AF342431

MULTIPLICATION OF RNA PLANT VIRUSES

Multiplication of RNA Plant Viruses

by

C. L. MANDAHAR

Botany Department, Panjab University, Chandigarh, India

 Springer

A C.I.P. Catalogue record for this book is available from the Library of Congress.

ISBN-10 1-4020-4724-X (HB)
ISBN-13 978-1-4020-4724-4 (HB)
ISBN-10 1-4020-4725-8 (e-book)
ISBN-13 978-1-4020-4725-1 (e-book)

Published by Springer,
P.O. Box 17, 3300 AA Dordrecht, The Netherlands.

www.springer.com

Figure on the front cover is reproduced from the *The Journal of Cell Biology*, November 1999, Vol. 147(5), p. 945-958 by copyright permission of The Rockefeller University Press, New York, USA

Printed on acid-free paper

Printed in the Netherlands.

Photograph on the Front Cover

The figure on the front cover presents a model to describe infection and replication of *Tobacco mosaic virus* (TMV) in Y2 tobacco protoplasts. Following introduction of virus to the cell, viral RNA is transported to the perinuclear endoplasmic reticulum (ER) on elements of the protein cytoskeleton, most likely microtubules (Fig. 1). The result ultimately leads to production of virus-specific proteins including viral replicase (R), and viral RNA (vRNA) in ER associated virus replication complexes. The ER-associated nascent viral RNAs function as mRNAs for the synthesis of movement protein (MP) that remains associated with the replication complexes (Fig. 2). Changes in different regions of TMV RNA-dependent RNA polymerase (RdRp/polymerase/replicase) may alter cell-to-cell virus movement (Hirashima and Watanabe, 2001); in fact, the polymerase protein has been implicated in movement of TMV and *Cucumber mosaic virus* (CMV) in infected plants (Deom *et al.*, 1997; Hirashima and Watanabe, 2001, 2003; Choi *et al.* 2005). Thus, both TMV MP and replicase are likely involved in TMV movement within and between cells. The MP remains associated with virus replication complexes (Kawakami *et al.*, 2004) that comprise large ER-derived structures (Fig. 3), which can be regarded as replication complexes. Formation and anchoring of large virus replication complexes is stabilized by MP and microfilament (MF) interactions (Fig. 3). Actin microfilaments and other cytoskeletal elements transport vRNA-MP complexes to the periphery of the cell to initiate cell-to-cell spread or toward other cellular sites where MP is degraded (Fig. 4). In late stages of infection in isolated protoplasts, vRNA and MP are localized in hair-like structures that protrude from the surface of the cell through plasma membrane (Fig. 5). These protrusions may be related to desmotubules. Alternatively, they may be a consequence of cell damage, which results in the extrusion of ER through the plasma membrane. Thus, at least two types of ER appear to be involved in TMV infection; one type is involved in vRNA replication and does not require the presence of MP while the second type corresponds to the filamentous protrusions that may be involved in intercellular spread of the virus and requires functional MP.

Notes

1. The figure is reproduced from Más and Beachy. 1999. *The Journal of Cell Biology* – November 1999, Vol. 147(5), p 945-958, by copyright permission of The Rockefeller University Press, New York, USA.

2. Kawakami, S., Watanabe, Y., and Beachy, R. N. 2004. *Tobacco mosaic virus* spreads cell-to-cell as intact replication complexes. *Proc. Natl. Acad. Sci. USA* 101: 6291-6296.

3. For other references, please see Chapter 5. RNA-Dependent RNA Polymerases and Replicases.

CONTENTS

List of Tables — xv
Preface — xvii
Acknowledgments — xix

1. Introduction — **1**

  I. Positive-Sense RNA Viruses — 1

  II. Putative 'Life/Replication Cycle' of Plant Viruses — 2

    A. Core Promoters — 7

    B. Replication Complexes — 8

    C. Switches — 8

  III. *cis*-Acting and *trans*-Acting Viral Nucleotide Sequences — 10

    A. *cis*-Acting Sequences — 10

      1. Internal Control Regions (Internal *cis*-Acting Sequences) — 11

      2. Role of *cis* Factors — 12

      3. *cis*-Acting Sequences in *Brome Mosaic Bromovirus* RNAs — 13

    B. *trans*-Acting Viral RNA Nucleotide Sequences — 14

  IV. Host Gene Shut-off and Hijacking of Host Cellular Machinery by Viruses — 15

    A. Mechanisms of Shut-off — 17

  V. Classification and Nomenclature — 19

    A. Classification — 19

    B. Nomenclature — 21

  VI. Abbreviations — 22

    A. Plant Viruses — 22

    B. Other Abbreviations — 23

  VII. References — 23

2. Positive-Sense Viral RNA — **29**

  I. Virus Genome — 29

  II. 5`-End — 34

    A. Cap or Cap-Like Structure — 35

      1. Mechanisms of Cap Formation — 36

        a. Co-transcriptional Capping — 36

        b. Cap Snatching — 39

2. Functions of Cap and Cap-Like Structure 39

B. Genome-Linked Protein 41

1. Functions 42

III. 3`-End 44

A. tRNA-Like Structure 45

1. Structure 46

2. Aminoacylation 48

3. Pseudoknot 50

4. Functions 51

a. Role in Transcription (Negative-Strand RNA Synthesis) 51

b. Role in Viral RNA Replication 52

c. Roles of Pseudoknot 54

d. Other Functions 54

B. Poly(A) Tail 55

C. Repair Mechanism of 3`-End of Viral RNA 56

IV. Viral Genes 58

A. Genes of Positive-Sense RNA Viruses 58

V. Functions of Viral RNA 62

VI. References 63

3. Infection by and Uncoating of Virus Particles **71**

I. Infection 71

A. Entry of Virus Particle/Genome into a Cell 71

B. Cell-Infecting Unit 72

C. Nucleo-Cytoplasmic Shuttling 73

II. Uncoating of Virus Particles 75

A. Uncoating 76

B. Site(s) of Uncoating 77

1. Extracellular Sites 77

2. Intracellular Sites 78

C. Mechanism(s) of Uncoating 78

1. Role of Ca^{2+} Divalent Cations 78

2. Role of pH 79

3. Co-translational and Co-replicational Mechanisms 80

D. Direction of Uncoating   82

III. References   83

4. Replication of Plus-Sense Viral RNA   **87**

I. Introduction   87

II. Models of Viral RNA Replication   90

III. Negative-Strand RNA Synthesis   92

A. Model of Negative-Strand RNA Synthesis   94

IV. Double-Stranded Forms of Replicative RNAs   96

A. Replicative Form RNA   97

B. Replicative Intermediate RNA   97

C. Double-Stranded RNAs as Artifacts   98

V. Synthesis of Progeny Positive-Strand RNA   98

VI. Asymmetry in Negative-Strand and Positive-Strand Progeny RNA   99
Synthesis

VII. Time Course of Viral RNA, Viral Protein and Virus Particle Synthesis   100

VIII. Replication Promoters, Enhancers and Repressors   101

IX. Template Selection by Cognate Viral Replicases   104

X. Capsid Protein and Viral RNA Replication   105

A. Functions of Capsid Protein   106

1. *Alfalfa mosaic alfavirus*   107

a. Functions of Capsid Protein Bound to Inoculum Viral RNAs   107

b. Functions of Capsid Protein Expressed from Viral RNAs 3 and 4   110

c. Switch   110

2. Other Plant Viruses   111

XI. References   111

5. RNA-Dependent RNA Polymerases and Replicases   **119**

I. Introduction   119

A. Time Course of Replicase Production   123

B. Isolation and Purification of Viral Replicases   124

II. Plant Viral Polymerases   125

III. Specificity   129

A. Specificity of Viral Polymerase Action   129

  1. Host Factors and Viral Polymerase Specificity   129

B. Absence of Specificity of Viral Polymerase Action   130

IV. Structure   131

A. Conserved Domains   131

B. Shape and Structure   132

C. Miscellaneous   133

V. Functions   134

A. Replication of Viral RNA   135

  1. Binding of Replicase and Role of *cis*-Acting Elements   137

B. Transcription of Viral RNA   138

C. Recombination of Viral RNA   138

D. Evolution of Plant Viruses   139

E. Miscellaneous Functions   140

VI. Classification   141

VII. References   142

6. Helicases   **151**

I. Introduction   151

A. Occurrence   152

B. Characteristics of Viral RNA Helicases   154

C. Nucleoside Triphosphate Connection   155

II. Classification   157

A. Relationship between Helicase Superfamilies   160

III. Structure   161

IV. Functions   161

V. References   163

7. Proteinases   **167**

I. Introduction   167

A. Classification   168

  1. Chymotrypsin-Like Cysteine and Serine Proteinases   171

  2. Papain-Like Cysteine Proteinases   172

II. Serine and Serine-Like Proteinases 172
  A. Family *Potyviridae* 172
    1. P1 Proteinase 173
    2. NIa Proteinase 174
  B. Family *Comoviridae* 175
    1. Comovirus Proteinase 176
    2. Nepovirus Proteinase 177
      a. Proteolytic Processing 177
      b. Genome-Linked Protein (VPg) 179
  C. Family *Sequiviridae* 179
  D. Putative Serine Proteolytic Activities 180
  E. Conclusions 180
III. Papain-like Cysteine Proteinases 180
  A. 'Main' Papain-Like Cysteine Proteinases 182
  B. Leader Proteinases/Papain-Like Leader Proteinases 183
    1. Potyviruses 183
    2. Closteroviruses 185
IV. Aspartic Proteinases 186
V. Functions of Proteinases 187
VI. References 191

8. Subgenomic RNAs **195**

I. Introduction 195
  A. 5`-Coterminal Subgenomic RNAs 199
  B. 3`-Coterminal Subgenomic RNAs 200
II. Mechanisms of Subgenomic RNA Synthesis 202
  A. Transcription or Promotion Mechanism 202
  B. Termination Mechanism 204
  C. Mechanism in *Citrus tristeza closterovirus* 204
  D. Post-transcriptional Mechanism 205
III. Replication of Subgenomic RNA 205
IV. Subgenomic RNA Promoters 206
  A. Structure and Recognition by Viral Polymerase 209
  B. Subgenomic RNA Promoters of Some Plant Viruses 210
  C. Location 212
  D. Role 214

V. Subgenomic RNAs of Some Plant Viruses 214

VI. Expression of Subgenomic RNAs 216

VII. Functions of Subgenomic RNAs 217

 A. Temporal Regulation of Gene Expression 217

 B. Role in Translation 218

 C. Role in Infectivity 218

VIII. References 218

9. Gene Expression **223**

I. Introduction 223

 A. Translation Initiation Factors 224

  1. Translation Initiation Factors and Plant Viruses 226

II. Canonical Translation 226

 A. Translation Initiation 228

  1. Aberrant Translation Initiation 228

   a. Leaky Scanning 228

   b. Ribosomal Shunting and Internal Translation Initiation 229

  2. Scanning 230

 B. Elongation Step of Translation 230

 C. Termination of Translation 231

 D. Interference by Viral RNAs 231

 E. Translation of Viral RNAs 232

III. Translation Strategies of Viral RNAs 234

 A. Regulation of Gene Expression at the Level of Genome Segments 235

 B. Regulation of Gene Expression at Transcription Level 235

 C. Regulation of Gene Expression at Translational Level 237

  1. Regulation of Gene Expression at Translation Initiation Level 237

   a. Factors Affecting Translation Initiation 237

   b. Unconventional Translation Initiation 239

  2. Regulation of Gene Expression at Translation Elongation Level

   (Frameshifting) 242

   a. Types of Frameshifting 244

  3. Regulation Gene Expression at Translation Termination

   (Readthrough) 246

   a. Readthrough Mechanisms 248

   b. Functions of Readthrough Mechanism 249

  4. Regulation of Gene Expression at Post-translational Level

   (Proteolytic Processing of Polyproteins) 250

IV. Cap-Independent Translation 251

    A. Cap-Independent Translation Enhancer (CITE) at 3`-End of Viral RNA 253

        1. Functions of 3`-CITE Domain 255

    B. Cap-Independent Translation Enhancer at 5`-End of Viral RNA 256

    C. Internal Ribosome Entry Sites (IRES) 256

        1. Translation Mechanism 257

        2. Advantages of IRES 258

    D. Plant Viruses Showing Cap-Independent Translation 258

        1. Plant Viruses Lacking Both 5`-Cap Structure and 3`-Poly(A) Tail 258

        2. Plant Viruses Lacking 5`-Cap Structure but 3`-Poly(A) Tail Present 259

        3. Plant Viruses Lacking 3`-Poly(A) Tail but 5`-Cap Structure Present 260

    E. Conclusions 261

V. References 263

10. Assembly of Virus Particles **271**

  I. Introduction 271

  II. Assembly of Rod-Like Virus Particles 272

    A. *Tobacco mosaic tobamovirus* 272

        1. Assembly of Capsid Protein *in vitro* 273

        2. Assembly of Virus Particles *in vitro* 274

            a. Nucleation (Assembly Initiation) 275

            b. Elongation Mechanism 277

            c. Conclusions 278

        3. Assembly of Virus Particles *in vivo* 278

    B. Assembly of Other Rod-Like Viruses 279

  III. Assembly of Flexuous Virus Particles 280

  IV. Assembly of Icosahedral Virus Particles 282

    A. *Southern bean mosaic sobemovirus*, *Tomato bushy stunt tombusvirus*, and *Satellite tobacco necrosis virus* 283

    B. *Alfalfa mosaic alfamovirus* 285

    C. Bromoviruses 286

        1. *Brome mosaic bromovirus* 287

            a. Polymorphic Capsids 288

            b. Capsid Protein-RNA Interactions 289

        2. *Cowpea chlorotic mottle bromovirus* 291

    D. Other Spherical Plant Viruses 291

V. Specificity of Virus Assembly — 292

VI. Conclusions — 294

VII. References — 295

11. Host Factors and Virus Multiplication — 299

I. Introduction — 299

II. Host Proteins and Membranes — 299

III. Functions of Cellular Factors — 302

IV. Sites of Viral RNA Replication — 304

  A. Cytoplasm — 304

    1. Cytoplasmic Inclusions (Viroplasms) — 305

    2. Endoplasmic Reticulum — 305

  B. Chloroplasts — 307

  C. Other Cellular Organelles — 307

V. Replication Complexes — 309

  A. Replication Complex of *Brome mosaic bromovirus* — 311

  B. Replication Complex of *Cucumber mosaic virus* — 313

  C. Replication Complex of *Tobacco mosaic virus* — 313

  D. Replication Complex of *Cucumber necrosis tombusvirus* — 315

  E. Replication Complexes of Other Plant Viruses — 316

  F. Structure of Replication Complexes — 316

VI. Membrane-Targeting and Anchoring of Replication Complexes — 318

VII. Vesiculation — 319

  A. Functions of Vesicles — 321

  B. Viral Signals Inducing Formation of Vesicles — 322

VIII. References — 323

Subject Index — 329

LIST OF TABLES

S. No.	Heading	Page No.
1	Characteristics of single-stranded positive-sense RNA genomes of plant viruses	32
2	Genes of positive-sense RNA genomes of plant viruses	59
3	Classification, lineage and mode of expression of RNA-dependent RNA polymerases of families and genera of positive-sense RNA plant viruses	121
4	RNA-dependent RNA polymerases of some positive sense RNA plant viruses	126
5	Plant viruses belonging to different helicase superfamilies and lineages	158
6	Plant virus groups and genera encoding different types of demonstrated or putative proteinases	169
7	Polyprotein precursors, cleavage products and proteinases of some positive-sense RNA plant viruses	170
8	Positive-sense RNA plant viruses that produce subgenomic RNAs	196
9	Positive sense RNA plant viruses showing frameshifting mechanism	243
10	Positive-sense RNA plant viruses showing readthrough mechanism	247
11	Proteins of host plants of *Arabidopsis thaliana* and of *Saccharomyces cerevisiae* that interact with viral RNAs or RdRps subunits of plant alpha-like plant viruses	301

PREFACE

Multiplication is the basic biological function of all organisms and is primarily dependent upon replication of genome. This is true of plant viruses also. Huge literature has been published concerning various aspects of multiplication of plant viruses. This can be imagined from the fact that some excellent reviews were published as far back as 1966-67 (Bald, 1966; Schlegel *et al.*, 1967; Ralph, 1969). Since then, more than one hundred reviews have been published while about 60 of them have been published since 1995. Research papers are obviously innumerable. It is perhaps because of this stupefying fact that no book on plant virus multiplication has been published so far. I could take up this assignment since, because of my involvement with writing and editing books on plant viruses (Mandahar, 1978, 1989, 1990, 1999), I had already scanned much of the relevant literature. Even then this was a real daunting task.

The published work on multiplication of positive-sense RNA plant viruses has passed through different stages till now it is the genetic, molecular and biochemical information that is appearing in torrents. Despite the importance of RNA replication in viral life cycle, biochemical studies on virus replication were in their earlier stages even in 1999 (Nagy *et al.*, 1999) so that not much was known about viral RNA replication (Teycheney *et al.*, 2000) and its initiation by 2000 (Deiman *et al.*, 2000). The picture has improved much over the last few years because structural and sequence requirements of viral RNA synthesis are beginning to be elucidated (Sivakumaran *et al.*, 2003; Hema and Kao, 2004), due primarily to molecular and biochemical studies. Such studies have thrown much new light on replication of positive-sense RNA plant viruses so that their multiplication has now emerged as a rapidly developing field. But much distance has still to be covered and traversing this distance could prove difficult because no one organised corpus of knowledge is available. Hopefully, this book fills the niche and generates understanding of multiplication of positive-sense RNA plant viruses, especially at molecular level. This understanding is important because advances in RNA virus replication are being powered by the urgent requirement for improved disease control strategies and by the ability of viruses to illuminate key cellular pathways (Ahlquist, 2002). Moreover, viruses are rightly regarded as exquisite models for study of cell strategies (Bernardi and Haenni, 1998).

Comments and suggestions about this book may please be sent to the author's residential address (H. No. 223A, Sector 2, Panchkula - 134112, Haryana, India).

REFERENCES

Ahlquist, P. 2002. RNA-dependent RNA polymerases, viruses and gene silencing. *Science* 296: 1270-1273.

Bald, J. G. 1966. Cytology of plant virus infections. *Adv. Virus Res.* 12: 103-125.

Bernardi, F., and Haenni, A.-L. 1998. Viruses: Exquisite models for cell strategies. *Biochemie* 80: 1035-1041.

Deiman, B. A. L. M., Verlaan, P. W. G., and Pleij, C. W. A. 2000. *In vitro* transcription by the turnip yellow mosaic virus RNA polymerase: A comparison with alfalfa mosaic virus and brome mosaic virus replicases. *J. Virol.* 74: 264-271.

Hema, M., and Kao, C. C. 2004. Template sequence near the initiation nucleotide can modulate brome mosaic virus RNA accumulation in plant protoplasts. *J. Virol.* 78: 1169-1180.

Mandahar, C. L. 1978 . Introduction to Plant Viruses. S. Chand & Co., New Delhi. Second edition. pp. 568.

Mandahar, C. L. (Editor). 1989. Plant Viruses. Volume 1. Structure and Replication. CRC Press, Boca Raton, Florida, USA. pp. 366.

Mandahar, C. L. (Editor). 1990. Plant Viruses. Volume 2. Pathology. CRC Press, Boca Raton, Florida, USA. pp. 371.

Mandahar, C. L. (Editor). 1999. Molecular Biology of Plant Viruses. Kluwer Academic Publishers, Boston/Dordrecht/London. pp. 281.

Nagy, P. D., Pogany, J., and Simon, A. E. 1999. RNA elements required for RNA recombination function as replication enhancers *in vitro* and *in vivo* in a plus strand RNA virus. *EMBO J.* 18: 5653-5665.

Ralph, R. K. 1969. Double-stranded viral RNA. *Adv. Virus Res.* 15: 61-158.

Schlegel, D. E., Smith, S. H., and de Zoeten, G. A. 1967. Sites of virus synthesis within cells. *Annu. Rev. Phytopathol.* 5: 223-246.

Sivakumaran, K., Hema, M., and Kao, C. C. 2003. Brome mosaic virus RNA synthesis *in vitro* and in barley protoplasts. *J. Virol.* 77: 5703-5711.

Teycheney, P.-Y., Aaziz, R., Dinant, S., Salánki, K., Tourneur, C., Balázs, E., Jacquemond, M., and Tepfer, M. 2000. Synthesis of (-)-strand RNA from the 3`-untranslated region of plant viral genomes expressed in transgenic plants upon infection with related viruses. *J. Gen Virol.* 81: 1121-1126.

ACKNOWLEDGEMENTS

This book could not have been possible without the essential inputs from four quarters: the library of the Institute of Microbial Technology (IMTECH), Chandigarh; the library of the Central Potato Research Institute (CPRI), Shimla; my family; and the publishers. I am thankful to the Director, to Dr. Naresh, and the library staff of the extremely well stocked, maintained, and managed library of the IMTECH. Equally important help was provided by the then Director Dr. S. M. Paul Khurana but now the Vice Chancellor of Rani Durgavati University, Jabalpur, and by Dr. I. D. Garg, Head of Plant Pathology/Plant Virology and the library staff of the equally well stocked, maintained, and managed library of CPRI, Shimla.

Without the help of these two sources, this book could not have been even thought of. Dr. Khurana, in addition, always kept me well informed about new developments and with relevant literature. His help was always forthcoming whenever needed; so was that of Dr. Garg. I am thankful to Dr. J. A. Khan, Molecular Virology Laboratory, National Botanical Research Institute, Lucknow, India, for the reprints provided by him.

I gratefully acknowledge the help of Drs. Ahlquist, A., Beachy, R. N., Bol, J. F., Dolja, V. V., Garcia, J. A., Haenni, A.-L., Miller, W. A., Nagy, P. D., Sanfaçon, H., Simon, A. E., and several others for providing me the reprints of their publications whenever I requested. Without these reprints, my efforts would have increased manifold. My gratitude to Dr. R. N. Beachy, President, Donald Danforth Plant Science Center, 975 North Warson Road, St. Louis, Missouri 63132, USA, and The Rockefeller University Press, New York, USA, is immense for permitting me to use the photograph printed on the front cover of this Book. This photograph is "Reproduced from Más and Beachy. 1999. *The Journal Of Cell Biology* – November 1999, Vol. 147(5), p 945-958, by copyright permission of The Rockefeller University Press, New York, USA."

My thanks to Ms Ing. Zuzana Bernhart, the Publishing Editor, and Ms Ineke Ravesloot for always solving my problems as and when they arose.

I am thankful to Dr. Karl Maramorosch for constant encouragement; to Dr. A-.L. Haenni for ready and immediate help that removed many types of roadblocks during preparation of the manuscript; to Drs. A. E. Gorbalenya, E. V. Koonin, and V. V. Dolja, whose comparative and statistical work gave new direction and insights to understanding of helicases, polymerases, and proteinases; to Dr. P. Ahlquist whose extensive biochemical and molecular studies laid the foundations of molecular aspect of plant virus replication; and to many plant virologists whose researches helped the subject of multiplication of RNA plant viruses to arrive at the stage at which it stands now.

Particular thanks to Mrs. Aruna Mandahar, my wife, for always providing me the required time at the cost of her genuine desires and needs and for her taking over several of my responsibilities; to son Atul Mandahar who went beyond the call of his filial duties to make his parents feel secure and wanted, for his constant affectionate support, and for always providing space and staff whenever the need arose; to son-in-law Venod Sharma, daughter-in-law Shelly Mandahar, daughter Abha Sharma, and grandchildren Akash, Abhishek, and Arshya – who together gave me the motivation and strength to complete this task.

Thanks to Mr. Tarun Chugh whose computer expertise was of great help in several ways and without which my task would have become far more arduous; to Mr. Rajiv Sharma and Mr. Suresh who always provided me the support whenever I needed; and to Dr. S. S. Kumar, friend and colleague, who always stood by my side during my endeavours.

1 INTRODUCTION

I. POSITIVE-SENSE RNA VIRUSES

RNA is labile; still it carries genetic information and multiplies efficiently. About 80% of viruses possess single-stranded RNA genome of positive polarity (positive-strand/ positive-sense/plus-sense/plus-strand/messenger-sense) so that it is understandable that a great majority of studies on replication of viruses pertain to these viruses. Positive-sense RNA viruses encompass over one-third of all virus genera and include numerous (plant, animal, and human) pathogens, such as the *Severe acute respiratory syndrome coronavirus* (SARS), *Hepatitis C virus*, and many of the viruses on the US Health and Human Services Department Select List of potential bioterrorism agents (Ahlquist *et al.*, 2003). Early work on computer-assisted comparative studies for understanding the relationships within these positive-strand RNA viruses (Franssen *et al.*, 1984; Haseloff *et al.*, 1984; Argos *et al.*, 1984; Kamer and Argos, 1984; Ahlquist *et al.*, 1985) demonstrated the presence of non-trivial links between these viruses of plants and animals (Koonin and Dolja, 1993). In fact, this was the least expected outcome of the comparative analysis of their RNA genomes (Goldbach, 1987). Many of the characteristics of positive-sense RNA viruses of animals and plants, particularly those concerned with RNA replication, are the same or nearly the same so that these viruses are now regarded to form a single unit.

Vast differences exist in morphology of virus particles, organization of genomic RNAs, host ranges, symptoms, and some other characters of positive-sense RNA viruses of plants (like *Brome mosaic virus*, *Cucumber mosaic virus*, *Tobacco mosaic virus* and others) and animals (like *Sindbis virus* and *Poliovirus*). Still these viruses exhibit striking similarities in replication strategies (Pogue *et al.*, 1994). For example, amino acid sequences of their virus-encoded RNA-dependent RNA polymerases (RdRps) and several other functional domains within other virus-encoded non-structural proteins (helicases, nucleoside triphosphate motifs, proteinases) of these viruses are conserved (Koonin, 1991a, 1991b; Koonin and Dolja, 1993; Dougherty and Semler, 1993; Zanotto *et al.*, 1996; Buck, 1996; Kadaré and Haenni, 1997). The above similarities clearly establish that these completely different plant and animal viruses and virus groups have identical RNA replication mechanisms although the nature of their replication proteins (like amino acid sequences of viral RdRps) and their interacting host proteins are different. The similarities between these positive-sense animal and plant viruses extend even to their having related genes and occasionally even possessing identical gene organization and gene expression. Thus, the replication of plus-sense viral RNA *per se* or its different stages has been reviewed innumerable times in viruses in general and/or in different groups of viruses since the sixties of the last century (Hofschneider and Husen, 1968; Montagnier, 1968) till date (White and

Nagy, 2004; and many other references concerning plant viruses mentioned in text at relevant places in different chapters).

Some of the main principles of evolution of positive-sense RNA viruses, as formulated by Koonin and Dolja (1993) are: RNA viruses evolve rapidly so that only very important functional motifs are conserved in a wide range of virus groups; the universal building blocks of positive-sense RNA viruses are the genes for RdRp and capsid protein, out of which only RdRp sequence contains universal motifs that are conserved in all positive-sense RNA viruses with known genome sequences; evolution of positive-sense RNA viruses is shaped by two opposing trends - conservation of distinct arrays of genes mainly those encoding viral replication proteins and recombinational shuffling of genes and gene blocks; widespread recombination, even among distantly related viruses, has made it impossible to depict evolutionary history of positive-strand RNA viruses as a single phylogenetic tree; and correlation between virus phylogeny and strategy of genome replication and expression is only limited - suggesting that fundamental expression and replication mechanisms could have evolved more than once.

The positive-sense RNA viruses exhibit great genetic compactness so much so that individual nucleotides often perform multiple functions during their life cycle. The virus genomic RNA of positive-sense RNA viruses nearly always serves two essential functions at the start of viral replication cycle. It acts at two independent levels: it acts as messenger RNA encoding the various virus-specific proteins [coat protein, replication proteins (polymerase, helicase, proteinase), movement protein, and still other proteins involved in viral RNA replication]. Then, it also acts as template for transcription into minus-strand RNA, which is the starting point of RNA replication process. But, there is no interference between these two functions of a single genomic RNA molecule. This means that there is no clash between the ribosomes reading the RNA from its 5`- to 3`-end and the replicase reading the genome from its 3`- to 5`-end. Imagine how a single RNA molecule regulates all these processes sometime simultaneously and sometime at different times, sometime at a one place and sometime at different places; how all these and other functions and processes are regulated and controlled in time and space. The mystique appears to become more mysterious. Since the positive-sense RNA genomes are templates for both translation and replication, it results in interactions between host translation factors and viral RNA replication at several levels. Plus-sense RNA of plant viruses thus acts as a substrate for transcripttion, RNA replication, translation, and encapsidation.

II. PUTATIVE 'LIFE/REPLICATION CYCLE' OF PLANT VIRUSES

Being parasites, viruses must enter and infect the host cells and exploit the cellular macromolecular machinery and sources of energy supply for their reproduction. Viral RNA replication involves both virus-encoded and cell-encoded proteins and is a very complex process. The ability of a virus to multiply requires successful virus-host interactions and certain definite host factors; involves a series of distinct and obligatory stages that may overlap chronologically; and requires a 3`-proximal core replication promoter or simply promoter (that promotes or initiates replication), and increasingly diverse inventory of other supporting elements like replication enhancers or simply enhancer (that enhances or increases replication), replication repressors or simply repressor (that represses replication), and 5`-terminal sequences. A putative 'life cycle' of plant viruses can be built and regarded to consist of seven steps/stages, which at one time or the other overlap each other. Infection process is the **first step**. It consists of viral attachment and ingress of viral genome into the cytoplasm of host cell. This involves delivery of the virion to some extra- or intra-cellular location of the cell or initial interaction and attachment of the virus particle with a host cell and targeting of virus particles to reproductive site. The infecting virus molecule is partially uncoated at the 5`-end during the process of virus entry into a cell. Decapsidation (uncoating of viral RNA) is the **second step**. It leads to unmasking of genomic RNA by removal of coat protein by co-translational and co-replicational disassembly mechanisms thereby releasing viral genome within the cell and making it available for other functions. The genomic RNA is recruited by ribosomes and translated to produce components of viral replicase. The process of uncoating implies that second, third (synthesis of viral replicase), fourth (RNA replication), and fifth (translation of viral RNA) steps partly overlap.

The **third** step is the formation of virus-specific RNA replication proteins, including RdRp, by translation of the viral genome by host ribosomal system. It is an early event of viral replication cycle and occurs nearly at the beginning of infection. Since no pre-existing cellular machinery for replication of RNA viral genomes (that is, no RdRp) exists in host cells, therefore, all RNA viruses encode an RNA polymerase and possess a relevant gene for it. This enzyme synthesizes new RNA or DNA as well as subgenomic RNA, acts as the transcribing enzyme for transcription of RNA from DNA or RNA and has also to recruit genomic RNA (out of the translation mode) for 3`- to 5`-copying to form negative-sense RNA. This happens at RNA replication complexes associated with intracellular membranes. The replicase complex binds with the 3`-untranslated region (UTR) of genomic RNA, and the RdRp component of replicase complex employs the full-length genomic RNA as a template for synthesis of a complementary full-length minus-strand viral RNA. In fact, the study of RNA replication was hampered for a long time mainly because the active viral RdRp could not be dissociated from the host membrane and the inducible endogenous host-encoded RdRp, which is found in plant systems, could not be distinguished from the viral RNA polymerase. It was only about fifteen years back that it became possible to isolate viral RNA polymerase from infected plant tissue. This viral polymerase, in most cases, can only synthesize the RNA of negative polarity and depends on the addition of an exogenous viral template for RNA synthesis. Such RNA polymerases had been obtained for 5 viruses by 1992-1993 (David *et al.*, 1992; Ishihama and Barbier, 1994). This made it possible to investigate as to which viral proteins and host proteins participate in viral RNA replication. The availability of *in vitro* active RdRp, expressed in/obtained from heterologous systems like *Escherichia coli* (Hong and Hunt, 1996; Anindya *et al.*, 2005) and *Saccharomyces cerevisiae* (Janda and Ahlquist, 1993; Quadt *et al.*, 1995), greatly facilitated research into the mechanisms of replication of positive-sense RNA viruses. Mutant RNA polymerases could be expressed in these systems even when they did not have to support viral RNA replication. These heterologous systems have contributed enormously to the study of plant virus replication.

The **fourth** step is the replication of viral RNA. It has several sub-steps. Firstly, initiation of RNA replication at the right template and at the right position requires specific recognition of *cis*-acting RNA elements (called replication promoters) by viral RdRp and/or associated host factors in plus-strand genomic RNA. For replication to begin, the genomic viral RNA switches (the first switch) from translation mode (during RdRp synthesis) to replication mode, indicating that viral RNAs must contain the necessary information to switch between translation and replication (Gamarink and Andino, 1998). This switch is necessary because during translation of viral genomic RNA, 5`- to 3`-ribosome trafficking blocks 3`- to 5`-polymerase copying of viral RNA. BMV genomic RNAs are recruited from translation to replication mode by a viral protein (1a, a helicase-like replication factor), which acts through *cis*-acting recognition elements of each of the viral genomic RNA (Chen *et al.*, 2001). In addition, internal locations can also bind RdRp, with RNA bridges directing the RdRp to 3`-end.

Secondly, replication of viral RNA yields progeny plus-sense RNA molecules. Viral replication involves reiterative transcription of the infecting virus plus-sense RNA template into the complementary negative-sense chain by replicase system, leading to formation of a double-stranded RNA molecule (the replicative form or RF RNA). The RF RNA is unwound by viral helicase activity, releasing the template and the single-stranded minus-sense RNA, and reiterative formation of progeny positive-sense RNA molecules by using the complementary negative-strands as templates and so leading again to the formation of a double-stranded RNA structure called replicative intermediate (RI) RNA.

Since RdRp must be recruited to 3`-ends of plus strands to initiate synthesis of full-length complementary RNAs, *cis*-acting elements thought to recruit RdRp are usually located within the 3`-noncoding regions (Duggal *et al.*, 1994). An interaction between viral polymerase and certain viral sequences that promote the minus-strand RNA synthesis begins the replication of positive-strand RNA viruses. The elements promoting negative-strand RNA synthesis are usually contained within the 3`-terminal 200 bases of many viral RNAs [as in *Brome mosaic bromovirus* (BMV), *Cucumber mosaic cucumovirus* (CMV), *Cowpea mosaic comovirus* (CPMV), *Cymbidium ringspot tombusvirus*, *Tobacco mosaic tobamovirus* (TMV) and in many other plant viruses], and include diverse forms such as tRNA-like structures, poly(A) tails, pseudoknots, small hairpins, and short primary sequences without high-order structures that have been identified as core promoters for minus-strand synthesis (Dreher, 1999). Thus, many sequences and/or structures required for negative-strand synthesis have been characterized for many RNA plant viruses. Transcription initiation site of TMV is located near extreme 3`-termini of RNA templates so that the replication promoter on positive-strand TMV RNA is located within 249-base sequence and the extreme 3`-terminal sequence of -CCCGGUAGGGGCCCA(3`) (Watanabe *et al.*, 1999). Involvement of 5`-proximal sequences and structural elements in minus-strand synthesis has also been found for an increasing number of viruses (Herold and Andino, 2001; Wu *et al.*, 2001; Vlot and Bol, 2003).

Thirdly, if RNA synthesis is initiated at internal promoters on minus-strand RNA, it leads to the formation of subgenomic mRNAs. Thus, minus strands can also act as templates for the synthesis of 3`-coterminal subgenomic RNAs by using internal promoters.

Fourthly, the synthesis of progeny positive-strand RNA occurs through asymmetric mechanism, which enables simultaneous synthesis of multiple progeny viral RNA strands that are at different stages of completion and appear to hang as tails. As many as five partially completed plus strands may be associated with RI; moreover, each full-length minus-strand RNA is used multiple times as a template for genomic or subgenomic RNA. This process results in formation of much greater number of progeny positive RNA strands so much so that the ratio of plus to minus strands at the exponential stage of replication cycle can range from 10:1 to as high as 1000:1 (Buck, 1996). This asymmetric accumulation of the minus- and plus-sense RNAs suggests the presence of a second switch, which represses minus-strand synthesis. The switch from minus-strand to plus-strand synthesis is in all probability mediated by *cis*-acting sequences that may or may not permit RdRp access to plus-strand promoter or 3`-terminal sequences. These *cis*-acting sequences and structural elements in untranslated regions of genomes allow viral RdRps to correctly initiate and transcribe asymmetric levels of plus and minus strands during replication of plus-sense RNA viruses. Such elements include promoters, enhancers, and transcriptional repressors that may require interactions with distal RNA sequences for function. The new genomic and subgenomic RNAs are released from minus-sense template by helicase activity. Later, the methyltransferase activity of the replicase complex caps new molecules of the genomic RNA and the subgenomic mRNAs. Progeny positive RNA strands generated as above may further enter into any one the three routes of virus replication cycle: act as mRNA and code for coat protein synthesis (the translation step), be converted to double-stranded RF RNA to provide more templates for greater progeny positive RNA synthesis, or be encapsidated to produce the virions (the encapsidation step). Pulse chase experiments and kinetics of labeling were the tools of choice for these studies.

For asymmetric viral RNA replication, viral RNAs must have information to support asymmetric synthesis of plus and minus strands. Requirements for this include transcriptional enhancers (Nagy *et al.*, 1999; Pogany *et al.*, 2003; Ray and White, 2003) and repressors that function via RNA-RNA (Pogany *et al.*, 2003) or protein-RNA (Dreher, 1999; Zhang *et al.*, 2004a, 2004b) interactions.

The promoter regions necessary for minus-strand RNA and subgenomic RNA synthesis have been mapped in several positive-strand RNA viruses. These two types of promoters generally do not share common structural characters despite their being recognised by the same viral polymerase. In fact, the single type of polymerase produced by a particular virus possesses the ability to recognise different types of promoters of that virus even though little similarity occurs in sequence and structure between these promoters located on a viral genome (Miller and Koev, 2000). The subgenomic RNA promoter is generally situated 3`-with respect to the transcription start site while the plus- and minus-strand genomic RNA promoters, so as to prevent loss of genetic information, are located 5`-of their start site. Enhancers and repressors regulate promoter efficiency and may be located either proximal or distal to core promoter (Nagy *et al.*, 1999, 2001; Panavas and Nagy, 2003; Pogany *et al.*, 2003; Ray and White, 1999, 2003).

The progeny positive RNA strands generated as above may further enter into any of the three routes of virus replication cycle: act as mRNA and code for viral proteins (the translation step), be converted to double-stranded RF RNA to provide more templates for greater progeny positive RNA synthesis, or be encapsidated to produce the virus particles (the encapsidation step).

The **fifth** step is the translation of viral genomic RNA when it directly serves as mRNA and produces structural and non-structural proteins by exploiting cellular functions. Virus-specific capsid is the structural protein while various enzymes are the non-structural proteins. To compensate for the limited amount of genetic information and to maximize their coding capacity, plant viruses have developed a great number of highly sophisticated strategies to synthesize their proteins. These strategies are: synthesis of subgenomic RNAs, use of polycistronic RNAs or of ambisense RNAs, RNA splicing, internal in-phase initiation, gene overlap, readthrough of termination codons, shift in reading frame and post-translational cleavage of a polyprotein. Translational programming permits the viruses to conserve the functional genome size by making efficient use of genome coding capacity.

Viral assembly/encapsidation of progeny plus-sense RNA molecules by capsid protein subunits to produce progeny virus particles is the **sixth** step. It involves recognition of the specific progeny genomic RNA by the specific capsid protein subunits and the assembly of viral RNA with its specific proteins to form complete virus particles. Empty capsids are also formed in some cases. Maturation of virus particles is the **seventh step**. However, it does not occur in plus-sense RNA viruses but occurs in enveloped viruses, which acquire the envelope by budding through some membranous system. Progeny virus particles, subsequent to their formation, spread within the host followed by release outside the host and/or transmission of virions from infected cells to next host in which the infection cycle is repeated. Successful infection involves formation of 'virus replication factories' on cell membranes and transport of such factories within and between cells.

Each of the above steps can occur at one or more specific sites and is a complex phenomenon requiring specific conditions. Some of these stages overlap chronologically. No natural DNA phase is involved in replication of positive-sense RNA plant viruses, this replication process shares fundamental similarities with replication phenomenon of cellular RNA, and the positive-strand RNA viruses possess a common replication strategy (Pogue *et al.*, 1994).

Synthesis of cellular RNA is DNA-dependent, and hence is called DNA-dependent RNA synthesis. However, most of the eukaryotic cells do not contain pre-existing machinery for conducting RNA-dependent RNA synthesis and so are unable to replicate RNA viral genome. To overcome this inability, all RNA viruses encode an RNA polymerase within the invaded cells so as to make the viruses competent for multiplication of their RNA genome. This enables the viruses to replicate and transcribe their genomes in eukaryotic cells by frequently subverting cellular factors.

These subverted factors play an integral and regulatory role in replication and transcription of viral RNA.

Viral RNA synthesis has been conducted both *in vitro* and *in vivo*. Generally, an *in vitro* system of RNA synthesis can be used to dissect the mechanism and roles of proteins involved in RNA replication. However an *in vitro* system may lack important properties that can only be found *in vivo*, since some host proteins are involved in the replication process (Lai, 1998; Ahlquist *et al.*, 2003). While *in vivo* results could be predicted from the *in vitro* results, the reverse is not always true. Some differences exist in the influence of stem-loops on replication of *Hibiscus chlorotic ringspot virus* between the *in vitro* and *in vivo* systems (Wang and Wong, 2004). Deletion of U loop of stem-loop 2 slightly affected the replication efficiency *in vitro* but the mutant could not replicate in knead protoplasts. The requirements of stem-loops for minus-strand RNA synthesis of TMV carried out *in vitro* were the same as in tobacco protoplasts (Osman *et al.*, 2000). However, differences were noted in replication efficiency of some TMV mutants *in vivo* (Chandrika *et al.*, 2000). Similarly, insertion of 3 nucleotides in stem-loop C of BMV RNA3 showed reduced RNA synthesis *in vitro* but not *in vivo* (Sivakumaran *et al.*, 2003). Therefore, *cis* sequence elements and structural requirements for RNA synthesis need to be analysed both *in vitro* and *in vivo*.

Replication of viral RNA begins with specific recognition of *cis*-acting RNA elements on the infecting positive-sense RNA by viral RdRp and/or associated host factors leading to synthesis of complementary negative-sense intermediates, that later act as templates for synthesis of plus-sense RNA progeny molecules. Therefore, RNA synthesis is regulated by sequences present on both positive- and negative-stranded RNA templates and the essential *cis*-acting motifs have to include the promoters for synthesis of the negative strand from the infecting positive-strand RNA and synthesis of the positive-strand progeny RNA molecules from the negative-strand templates.

A. Core Promoters

Core promoters, present in viral RNAs, contain multiple sequence and structural features that are essential for efficient recognition of viral RNA by cognate RdRp (Kim *et al.*, 2000). They recruit viral RdRp to transcription site, are located at the 3`-ends of positive-strand RNA (and also negative-strand RNAs), are generally constituted by one or a few 3`-proximal hairpins that specifically interact with polymerase for initiation of minus-strand RNA synthesis, are generally located proximal to 3`-terminal sequences, and are usually comprised of one or more hairpins with adjoining single-stranded sequence (Duggal *et al.*, 1994; Song and Simon, 1995; Buck, 1996; Chapman and Kao, 1999; Dreher, 1999; Sivakumaran *et al.*, 1999). However, some viral RNAs are able to replicate in the absence of large 3`- or 5`-terminal sequences (Wu and White, 1998), which suggests that promoter sequences may not be required for them. All core promoters permit basal levels of RNA transcription; however, efficient RNA synthesis requires additional viral elements like structures and sequences at 5`-ends that may be needed for genome circularization (Herold and Andino, 2001; Khromykh *et al.*, 2001) and internal elements such as enhancers, promoters, repressors, and RNA chaperons, which function either in *cis* (Nagy *et al.*, 1999, 2001; Ray and White, 1999, 2003; Vlot *et al.*, 2001; Panavas and Nagy, 2003; Pogany *et al.*, 2003; Zhang *et al.*, 2004a, 2004b) or in *trans* (Sit *et al.*, 1998; Eckerle and Ball, 2002). Genomic RNAs of all carmoviruses, except of *Galinsoga mosaic virus*, contain a 3`-proximal hairpin that comprises the core promoter for minus-strand RNA synthesis (Song and Simon, 1995). The best-characterized BMV core promoter is the one that directs initiation of minus-strand RNA synthesis.

B. Replication Complexes

It is being increasingly realized now that it is not simply the replicase molecules that are involved in viral RNA replication but rather a complex entity called replication complex. Replication complexes are known/regarded to be constituted by several of the following elements: virus-encoded proteins [the RdRp itself that is characterized by a conserved Gly-Asp-Asp (GDD) motif and the auxiliary proteins], an nucleoside 5`-triphosphate (NTP)-binding RNA helicase motif in the same or in another protein in nearly all cases, methyltransferase, host-derived proteins/factors and the RNA template (Buck, 1996, 1999; Osman and Buck, 1997; Lai, 1998; Kao *et al.*, 2001; Ahlquist *et al.*, 2003). In fact, replication complexes of some, if not of many positive-strand RNA plant viruses, are considered to sequester viral RNA templates along with replicases, putative RNA helicases and capping factors (Schwartz *et al.*, 2002). Thus, replication and transcription in positive-strand RNA viruses use a complex of viral and cellular proteins - the replication complex - that initiates RNA synthesis at 3`-end of genomic RNA. Such replication complexes have been detected in plants infected with BMV, CMV, *Cucumber necrosis virus* (CNV), TMV, and many other plant viruses. Specific and complex interactions between viral genomic RNAs and viral replicase are required for replication and/or transcription of viral RNAs. Viral replication complexes are treated in detail in a later chapter.

C. Switches

The changeover of one stage of RNA replication to the next is a controlled event, although we do not yet understand this controlled switchover. The first switch is regarded to be operative subsequent to formation of RdRp by ribosomes; it ensures that the genomic viral RNA switches from translation mode (during RdRp synthesis) to replication mode. For this, RdRp and possibly other viral or host factors form an initiation complex at the promoter and initiate complementary minus strand. The second switch is considered to be the one that ensures asymmetric accumulation of minus- and plus-RNA strands; it represses minus-strand synthesis, thereby constraining the RdRp to synthesize minus strands but favours synthesis of plus-strands of RNA. This switch from minus-strand to plus-strand synthesis is likely mediated by *cis*-acting elements that allow or deny access of RdRp to the plus-strand promoter or 3`-terminal sequences (Sun *et al.*, 2005).

Regulation of switches is carried out by some plant viruses by changing the conformation of 3`-proximal structures and this may be mediated by one or more unstable base pairs occurring between complementary short sequences located within and outside hairpins (Olsthoorn *et al.*, 1999; Koev *et al.*, 2002; Pogany *et al.*, 2003; Zhang *et al.*, 2004a; Sun *et al.*, 2005). For example, repression of minus-strand RNA synthesis by *Barley yellow dwarf virus* occurs by altering the conformation of its 3`-end to a 'pocket' structure in which the transcription initiation site is unavailable to RdRp since the site is embedded in a stem (Koev *et al.*, 2002). Moreover, apart from the *cis*-acting sequences, *trans*-acting cellular factors or virus-encoded proteins could also influence the balance between alternative structural conformations (Olsthoorn *et al.*, 1999). Thus, the presence of alternative structural conformation of some viral 3`-ends may be one mechanism to limit minus-strand synthesis and permit greater synthesis of progeny plus strands.

Another type of switch controlling virus replication has been found during studies on replication of BMV RNA. It is the mutual interference by and influence of different types of viral RNAs on each other during their synthesis. BMV RNA3 acts as a template for RNA replication as well as for sgRNA transcription. Grdzelishvili *et al.* (2005) studied replication of BMV RNA3 and formation of subgenomic RNA (sgRNA) in *Saccharomyces cerevisiae* expressing 1a and 2a replication proteins and which support full RNA3 replication cycle. They observed two main and opposing trends: firstly that blocking of sgRNA transcription (that is, inhibiting sgRNA synthesis) stimulated RNA3 replication by up to 350%, which shows that sgRNA transcription inhibits RNA3 replication. This inhibition only operated in *cis* and was independent of the product of sgRNA synthesis that is the capsid protein. Secondly, transcription/formation of sgRNA inhibited RNA3 replication at a step or steps after negative-strand RNA3 synthesis, implying competition with positive-strand RNA3 synthesis for negative-strand RNA3 templates, viral replication factors, or common host components. Accordingly, inhibition of positive-strand RNA3 synthesis stimulated sgRNA transcription by up to 400%. They concluded that BMV sub-genomic transcription and genomic RNA3 replication/synthesis mutually and strongly interfere with each other *in vivo*. It is the sgRNA synthesis itself and not sgRNA promoter sequences or products of sgRNA synthesis that inhibit RNA3 replication. Strikingly, this inhibition of sgRNA synthesis specifically targeted replication of its parental RNA3 but not separate RNA3 derivatives replicating in the same cell.

Hema and Kao (2004) and Sivakumaran *et al.* (2004) studied interaction of sgRNA and genomic RNA3 synthesis in barley protoplasts transfected with BMV RNAs 1 to 3. These studies also indicated interference of positive-strand RNA3 synthesis with sgRNA transcription (Hema and Kao, 2004) but there was one difference from the results of Grdzelishvili *et al.* (2005): inhibition of sgRNA transcription did not stimulate RNA3 accumulation in transfected protoplasts; rather, in many cases, inhibiting sgRNA transcription reduced RNA3 accumulation relative to that of wild type due largely to blocking capsid protein (CP) expression. Grdzelishvili *et al.* (2005), in line with the results of Hema and Kao (2004) and Sivakumaran *et al.* (2004), found that blocking sgRNA synthesis often has less effect on RNA3 accumulation in the barley protoplast system than in yeast.

Grdzelishvili *et al.* (2005) suggested that the mutual interference between sgRNA transcription and positive-strand RNA3 synthesis could be due to their competition for negative-strand RNA3 template, viral replication factors and for one or more common, limiting components of the BMV RNA synthesis machinery, or limiting host components. This competition is understandable since synthesis of both sgRNA and positive-strand RNA3 occurs in 50- to 70-nm-diameter endoplasmic reticulum (ER) membrane-bound RNA replication compartments, use the same negative-strand RNA3 template, requirement for BMV replication factors 1a and 2a, and have similar replication kinetics (Ishikawa *et al.*, 1997).

III. *CIS*-ACTING AND *TRANS*-ACTING VIRAL NUCLEOTIDE SEQUENCES

Positive-strand RNA viruses contain *cis*- and *trans*-acting nucleotide sequences, which affect and perform indispensable roles in viral RNA replication, in their 5`- and 3`-UTRs (Duggal *et al.*, 1994; Buck, 1996; Lai, 1998; Héricourt *et al.*, 1999). These nucleotide sequences have been referred to in literature by several interchangeable terms: elements/factors/motifs/sequences/signals/structures. In *cis* reaction, the nucleotide sequences and their site of action are contained in the same molecule; hence such a reaction is monomolecular. In *trans* reaction, the nucleotide sequences and their site of action are contained in separate molecules; hence this reaction is bimolecular.

A. *cis*-Acting Sequences

The site of action of the *cis*-acting motif may be near to or distal from the site where this motif is located. The *cis* reactions are generally very fast, occur cotranslationally, are also called 'autocatalytic' or 'autoproteolytic', and are functional in all stages of virus replication so that different stages of viral RNA replication have their own specific *cis*-acting sequences. The best-known *cis*-acting elements are the positive-strand viral RNA replication and transcription promoters, which are essential for initiation of RNA synthesis and formation of negative-strand RNA synthesis by specific RdRps. Many such *cis* sequences have been characterized in several RNA plant viruses (de Graaff and Jaspars, 1994; Buck, 1996; Lai, 1998; Héricourt *et al.*, 1999; Mandahar, 1999). The *cis* factors needed for formation of plus-strand progeny RNA and replication complexes have also been distinguished/predicted.

The *cis*-acting signals needed for RNA replication are always located on viral RNA genomes and are generally found at both the 3`- and 5`-UTRs of the RNA genomes. The 3`-terminal *cis*-acting elements are essentially required for faithful recognition and replication of only the cognate RNA by specific viral polymerases; have been characterized in many plant viruses; and contain poly(A) tails, pseudoknots, tRNA-like structures, stem-loop structures, hairpins or short primary sequences without high order structures. However, the *cis*-acting replication elements also exist in protein-coding regions as in *Tomato bushy stunt virus* (TBSV) (Ray and White, 2003). Several internally placed *cis*-acting elements are also known as in BMV and in other plant viruses (Duggal *et al.*, 1994; Buck, 1996; Dreher, 1999). More complicated cases of *cis* factors also exist where internal elements have a long-distance interaction with terminal sequences for efficient positive-strand synthesis as in *Potato virus X* (PVX) genomic RNA (Kim and Hemenway, 1999). Moreover, some *cis*-acting elements function as RNA chaperons and assist in replicase maturation (Quadt *et al.*, 1995; Vlot *et al.*, 2001). Thus, a large variety of *cis*-acting elements occur ranging from short linear nucleotide sequences to extensive multiple stem-loop structures. Plant viruses may contain several of these sequences. For example, BMV possesses three different types of promoters for different functions. One is the tRNA-like structure at 3`-end of genomic RNAs and contains motifs required for negative-strand synthesis. Second promoter is located at the 3`-terminus of negative-strand intermediates and is required for synthesis of genomic positive-strand RNA molecules. Third is the subgenomic RNA promoter, which is located internally on the genomic RNA. Similar replication enhancers also exist in several other plant viruses like *Alfalfa mosaic virus* (AMV), *Turnip crinkle virus* (TCV), TBSV, etc. and must be existing, but not yet identified, in many other plant viruses. Apart from the above well-defined and definite *cis*-acting structures, other *cis*-acting elements (involved in RNA replication) occur in many plant viruses and do not seem to possess easily defined sequence motifs.

It seems that RNA replication promoters (in TCV) are organized in a 'modular fashion', contain an RNA replication enhancer sequence and a replication initiation sequence. The segment A (a sub-element of region III of TBSV RNA) also behaves as a modular element (Ray and White, 2003). Both the activator proteins (of transcription and splicing) possess a common modular organization that has separate nucleic acid binding regions as well as regions that bind proteins by specific protein-protein interactions.

1. Internal Control Regions (Internal *cis*-Acting Sequences)

The internal control regions (ICR)-like motifs, frequently found in RNA genomes of plant viruses, act as promoters of viral positive-strand RNA synthesis. Marsh and Hall (1987) and French and Ahlquist (1987) discovered that 5`-untranslated regions of each BMV RNA and the intercistronic region of BMV RNA3 contain sequences that closely resemble the internal control regions (ICRs) 1 and 2 of tRNA genes. These sequences have been called by various names in literature - internal sequences/intergenic motifs or sequences/intercistronic regions/internal *cis*-acting sequences/internal control regions or motifs, The ICR-like sequences are present among BMV RNA 1; CMV RNAs 1, 2, and 3; TMV RNA; *Turnip yellow mosaic tymovirus* (TYMV) RNA and other plant virus genomes. These motifs are always placed within the known or putative promoter sequences involved in viral RNA replication.

The intercistronic region of BMV RNA3 occurs between genes for 3a protein and capsid protein, and contains one ICR-1 motif, three ICR-2 motifs and the intercistronic *cis* elements. The ICR-2-like motif starts at base 1100 in the intercistronic region and is almost identical to those present at the 5`-termini of BMV RNAs 1 and 2 and the tRNA-like consensus. Also, the 3`-most ICR-2 motif of the intercistronic region bears core promoters for subgenomic RNA formation. Pogue and Hall (1992) proposed that the ICR motifs at the 5`-end of RNA2 are part of a stem-loop structure such that the ICR2 domain is present in the loop region while the ICR1 domain is part of the stem.

Multiple consecutive C residues, contained in all the above sequences (Song and Simon, 1995; Wang and Simon, 1997), are arranged in a linear fashion and also occur in a hairpin replication enhancer motif located on negative strands of TCV satellite RNA C. Perhaps, this 5`-proximal linear replication enhancer element attracts RdRp to template (the negative-strand of TCV satellites RNA C) (Nagy *et al.*, 1998, 1999) and is required for synthesis of positive-strand of the satellite RNA C (Guan *et al.*, 2000). The above linear RNA element resembles BMV subgenomic RNA core promoter in which four (at positions -17, -14, -13, and -11) of the 22 nucleotides are essential for recognition by BMV RdRp *in vitro* (Siegel *et al.*, 1997, 1998). Two motifs (ICR1 and ICR2), located at 5`-terminal region of genomic RNAs of *Bromoviridae*, bind to cell transcription factor that has significant role in formation of initiation complex for synthesizing positive-strand RNA.

A 351-nucleotide segment, located upstream of an AUG that is distal to the first initiation codon in cowpea mosaic virus M component RNA, permits internal initiation at the distal AUG. Similarly, the 5`-leader of turnip mosaic potyvirus RNA or the nucleotide sequence located upstream of the 3`-end capsid protein subgenomic mRNA of a crucifer-infecting tobamovirus also permit internal initiation (Basso *et al.*, 1994; Ivanov *et al.*, 1999).

A *cis*-acting replication motif is located internally within the 3`-UTR of tombusviruses. It is called region III (Ray and White, 2003), has been characterized, and is postulated to have enhancer-like function during tombusvirus replication. The negative-stranded region III functions as a 'strong' replication enhancer and it increases RNA synthesis by 10-to-20-fold as compared to other viral sequences (Panavas and Nagy, 2003). These workers concluded that the replicase enhancer located in region III functions as a strong enhancer in negative-stranded RNA and a weak enhancer in positive-stranded RNA of tombusviruses. Ray and White (2003) also arrived at the same results concerning the role of region III in enhancing tombusvirus replication. Panavas and Nagy (2003) establish that region III of tombusviruses contains a replication enhancer that directly facilitates *de novo* RNA synthesis from minimal promoters; that negative-strand region III is a stronger replication enhancer than positive-strand region III; and that their *in vitro* studies support the results of *in vivo* studies of Ray and White (2003). Intergenic regions of RNA genomes of *Barley stripe mosaic virus* (Zhou and Jackson, 1996), and *Cucumber mosaic virus* (Boccard and Baulcombe, 1993) contain *cis* elements.

2. Role of *cis* Factors

Three functions have been attributed to the *cis*-acting elements: to recognize and then replicate cognate RNA, to recruit viral RdRp, and to facilitate *de novo* initiation of complimentary strand RNA synthesis. In addition, the RNA *cis*-acting sequences may also perform important roles during RNA recombination, viral evolution and virus adaptation. The role of the internal *cis*-acting motifs has not been well established but

(Buck, 1996; Kim and Hemenway, 1999). Mostly, the *cis*-acting elements seem to function through direct interaction with other sequence elements or viral or cellular proteins (Duggal *et al.*, 1994; Dreher, 1999; Sivakumaran *et al.*, 1999; Fabian *et al.*, 2003).

The distal *cis* sequences and/or structural elements affect ribosomal frameshifting, transcription, and translation in *Barley yellow dwarf virus* (Koev *et al.*, 1999; Guo *et al.*, 2000; Paul *et al.*, 2001). Stem-loop structures have great importance for transcription of TMV subgenomic RNA (Grdzelishvili *et al.*, 2000).

Mechanisms for transcription and splicing regulation require special *cis*-acting RNA sequences that promote and regulate activities of transcription or splice sites during transcription and RNA splicing, respectively. The promoters get associated with regulatory proteins to constitute multicomponent promoter/enhancer complexes that recruit the relevant enzymatic machinery to sequences that recognize the promoter or splice sites. During recruitment, direct interactions occur between the regulatory proteins in the enhancer complexes and the components of the basic enzymatic machineries. The high degree of regulatory specificity during transcription and splicing is largely due to the multicomponent nature of the enhancer complexes and to their cooperative assembly. Specific activator proteins recognize transcription enhancers while members of serine-arginine-rich protein family mostly recognise the splice enhancers.

Exact mechanisms of functioning of the enhancers during transcription and splicing phenomena are not well understood. However, the two processes have certain common features: cooperative binding of regulatory proteins to promoters forms highly stable complexes that recruit the basic components of transcription or splicing machinery to the nearby recognition motifs; regulation of transcription and splicing occurs by similar ways which include multicomponent nature of enhancer complexes, and modular organization of the enhancer-binding proteins.

The ICR-like motifs found in RNA genomes of plant viruses possibly act as promoters of viral positive-strand synthesis.

A role of RNA polymerase III involved in tRNA synthesis can be suggested in viral RNA replication.

Thus, some of the different steps in viral replication regulated by specific *cis*-acting elements (present in both plus- and negative-strand RNA templates) are: assembly of a replicase activity on inoculum RNAs, formation and recognition of replication complexes, transcription of the incoming infecting viral RNA genome, formation of positive-sense RNA from the negative-sense transcripts, increased translation of RNA, splicing of the translated polyproteins, cap-independent translation, assembly of virus particles, recognition of site for replication proteins, synthesis of genomic and subgenomic RNAs; and in even still other stages.

3. *cis*-Acting Sequences in Brome mosaic bromovirus RNAs

The *cis*-acting elements for efficient BMV genomic plus-strand, minus-strand, and subgenomic RNA synthesis have been characterized and a number of required *cis* elements have been identified. Choi *et al.* (2004) have given a summary of the known *cis*-acting sequences in BMV RNAs and the following information is based on this paper.

Plus-sense RNA1 contains the regulatory element called B-box located at +17 to +30 nucleotides, and stem-loop C (SLC) (within tRNA-like structure) located at +3156 to +3186 nucleotides. The 3`-noncoding regions of BMV genomic plus-strand RNAs form a tRNA-like structure that directs the initiation of minus-strand RNA synthesis *in vitro* and *in vivo*. The SLC of RNA3 binds BMV replicase through an RNA structure known as clamped adenine motif. Highly similar structures are predicted for RNA1 and RNA2, so that most likely the same structures are required to direct their minus-strand replication.

Negative-sense RNA1 contains complementary B-box (cB-box) located at –3204 to –3217 nucleotides. The core of cB box is a 4-nucleotide sequence, CCAA, which is conserved at comparable positions in RNA1 and RNA2 of other members of *Bromoviridae*. The cB box is the sequence complementary to the previously identified box B regulatory element.

Plus RNA2 contains B-box located at +17 to +30 nucleotides, and SLC located at +2787 to +2817 nucleotides.

Negative RNA2 bears cB-box located at –2835 to –2848 nucleotide.

Plus RNA3 bears a B-box-like structure located at +23 to +36 nucleotides, replication enhancer located at +1012 to +1221 nucleotides, replicase assembly site located at +1003 to +1217 nucleotides, B-box located at +1100 to +1113 nucleotides, and SLC located at +2039 to +2069 nucleotides.

Negative RNA3 contains minimal promoter located at –2092 to –2117 nucleotides, and subgenomic core promoter located at –876 to –895 nucleotides.

B. *trans*-Acting Viral RNA Nucleotide Sequences

Four satellite viruses exist in association with their helper viruses in natural plant infections: *Satellite maize white line mosaic virus* (SMWMV), *Satellite panicum mosaic virus* (SPMV), *Satellite tobacco mosaic virus* (STMV), and *Satellite tobacco necrosis virus* (STNV) (Dodds, 1998; Scholthof *et al.*, 1999). Satellite RNAs are dependent upon the respective helper viruses for replication, encapsidation and dissemination (Taliansky and Palukaitis, 1999). Helper viruses recognise the respective satellite viruses/RNAs in *trans* so that the helper viruses and the host plants have to provide all *trans*-acting factors necessary for replication of satellites (Qiu and Schlothof, 2000; Song and Miller, 2004).

This recognition of satRNAs in *trans* by their respective helper virus replicase complex is ensured mainly by sequences/signals present in their 5`- and 3`-UTRs (Routh *et al.*, 1997; Bringloe *et al.*, 1998). The *cis*-acting elements essential for STNV replication are embedded in three hairpin-like structures on 5`-UTR, two discrete regions identified as 3`-proximal terminal sequences, and the sequences immediately adjacent to capsid protein open reading frame (ORF) in 3`-UTR (Bringloe *et al.*, 1998). Similarly, the 3`-UTR sequence immediately downstream of capsid protein ORF is critical for accumulation of STMV (Routh *et al.*, 1995, 1997). On the other hand, capsid protein genes of both STNV and STMV are not essential for their accumulation and movement (Bringloe *et al.*, 1998; Routh *et al.*, 1995, 1997). Signals for replication and movement of SPMV RNA are located in 5`-proximal nucleotides - namely nucleotides numbers 62 to 110 in 5`-end, the entire 3`-UTR, and 73 nucleotides upstream of capsid protein stop codon.

The RNA1 of bipartite *Lettuce infectious yellows virus* (LIYV) encodes a *trans* enhancer for RNA2 accumulation (Yeh *et al.*, 2000). The RNA1 3`-terminal ORF2 encodes a protein p32 that acts as the *trans* enhancer of LIYV RNA2 accumulation. This LIYV RNA1-mediated *trans* replication enhancer activity (through p32) affecting accumulation of LIYV RNA2 is unusual among RNA plant viruses.

Red clover necrotic mosaic virus (RCNMV) is a bipartite virus and its genome consists of RNA1 and RNA2. Capsid protein is translated from a subgenomic RNA transcribed from RNA1. Sit *et al.* (1998) showed that a 34-nucleotide sequence in RNA2 is required for transcription of subgenomic RNA (from RNA1 of RCNMV) and they proposed a model in which RNA2 directly binds to RNA1 and *trans*-activates synthesis of subgenomic RNA. This regulation of transcription by RNA mediation is not a usual process in RNA viruses that normally rely on protein regulators. The *trans*-acting RNA element required for subgenomic RNA synthesis from RNA1 is located in protein-coding region of RNA2; however, RNA2 is not required for repliation of RNA1. It is the stem-loop structure SL2 of 20 nucleotides (of RNA2) that functions as a *trans*-activator for subgenomic RNA synthesis (from RNA1) (Sit *et al.*, 1998; Guenther *et al.*, 2004; Tatsuta *et al.*, 2005). Thus, SL2 is responsible for RNA-mediated *trans*-activation of subgenomic RNA transcription from RNA1 through direct pairing between RNA1 and RNA2. Simultaneously, SL2 is also a *cis*-acting RNA2 replication element; in fact, SL2 of RCNMV RNA2 has at least three different functions: as a *cis*-acting sequence for replication of RNA2, as a *trans*-acting sequence for capsid protein expression through production of subgenomic RNA, and as a coding sequence for movement protein (Tatsuta *et al.*, 2005).

IV. HOST GENE SHUT-OFF AND HIJACKING OF HOST CELLULAR MACHINERY BY VIRUSES

Viruses have developed the capacity to use machinery of the host cell for synthesis of their own DNA and RNA (during the viral nucleic acid replication process), and proteins (during the translation process) and utilize the cellular metabolic machinery to produce viral proteins and viral genome. DNA viruses have an inherent advantage here; they can use the already existing host cell machinery, which replicates and transcribes the host DNA, for replication and transcription of their DNA. In contrast, since no pre-existing cellular machinery exists in host cells for replication of RNA viral genomes, all RNA viruses encode an RNA polymerase. In fact, close integration between viral and host factors suggests the virus-infected cell to be a unified entity that constitutes the functional unit of infection (Ahlquist *et al.*, 2003).

There is some evidence about the way plant viruses out-compete cellular processes of the infected cells. The 3`-untranslated region of AMV RNA4 has a role in translation initiation and is also required for efficient competition of virus RNA4 with cellular messenger RNAs in wheat germ extracts and oocytes (Hann *et al.*, 1997). Moreover, many viral mRNAs have developed strategies of translation enhancement that are different from those used by most cellular mRNAs. Typically, cellular mRNAs have a terminal cap and a poly(A) tail which interact synergistically and function as co-dependent regulators of translation by interaction between these termini of mRNAs (Gallie, 1996, 1998). This happens in many of the plant viruses also but cap is absent in a significant number of plant viruses so that they exhibit cap-independent translation mechanism. Enhanced translation of specific viral mRNAs leading to high levels of protein synthesis of specific viral genes in plants may be a fundamental mechanism by which viral mRNAs out-compete their cellular counterparts (Qu and Morris, 2000).

Viruses can down-regulate the expression of cellular genes, an effect that has been called host gene 'shut-off' by Aranda and Maule (1998). Host shut-off is the process that results in suppression of cellular macromolecular synthesis after infection by many but not all viruses, is caused by viral domination of host metabolism and so is a virus-directed process, favours selective translation of viral mRNAs over that of the endogenous mRNAs, is partial and never absolute, and is not always required for virus replication. This selective translation of viral mRNAs during host shut-off is due to a general competition between viral and host mRNAs for host translation machinery. There was a complete transient inhibition of expression of at least 10 different cellular genes in host cells associated with replication of *Pea seed-borne mosaic potyvirus* (PSbMV) (Wang and Maule, 1995). PSbMV transiently suppressed the expression of host genes in cells most active in viral replication and these cells formed a band encompassing between four to eight cell layers. Restrepo-Hartwig and Carrington (1992) had suggested, but not proven, that nuclear inclusion proteins a and b (NIa and NIb) of *Tobacco etch virus* (TEV) may affect host gene expression, ribosome assembly, or some other nuclear event. In fact, host shut-off is now considered a common phenomenon associated with animal and plant virus replication. It is not yet certain whether host shut-off is a prerequisite for virus multiplication or is an indirect consequence of viral protein function. Simultaneously, there is more than a compensatory increase in protein in cells just behind the infection front (Wang and Maule, 1995).

The host shut-off process is selective and never all-inclusive because all viruses need some components of host pathways for genome replication and expression. Thus, despite the down-regulation of host gene expression, some host genes escape shut-off and may even show coordinate induction. In PSbMV-infected peas, many host mRNAs were removed from the infected cell while HSP70 and polyubiquitin mRNA accumulation increased; these host RNAs were translated to give increased protein in the shut-off zone. These induced host genes, like the viral RNA, have the ability to escape shut-off.

Host shut-off may have some consequences for the host cell as well. However, no obvious cytopathic effect was apparent in cells showing shut-off (Wang and Maule, 1995; Aranda *et al.*, 1996) but electron microscope studies showed a rapid but transient loss of ER linked to shut-off onset (Roberts *et al.*, 1998). The long-term effect of shut-off was neither cell lysis nor necrosis. However, transient shut-off in infected cells could act as a trigger for a series of physiological changes manifested later as virus-induced symptoms (Técsi *et al.*, 1996). Other proteins also increase during plant virus infection (Aranda *et al.*, 1996; Havelda and Maule, 2000). Increase in protein expression could prepare the cell for biosynthetic demands of viral infection.

A. Mechanisms of Shut-off

Host shut-off has been investigated in far more detail in animal than plant viruses (Aranda and Maule, 1998; Bushell and Sarnow, 2002); hence the examples given here more often pertain to the animal viruses. The precise mechanism(s) for achieving host shut-off vary widely in different virus infections and has been attributed to various factors including perturbation of intracellular ion concentration, altered nucleotide metabolism, alterations in RNA stability, changes in processing and recruitment of specific host factors, alterations in nucleocytoplasmic transport of mRNAs, modifications in cellular transcription process, and alterations in cellular translation machinery and mechanism. Host shut-off can function at different points of the eukaryotic gene expression pathway and any of the key control steps on the pathway can serve as targets for shut-off.

Poliovirus replicates in cytoplasm but causes shut-off of nuclear functions since it blocks cellular transcription by inhibiting RNA pol II transcription by its protease $3C^{pro}$. Thus, decrease in host RNA synthesis in *Poliovirus*-infected cells is correlated with degradation of transcription factors by viral serine protease 3C, which causes a cleavage of the cellular translational component p220. The mRNA maturation is another method. *Herpes simplex virus-1* (HSV-1) replicates in nucleus but influences host mRNA maturation through the action of its IE63 (also called ICP27) gene product. This results in reduced accumulation of host spliced mRNA. Thus, HSV-1 disrupts the cellular splicing machinery so as to promote preferential maturation of its own transcripts.

Modification of nucleocytoplasmic transport of mRNA is still another method by which viruses affect host shut-off. *Vesicular stomatitis virus* (VSV) (a negative-sense RNA virus) controls nucleocytoplasmic transport of host mRNAs. The virus replicates in cytoplasm but its matrix (M) protein localizes to nucleus of infected cell and effectively blocks export of host mRNAs and proteins from nucleus in *Xenopus laevis* oocytes. The M protein interferes with transport mechanisms dependent on guanosine triphosphatase (GTPase) and its associated factors. In contrast, subgroup C human adenoviruses, which replicate in nucleus, act in another way. Most cellular mRNAs, although transcribed, fail to enter cytoplasm efficiently during the late phase of infection, whereas viral mRNAs and some cellular RNA species are translocated. This selective block requires the synthesis of viral E1B-55K and E4 proteins.

Influence of viruses on stability of host mRNA(s), like decreasing the stability of mRNA by HSV-1, causes host shut-off. Such degradation of host mRNAs is widespread in animal viruses and has also been detected in a plant-infecting potyvirus, PSbMV. Many host mRNAs, presumably by degradation, are lost very rapidly following the onset of viral RNA replication (Wang and Maule, 1995). There is some selectivity in this process since some host mRNAs accumulate and are translated.

Deleterious effect of virus infection on translation process of the host cell is the most important cause of host shut-off. Translation of viral mRNAs by host machinery occurs essentially as that of cellular mRNAs (Gale *et al.*, 2000; Thompson and Sarnow, 2000) and one of the first steps is the recruitment of mRNAs by translation initiation factor eIF4F complex. But, an invading virus does not contain functional ribosomes so that for successful amplification of its genomes, a virus mRNA has to compete with cellular mRNAs for host cell translation machinery and for other functions. Several RNA viruses have developed remarkable strategies to do so. Bushell and Sarnow (2002) review the mechanisms by which viruses take over the translational machinery of the host cell. The most studied example of a block in translation involves animal picornaviruses and is reviewed by Bushell and Sarnow (2002). Recruiting the host translation initiation factors for translation initiation by the invading virus is one such step. This is done in three different ways. Cleavage of poly(A)-binding protein (PABP) and disruption of closed-loop translation complex is one such method. This happens by proteolysis/ cleavage of PABP and eukaryotic initiation factor 4G (eIF4G that is a subunit of eIF4F) and results in cessation of translation initiation. This process is adopted by picornaviruses. Poliovirus 2A^{pro} causes this disruption. Host mRNA translation through modulation of PABP, which is a 70-kDa RNA-binding protein that plays a direct role in mRNA stability through its interaction with poly(A) tails of mRNAs. The circularization of translation initiation complex is facilitated through PABP interaction with mRNA poly(A) tail and eIF4G subunit. The resulting 'closed loop' translation initiation complex is thought to stabilize assembled initiation factors and increase translation efficiency. Thus, viral disruption of PABP function can be expected to alter mRNA stability and reduce the overall rate and efficiency of cap-dependent mRNA translation. Moreover, the modified eIF4F becomes inactive for cap-dependent translation of cellular mRNAs but remains functional for translation of uncapped viral RNAs by internal ribosome entry as in picornaviruses. Thus, decreasing the cap-binding protein complex either by cleavage of eIF4GI and eIF4GII or by sequestration of eIF4E selectively inhibits translation of capped host cell mRNAs without inhibiting the translation of picornaviral mRNAs. Léonard *et al.* (2004) found that viral protein linked to genome (VPg) of *Turnip mosaic potyvirus* (TuMV) interacts *in vitro* with translation eukaryotic translation initiation factor (eIF) 4E and also showed direct interaction between VPg-Pro (a precursor of VPg) and PABP and between VPg-Pro and eIF4E. Association of PABP with eIF4F results in circularization of mRNAs. Thus, interaction between VPg-Pro and eIF4E as well as PABP could possibly promote viral RNA circularization during translation.

Gallie and Browning (2001) propose that eIF4F (a higher order protein complex that also contains eIF4E besides eIF4G) may promote translation under cellular conditions when cap-dependent translation is inhibited. Viral infection often leads to cap-dependent inhibition of host mRNA translation (Gale *et al.*, 2000; Bushell and Sarnow, 2002). Léonard *et al.* (2004) suggest that TuMV infection of *Brassica pervirdis* leads to inactivation of eIF4E or to its monopolization for viral protein synthesis.

Second method is by substitution of PABP as performed by rotaviruses of *Reoviridae*. Interaction between PABP and eIF4G is disrupted in rotavirus-infected cells by viral protein NSP3 - leading to reduced efficiency of host mRNA translation and circularization-mediated translational enhancement of rotavirus mRNAs. This also demonstrates how viral nonpolyadenylated mRNAs can usurp the host cell translation apparatus by encoding a protein that binds to the 3`-ends of viral mRNAs, evicting PABP from eIF4G, which is the most important factor involved in recruitment of ribosomes to mRNAs.

Third method is by bypassing of initiator tRNA as practiced by cricket paralysis-like viruses. In *Cricket paralysis virus*, the intergenic region (IGR)-internal ribosome entry site (IRES) sequences themselves occupy the ribosomal P-site; a CCU triplet at the start site base pairs directly with upstream IGR-IRES sequences. This IRES can recruit both 40S and 60S subunits without any known canonical eIFs to form an 80S ribosome that can start protein synthesis from the next codon, a GCU, which is located in the ribosomal A-site. Thus, the first amino acid in the protein is alanine encoded by the A-site located GCU codon. Above findings show that IGR-IRES element propels the ribosome into elongation mode without prior formation of a peptide bond.

V. CLASSIFICATION AND NOMENCLATURE

The various reports of the International Committee on Taxonomy of Viruses (ICTV) have removed much of the confusion and controversy about classification and nomenclature of viruses. In fact, ICTV has evolved a universal system of virus classification on the basis of principles and rules agreed upon by all branches of virology (Mayo, 2002). Its latest report is the Seventh Report published in 2000 (Van Regenmortel *et al.*, 2000) while Mayo (2002) and Van Regenmortel (2002) have summarized the relevant fields of plant viruses. The number of plant virus genera is given below on the basis of their genome types as per Van Regenmortel *et al.* (2000) as mentioned by Mayo (2002): plant virus genera having double-stranded DNA genome - none; plant virus genera having single-stranded DNA genome - 4; plant virus genera having retroid DNA genome - 6; plant virus genera having retroid RNA genome - 2; plant virus genera having double-stranded RNA genome - 6; plant virus genera having single-stranded RNA genome (negative-sense RNA) - 5; plant virus genera having single-stranded RNA genome (positive-sense RNA) - 49; the above numbers together bring the total number of plant virus genera to 72.

A. Classification

The current classification of plant virus genera containing positive-sense RNA genome, shown below, is as per Mayo (2002) [except that the genera have been arranged here in alphabetic order; that the words 'Unassigned genera' are included here; and that assignment of various genera to superfamilies, of genera of family *Luteoviridae*, and of 'Unassigned Genera' is after Morozov and Solovyev (1999)]. The name given in parenthesis, after the name of genus, is the type species of that particular genus.

1. Family *Bromoviridae* (Alphavirus-like supergroup)
    Genus *Alfamovirus* (*Alfalfa mosaic alfamovirus*)
    Genus *Bromovirus* (*Brome mosaic bromovirus*)
    Genus *Cucumovirus* (*Cucumber mosaic cucumovirus*)
    Genus *Ilarvirus* (*Tobacco streak ilarvirus*)
    Genus *Oleavirus* (*Olive latent oleavirus 2*)
2. Family *Closteroviridae* (Alphavirus-like supergroup)
    Genus *Closterovirus* (*Beet yellows closterovirus*)
    Genus *Crinivirus* (*Lettuce infectious yellows crinivirus*)
3. Family *Comoviridae* (Picorna-like supergroup)
    Genus *Comovirus* (*Cowpea mosaic comovirus*)
    Genus *Fabavirus* (*Broad bean wilt fabavirus*)
    Genus *Nepovirus* (*Tobacco ringspot nepovirus*)
4. Family *Luteoviridae*
    Genus *Enamovirus* (*Pea enation mosaic enamovirus1*) (Sobemo-like supergroup)
    Genus *Luteovirus* (*Barley yellow dwarf luteovirus*-PAV) (Carmo-like supergroup)
    Genus *Polerovirus* (*Potato leafroll polerovirus*) (Sobemo-like supergroup)
5. Family *Potyviridae* (Picorna-like supergroup)
    Genus *Bymovirus* (*Barley yellow mosaic bymovirus*)
    Genus *Ipomovirus* (*Sweet potato mild mottle ipomovirus*)
    Genus *Macluravirus* (*Maclura mosaic macluravirus*)
    Genus *Potyvirus* (*Potato potyvirus Y*)
    Genus *Rymovirus* (*Ryegrass mosaic rymovirus*)
    Genus *Tritimovirus* (*Wheat streak mosaic tritimovirus*)
6. Family *Sequiviridae* (Picorna-like supergroup)
    Genus *Sequivirus* (*Parsnip yellow fleck sequivirus*)
    Genus *Waikavirus* (*Rice tungro spherical waikavirus*)
7. Family *Tombusviridae* (Carmo-like supergroup)
    Genus *Aureusvirus* (*Pothos latent aureusvirus*)
    Genus *Avenavirus* (*Oat chlorotic stunt avenavirus*)
    Genus *Carmovirus* (*Carnation mottle carmovirus*)
    Genus *Dianthovirus* (*Carnation ringspot dianthovirus*)
    Genus *Machlomovirus* (*Maize chlorotic mottle machlomovirus*)
    Genus *Necrovirus* (*Tobacco necrosis necrovirus*)
    Genus *Panicovirus* (*Panicum mosaic panicovirus*)
    Genus *Tombusvirus* (*Tomato bushy stunt tombusvirus*)
8. Unassigned Genera (Genera not assigned to any family by Mayo, 2002)
    Genus *Allexvirus* (*Shallot allexvirus X*) (Family *Potexviridae*)
    Genus *Benyvirus* (*Beet necrotic yellow vein benyvirus*) (Family '*Tubiviridae*')
    Genus *Capillovirus* (*Apple stem grooving capillovirus*) (Proposed order *Tymovirales* of alphavirus-like supergroup)

Genus *Carlavirus* (*Carnation latent carlavirus*) (Family *Potexviridae*)
Genus *Foveavirus* (*Apple stem pitting foveavirus*) (Family *Potexviridae*)
Genus *Furovirus* (*Soil-borne wheat mosaic furovirus*) (Family '*Tubiviridae*')
Genus *Hordeivirus* (*Barley stripe mosaic hodeivirus*) (Family '*Tubiviridae*')
Genus *Idaeovirus* (*Raspberry bushy dwarf idaeovirus*) (Family *Bromoviridae* of alphavirus-like superfamily)
Genus *Marafivirus* (*Maize rayado fino marafivirus*) (Proposed order *Tymovirales* of alphavirus-like supergroup)
Genus *Ourmiavirus* (*Ourmia melon ourmiavirus*)
Genus *Pecluvirus* (*Peanut clump pecluvirus*) (Family '*Tubiviridae*')
Genus *Pomovirus* (*Potato mop-top pomovirus*) (Family '*Tubiviridae*')
Genus *Potexvirus* (*Potato potexvirus X*) (Family *Potexviridae*)
Genus *Sobemovirus* (*Southern bean mosaic sobemovirus*) (Sobemo-like supergroup)
Genus *Tobamovirus* (*Tobacco mosaic tobamovirus*) (Family '*Tubiviridae*')
Genus *Tobravirus* (*Tobacco rattle tobravirus*) (Family '*Tubiviridae*')
Genus *Trichovirus* (*Apple chlorotic leaf spot trichovirus*) (Proposed order *Tymovirales* of alphavirus-like supergroup)
Genus *Tymovirus* (*Turnip yellow mosaic tymovirus*) (Proposed order *Tymovirales* of alphavirus-like supergroup)
Genus *Umbravirus* (*Carrot mottle umbravirus*) (Carmo-like supergroup)
Genus *Vitivirus* (*Grapevine vitivirus A*) (Proposed order *Tymovirales* of alphavirus-like supergroup)

B. Nomenclature

Virologists are strongly opposed to the introduction of Latin binomials for naming viruses (Van Regenmortel, 1989), and the nomenclature system developed by ICTV reflects this position clearly (Van Regenmortel *et al.*, 2000). In formal virus taxonomy, names of orders, families, subfamilies, genera, and species are always printed in italics and the first letters of the names are capitalized, except in case of species where the first letter is also capitalized if they are proper nouns, or parts of proper names. Use of italics for the names of species taxa clearly indicates that the species name has been approved as the official, internationally recognized name (Pringle, 1998). It should be stressed that italics and an initial capital letter need to be used only if the species name refers to a taxonomic category (Van Regenmortel *et al.*, 2000); thus, use of italics is necessary if the name of a species is used as a taxonomic entity.

However, italicized taxonomic names and initial capital letters are not appropriate but the names are written in lower case roman script and correspond to informal vernacular in the following situations. Firstly, when the names refer to physical entities such as the virions found in a virus preparation or seen in an electron micrograph since it is not possible to centrifuge or visualize a virus family (like *Tombusviridae*) or a genus (like *Tobamovirus*), or a species (like *Poliovirus*). Secondly, no italics are needed when the virus names are used in adjectival form, for instance, tobacco mosaic virus polymerase. Thirdly, when the taxonomic status of a new putative species is not certain or its positioning within an established genus has not been clarified, it is considered as a 'tentative' species and its name is not to be given in italics, although its initial letter will be capitalized.

VI. ABBREVIATIONS

A. Plant Viruses

ACLSV - *Apple chlorotic leafspot trichovirus*
AMV - *Alfalfa mosaic alfamovirus*
ASGV - *Apple stem grooving capillovirus*
BaMV - *Bamboo mosaic potexvirus*
BaYMV - *Barley yellow mosaic bymovirus*
BBMV - *Broad bean mottle bromovirus*
BCMV - *Bean common mosaic bymovirus*
BCV1 - *Beet cryptic alphacryptovirus 1*
BCV3 - *Beet cryptic alphacryptovirus 3*
BlSV - *Bluberry scorch carlavirus*
BMV - *Brome mosaic bromovirus*
BNYVV - *Beet necrotic yellow vein benyvirus*
BrSMV - *Brome streak mosaic rymovirus*
BSMV - *Barley stripe mosaic hordeivirus*
BVQ - *Beet furo-like virus Q*
BWYV - *Beet western yellows luteovirus*
BYDV - *Barley yellow dwarf luteovirus*
BYMV - *Barley yellow mosaic bymovirus*
BYV - *Beet yellows closterovirus*
CaMV - *Cauliflower mosaic caulimovirus*
CarMV - *Carnation mottle carmovirus*
CCMV - *Cowpea chlorotic mottle bromovirus*
CCSV- *Cucumber chlorotic spot closterovirus*
CfMV - *Cocksfoot mottle sobemovirus*
ClYMV - *Clover yellow mosaic potexvirus*
CLRV - *Cherry leafroll nepovirus*
CMV - *Cucumber mosaic cucumovirus*
CNV - *Cucumber necrosis tombusvirus*
CPMV - *Cowpea mosaic comovirus*
CRSV - *Carnation ringspot dianthovirus*
CTLV - *Citrus tatter leaf capillovirus*
CTV - *Citrus tristeza closterovirus*
CymRSV - *Cymbidium ringspot tombusvirus*
GFLV - *Grapevine fanleaf nepovirus*
GRV - *Groundnut rosette umbravirus*
GVA - *Grapevine virus A*
GVB - *Grapevine trichovirus B*
LIYV - *Lettuce infectious yellows closterovirus*
MCMV - *Maize chlorotic mottle machlomovirus*
MFV - *Maize rayado fino virus*
MRDV - *Maize rough dwarf fijivirus*
MStV - *Maize stripe tenuivirus*

OBDV - *Oat blue dwarf marafivirus*
PCV - *Peanut clump pecluvirus*
PDV - *Prune dwarf ilarvirus*
PEBV - *Pea early browning tobravirus*
PEMV - *Pea enation mosaic enamovirus*
PLRV - *Potato leaf roll polerovirus*
PMTV - *Potato mop top pomovirus*
PMV - *Panicum mosaic sobemovirus*
PPV - *Plum pox potyvirus*
PSbMV - *Pea seedborne mosaic potyvirus*
PVM - *Potato carlavirus M*
PVX - *Potato potexvirus X*
PVY - *Potato potyvirus Y*
PYFV - *Parsnip yellow fleck Sequivirus*
RBDV - *Rasberry bushy dwarf idaeovirus*
RCNMV - *Red clover necrotic mosaic dianthovirus*
RDV - *Rice dwarf phytoreovirus*
RGSV - *Rice grassy stunt tenuivirus*
RRSV - *Rice ragged stunt oryzavirus*
RStV - *Rice stripe tenuivirus*
RTSV - *Rice tungro spherical waikavirus*
RYMV - *Rice yellow mottle sobemovirus*
SbDV - *Soybean dwarf luteovirus*
SBMV - *Southern bean mosaic sobemovirus*
SBWMV - *Soil-borne wheat mosaic furovirus*
STNV - *Satellite tobacco necrosis virus*
TBRV - *Tomato black ring nepovirus*
TBSV - *Tomato bushy stunt tombusvins*
TCV - *Turnip crinkle carmovirus*
TEV - *Tobacco etch potyvirus*
TMV - *Tobacco mosaic tobamovirus*
TNV - *Tobacco necrosis necrovirus*
TomRSV - *Tomato ringspot nepovirus*
TRV - *Tobacco rattle tobravirus*
TSV - *Tobacco streak ilarvirus*
TSWV - *Tomato spotted wilt tospovirus*
TuMV - *Turnip mosaic potyvirus*
TVMV - *Tobacco vein mottling potyvirus*
TYMV - *Turnip yellow mosaic tymovirus*
WClMV - *White clover mosaic potexvirus*
WTV - *Wound tumor phytoreovirus*

B. Other Abbreviations

CI - Cytoplasmic inclusion
CIT - Cap-independent translation
CP - Coat/capsid protein
DI - Defective interfering RNA
ds - Double-strand(ed)
ER - Endoplasmic reticulum
f/s - Frameshift(ing)
HC-Pro - Helper component proteinase
HEL - Helicase
HSP - Heat shock protein
ICR - Internal control region
IRES - Internal ribosome entry site
K/kDa - Kilodalton
MP - Movement protein
MT - Methyltransferase
NI - Nuclear inclusion

NMR - Nuclear magnetic resonance
ns - Non structural
nt - Nucleotide
NTP - Nucleotide triphosphate
NTR - Nontranslated region
ORF(s) - Open reading frame(s)
PABP - Poly(A) binding protein
POL - Polymerase
RdRp - RNA-dependent RNA polymerase
sgRNA - Subgenomic RNA
ss - Single-strand(ed)
TGB - Triple gene block
TLS - tRNA-like structure
UTR - Untranslated region
VPg - Genome-linked protein

VII. REFERENCES

Ahlquist, P., Noueiry, A. O., Lee, W.-M., Kushner, D. B., and Dye, B. T. 2003. Host factors in positive-strand RNA virus genome replication. *J. Virol.* 77: 8181-8186.

Ahlquist, P., Strauss, E. G., Rice, C. M., Strauss, J. H., Haseloff, J., and Zimmern, D. 1985. Sindbis virus proteins nsP1 and nsP2 contain homology to nonstructural proteins from several RNA plant viruses. *J. Virol.* 53: 536-542.

Anindya, R., Chittori, S., and Savithri, H. S. 2005. Tyrosine 66 of pepper vein banding virus genome-linked protein is uridylylated by RNA-dependent RNA polymerase. *Virology* 336: 154-162.

Aranda, M., and Maule, A. 1998. Virus-induced host gene shutoff in animals and plants. *Virology* 243: 261-267.

Aranda, M. A., Escaler, M., Wang, D. W., and Maule, A. J. 1996. Induction of HSP70 and polyubiquitin expression associated with plant virus replication. *Proc. Nat. Acad. Sci. USA* 93: 15289-15293.

Argos, P., Kamer, P., Nicklin, M. J. H., and Wimmer, E. 1984. Similarity in gene organisation and homology between proteins of animal picornaviruses and plant comoviruses suggest common ancestry of these virus families. *Nucl. Acids Res.* 12: 7251-7267.

Basso, J., Dellaire, P., Charest, P. J., Devantier, Y., and Laliberte, J. -F. 1994. Evidence for an internal ribosome entry site within the 5` nontranslated region of turnip mosaic potyvirus RNA. *J. Gen. Virol.* 75: 3157-3165.

Boccard, F., and Baulcombe, D. C. 1993. Mutational analysis of *cis*-acting sequences and gene function in RNA3 of *Cucumber mosaic virus*. *Virology* 193: 563-578.

Bringloe, D. H., Gultyaev, A. P., Pelpel, M., Pleij, C. W., and Coutts, R. H. 1998. The nucleotide sequence of *Satellite tobacco necrosis virus* strain C and helper-assisted replication of wild-type and mutant clones of the virus. *J. Gen. Virol.* 79: 1539-1546.

Buck, K. W. 1996. Comparison of the replication of positive-stranded RNA viruses of plants and animals. *Adv. Virus Res.* 47: 159-251.

Buck, K. W. 1999. Replication of tobacco mosaic virus RNA. *Phil. Trans. Roy. Soc., London* 354: 613-627.

Bushell, M., and Sarnow, P. 2002. Hijacking the translation apparatus by RNA viruses. *J. Cell Biol.* 158: 395-399.

Chandrika, R., Ranbidran, S., Lewandowski, D. J., Manjunath, K. L., and Dawson, W. O. 2000. Full tobacco mosaic virus RNAs and defective RNAs have different 3`-replication signals. *Virology* 273: 198-203.

Chapman, M. R., and Kao, C. C. 1999. A minimal RNA promoter for minus-strand RNA synthesis by the brome mosaic virus polymerase complex. *J. Mol. Biol.* 286: 709-720.

Chen, J., Noueiry, A., and Ahlquist, P. 2001. Brome mosaic virus protein 1a recruits viral RNA2 to RNA replication through a 5`-proximal RNA2 signal. *J. Virol.* 75: 3207-3219.

Choi, S.-K., Hema, M., Gopinath, K., Santos, J., and Kao, C. C. 2004. Replicase-binding sites on plus- and minus-strand brome mosaic virus RNAs and their roles in RNA replication in plant cells. *J. Virol.* 78: 13420-13429.

David, C., Gargouri-Bouzid, R., and Harnni, A.-L. 1992. RNA replication of plant viruses containing an RNA genome. *Prog. Nucleic Acids Res. & Mol. Biol.* 42: 157-227.

de Graaff, M., and Jaspars, E. M. J. 1994. Plant viral RNA synthesis in cell-free systems. *Annu. Rev. Phytopathol.* 32: 311-335.

Diez, J., Ishikawa, M., Kaido, M., and Ahlquist, P. 2000. Identification and characterization of a host protein required for efficient template selection in viral RNA replication. *Proc. Nat. Acad. Sci. USA* 97: 3913-3918.

Dodds, J. A. 1998. Satellite tobacco mosaic virus. *Annu. Rev. Phytopathol.* 36: 295-310.

Dougherty, W. G., and Semler, B. L. 1993. Expression of virus-encoded proteinases: Functional and structural similarities with cellular enzymes. *Microbiol. Rev.* 57: 781-822.

Dreher, T. W. 1999. Functions of 3`-untranslated regions of positive-stranded RNA viral genomes. *Annu. Rev. Phytopathol.* 37: 151-174.

Duggal, R., Lahser, F. C., and Hall, T. C. 1994. *cis*-Acting sequences in the replication of plant viruses with plus-sense RNA genomes. *Annu. Rev. Phytopathol.* 32: 287-309.

Eckerle, L. D., and Ball, L. A. 2002. Replication of the RNA segments of a bipartite viral genome is coordinated by a transactivating subgenomic RNA. *Virology* 296: 165-176.

Fabian, M. R., Na, H., Ray, D., and White, K. A. 2003. 3`-Terminal RNA secondary structures are important for accumulation of tomato bushy stunt virus DI RNAs. *Virology* 313: 567-580.

Franssen, H., Leunissen, J., Goldbach, R., Lomonossoff, G. P., and Zimmern, D. 1984. Homologous sequences in nonstructural proteins from *Cowpea mosaic virus* and picornaviruses. *EMBO J.* 3: 855-861.

French, R., and Ahlquist, P. 1987. Intercistronic as well as terminal sequences are required for efficient amplification of brome mosaic virus RNA3. *J. Virol.* 61: 1457-1465.

Gale, M., Jr., Tan, S.-L., and Katze, M. G. 2000. Translational control of viral gene expression in eukaryotes. *Microbiol Mol. Biol. Rev.* 64: 239-280.

Gallie, D. R. 1996. Translational control of cellular and viral mRNAs. *Plant Mol. Biol.* 32: 145-158.

Gallie, D. R. 1998. A tail of two termini: A functional interaction between the termini of an mRNA is a prerequisite for efficient translation initiation. *Gene* 216: 1-11.

Gallie, D. R., and Browning, K. S. 2001. eIF4G functionally differs from eIFiso4G in promoting internal initiation, cap-independent translation and translation of structured mRNAs. *J. Biol. Chem.* 276: 36951-36960.

Gamarnik, A. V., and Andino, R. 1998. Switch from translation to RNA replication in a positive-stranded RNA virus. *Genes Dev.* 12: 2293-2304.

Goldbach, R. 1987. Genome similarities between plant and animal RNA viruses. *Microbiol. Sci.* 4: 197-202.

Grdzelishvili, V. Z., Chapman, S. N., Dawson, W. O., and Lewandowski, D. J. 2000. Mapping of the tobacco mosaic virus movement protein and coat protein subgenomic RNA promoters *in vivo*. *Virology* 275: 177-192.

Grdzelishvili, V. Z., Garcia-Ruiz, H., Watanabe, T., and Ahlquist, P. 2005. Mutual interference between genomic RNA replication and subgenomic mRNA transcription in *Brome mosaic virus*. *J. Virol.* 79: 1438-1451.

Guan, H., Carpenter, C. D., and Simon, A. E. 2000. Analysis of *cis*-acting sequences involved in plus-strand synthesis of a *Turnip crinkle virus*-associated satellite RNA identifies a new carmovirus replication element. *Virology* 268: 345-354.

Guenther, R. H., Sit, T. L., Gracz, H. S., Dolan, M. A., Townsend, H.-L., Liu, G., Newman, W. H., Agris, P. F., and Lommel, S. A. 2004. Structural characterization of an intermolecular RNA-RNA interaction involved in the transcription regulation element of a bipartite plant virus. *Nucl. Acids Res.* 32: 2819-2828.

Guo, L., Allen, E. M., and Miller, W. A. 2000. Structure and function of a cap-independent translation element that functions in either the 3`- or the 5`-untranslated region. *RNA* 6: 1808-1820.

Hann, L. E., Webb, A. C., Cai, J. M., and Gehrke, L. 1997. Identification of a competitive translation determinant in the 3`-untranslated region of alfalfa mosaic virus coat protein mRNA. *Mol. Cell. Biol.* 17: 2005-2013.

Haseloff, J., Goelet, P., Zimmern, D., Ahlquist, P., Dasgupta, R., and Kaesberg, P. 1984. Striking similarities in amino acid sequence among non-structural proteins encoded by RNA viruses that have dissimilar genomic organisation. *Proc. Nat. Acad. Sci. USA* 81: 4358-4362.

Havelda, Z., and Maule, A. J. 2000. Complex spatial responses to *Cucumber mosaic virus* infection in susceptible *Cucurbita pepo* cotyledons. *Plant Cell* 12: 1975-1986.

Hema, M., and Kao, C. C. 2004. Template sequence near the initiation nucleotide can modulate brome mosaic virus RNA accumulation in plant protoplasts. *J. Virol.* 78: 1169-1180.

Héricourt, F., Jupin, I., and Haenni, A.-L. 1999. Genome of RNA viruses. In: Mandahar, C. L. (Ed.). Molecular Biology of Plant Viruses. Kluwer Academic Publishers, Boston/Dordrecht/London. p. 1-28.

Herold, J., and Andino, R. 2001. Poliovirus RNA replication requires genome circularization through a protein-protein bridge. *Mol. Cell* 7: 581-591.

Hofschneider, P. H., and Husen, P. 1968. The small RNA viruses of plants, animals and bacteria. C. The replicative cycle. In: Fraenkel-Conrat, H. (Ed.). Molecular Basis of Virology. Van Nostrand Reinhold, New York. pp. 169-208.

Hong, Y., and Hunt, A. G. 1996. RNA polymerase activity by a potyvirus-encoded RNA-dependent RNA polymerase. *Virology* 226: 146-151.

Ishihama, A., and Barbier, P. 1994. Molecular anatomy of viral RNA-directed RNA polymerases. *Arch. Virol.* 134: 235-258.

Ishikawa, M., Janda, M., Krol, M. A., and Ahlquist, P. 1997. *In vivo* DNA expression of functional brome mosaic virus RNA replicons in *Saccharomyces cerevisiae.. J. Virol.* 71: 7781-7790.

Ivanov, P. A., Karpova, O. V., Skulachev, M. V., Tomashevskaya, O. L., Rodionova, N. P., Dorokhov, Yu. L., and Atabekov, J. G. 1997. A tobamovirus genome that contains an internal ribosome entry site functions *in vitro. Virology* 232: 32-43.

Janda, M., and Ahlquist, P. 1993. RNA-dependent replication, transcription and persistence of brome mosaic virus RNA replicons in *S. cerevisiae. Cell* 72: 961-970.

Kadaré, G., and Haenni, A.-L. 1997. Virus-encoded helicases. *J. Virol.* 71: 2583-2590.

Kamer, G., and Argos, P. 1984. Primary structural comparisons of RNA-dependent polymerases from plant, animal and bacterial viruses. *Nucl. Acids Res.* 12: 7269-7282.

Kao, C. C., Singh, P., and Ecker, D. J. 2001. *De novo* initiation of viral RNA-dependent RNA synthesis. *Virology* 287: 251-260.

Khromykh, A. A., Meka, H., Guyatt, K. J., and Westaway, E. G. 2001. Essential role of cyclization in flavivirus RNA replication. *J. Virol.* 75: 6719-6728.

Kim, C. H., Kao, C. C., and Tinoco, I. 2000. RNA motifs that determine specificity between a viral replicase and its promoter. *Nat. Struct. Biol.* 7: 415-423.

Kim, K.-H., and Hemenway, C. 1999. Long-distance RNA-RNA interactions and conserved sequence elements affect potato virus X plus-strand RNA accumulation. *RNA* 5: 636-645.

Kim, M., Zhong, W., Hong, Z., and Kao, C. C. 2000. Template nucleotide moieties required for *de novo* initiation of RNA synthesis by a recombinant viral RNA-dependent RNA polymerase *J. Virol.* 74: 10312-10322.

Koev, G., Mohan, B. R., and Miller, W. A. 1999. Primary and secondary structural elements required for synthesis of barley yellow dwarf subgenomic RNA1. *J. Virol.* 73: 2876-2885.

Koev, G., Liu, S., Beckett, R., and Miller, W. A. 2002. The 3'-terminal structure required for replication of barley yellow dwarf virus RNA contains an embedded 3'-end. *Virology* 292: 114-126.

Koonin, E. V. 1991a. The phylogeny of RNA-dependent RNA polymerases of positive-strand RNA viruses. *J. Gen. Virol.* 72: 2197-2206.

Koonin, E. V. 1991b. Similarities in RNA helicases. *Nature* 352: 290.

Koonin, E. V., and Dolja, V. V. 1993. Evolution and taxonomy of positive-strand RNA viruses: Implications of comparative analysis of amino acid sequences. *Crit. Rev. Biochem. Mol. Biol.* 28: 375-430.

Lai, M. M. C. 1998. Cellular factors in the transcription and replication of viral RNA genomes: A parallel to DNA-dependent RNA transcription. *Virology* 244: 1-12.

Léonard, S., Viel, C., Beauchemin, C., Daigneault, N., Fortin, M, G., and Laliberte, J.-F. 2004. Interaction of VPg-Pro of *Turnip mosaic virus* with the translation initiation factor 4E and the poly(A)-binding protein *in planta. J. Gen. Virol.* 85: 1055-1063.

Mandahar, C. L. (Ed.). 1999. Molecular Biology of Plant Viruses. Kluwer Academic Publishers, Boston/Dordrecht/London. pp. 281.

Marsh, L. E., and Hall, T. C. 1987. Evidence implicating a tRNA heritage for the promoters of positive-strand RNA synthesis in *Brome mosaic virus* and related viruses. *Cold Spring Harbor Symp. Quant. Biol.* 52: 331-341.

Mayo, M. A. 2002. The principles and current practice of plant virus taxonomy. *In*: Khan, J. A., and Dijkstra, J. (Eds.). Plant Viruses as Molecular Pathogens. Howarth Press, Inc., New York, London, Oxford. pp. 3- 23.

Miller, W. A., and Koev, G. 2000. Synthesis of subgenomic RNAs by positive-strand RNA viruses. *Virology* 273: 1-8.

Montagnier, L. 1968. The replication of viral RNA. 18th Symp. Soc. Gen. Microbiol. (Molecular Biology of Viruses), London. pp. 125-147.

Nagy, P. D., Zhang, C., and Simon, A. E. 1998. Dissecting RNA recombination *in vitro*: Role of RNA sequences and the viral replicase. *EMBO J.* 17: 2392-2403.

Nagy, P. D., Pogany, J., and Simon, A. E. 1999. RNA elements required for RNA recombination function as replication enhancers *in vitro* and *in vivo* in a plus strand RNA virus. *EMBO J.* 18: 5653-5665.

Nagy, P. D., Pogany, J., and Simon, A. E. 2001. *In vivo* and *in vitro* characterization of an RNA replication enhancer in a satellite RNA associated with *Turnip crinkle virus*. *Virology* 288: 315-325.

Olsthoorn, R. C. L., Martens, S., Brederode , F. T., and Bol, J. F. 1999. A conformational switch at the 3` end of a plant virus RNA regulates virus replication. *EMBO J.* 18: 4856-4864.

Osman, T. A. M., and Buck, K. W. 1997. The tobacco mosaic virus RNA polymerase complex contains a plant protein related to RNA-binding subunit of yeast elF-3. *J. Virol.* 71: 6075-6082.

Osman, T. A. M., Hemenway, C. L., and Buck, K. W. 2000. Role of the 3`-tRNA-like structure in tobacco mosaic virus minus-strand RNA synthesis by the viral RNA-dependent RNA polymerase. *J. Virol.* 74: 11671-11680.

Panavas, T., and Nagy, P. D. 2003. The RNA replication enhancer element of tombusviruses contains two interchangeable hairpins that are functional during plus-strand synthesis. *J. Virol.* 77: 258-269.

Paul, C. P., Barry, J. K., Dinesh-Kumar, S. P., Brault, V., and Miller, W. A. 2001. A sequence required for – 1 ribosomal frameshifting located four kilobase downstream of the frameshift site. *J. Mol. Biol.* 310: 987- 999.

Pogany, J., Fabian, M. R., White, K. A., and Nagy, P. D. 2003. Functions of novel replication enhancer and silencer elements in tombusvirus replication. *EMBO J.* 22: 5602-5611.

Pogue, G. P., and Hall, T. C. 1992. The requirement for a 5` stem-loop structure in brome mosaic virus replication supports a new model for viral positive-strand RNA initiation. *J. Virol.* 66: 674-684.

Pogue, G. P., Huntley, C. C., and Hall, T. C. 1994. Common replication strategies emerging from the study of diverse groups of positive-strand virus. *Arch. Virol.* 9: 181-194.

Pringle, C. R. 1998. Virus taxonomy - San Diego 1998. *Arch. Virol.* 143: 1449-1459.

Quadt, R., Ishikawa, M., Janda, M., and Ahlquist, P. 1995. Formation of brome mosaic virus RNA-dependent RNA polymerase in yeast requires coexpression of viral proteins and viral RNA. *Proc. Nat. Acad. Sci. USA* 92: 4892-4896.

Qiu, W. P., and Scholthof, K.-B. G. 2000. *In vitro*- and *in vivo*-generated defective RNAs of *Satellite panicum mosaic virus* define *cis*-acting RNA elements required for replication and movement. *J. Virol.* 74: 2247-2254.

Qu, F., and Morris. T. J. 2000. Cap-independent translational enhancement of *Turnip crinkle virus* genomic and subgenomic RNAs. *J. Virol.* 74: 1085-1093.

Ray, D., and White, K. A. 1999. Enhancer-like properties of an RNA element that modulates tombusvirus RNA accumulation. *Virology* 256: 162-171.

Ray, D., and White, K. A. 2003. An internally located RNA hairpin enhances replication of tomato bushy stunt virus RNAs. *J. Virol.* 77: 245-257.

Restrepo-Hartwig, M. A., and P. Ahlquist. 1996. Brome mosaic virus helicase- and polymerase-like proteins colocalize on the endoplasmic reticulum at sites of viral RNA synthesis. *J. Virol.* 70: 8908-8916.

Restrepo-Hartwig, M. A., and Ahlquist, P. 1999. Brome mosaic virus RNA replication proteins 1a and 2a colocalize and 1a independently localizes on the yeast endoplasmic reticulum *J. Virol.* 73: 10303-10309.

Restrepo-Hartwig, M. A., and Carrington, J. C. 1992. Regulation of nuclear transport of a plant, potyvirus protein by autoproteolysis. *J. Virol.* 66: 5662-5666.

Roberts, I. M., Wang, D., Findlay, K., and Maule, A. J. 1998. Ultrastructural and temporal observations of the potyvirus cylindrical inclusions (CIs) show that the CI protein acts transiently in aiding virus movement. *Virology* 245: 173-181.

Routh, G., Dodds, J. A., Fitzmaurice, L., and Mirkov, T. E. 1995. Characterization of deletion and frameshift mutants of *Satellite tobacco mosaic virus. Virology* 212: 121-127.

Routh, G., Yassi, M. N. A., Rao, A. L. N., Mirkov, T. E., and Dodds, J. A. 1997. Replication of wild type and mutant clones of *Satellite tobacco mosaic virus* in *Nicotiana benthamiana* protoplasts. *J. Gen. Virol.* 78: 1271-1275.

Scholthof, K.-B. G., Jones, R. W., and Jackson, A. O. 1999. Biology and structure of plant satellite viruses activated by icosahedral helper viruses. *Curr. Top. Microbiol. Immunol.* 239: 123-143.

Schwartz, M., Chen, J., Janda, M., Sullivan, M., de Boon, J., and Ahlquist, P. 2002. A positive-strand RNA virus replication complex parallels form and function of retrovirus capsids. *Mol. Cell* 9: 505-514.

Siegel, R. W., Adkins, S., and Kao, C. C. 1997. Sequence-specific recognition of a subgenomic RNA promoter by a viral RNA polymerase. *Proc. Natl. Acad. Sci. USA* 94: 11238-11243.

Siegel, R. W., Bellon, L., Beigelman, L., and Kao, C. C. 1998. Moieties in an RNA promoter specifically recognised by a viral RNA-dependent RNA polymerase. *Proc. Nat. Acad. Sci. USA* 95:11613-11618.

Sit, T. L., Vaewhongs, A. A., and Lommel, S. A. 1998. RNA-mediated transactivation of transcription from a viral RNA. *Science* 281: 829-832.

Sivakumaran, K., Kim, C. H., Tayon, R., Jr., and Kao, C. C. 1999. Sequence and secondary structural determinants in minimal viral promoter that directs replicase recognition and initiation of genomic plus-strand RNA synthesis. *J. Mol. Biol.* 294: 667-682.

Sivakumaran, K., Hema, M., and Kao, C. C. 2003. Brome mosaic virus RNA synthesis *in vitro* and in barley protoplasts. *J. Virol.* 77: 5703-5711.

Sivakumaran, K., Choi, S. K., Hema, M., and Kao, C. C. 2004. Requirements for brome mosaic virus subgenomic RNA synthesis *in vivo* and replicase-core promoter interactions *in vitro. J. Virol.* 78: 6091-6101.

Song, C., and Simon, A. E. 1995. Requirement for a 3`-terminal stem-loop for *in vitro* transcription by an RNA-dependent RNA polymerase. *J. Mol. Biol.* 254: 6-14.

Song, S. I., and Miller, W. A. 2004. *cis* and *trans* requirements for rolling circle replication of a satellite RNA. *J. Virol.* 78: 3072-3082.

Sun, X., Zhang, G., and Simon, A. E. 2005. Short internal sequences involved in replication and virion accumulation in a subviral RNA of *Turnip crinkle virus. J. Virol.* 79: 512-524.

Taliansky, M. E., and Palukaitis, P. F. 1999. Satellite RNAs and satellite viruses. In: Granoff, A., and Webster, R. G. (Eds.). Encyclopedia of Virology, Second ed., Vol. 3. Academic Press, San Diego, California. pp. 1607-1615.

Tatsuta, M., Mizumoto, H., Kaido, M., Mise, K., and Okuno, T. 2005. The *Red clover necrotic mosaic virus* RNA2 *trans*-activator is also a *cis*-acting RNA2 replication element. *J. Virol.* 79: 978-986.

Técsi, L. I., Smith, A. M., Maule, A. J., and Leegood, R. C. 1996. A spatial analysis of physiological changes associated with infection of cotyledons of marrow plants with *Cucumber mosaic virus. Plant Physiol.* 111: 975-986.

Van Regenmortel, M H. V. 1989. Applying the species concept to plant viruses. *Arch. Virol.* 104: 1-17.

Van Regenmortel, M H. V. 2002. How to write the names of virus species. *In*: Khan, J. A., and Dijkstra, J. (Eds.). Plant Viruses as Molecular Pathogens. Howarth Press, Inc., New York, London, Oxford. pp. 25-27.

Van Regenmortel, M H. V., Fauquet, C. M., Bishop, D. H. L., Carstens, E. B., Estes, M. K., Lemon, S. M., Maniloff, J., Mayo, M. A., McGeoch, D., Pringle, C. R., and Wickner, R. B. 2000. Virus Taxonomy. Seventh Report of the International Committee on Taxonomy of Viruses. Academic Press, New York, San Diego. 1162 pp.

Vlot, A. C., and Bol, J. F. 2003. The 5`-untranslated region of alfalfa mosaic virus RNA1 is involved in negative-strand RNA synthesis. *J. Virol.* 77: 11284-11289.

Vlot, A. C., Neeleman, L., Linthorst, H. J. M., and Bol, J. F. 2001. Role of the 3`-untranslated regions of alfalfa mosaic virus RNAs in the formation of a transiently expressed replicase in plants and in the assembly of virions. *J. Virol.* 75: 6440-6449.

Wang, D., and Maule, A. J. 1995. Inhibition of host gene expression associated with plant virus replication. *Science* 267: 229-231.

Wang, J., and Simon, A. E. 1997. Analysis of the two subgenomic RNA promoters for *Turnip crinkle virus in vivo* and *in vitro*. *Virology* 232: 174-186.

Wang, K. -H., and S. -M. Wong. 2004. Significance of the 3`-terminal region in minus-strand RNA synthesis of *Hibiscus chlorotic ringspot virus*. *J. Gen. Virol.* 85: 1763-1776.

Watanabe, T., Honda, A., Iwata, A., Ueda, S., Hibi, T., and Ishihama, A. 1999. Isolation from tobacco mosaic virus-infected tobacco of a solubilized template-specific RNA-dependent RNA polymerase containing a 126/183K protein heterodimer. *J. Virol.* 73: 2633-2640.

White, K. A., and Nagy, P. D. 2004. Advances in the molecular biology of tombusviruses: Gene expression, genome replication, and recombination. *Prog. Nucl. Acid Res. Mol. Biol.* 78: 187-226.

Wu, B., Vanti, W. B., and White, K. A. 2001. An RNA domain within the 5`-untranslated region of the tomato bushy stunt virus genome modulates viral RNA replication. *J. Mol. Biol.* 305: 741-756.

Wu, B. D., and White, K. A. 1998. Formation and amplification of a novel tombusvirus defective RNA which lacks the 5`-nontranslated region of the viral genome. *J. Virol.* 72: 9897-9905.

Yeh, H.-H., Tian, T., Rubio, L., Crawford, B., and Falk, B.-W. 2000. Asynchronous accumulation of lettuce infectious yellows virus RNAs 1 and 2 and identification of an RNA1 *trans* enhancer of RNA2 accumulation. *J. Virol.* 74: 5762-5768.

Zanotto, P. M., Gibbs, M. J., Gould, E. A., and Holmes, E. C. 1996. A reevaluation of the higher taxonomy of viruses based on RNA polymerases. *J. Virol.* 70: 6083-6096.

Zhang, G., Zhang, J., and Simon, A. E. 2004a. Repression and derepression of minus-strand synthesis in a plus-strand RNA virus replicon. *J. Virol.* 78: 7619-7633.

Zhang, J., Stuntz, R., and Simon, A. E. 2004b. Analysis of a viral replication repressor: Sequence requirements in the large symmetrical loop *Virology* 326: 90-102.

Zhou, H., and Jackson, A. O. 1996. Expression of the barley stripe mosaic virus RNA β triple gene block. *Virology* 216: 367-379.

2 POSITIVE-SENSE VIRAL RNA

I. VIRUS GENOME

Much information is available about structure, organisation and functions of positive-sense single-stranded RNA genomes of plant viruses. This information has been published in many reviews starting from Klug and Caspar (1960) to those published from 1990s onwards (ten Dam *et al.*, 1990; Mans *et al.*, 1991; Koonin and Dolja, 1993; Florentz and Giegé, 1995; Zaccomer *et al.*, 1995; Söll and RajBhandary, 1995; Giege, 1996; Buck, 1996; Deiman and Pleij, 1997; Haenni and Chapeville, 1997; Giege *et al.*, 1998a, 1998b; Dreher, 1999; Héricourt *et al.*, 1999; Furuichi and Shatkin, 2000; Fechter *et al.*, 2001a; Carr, 2004). More review papers are listed under the appropriate heads and subheads.

Single-stranded or double-stranded RNA or DNA constitutes the virus genome. The RNA viruses are very successful in nature so that the vast majority of plant viruses, about 80% of them, contain an RNA genome. The single-stranded RNA genome may be of positive/plus/messenger-sense/strand/polarity/type or of negative-sense. The positive-sense RNA virus genomes play multiple roles during the infection cycle: they act as mRNAs and direct virus-specific protein synthesis from their genes; they act as templates for transcription into negative-sense RNA copies which are the starting point of all the subsequent stages of virus genome replication followed by encapsidation of progeny positive-sense RNA molecules leading ultimately to formation of progeny virions; they serve as templates for subgenomic RNA synthesis in many plant viruses; and they act as regulators of gene expression. Viral genomic RNA functions as *cis*-acting element in the above processes and recruits the required factors like translation factors, RNA replicase components, and structural proteins (Buck, 1996; Dreher, 1999; Gale *et al.*, 2000). These processes could function simultaneously or become mutually exclusive at some stage of viral life cycle; virus itself controls and regulates the direction towards which progeny RNA is directed at a particular time (Gamarnik and Andino, 1998; Olsthoorn *et al.*, 1999; Butsch and Boris-Lawrie, 2002). The negative-sense RNA cannot directly act as mRNA and fails to act as a template for protein synthesis.

The viral RNA genome contains both the coding and non-coding functions/regions; the latter regions contain information in the form of *cis*-acting regulatory sequences that control a host of viral functions like translation of viral proteins, synthesis of full-length plus- and minus-strand RNAs, transcription of subgenomic RNAs, and still other functions. Viral RNA genome may be either monopartite when a single RNA molecule contains all the genetic information or multipartite when the genetic information is spread over two or more than two RNA segments all of which must be present together for infectivity of a multipartite virus. Each nucleic acid has two ends: the 3`-end/terminus/untranslated sequence or region (3`-UTR) and the 5`-end/terminus/untranslated sequence or region (5`-UTR) and proper maintenance of both these ends in functional condition is a priority of the viruses for their own fitness. The 3`-end refers to that end at which the ribose of terminal nucleotide of the nucleic acid bears a free 3`-hydroxyl group and 5`-end to the end bearing a free 5`-hydroxyl group. The 5`-untranslated sequence of RNAs is of short length, is constituted by only a few nucleotides and can bear any one of the following three structures: cap-like structure, genome-linked protein (VPg), or di- or tri-phosphate. The 3`-untranslated sequences are generally very long, are composed of about 100 or even up to 300 bases, and can carry any one of the following three structures: tRNA-like structure, poly(A) tail or lacks both and culminates in pX_{OH}. Some of the permutations and combinations of various structures found at 5`- and 3`-ends and the virus supergroups/families/ genera containing various combinations of these structures are shown in Table 1.

Most RNA preparations are usually much less infectious than the intact virus on equal RNA basis, and this varies with the method of preparation. Gierer and Schramm (1956), who first demonstrated the infectious nature of the intact RNA isolated by phenol extraction from TMV, found that infectivity of the isolated RNA was about 0.1% compared to the same amount of RNA in intact virus. Complete nucleotide sequence (primary structure) of RNA genomes of many plant viruses is known but nucleotide sequence of only the regions in vicinity of 5`- and/or 3`-termini has been reported in many other cases.

It is quite common in plant viruses to have 'compressed' genomes, i.e., to bear extensively overlapping ORFs. *Grapevine fleck virus* has a monopartite RNA genome of about 8800 nucleotides containing five ORFs out of which three ORFs overlap and occur over a range of 1000 nucleotides. However, occurrence of overlapping ORFs is not an inviolable condition; no overlapping ORFs occur in BMV RNA. It is exceptional if any nucleotide sequence in viral genomes does not have a function to perform. Non-coding intergenic areas exist between several plant viral genes in many plant viruses. These intergenic regions generally have great significance for regulation of replication/transcription of viral genome and for translation of viral proteins.

All viral genomic RNAs are polycistronic mRNAs although all genes are not 'open' to translation by ribosomes; some genes are not accessible to ribosomes for translation since they are 'closed or masked'. In fact, many plant virus RNAs contain only one open initiation site so that they function as monocistronic mRNAs. The triplet AUG is the translation initiation codon and is generally situated near but a few nucleotides away from the 5`-end. Hence, only the 5`-proximal cistron of virus RNAs is translatable (Kozak 1978, 1980). The initiator codon in RNA4 of BMV starts from 10th nucleotide from 5`-end. However, RNAs of some plant viruses bear at least two different translation initiation sites as in middle and bottom component RNAs of CPMV and RNA3 of AMV. Nevertheless, only one of these translation initiation sites is preferentially utilized. Thus, normally only a single polypeptide is synthesized from a polycistronic mRNA and this is valid for both monopartite and multipartite plant viruses.

The size of positive-sense RNA viral genome (including animal viruses) ranges from about 3.5 to about 30 kb (Koonin and Dolja, 1993) but the positive-sense RNA genomes of plant viruses are smaller, generally range from 4 to 15 kb, carry extremely compact genetic information, contain limited number of genes, and encode 4 to 12 proteins (Koonin and Dolja, 1993; Héricourt *et al.*, 1999; Morozov and Solovyev, 1999). The size of RNA genome of positive-sense RNA plant viruses, in terms of number of nucleotides, varies from 1394 nucleotides in RNA2 of *Carnation ringspot dianthovirus* (family *Tombusviridae*) to 15,480 nucleotides in RNA genome of BYV to 20,000 nucleotides (only partial) in CTV (Table 1).

In fact, closteroviruses have the largest genomes among the plant positive-sense RNA viruses; the genome size ranges from 15.5 kb (15480 nucleotides) in BYV to 19.3 kb (around 2000 nucleotides with 12 ORFs) in CTV. The CTV is the largest single-stranded RNA plant virus and at least its ten 3`-ORFs are expressed through a nested set of 3`-coterminal subgenomic RNAs. In contrast, RNA genome of *Turnip rosette sobemovirus* and of STNV appears to be among the smallest in plant viruses. The limitation of genome size is imposed by constraints of packaging and replication.

Like cellular mRNAs, viral mRNAs also contain elements that facilitate and/or regulate viral protein synthesis in host cells (Gallie, 1996) so that many positive-sense RNA plant viruses have a 5`-cap structure and a 3`-poly(A) tail in their genomes (Table 1). Some such plant viruses are: family *Potexviridae* (genera *Allexvirus, Carlavirus, Foveavirus* and *Potexvirus*), genus *Benyvirus* (BNYVV), genus *Marafivirus* (OBDV), and genus *Trichovirus*.

But other plant viral RNAs do not contain either or both these structures (Buck, 1996; Héricourt *et al.*, 1999) (Table 1). RNAs of potyviruses lack a 5`-cap structure but contain 3`-poly(A) tail so that translation initiation of these plant viruses starts from IRES in 5`-UTR. An interaction between eukaryotic translation initiation factor 4G (eIF4G) (bound to IRES) and PABP of TEV appears to be required for efficient translation of viral RNAs (Gallie, 2001). On the other hand, TMV has a capped genome but lacks poly(A) sequence so that TMV 3`-UTRs are functionally equivalent to poly(A) tail of cellular mRNAs (Gallie and Kobayashi, 1994). The pseudoknot domain located upstream of 3`-tRNA-like structure of TMV UTR appears to substitute functionally for poly(A) tail and is particularly responsible for increasing translation expression (Gallie and Walbot, 1990; Leathers *et al.*, 1993). Similarly, RNAs 1 and 2 of RCNMV are capped but lack a poly(A) tail. However, uncapped *in vitro* transcripts of RCNMV RNA1 and RNA2 show high infectivity, which implies that this virus possesses cap-independent translation mechanism of proteins (Mizumoto *et al.*, 2003). BMV RNAs also lack the poly(A) tail. Genomic RNAs of many other plant viruses also lack a poly(A) tail and elements in 5`- and/or 3`-UTR are believed to compensate for absence of the poly(A) tail (Gallie and Kabayashi, 1994; Wells *et al.*, 1998). Plant viruses lacking both the cap structure as well as the poly(A) tail belong to genera *Luteovirus* and *Necrovirus*. Translation of BYDV requires base pairing between a stem-loop in 5`-UTR and a stem-loop in a 100-nucleotide translation element in 3`-UTR, presumably to deliver translation factors and/or ribosomes to 5`-end (Guo, L., *et al.*, 2001). Carnation Italian ringspot tombusvirus RNA is not polyadenylated and lacks 5`-cap structure. Genomic RNAs of BYDV, STNV, and TBSV harbour sequences in 3`-UTR which confer cap-independent translation (Meulewaeter *et al.*, 1998; Wang, S. P. *et al.*, 1997; Wu and White, 1999).

The mRNAs, whether cellular or viral, must be protected from degradation by 5`- and 3`-exonucleases as well as other cellular degradation pathways. They are, therefore, never single-stranded and have developed strategies that could prevent degradation; these strategies include the formation of base paired structures, the binding of cellular proteins as in phage Qβ, or covalent linkage of viral proteins to the ends of viral genomes. The structures present at both the 5`- and 3`-termini of RNA genomes ensure this protection. This also explains the presence of homologous and conserved nucleotide sequences at the 3`-termini of all the four RNA segments of BMV and of CCMV.

General features of viral RNA genomes are listed in Table 1.

TABLE 1

Characteristics of single-stranded positive-sense RNA genome of plant viruses

(Based on Tables 1 and 2 of Héricourt *et al.*, 1999, with additions and corrections)

Genus	Virus	Structure		RNA	
		5`-end	3`-end	Segments	Nucleotides [1]
Alfamovirus	AMV	Cap	tRNA-like[2]	3	1:3644
					2:2593
					3:2142
Bromovirus	BMV	Cap	tRNA-like	3	1:3234
					2:2865
					3:2117
Bymovirus	Ba YMV	VPg proposed	Poly(A)	2	1:7632
					2:3585
Capillovirus	ASGV	Cap prop osed in CTLV	Poly(A)	1	6496
Carlavirus	PVM	Cap in BISV	Poly(A)	1	8535
Carmovirus	CarMV	Cap	No poly(A) No tRNA-like	1	4003
Closterovirus	BYV	Cap prop osed	No pol y(A) No tRNA-like	1	15480
Comovirus	CPMV	VPg	Poly(A)	2	B:5889 M:3481
Cucumovirus	CMV	Cap	tRNA-like	3	1:3357 2:3050 3:2216
Dianthovirus	CRSV	Cap in RCNMV	No pol y(A)	2	1:3756 2:1394
Enamovirus	PEMV	VPg	No poly(A) No tRNA-like	2	1:5706 2:4253
Furovirus	SBWMV	Cap	tRNA-like	2	1:7099 2:3593
	BNYVV	Cap	Poly(A)	4	1:6746 2:4612

					3:1774 4:1108
Hordeivirus	BSMV	Cap	tRNA-like	3	1:3768 (α) 2:3289 (β) 3:3164 (χ)
Idaeovirus	RBDV	Cap prop osed in RNA 1	No pol y(A) No tRNA-like	2	1:5449 2:2231
Ilarvirus	TSV		tRNA-like	3	1:3491 2:2926 3:2205
Luteovirus	BYDV	VPg;No VPg in BYDV (PAV isolate)	No pol y(A) No tRNA-like	1	5677
Machlomovirus	MCMV	Cap	No poly(A)	1	4437
Marafivirus	OBDV	Cap prop osed	Poly(A)	1	6509
Necrovirus	TNV	PpA	No pol y(A) No tRNA-like	1	3759
Nepovirus	TBRV	VPg	Poly(A)	2	1:7356 2:4662
Potexvirus	PVX	Cap	Poly(A)	1	6435
Potyvirus	PVY	VPg	Poly(A)	1	9704
Rymovirus	BrSMV		Poly(A)	1	9672
Sequivirus	P YFV	Vpg proposed	No pol y(A) No tRNA-like	1	9871
Sobemovirus	SBMV	VPg	No pol y(A) No tRNA-like	1	4194
Tobamovirus	TMV	Cap	tRNA-like	1	6395
Tobravirus	TRV	Cap	tRNA-like folding but can not be aminoac ylated	2	1:6791 2:1799 1905 3389
Tom busvir us	TBSV		No pol y(A) No tRNA-like	1	4776
Trichovirus	ACLSV	Cap prop osed; Cap in GVB	Poly(A)	1	7555
Tym ovirus	TYMV	Cap	tRNA-like	1	6318
Um bravirus	GRV		No pol y(A) No tRNA-like	1	4019
W aikavirus	RTSV		Poly(A)	1	12484

1 - Nucleotides excludes poly(A) tails.

2 - AMV RNA bears a tRNA-like structure at 3` terminus. This tRNA-like structure contains binding sites for capsid protein which, upon binding to these sites, favours formation of a linear confirmation at the expe nse of tRNA -like structure. Thus capsid protein appears to have a regulatory role (Olsthoorn *et al.*, 2004).

The various combinations of terminal structures present in viral RNAs of different plant viruses can be as below:

- 5`-cap-like structure and 3`-poly(A) tail: *Potexviridae, Benyvirus, Marafivirus, Trichovirus*
- 5` cap-like structure and 3`-tRNA-like structure: *Closteroviridae, Bromoviridae,* carmo-like supergroup, *Tobamovirus, Tobravirus, Hordeivirus, Furovirus, Tymovirus*
- 5` cap-like structure and 3`-pX$_{OH}$: *Tombusviridae, Machlomovirus, Dianthovirus*
- 5` genome-linked protein and 3`-poly(A) tail: *Potyviridae, Comoviridae*
- 5` genome-linked protein and 3`-tRNA-like structure: Sobemo-like supergroup
- 5` genome-linked protein and 3`-pX$_{OH}$: *Sobemovirus, Luteovirus* (subgroup II)
- 5` cap-like structure as well as VPg absent and 3`-tRNA-like structure as well as poly(A) tail absent: Genus *Luteovirus* (subgroup I)

II. 5`-END

As already mentioned, the 5`-end of plant virus RNAs can bear any one of the three structures - cap-like structure, genome-linked protein (VPg), or di- or tri-phosphate. A di- or tri-phosphate (ppA) is present at 5`-terminus of STNV and TNV. The 5`-termini of all RNAs of a multipartite virus carry identical structures. The nucleotide adjacent to cap-like structure or VPg in all virus RNAs is labeled as number 1.

Most of the small RNA plant viruses possess guanylates as 5`-termini of genomic and subgenomic RNAs (*Artichoke mottle crinkle virus*, BYDV, BMV, CarMV, *Citrus leaf blotch virus*, *Clover yellow mosaic virus*, MCMV, *Panicum mosaic virus*, PVX, *Saguaro cactus virus*, TMV, TBSV, TCV, and others). Adenylate is found in 5`-terminus of RNAs of several plant viruses like BaMV, BBMV, CTV and RCNMV. The 5`-termini of all CTV subgenomic RNAs mapped so far and of genomic RNA have an adenylate (Karasev *et al.*, 1995, 1997; Ayllón *et al.*, 2003). An adenylate is also at 5`-terminus of several subgenomic RNAs of other closteroviruses like BYV and *Beet yellow stunt virus* (Agranovsky *et al.*, 1994). However, BYV subgenomic RNA of protein 6 (p6) contains a guanylate at 5`-terminus (Peremyslov and Dolja, 2002). Uridylate and cytidylate as 5`-termini are uncommon in plant (and also animal) viruses but one of these is reported to exist in *Oat chlorotic stunt avenavirus* (Boonham *et al.*, 1998). Some interesting situations are also reported. The 5`-termini of subgenomic RNAs of *Sweet potato chlorotic stunt virus* contain adenylate, guanylate, or even uridylate while 5`-termini of both genomic RNAs of this virus have guanylates.

Secondary structure of 5`-UTR of BMV RNAs 1 and 2 consists of a modular arrangement of three subdivisions A, B, and C. Subdivisions A and C are common to both RNAs while subdivision B is unique to RNA2. Subdivision A is suggested to be involved in RNA replication and also in translation inhibition. It is composed of 5`-terminal stem-loop structure, which is conserved in the two RNAs of all sequenced bromoviruses. Subdivision B attenuates replication of RNA2 while subdivision C possibly overcomes inhibitory effect of subdivision A on translation. The 5`-UTRs of RNA genomes of all or a large majority of plus-sense RNA viruses may also show identical subdivisions with the same functions as given above.

The 5`-UTR in plant virus mRNAs can affect translation both directly and indirectly. The direct influence depends upon the length and degree of secondary structure of 5`-UTR while indirect effect is due to the presence of specific elements constituting the binding sites for protein factors (Gallie, 1996). The elements in 5`-UTR mediate several steps in virus replication process. These steps are: targeting of templates to replication complexes as in BMV in which targeting of RNA2 to replication complexes requires a hairpin in its 5`-UTR (Chen *et al.*, 2001); promoting negative-strand RNA synthesis as in AMV (Vlot and Bol, 2003); and enhancing synthesis of positive-strand RNA as in PVX where an RNA structure in its 5`-UTR facilitates positive-strand synthesis (Miller *et al.*, 1998). Although none of the RNA elements in 5`-UTR are essential in *cis* for TBSV RNA replication, this region clearly facilitates one or more steps during RNA replication. Ray *et al.* (2004) demonstrated that various sub-elements of T-shaped domain within the 5`-UTR of TBSV RNA contribute to the efficiency of viral replication. The stabilities of all the three stems (S1, S2, and S3) forming the 3-helix junction are particularly important, while stem-loop 3 (a terminal extension of S2) is largely dispensable.

A. Cap or Cap-Like Structure

The cap-like structure is a 7 methylguanosine base which is connected to the next sugar through a 5`- to 5`-triphosphate linkage via its 5`-OH group and is represented as $m^7G^5ppp^5$ (or $M^7G^5ppp^5$) or a cap structure consists of 7-methylguanosine linked to 5`-end of transcripts by a 5`-5`-triphosphate bridge; the methyl group may be represented by capital M or small m. Cap-like structure has been variously represented in literature as M^7GpppN, $M^7G(5`-)ppp(5`-)N$, $M^7G(5`)ppp(5`)G$, 7-methyl-G(5`-)ppp, and M^7GpppG; N in all these cases usually represents A or G. Most, if not all, capped mRNAs possess a single methyl group on terminal G residue. Eukaryotic viral mRNAs contain the same cap structure as cellular mRNAs, whatever be their genomic structures and replication strategies, possibly because capped mRNAs are the functional form in host cells.

Nearly all eukaryotic mRNAs and many viral RNAs possess a cap structure. Capped genomic RNAs are present in many positive-sense plant viruses including members of alpha-like and carmo-like virus supergroups. The plant viruses of alpha-like supergroup having capped genomes belong to various genera like *Alfamovirus, Bromovirus, Carmovirus, Closterovirus, Cucumovirus, Furovirus, Hordeivirus, Machlomovirus, Potexvirus, Tobamovirus, Tobravirus, Tymovirus* and to still other genera (Morozov and Solovyev, 1999). Other plant viruses possessing capped mRNAs are DNA viruses, negative-strand RNA viruses, pararetroviruses, and some double-strand RNA viruses. In fact, there are very few plant viruses that lack capped mRNAs; some of these are *Barley yellow dwarf virus*, CPMV, STNV, possibly *Velvet tobacco mottle virus* (VTMoV) and viruses of family *Picornaviridae*. The STNV RNA has a 5`-terminal AGUAAAGACAGGAAACUUUACUGACUAACAUGGCAAAACAAC.

This structure is able to form a perfect Watson-Crick base-pair structure between the 5`-nucleotides 1-7 and 16-22 to give a stick-like secondary structure and a potential hairpin loop. This reminds the base-paired 3`- and 5`-sequences of tRNA that may protect against hydrolysis by cellular 5`- to 3`-exonucleases. The 'capless' viral genomic RNAs may contain a small protein, VPg, covalently attached to 5`-end.

1. Mechanisms of Cap Formation

Addition of a cap structure at 5`-end of cellular mRNAs and most of the DNA viruses, retroviruses, and pararetroviruses occurs in nucleus (by employing cellular capping enzymes) and is closely associated with RNA polymerase II transcription. The cap is added to transcripts of viruses, which replicate in nucleus, during initial phases of transcription. The capping of mRNAs of some of negative-strand RNA viruses (like rhabdoviruses), of double-stranded RNA viruses, of those positive-strand RNA viruses that possess a cap, and of some DNA viruses (like pox viruses) occurs in cytoplasm. The RNA polymerase II transcription system is not available to cytoplasmic viruses for capping their RNAs. Majority of these viruses, therefore, encode their own capping systems and capping enzymes. In fact, various groups of RNA viruses have developed their own different pathways for capping their RNAs. However, some of these viruses conserve the cellular capping reactions.

Capping mechanism is of two types: the co-transcriptional capping and the addition of preformed capped 5`-end to cellular and viral RNAs (cap-snatching) (Furuichi and Shatkin, 2000).

a. Co-transcriptional Capping

Capping of cellular mRNAs occurs along with the transcription process; hence the name - co-transcriptional capping. Cellular co-transcriptional capping mechanism, whether it occurs in nucleus or in cytoplasm, is essentially the same involving the same steps and the same enzymes. This capping reaction can be divided into four steps (Furuichi and Shatkin, 2000; Shuman, 1995, 2002). In first step, the nascent transcripts of RNA polymerase II are first modified by removal of 5`-$\gamma$-phosphate by (mRNA) triphosphatase causing dephosphorylation of viral RNA 5`-pppN end. This results in a 5`-ppN terminus to give a diphosphorylated end (ppNpNp). This is the RNA 5`-triphosphatase activity. In second step, the diphosphorylated end is then converted to GpppNpNp by addition/transfer of GMP moiety (from GTP) by (mRNA) guanylyltransferase resulting in the formation of a 5`-5`-triphosphate bridge by a covalent enzyme called GMP intermediate. This is the guanylyltransferase step (guanylyltransferase activity). In simpler terms, the 5`-diphosphate end of RNA is capped first by GTP:mRNA guanylyltransferase. In third step, the capped guanosine moiety (G5`-ppp5`-G) undergoes methylation by mRNA guanine-7-methyltransferase to ultimately give a complex/structure $M^7G(5`-)ppp(5`-)$ which is also sometime written as $m^7GTP$ or as $m^7GTP$-MT complex. The $N^7$-methylation on guanine occurs co-transcriptionally by action of methyltransferase. This is the methyltransferase step (methyltransferase activity). In other words: the G5`-ppp5`-G cap of RNA is then methylated by RNA (guanine-N7) methyltransferase (Shamun, 1995). In the fourth and last step, there is transfer of m$^7$GTP moiety of m$^7$GTP-MT to 5`-ppN terminus of RNA, resulting in m^{7}GpppN(pN). Thus, three enzymes (RNA triphosphatase, guanylyltransferase and methyltransferase) are involved in a sequential manner during cellular capping process. Both cellular guanylyltransferase and methyltransferase are structurally conserved during evolution (Shuman, 2002).

Capped genomic RNAs of viruses of alphavirus-like superfamily replicate in cytoplasm of infected cells and, consequently, these viruses encode their own capping enzymes. Similarly, capping of great majority of plant viruses takes place in cytoplasm so that these plant viruses have also to synthesize their own capping enzymes. Ample available evidence shows that the cellular capping mechanism, outlined above, also operates in cytoplasm of virus infected plant cells. Presence of the proposed capping domains in the relevant proteins of TMV (Merits *et al.*, 1999), BaMV (Li *et al.*, 2001a), and BMV (Ahola and Ahlquist, 1999; Kong *et al.*, 1999) support the *in vitro* operation of methyltransferase and guanylyltransferase reactions in these viruses. Thus, the 126-kDa TMV replication protein contains a methyltransferase-like domain at N-terminus, a helicase domain at C-terminus, and also acts as a guanylyltransferase-like enzyme. The RNA 5`-triphosphtase activity has been assigned to helicase-like protein of BaMV and some animal viruses like *Semliki Forest virus* (SFV) and *Sindbis virus* (SIV). The BaMV RNA can also be capped *in vitro* by sequential treatment of the template with BaMV helicase-like domain (that exhibits triphosphatase activity) and methyltransferase domain (Li *et al.*, 2001b). A methyltransferase domain is present near N-terminus of replication proteins of all members of alpha-like virus supergroup. Additional conserved motifs also exist. Viruses having methyltransferase-like domain have been divided into two groups: the 'altovirus' group includes *Bromoviridae* and *Togaviridae* families and *Tobamovirus, Tobravirus, Hordeivirus, Furovirus* genera while 'tymovirus' group includes *Carlavirus, Trichovirus,* and *Tymovirus* genera.

Methyltransferase activity had been clearly demonstrated in SIV by genetic and biochemical analysis in 1989-1991. Identical methyltransferase-like domain was tentatively detected in several other viruses, belonging to related virus groups, by comparing their amino acid sequences; these domains were found to possess unique conserved motifs different from those found in cellular methyltransferases (Rozanov *et al.*, 1992). Some of the plant viruses in which these sequences were detected are: ACLV, AMV, ASGV, BMV, BNYVV, BSMV, CMV, *Kennedya yellow mosaic tymovirus* (KYMV), *Narcissus mosaic virus* (NMV), PVM, PVX, RBDV, *Shallot virus X* (ShVX), TMV, TRV, TYMV, WClMV, and some other plant viruses (Koonin and Dolja, 1993). The tentative phylogenetic tree of the methyltransferases-like structures of these viruses separated into two clear lineages - the tymo lineage (ACLV, ASGV, PMV, PVM, TYMV, KYMV, NMV, PVX, and ShVX) and tobamo lineage (BNYVV, BYV, TRV, SBWMV, BSMV, TMV, RBDV, AMV, CMV, and BMV) (Rozanov *et al.*, 1992; Koonin and Dolja, 1993). The methyltransferase domain is always associated with supergroup 3 polymerases and superfamily 1 helicases (Koonin and Dolja, 1993).

The amino-terminal domain of BMV 1a protein is the putative capping domain while its carboxy-terminal domain is the helicase-like domain. The two domains conceivably function in some interdependent or cooperative manner. Ahola and

Ahlquist (1999) and Ahola *et al.* (2000) show that BMV 1a protein has a guanine-7-methyltrasferase and covalent m^{7}GMP binding activities; that it could specifically methylate by adding a single methyl group at position 7 of guanosine ring; and that 1a is highly specific for GTP. Thus, methylation proceeds or coincides with covalent binding in the capping reactions catalyzed by the replicase proteins in alphavirus-like superfamily - indicating conservation of reactions involved in capping of viral mRNAs in this superfamily. It seems that possibly the amino-terminal domain caps the nascent viral RNAs immediately after the synthesis of 5`-ends by a 2a polymerase-1a helicase complex or soon after the newly formed 5`-ends are released from the template by helicase-like 1a domain. This capping of nascent RNA strands in conjunction with their synthesis is likely to be an efficient process. The BMV 1a capping domain also functions in capping of subgenomic RNA4 *in vivo* (Ahola *et al.*, 2000).

Huang *et al.* (2004) studied cap formation by BaMV in detail and proposed a model that has four steps: step 1 is binding of TGP and *S*-adenosylmethionine (AdoMet); step 2 is transfer of a methyl group from AdoMet to position N7 of GTP by methyltransferase; step 3 is formation of a phosphoamide bond between m^{7}GMP and an active-site residue; and step 4 is guanylyltransferase activity to transfer of m^{7}GMP moiety of m7GTP to 5`-diphosphate end of viral RNAs. Cap formation in RNA transcripts of BaMV requires sequential enzyme actions of methyltransferase and guanylyltransferase. The 4.1 kb ORF1-encoded 155-kDa polypeptide of BaMV is postulated to be involved in replication of viral genome and formation of a cap structure at 5` -ends of BaMV transcripts. The N-terminal domain (consisting of 442 amino acids) of viral replicase (155-kDa viral protein) functions as an mRNA capping enzyme and possesses both enzymatic activities of GTP methyltransferase and *S*-adenosylmethionine (AdoMet)-dependet guanylyltransferase (Li *et al.*, 2001a, 2001b). Thus, both the methyltransferase and guanylyltransferase activities are executed by a single capping enzyme domain located at N terminus of 155-kDa viral protein. Residues 514 to 892 contain nucleoside triphosphate-binding and helicase-like motifs and so has nucleoside triphosphatase activities and an RNA 5`-triphosphatase activity that specifically cleaves the γ phosphate off from 5`-end of nascent RNA (Li *et al.*, 2001b). The RNA 5`-triphosphatase activity of BaMV is recruited from central helicase-like domain.

The order of cap formation found in BaMV exists in other members of alphavirus-like superfamily [such as *Semliki Forest Virus* (SFV), *Hepatitis E virus,* TMV (Merits *et al.*, 1999), and BMV (Ahola and Ahlquist, 1999; Kong *et al.*, 1999)]. Therefore, it is likely that capping enzymes within alphavirus-like superfamily have the same catalytic mechanism. The cap formation in this alphavirus-like plant virus is distinct from that of eukaryotic cellular systems in chronology of action of different participating enzymes and also with regard to protein primary sequence and genetic organisation. Thus, sequence of action of three capping enzymes during cellular capping process is - triphosphatase, guanylyltransferase and methyltransferase - and each enzyme is a separate entity. In contrast, sequence of action of the three capping enzymes in BaMV is triphosphatase, methyltransferase and guanylyltransferase and both the methyltransferase and guanylyltransferase activities are executed by a single capping enzyme domain located at N terminus of 155-kDa viral protein.

b. Cap-Snatching (Addition of Preformed Capped 5`-End to Viral RNAs)

The negative-strand RNA viruses replicating either in nucleus, like the *Influenza orthomyxovirus*, or in cytoplasm, like bunyaviruses, do not have capping enzymes. Instead they have evolved a mechanism to 'snatch' the cap structure from host cell mRNAs. During cap-snatching, viral polymerase binds to cellular capped mRNA in infected cells and an endonuclease activity associated with polymerase then cleaves the capped oligonucleotides. This cleaving of the host mRNA generally is at 10 to 20 nucleotides from its 5`-capped end, preferentially at a purine residue. The capped oligonucleotides are later employed as primers by virion-associated RdRp to initiate transcription of single-stranded negative genomic RNA (Furuichi and Shatkin, 2000). Thus, the negative-strand RNA viruses employ a preformed cap to produce 5`-capped mRNAs; in fact, transcription of negative-strand RNA viruses is initiated by a cap-snatching mechanism. Synthesis of influenza virus mRNAs occurs with initiation by host cell m^7GpppNm-containing RNA fragments derived from RNA polymerase II transcripts. Viral mRNA synthesis requires continuous functioning of RNA polymerase II The nucleus of infected cells is the site of this reaction. Capping of mRNAs of TSWV, the type species of genus *Tospovirus* of *Bunyaviridae*, also occurs by cap-snatching mechanism. The biological importance of cap-snatching is not yet known but, according to Furuichi and Shatkin (2000), one possibility is that the limited genomic capacity, for example of small viruses, is enhanced by use of a capped 5`-sequence from host cell transcripts or a preformed cap-oligonucleotide for mRNA synthesis.

2. Functions of Cap and Cap-like Structure

The precise role(s) of the cap-like structure is not yet known. However, several important leads exist that point to the significance of cap/cap-like structure in viral RNA replication and in other viral functions. The important biological functions of a cap seem to be related to N^7-methyl on 5`-terminal G and to the two pyrophosphoryl bonds that connect $m^7$G in a 5`-5`-configuration to the first nucleotide of mRNA.

The cap/cap-like structure has been suggested to protect mRNAs from degradation by cellular enzymes like 5`-exonucleases and phosphatases. Cap, therefore, maintains/ increases integrity and stability of mRNAs so as to promote efficient virus replication and translation. It is, thus, widely recognised that capped RNAs are more stable while uncapped mRNAs are very unstable in plant and animal cytoplasm. Then, the cap-like structure may be involved in synthesis of negative-strand RNA. Vlot *et al.* (2002) found that methyltransferase-like domain of AMV p1 protein, in addition to its putative role in RNA capping, has distinct roles in replication-associated functions necessary for negative-strand RNA synthesis. They proposed that p1 protein supports negative-strand synthesis of RNA1 in *trans*. But opinion also exists that presence or absence of a 5`-cap may be independent of replication mechanism.

Presence of 5`-cap structure enhances accuracy as well as efficiency of pre-mRNA splicing in mammalian cell extracts and in *Xenopus* oocytes. The cap on pre-mRNA interacts with splicing components at the adjacent 5`-splice site. This interaction is mediated by a nuclear cap-binding complex (CBC) that contains two associated proteins (CB20 and CB80). The CBC is specifically bound to the monomethylated 5`-cap on pre-mRNA, remains bound to the cap during processing of RNA, and is actively involved in splicing as well as RNA export. The CBC/capped RNA complex is also involved in 3`-polyadenylation and nucleocytoplasmic transport so that the 5`-terminal m7GpppN could also be involved in transport of RNA from nucleus to cytoplasm.

Presence of cap/cap-like structure stimulates addition of poly(A) tail to 3`-end of viral RNA.

Like BMV 1a and 2a proteins, C-terminus of AMV p1 protein interacts with N-terminus of p2 (van der Heijden *et al.*, 2001). For 1a proteins of BMV and CMV, interactions exist between capping and helicase-like domains. Moreover, sequences in capping domains of BMV and SFV (and flanking sequences) mediate membrane association and it has been proposed that these domains guide RdRp/template complexes to intracellular membranes, where RNA replication occurs (Ahola *et al.*, 1999; den Boon *et al.*, 2001; Schwartz *et al.*, 2002). In fact, mRNA capping enzyme of SFV requires association with membrane phospholipids for activity (Ahola *et al.*, 1999).

Viruses disrupt the various translational cap functions and re-programme the host cell towards synthesis of viral proteins. Viral disruption of cap-dependent translation contributes to host shut-off that often occurs during productive virus infection (Aranda and Maule, 1998). The RNA molecules having 5`-terminal cap are also preferentially translated in extracts of plant or animal cells. Moreover, capped RNAs are recognized and selected, in preference to uncapped RNAs, during binding to small subunits of ribosomes and also during early stages of protein synthesis. The eIF4A is an ATPase/ RNA helicase. The RNA4 of AMV is uncapped but its capped transcripts are translated efficiently *in vitro* and cap structure is also essential for its *in vivo* translation. Thus cap appears to be essential for translation of this RNA while the uncapped RNAs 1 and 2 are non-infectious.

The cap structure performs following functions during the eukaryotic translation initiation (Kozak, 1989; Merrick and Hershey, 1996; Hershey and Merrick, 2000). Firstly, cap structure promotes mRNA export. Secondly, cap serves as the binding site for eukaryotic translation initiation factor 4F (eIF4F), which specifically recognises 5`-cap structure. More specifically, the cap structure interacts with a subunit (eIF4E) of the heterotrimeric eIF4F. Thirdly, the eIF4F initiation factor also assists the binding of 40S ribosomes to mRNAs and, later, with the help of other initiation factors such as eIF3, recruits 43S pre-initiation complex that scans mRNA along 5`- to 3`-direction. Fourthly, cap-dependent control of mRNA translation confers several advantages to the cell. Cap dependence allows the cell to have a mechanism through which to control gene expression by modulating the assembly and activity of cap-binding complex components. Moreover, cap dependency provides selectivity of translation through translational regulatory properties of a specific mRNA. Fifthly, translational control (through functional activities of cap structure) allows a cell to fine-tune gene expression by stimulating or repressing translation of specific mRNAs, usually through reversible phosphorylation of translation factor attached to cap structure. In short, cap/cap-like structure is possibly required for efficient initiation and regulation of mRNA translation. Then, multiple rounds of translation are facilitated by eIF4G/poly A-binding protein interaction that increases the proximity of the 5`-capped and 3`-polyadenylated mRNA ends.

B. Genome-linked Protein

Virus protein linked to genome (VPg)/genome-linked protein is a small, virus RNA-encoded protein of 3500 to 24,000 molecular weight which is covalently linked by a phosphodiester bond to the 5`-terminal nucleotide of genomic RNA of many plant viruses and thus blocks 5`-terminus of viral RNAs. VPg, if present, occurs in all RNA segments of a multipartite virus. The VPg occurs at 5`-end of RNA genome of following genera of plant viruses of picorna-like and sobemo-like supergroups: *Bymovirus* (BaMV); *Comovirus* (CPMV); *Enamovirus* (Pea enation mosaic virus RNA1); *Luteovirus* [BYDV (RGV, RMV, and RPV strains), and BWYV]; *Nepovirus* (GFLV, TBRV, and *Tomato ringspot nepovirus* - TomRSV); *Polerovirus* (PLRV); *Potyvirus* [*Clover yellow vein virus* (Yambao *et al.*, 2003), *Pea seed-borne mosaic virus* (Guo, D., *et al.*, 2001); *Pepper vein banding virus* (Anindya *et al.*, 2005), *Potato virus A* (PVA), PVY, PPV, TEV, TVMV, and TuMV]; *Rymovirus* (*Ryegrass mosaic virus*); *Sequivirus* (PYFV); *Sobemovirus* (CfMV, SBMV, and *Turnip rosette sobemovirus*); and *Waikavirus* (RTSV). VPg is processed from a polyprotein in several plant viruses including BaYMV, GFLV, PLRV, *Ryegrass mosaic rymovirus*, TEV, TVMV, and seemingly in other plant viruses also.

VPg is the N-proximal portion of nuclear inclusion protein a (NIa protein) synthesized as part of the potyvirus polyprotein. Subsequently, NIa is cleaved from polyprotein by its C-proximal proteinase domain (NIa-Pro) and VPg domain is separated from proteinase by processing of a suboptimal cleavage site between these two domains (Merits *et al.*, 2002). Virions of TVMV bear a 24-kDa VPg, which is linked to 5`-end through a tyrosine residue that is conserved in VPg domains of polyproteins of other potyviruses. TEV has a VPg of 24-kDa mass. VPg of PVA RNA is about 25-kDa and is phosphorylated by the activity of a cellular kinase from tobacco (Puustinen *et al.*, 2002) indicating that VPg bound to virions can be phosphorylated. VPg of PYV has a molecular weight of 24 to 27-kDa, is a part of NIa, and is covalently linked to 5`-end of viral RNA through the tyrosine residue. Molecular mass of VPg of other potyviruses varies from 24 to 27-kDa.

VPg occurs in two forms in potyviruses: as a fully processed protein and as precursor forms with each form possibly having different functions during replication (Riechmann *et al.*, 1992). Virions of TVMV carry only VPg; so does RNA of PVA. Thus, VPg is found in virions of these potyviruses only in the processed form (Oruetxebarria *et al.*, 2001). In contrast, virions of TEV may carry either unprocessed viral NIa protein or only its N-terminal domain, which is the processed 24-kDa VPg. Thus, TEV RNA in virions carries both the processed VPg and the unprocessed NIa. In PVA, several high molecular weight mass precursor forms of VPg were detected in insect and plant cells (Merits *et al.*, 2002) but not in TuMV. Plochocka *et al.* (1996) gave the three-dimensional structure of PVY VPg.

VPg of CPMV is made up of 28 amino acid residues while a serine phosphodiester linkage binds it to the 5`-end of viral RNA. Encapsidated RNAs 1 and 2 of pea enation mosaic disease [caused by an obligatory association between *Pea enation mosaic virus*-1 (PEMV-1) (an enamovirus of family *Luteoviridae*) and *Pea enation mosaic virus*-2 (PEMV-2) (an umbravirus)] are covalently linked to a VPg encoded by PEMV-1 (Skaf *et al.*, 2000). In some plant viruses, a portion of VPg becomes covalently linked to the newly synthesized viral genomic RNA strands so that it remains bound to the encapsidated genome. On the other hand, Nicolas *et al.* (1997) suggested that VPg might be exposed in virions.

VPg can interact with other viral proteins. It shows VPg-VPg self-interaction in PVA (Oruetexebarria *et al.*, 2001), PSbMV (Guo, D., *et al.*, 2001), and in *Clover yellow vein potyvirus* (Yambao *et al.*, 2003). A novel protein-protein interaction, not reported in any other potyvirus so far, occurs between HC-Pro and VPg (HC-Pro-VPg interaction) of *Clover yellow vein potyvirus* (Yambao *et al.*, 2003). The C-terminal region (38 amino acids) of VPg is important for VPg-VPg interaction while central 19 amino acids are important for HC-Pro-VPg interaction.

1. Functions

The precise function(s) of VPg in many cases is unknown; however, its role in several events of virus life cycle has been detected or suggested. VPg is needed for virus RNA replication. Various demonstrated/suggested functions of VPg during viral RNA replication are: contains a nuclear localization signal; probably involved in formation of virus-specific polymerase; interacts with viral polymerase (NIb) of TVMV, stimulates its activity *in vitro* (Li, *et al.*,1997; Fellers *et al.*, 1998; Daros *et al.*, 1999) and in yeast (Hong *et al.*, 1995), and is likely to be involved in initiation of both the negative- and positive-strand RNA synthesis of TVMV and CPMV; contains a sequence-non-specific RNA-binding domain in PVA (Merits *et al.*, 1998); is a host-genotype-specific determinant for long-distance movement of TEV (Schaad *et al.*, 1997); interacts with eIF(iso)4F and the binding domain of eIF(iso)4E is located within VPg (Wittmann *et al.*, 1997; Léonard *et al.*, 2000); possibly mediates RNA translation and protein cleavage (Wimmer, 1982; Wittmann *et al.*, 1997; Schaad *et al.*, 2000); could interact with host proteins and undergoes phosphorylation by host kinases (Puustinen *et al.*, 2002); participates in formation of replication complex; and seems to act as an encapsidation signal for RNA of PEMV-2 since only RNA covalently attached to VPg is encapsidated and regulates replication of PEMV-1 (Skaf *et al.*, 2000). VPg is thus an integral part of viral life/infection cycle of such viruses. Data of Puustinen *et al.* (2002) imply that phosphorylation of VPg may play a role in VPg-mediated functions during infection cycle of potyviruses. The VPg of potyviruses may be the putative primer for potyvirus replication (Oruetxebarria *et al.*, 2001).

RNA of potyviruses is not capped so that translation initiation, including binding of initiation factor eIF4E to m^{7}G cap, cannot take place in the conventional manner. Instead, the plant cap-binding proteins eIF4E and its isomer eIF(iso)4E bind to and interact with VPg of potyviruses (Wittmann *et al.*, 1997; Léonard *et al.*, 2000; Schaad *et al.*, 2000). This interaction is important for replication and infectivity of TuMV, *Lettuce mosaic virus*, PVY, and TEV in *Arabidopsis thaliana* and pepper (Wittmann *et al.*, 1997; Lellis *et al.*, 2002; Ruffel *et al.*, 2002). It has been speculated that VPg might have a role in polyprotein translation since it interacts with cap-binding translation initiation factor eIF4E (Wittmann *et al.*, 1997). But potyviruses are capable of cap-independent translation (Niepel and Gallie, 1999). This makes the role of VPg in translation doubtful.

Presence of VPg on polyprotein precursor causes an increase in efficiency of proteolytic processing at Pro-RdRp site in TomRSV (Wang *et al.*, 1999). Similarly, presence of VPg on CPMV B RNA-encoded polyprotein precursors enhanced processing at Pro-RdRp site.

VPg is involved in transport of TEV, TVMV, PSbMV and PVA potyviruses (Schaad *et al.*, 1997; Hämäläinen *et al.*, 2000; Rajamäki and Volkenen, 1999, 2002) in infected plants. VPg forms a 'movement complex' with another viral protein and/or host factor(s) in *Clover yellow vein potyvirus*-infected host plants (Yambao *et al.*, 2003); central domain of VPg is particularly important for potyvirus movement (Rajamäki and Volkenen, 1999, 2002). The ability of VPg to support vascular transport and accumulation of PVA in tobacco and in a wild potato species is more effective in one host than in the other host (Puustinen *et al.*, 2002).

Phosphorylation of VPg is hypothesized to regulate infection in cells (Puustinen *et al.*, 2002). It could trigger virion disassembly and the subsequent translation process in potyviruses and it could have some functional significance for virus-host interactions since host factors in *Solanum commersonii* recognise VPg differently from those in tobacco plants. Moreover, phosphorylation of different parts of a virus can induce different effects. Thus, phosphorylation of CP (of encapsidated) PVX results in translational activation of the encapsidated virus RNA genome (Atabekov *et al.*, 2001) and phosphorylation of MP of TMV causes regulation of plasmodesmatal transport of the virus (Waigmann *et al.*, 2000) as well as affects intracellular stability and localization of virus MP (Kawakami *et al.*, 1999). So it is conceivable that phosphorylation of VPg of a virus (PVA) may effect regulation of specific interactions between VPg and host factors or regulation of viral factors taking part in virus replication and/or transport or causes virion disassembly (Puustinen *et al.*, 2002).

VPg is involved in overcoming viral resistance in plants with consequent increased susceptibility of host plants. All the reported cases pertain to potyviruses. The recessive resistance gene of a plant host encodes eIF4E, which upon interaction with VPg, leads to susceptibility (by overcoming the recessive resistance) of a plant host to the respective plant virus - *Lettuce mosaic virus* (Nicaise *et al.*, 2003), PVY (Ruffel *et al.*, 2002), TEV (Schaad *et al.*, 2000), TuMV (Wittmann *et al.*, 1997; Léonard *et al.*, 2000). Conversely, the recessive *Arabidopsis thaliana* mutant plants, that do not express eIF(iso)4E (an isomer of eIF4E), were immune to TuMV and TEV (Lellis *et al.*, 2002) indicating a relationship between VPg-eIF4E interaction and plant resistance to potyviruses.

Léonard *et al.* (2004) found that VPg of TuMV interacts *in vitro* with eIF4E; that direct interaction occurred between VPg [actually between polyprotein precursor forms of VPg (VPg-Pro and/or $6K_2$-VPg-Pro)] and PABP; that $6K_2$-VPg-Pro/VPg-Pro polyproteins were associated with ER membranes and were the actual viral forms that also interacted with eIF(iso)4E and eIF4E *in planta*; and that VPg-Pro also interacts with PABP *in planta*. Interaction between VPg-Pro and eIF(iso)4E/eIF4E as well as PABP could possibly promote RNA circularization (because of formation of the VPg-Pro-PABP complex) that could bring both ends of viral RNA in close proximity and lead to translation and replication of the virus. This is the way VPg overcomes viral resistance in plants.

The VPg plays some role in viral infectivity and pathogenicity of some plant viruses. Linkage of VPg to 5`-end of viral RNA in potyviruses is essential for virus infectivity, as in TVMV (Murphy *et al.*, 1996) so that degradation of VPg linkage to RNA in virions (of TVMV, TEV, and PPV) alters virus infectivity (Riechmann *et al.*, 1989). The linkage between VPg and TVMV RNA is mediated by tyrosine amino acid residue (Murphy *et al.*, 1996).

Modification of VPg protein leads to breaking of plant resistance in many *Potyviridae*/plant interaction systems. The VPg in bymoviruses (genus *Bymovirus* of family *Potyviridae*) is required for pathogenicity of genotypes carrying recessive resistance genes in several potyvirus/dicotyledonous plant pathosystems (Kuhne *et al.*, 2003). The ability of a pathotype of BYMV to overcome the host gene *rym4*-mediated resistance in winter barley correlates with a codon change (lysine at position 1307 changed to asparagine) in VPg coding region of RNA1 (Kuhne *et al.*, 2003). Thus, a single amino acid change in VPg protein of BYMV does the trick.

Both the recessive resistance gene(s) and the corresponding virus component that determines pathogenicity have been characterized in several potyviruses. Variations in VPg protein permit TVMV to overcome *va* gene resistance in tobacco (Nicolas *et al.*, 1997) and TEV to break resistance controlled by two unlinked recessive genes in tobacco cv. V20 (Schaad *et al.*, 1997). The ability of *Lettuce mosaic virus* to overcome two resistance genes of *Lactuca sativa* mapped to 3`-half of viral genome, which contains the VPg cistron (Redondo *et al.*, 2001).

VPg is suggested to be part of a transmembrane protein. Wang, A. *et al.* (2004) propose that the putative NTP-binding protein (NTB) interacts with VPg, that NTB-VPg is a transmembrane protein with VPg domain present on luminal face of membranes, and that NTB-VPg complex may serve as a membrane anchor for TomRSV replication complex. Han and Sanfaçon (2003) suggest that NTB-VPg is an integral membrane protein in plants. Similarly, in CPMV, both the 60-kDa (NTB-VPg) and 32-kDa proteins are associated with and modify membranes of ER (Carette *et al.*, 2002). Replication complexes of plant viruses, whose RNA genome possesses a VPg at 5`-end, could be anchored at host endoplasmic reticulum membranes in this way.

III. 3`-End

The 3`-terminus (3`-UTR) of plant virus RNAs can carry any one of the following three structures: tRNA-like structure, poly(A) tail or lacks both and culminates in pX_{OH}. The 3`-end of CarMV, PVX, SBMV, TRV, TSV, and *Turnip rosette virus* is represented by $pX_{OH}$ since it can neither be charged by an amino acid nor it carries a poly(A) tract. The 3`-end is highly structured when it carries a tRNA-like structure.

The roles of 3`-end during messenger functions of viral genomic RNA could be to ensure stability, targeting of viral RNA to specific sites in cells, required for replication of the RNAs, required for proper translation, and modulation of translation mechanism/machinery. The 3`-end may also have a regulatory function like non-interference between translation and replication of a particular genomic RNA so that there is no clash between the ribosomes reading RNA from 5`- to its 3`-end and the replicase reading the genome from its 3`- to 5`-end. Then, 3`-end is involved during transcription of the initial positive-strand RNA genome to a limited number of negative-strand RNA copies, which, in turn, are transcribed several times into new progeny positive-strand RNA copies. The 3`-UTR in BMV contains a promoter element that directs negative-strand RNA synthesis while that of TMV contains a translational enhancer. The RCNMV 3`-UTR can function alone (without its 5`-UTR) as a cap-independent translational enhancer in cowpea protoplasts (Mizumoto *et al.*, 2003). The 3`-UTR of RNA1 of RCNMV contains the determinant for temperature-sensitivity (Mizumoto *et al.*, 2002). Optimum and most efficient temperature for SBWMV RNA1 replication is 17°C and is determined by p152 and p211 replicase proteins and/or 5`- and 3`-UTRs of RNA1 (Ohsato *et al.*, 2003).

The homologous 145 nucleotides at 3`-terminus of AMV RNAs (1, 2, 3, and 4) can adopt two alternative conformations - a linear array of hairpins A to E with flanking AUGC motifs 1 to 4 that represent high-affinity CP-binding site and a tRNA-like structure that represents the promoter for minus-strand RNA synthesis. PCV is composed of two plus-sense RNAs that exhibit little sequence homology except for 3`-terminal 273 nucleotides that are 95% identical between two RNAs.

A. tRNA-Like Structure

The tRNA-like structure is so-called because, like a true cellular tRNA, it can be folded into a clover leaf configuration, terminates in the hallmark CCA_{OH}, and can be aminoacylated *in vitro* by a specific amino acid by the relevant aminoacyl-tRNA synthetase (Söll and RajBhandary, 1995; Giegé, 1996; Haenni and Chapeville, 1997; Dreher, 1999). Moreover, the three-dimensional folding of tRNA-like structure has certain characters reminiscent of the three-dimensional structure of canonical tRNAs and, even if the tRNA-like domain has a somewhat different folding pattern, it always possesses the canonical tRNA function of aminoacylation. The CCA_{OH} is a major functional domain, is well separated from rest of the tRNA-like structure, permits aminoacylation, elongation factor binds to it, is the initiation site for RNA replication, and is recognised by tRNA nucleotidyl-transferase which helps to maintain an intact 3`-end of viral RNA so that nucleotidyl-transferase activity is regarded as a telomerase activity (Rao *et al.*, 1989). The tRNA-like structure has been reviewed several times by now but still the presence of tRNA-like structures with tRNA-like properties at the end of a number of plant viral genomic RNAs has been and is still intriguing (Fechter *et al.*, 2001a). In fact, tRNAs could be molecular fossils of an RNA world (Maizels and

Weiner, 1993) and may correspond to primitive telomers ensuring that 3`-end of viral genomes are not lost during several cycles of RNA replication. Rao *et al.* (1989) also suggested telomeric function of tRNA-like structure of BMV RNA.

The aminoacylation property of tRNAs also remains enigmatic (Haenni and Chapeville, 1997): the capacity of viral RNA genomes to be aminoacylated is restricted only to plant viruses and, even among them, only to a few families having tRNA-like structures at 3`-end of their RNA genomes; only one of the three specific amino acids (histidine, tyrosine and valine) can be employed for aminoacylating a viral RNA; and tRNAs and tRNA-like structures can take up different folding patterns in plant viral RNAs.

The tRNA-like structure is present at 3`-end of seven different positive-strand RNA plant virus genera: *Bromovirus, Cucumovirus, Furovirus* (Goodwin and Dreher, 1998), *Hordeivirus, Tobamovirus, Tobravirus,* and *Tymovirus* (Fechter *et al.*, 2001a). The first characterised tRNA-like domains were found at 3`-ends of genomic RNAs of plant viruses. By now, many tRNA mimics having structural and functional characters of tRNAs have been found in nature. The most studied tRNA-like structures are of RNAs of TYMV, TMV, and BMV, which are representatives of each of the three amino acid specificities.

All the four BMV RNAs (including CP subgenomic RNA) carry a 5`-cap and share a highly conserved sequence of about 200 nucleotides at 3`-end of each of the RNAs (Kao and Sivakumaran, 2000). This sequence is proposed to mimic a tRNA-like structure on the basis of computer modeling, chemical probing and phylogenetic analyses (Felden *et al.*, 1994; Kao and Sivakumaran, 2000; Fechter *et al.*, 2001a), and possesses the canonical CCA sequence at their ends. The tRNA-like structure is tyrosylated both *in vivo* and *in vitro* (Fechter *et al.*, 2001b) although it does not strongly mimic a canonical tNRA^{Tyr}. Moreover, a structure named stem-loop C (SLC) in BMV tRNA-like structure does not have a direct analog in canonical tRNA tertiary structure. In fact, the BMV tRNA-like structure (of the three BMV genomic RNAs and subgenomic RNAs) is one of the best investigated. It can interact with RNA-associated enzymes, including aminoacyl tRNA synthetase, elongation factor 1a, and nucleotidyl transferase.

1. Structure

The tRNA-like domains of plant viruses are highly structured elements and bear several tRNA-specific domains. It is suggested that tRNA-like structures of TYMV and TMV are architectural scaffolds that allow the correct presentation of valine and histidine identity elements towards the corresponding synthetases in a way similar to that in tRNAs. The most complicated and divergent tRNA-like motifs are present in RNAs that have to be tyrosylatable. It is even very difficult to assign an anticodon domain in this case. Structural properties of tRNA are thought to be important in at least two ways: these properties enable the viral RNA to distinguish specific -CCA$_{OH}$ sequence (at 3`-end) from the internal -CCA, and permit access of relevant enzymes to this -CCA$_{OH}$.

The tRNA molecules fold into a characteristic L-shaped configuration having an amino acid acceptor end and an anticodon end at the two extremities of the fold. The amino acid acceptor arm is formed by stacking of 12 base pairs while the anticodon arm consists of 10 base pairs and the two are connected by 26 to 44 base pairs. Presence of a pseudoknot is essential for making the equivalent of amino acid acceptor arm in all the tRNA structures. The three-dimensional architecture of a tRNA-like structure mimics L-shaped tRNA but has some deviations from the standard tRNA molecule. This L-shaped configuration can be constructed from sequences of about all known tRNA-like structures. The structure of tRNAs of three plant viruses (BMV, TMV and TYMV) is discussed below.

The tRNA-like structure TYMV is the simplest, the best studied (Rietveld *et al.*, 1982), and has the maximum resemblance with canonical tRNAs. It was the first precise three-dimensional model of a viral tRNA-like structure and was made possible by chemical and enzymatic probing experiments coupled with graphical and computer modeling. Its resembles an L-shaped structure and is strikingly identical to that of the canonical tRNAs, especially at the level of anticodon domain. Construction of an amino acid acceptor branch, equivalent to that of tRNAs, required a peculiar folding called the pseudoknot (Rietveld *et al.*, 1982). The aminoacyl acceptor domain is formed exclusively by 3`-half of tRNA-like structure and it is independent of the remaining part of this structure. The minimum length of viral RNA necessary for interaction with tRNA-nucleotidyl transferase (CCA-enzyme) is about 50 nucleotides and this length corresponds to the aminoacyl acceptor arm. The above proposed model involves the last 86 nucleotides at 3`-terminal end of valylable TYMV RNA and was established by computer modeling (Dumas *et al.*, 1987).

The TYMV was the first virus shown to be specifically valylated. The RNA genome of all the investigated tymoviruses can accept valine except *Erysimum latent virus* whose RNA is not valylated *in vitro*. Valine is covalently bound, in presence of valyl-tRNA synthetase, to the tRNA-like structure at 3`-end of RNA. Central position of anticodon in TYMV tRNA-like structure is important for efficient valylation. Two cytidylic acid residues terminate the 3`-terminal sequence of tymovirus viral RNA. The CCA-enzyme has to first bind an adenylic acid residue at RNA terminus before valylation can take place. Apart from CCA-enzyme and valyl-tRNA synthetase, some other enzymes like peptidyl-tRNA hydrolase, RNAse, etc. also interact with tRNA-like structure of TYMV RNA. Valylation of TYMV RNA also takes place in infected Chinese cabbage leaves. But the RNA within virus particle is not valylated and even lacks 3`-terminal adenylic acid residue; it appears from this observation that esterification *in vivo* is either transient or occurs in only a fraction of virus particles in a large population.

The three-dimensional model of tRNA-like structure of BMV RNA is more complicated. Felden *et al.* (1994) generated its refined model by computer modeling out of the last 201 nucleotides. This model is made up of eight intricate structural domains (A, B1, B2, B3, C, D, E, and F) and has an overall mimicry with L-shaped architecture of tRNAs. The domain A mimics the amino acid accepting branch, domain B1 is the possible mimic of anticodon branch, hairpin B2 is specified as the anticodon domain (Fechter *et al.*, 2001a) in contrast to the B3 hairpin which was earlier regarded as the anticodon domain, hairpin C directs synthesis of viral RNA and is constituted by two discrete domains that are separated by an internal loop.

The tRNA-like structure of BMV RNA has three different parts: a pseudoknot structure in aminoacyl acceptor arm, the C residue (arm C) which is adjacent to 3`-terminus, and a hairpin structure 49 to 79 nucleotides upstream of 3`-terminal arm (arm C). All these three parts of tRNA-like structure are essential for optimal promoter activity. The pseudoknot possibly has a structural importance while the loop regions of arm C could be the bases that are specifically involved in interaction with replicase.

The 3`-terminal part of TMV RNA can be folded into a tRNA-like structure, which  has three structural domains (D1, D2, and D3) connected by a central core C (Rietveld *et al.*, 1984). Felden *et al.* (1996) gave the model of 3`-terminal 182 nucleotides of TMV RNA. This model mimics L-shape of canonical tRNA and this is achieved by pseudoknotted domain D1 (which is equivalent to amino acid-accepting arm) and by anticodon branch (mimicked by domain D2). D3 is an upstream pseudoknotted region. Tertiary folding of central core of the RNA produces the structure and orientation of the entire tRNA-like structure. The TMV tRNA-like domain has extra-large anticodon branch of the L-shape structure.

2. Aminoacylation

A significant and striking feature of many plant viruses is the capacity of the 3`-end of their RNA genomes to be aminoacylated. The RNA of most members of a particular virus group can be aminoacylated with an amino acid that most likely is specific for that group. Nevertheless, exceptions are known. Thus, RNAs of most of the members of cucumoviruses are tyrosylated while those of tymoviruses are generally valylated. But *Tomato aspermy cucumovirus* RNA can be adenylated instead of being tyrosylated and *Erysimum latent tymovirus* RNA cannot be valylated. All RNAs of a multi-component virus accept the same amino acid. Majority of such reports are based on *in vitro* studies and it is still to be settled whether the RNAs of these viruses are actually able to acquire the relevant amino acid *in situ* within the infected cells. The RNAs of TRV cannot be aminoacylated, even though they harbour aminoacyl acceptor domain similar to that of tymovirus and bromovirus RNAs, because the TRV tRNA-like structure does not to appear to contain an anticodon domain. This also probably explains the fact that TRV RNA, instead of being aminoacylated, can be adenylated as in *Erysimum latent tymovirus*.

No relationship seems to exist between the role of tRNA-like structures and their aminoacylation properties. This appears to be so from the situation prevalent in the two RNAs of PCV: valylation of tRNA-like domain is present in RNA1 but absent in RNA2 although both RNAs have very similar tRNA-like ends. It is so because RNA2 is missing an essential anticodon valine identity element. Thus, aminoacylation may not perform the same roles in all viruses and in all RNAs of a given genus/virus.

The 3`-end of TYMV RNA genome carries a tRNA-like structure that is a highly efficient mimic of tRNA$^{Val}$, which interacts with at least the three tRNA-specific proteins as efficiently as tRNA$^{Val}$. These three proteins are: CCA nucleotidyl-transferase that adds the 3`-terminal adenosine to complete the -CCA 3`-end since most virion RNAs lack the terminal A; valyl-tRNA synthetase that covalently links valine to (that is, valylates) the 3`-end; and translational elongation factor eIF1A that binds the valylated RNA to form a valyl-tRNA1A.GTP ternary complex. Valylation of TYMV RNA demonstrated that aminoacylation is important for genome amplification but it is not essential to have valine at 3`-end of RNA. This indicates that functional attributes of tRNA-like motif are directly connected to viral replication, that tRNA-like domain bears replication promoter signals (Matsuda and Dreher, 2004), that tRNA-like domain contains the translation initiation site (- $CCA_{OH}$ at 3`-terminal sequence), it also allows regulation of minus strand transcription, and is probably important in regulating the transition from the early phase of viral gene expression (when viral RNA is used as mRNA from which early viral genes are expressed) to the subsequent phase of genome replication (Matsuda *et al.*, 2004).

Dreher (1999) in fact proposed that tRNA-like structure most probably plays a dual role. It is the initiation site for minus-strand RNA synthesis and simultaneously acts as a repressor towards synthesis of minus-strands. This adverse regulation of negative-strand RNA synthesis could delay the switch from RNA translation to RNA replication mode or promote a switch from minus- to plus-strand RNA synthesis. On the other hand, aminoacylation of BMV RNAs do not seem to promote negative-strand RNA synthesis so that aminoacylation of RNA is not needed for RNA replication in BMV. This is also true of TYMV (Dreher *et al.*, 1996; Goodwin *et al.*, 1997). The amino acid bound to 3`-end of viral RNA by the relevant aminoacyl-tRNA synthetase and the plant viruses whose RNA genome can be aminoacylated *in vitro* are discussed below.

Valine is present at 3`-end of viral RNA of the following plant viruses: *Eggplant mosaic virus*, *Okra mosaic virus*, TYMV, cowpea strain of TMV and other plant viruses mentioned below. Both viral (of TYMV) and canonical tRNAs are recognised and charged by aminoacyl-tRNA synthetases. Valine identity elements in both are identical.

Sequence comparison of tRNA-like motifs of RNAs of four tymoviruses [TYMV, *Erysimum latent virus* (ELV), *Kennedya yellow mosaic virus* and *Eggplant mosaic virus* (EMV)], of seven furoviruses or furoviral-like RNAs [SBWMV1 and SBWMV2, *Potato mop top virus* (PMTV2), PCV1 and PCV2, and *Indian Peanut clump virus* (IPCV1) and one more furovirus], and one tobamovirus [*Sunn-hemp mosaic virus* (SHMV)] brought out that only four nucleotides are strictly conserved in efficiently valylated RNA molecules. These are the four valine identity nucleotides of $tRNA^{Val}$ and are A_{35}, C_{36}, C_{38}, A_{73}. The first three nucleotides (A_{35}, C_{36}, C_{38}) are located in anticodon loop while the last nucleotide is the discriminator residue and is located at discriminator position. Valylation capacity of three of above tRNA-like motifs is less efficient because discriminator analogue (A_{73}) in SHMV RNA is a C residue, central anticodon analogue is absent in PCV2 RNA, and anticodon motif is absent in ELV RNA. The result is that SHMV RNA has weaker valylation efficiency while PCV2 RNA and ELV RNA are not valylated at all.

Tyrosine is carried by 3`-end of the following plant viruses, apart from the others mentioned below at relevant places: BSMV, BBMV, BMV, CCMV, CMV. Thus, plant viruses of bromo-, cucumo- and hordeiviruses have tyrosylable RNA genomes. In fact, RNAs of most members of other investigated viruses within bromoviruses and cucumoviruses can be tyrosylated *in vitro*. The tRNA-like structures carrying tyrosine are intricate - as shown by tRNA-like motif in BMV. The tyrosine identity nucleotides of tRNA^{Tyr} are: A_{73}, G_1-G_{72}, G_{35}, U_{35}. A_{36}. Significantly, BMV tRNA-like structure has neither a canonical seven nucleotide-anticodon loop nor the tyrosine anticodon triplet GUA. The proposed tRNA-like structure of BMV also contains a pseudoknot, and is completely different and far more complicated than tRNA-like structures of RNAs of tymoviruses and tobamovirus. On the other hand, RNAs of *Tomato aspermy cucumovirus*, instead of being aminoacylated, is adenylated.

Fechter *et al.* (2001b) reported the tyrosine identity elements of BMV tRNA-like structure. Tyrosine identity relies solely on nucleotides from the pseudoknotted acceptor stem but efficiency of tyrosylation depends on the presence of an additional hairpin, the B2 structure, which is likely to be located perpendicularly to acceptor branch. This is suggested to enhance anchoring of RNA on synthetase. The sole residue determining tyrosine identity of BMV tRNA-like structure are base pair C116-G5 and residue A4, both located in amino acid acceptor arm (Fechter *et al.*, 2001b). These residues mimic structurally and functionally the major yeast tyrosine identity elements C1-G72 and A73. No functional equivalents of yeast tRNA^{Tyr} anticodon residues exist in BMV RNAs. Thus, specific tyrosylation of the tRNA-like structure of BMV RNA relies solely on identity nucleotides present in its amino acid-accepting domain. It can be concluded that B2 and B3 loops are not involved in tyrosylation of BMV RNA and that tyrosylation of this molecule relies on the sole presence of C116-G5 base pair and of discriminator residue A4 in its amino acid-acceptor branch. This conclusion likely holds for other bromo-, cucumo, and hordeivirus RNAs. Fechter *et al.* (2001a) demonstrate the strict equivalence of the major tyrosine identity elements embedded in amino acid-acceptor branches of BMV tRNA-like domain and yeast tRNA^{Tyr}. A similar conclusion arose from studies on TYMV and TMV tRNA-like structures in which mimics of the major valine and histidine identity elements of tRNAs were found.

Histidine is the amino acid accepted by genome of most of the tested tobamoviruses but efficiency of aminoacylation is greatly variable in different tobamoviruses or in different tobamovirus strains. In contrast, cowpea strain of TMV (also called *Sunnhemp mosaic tobamovirus*) instead accepts valine, like TYMV, in place of histidine. TMV genome is histidinylated *in vitro* at 3`-CCA terminal. The minimum aminoacylatable size of TMV RNA is about 95 nucleotides long, which agree with the proposed folding pattern of tRNA-like structure. The aminoacyl acceptor arm of TMV RNA is also independent of the rest of tRNA-like structure. Formation of pseudoknot is identical to that of TYMV RNA. The aminoacyl acceptor arm suffices for this interaction. The histidine identity nucleotides of tRNA^{His} are G_{-1}, A_{73}, G_{34}, U_{35}. It is the phosphate group of nucleotide -1 that is mainly involved in specificity of histidylation in classical tRNAs. The histidylation of several tRNA-like structures was suggested to account for presence of mimics of nucleotides G-1 and N73, the major tRNA^{His} identity nucleotides, in viral RNAs (Felden and Giegé, 1998).

3. Pseudoknot

Pseudoknot is a type of special secondary/tertiary folding pattern present in 3`-tRNA-like structure of RNA and so is a part of tRNA-like structure but is located upstream of the tRNA-like structure in a few plant viruses, is formed by aminoacyl acceptor arm of tRNA-like structure, and occurs in many viral and cellular RNAs. It was discovered while constructing an amino acid acceptor branch of TYMV tRNA-like domain (Rietveld *et al.*, 1982). Since then, it has been observed that all the viral tRNA mimics have a pseudoknotted amino acid-accepting stem. Its presence is generally revealed through phylogenetic comparisons and enzymatic or chemical structure probing. Virus RNAs of SBWMV possess valylated tRNA-like structure that is preceded by a long pseudoknot.

Pseudoknot is an RNA hairpin in which loop residues base pair with nucleotides outside the loop (Pleij, 1994); that is, pseudoknots arise by forming base-pairs between nucleotides that occur within a loop and complementary nucleotides that are located outside of this loop. Simultaneously, pseudoknot results in coaxial stacking of two stem regions that are connected by two loops. Structure and exact folding of pseudoknot of TYMV RNA in solution has been determined by NMR spectroscopic studies (Kolk *et al.*, 1998). A pseudoknot has low stability. The pseudoknot conformation (of AMV and ilarviruses) resembles tRNA-like structures of bromo-, cucumo-, and hordeiviruses. In TMV, the 3` UTR comprises of a tRNA-like structure and three associated upstream pseudoknot domains (Okada, 1999).

4. Functions

The tRNA-like structures are multifunctional. Several possible functions of vital importance have been assigned to tRNA-like structures of RNAs of plant viruses although no unanimity exists yet with regard to any of these postulated functions. Some of the viral RNA functions may be dependent on aminoacylation properties of tRNA-like domains and/or structural properties within or outside these domains. As will be clear from below, 3`-tRNA-like structure and pseudoknot structures play roles in maintenance of RNA stability, in promoting initiation of viral RNA translation, as a promoter for synthesis of negative-sense RNA by viral replicase, and in increasing RNA repair by recruiting cellular enzymes that repair damaged cellular tRNAs (Rao *et al.*, 1989; Dreher, 1999; Barends *et al.*, 2004).

a. Role in Transcription (Negative-Strand RNA Synthesis)

The tRNA-like structures are involved in transcription and negative-strand RNA synthesis in AMV, BMV, TMV, TYMV, and possibly also in other plant viruses. However, the entire tRNA-like structure (of TYMV RNA) is not required for this purpose under *in vitro* conditions (Deiman *et al.*, 2000); a tRNA fragment is sufficient (Deiman *et al.*, 1997, 1998; Singh and Dreher, 1997, 1998) so that only a short specific non-base-paired sequence is required for *in vitro* transcription by RdRp in AMV, BMV and TYMV. Thus, the pseudoknot and arm C of tRNA-like structure of BMV RNA are not required for transcription initiation (of the infecting positive-strand RNA for producing complementary negative-strand) *in vitro* (Chapman and Kao, 1999; Deiman *et al.*, 2000). The minimum template requirements for negative-strand RNA synthesis of AMV, BMV and TYMV resemble each other.

The main determinant in tRNA-like structure for initiating BMV and TYMV negative-strand RNA synthesis (by RdRp) *in vitro* is 3`-ACCA end. Only a non-base-paired ACCA end (of 3`-terminal hairpin) is functional, the minimum template length is 9 nucleotides and transcription improves with increasing the length of template and that proper base stacking contributes to efficient transcription initiation (Deiman *et al.*, 2000). Sequence domains within the 3`-region of BMV RNA that direct negative-strand synthesis have been extensively characterized by using *in vitro* and *in vivo* replication systems. The tRNA-like domains most likely act as initiation sites, rather than promoters, of minus-strand RNA synthesis.

Position of transcription site determines the efficiency of transcription. The best transcription of BMV RNA is obtained from site 13 to 16 nucleotides downstream of 5`-end of template. Sequences upstream and downstream of internal initiation site contribute to increased transcription efficiencies and efficient internal initiation (Deiman *et al.*, 2000). The tRNA-like structures are also now regarded as suppressors of minus-strand RNA synthesis. Low stability of a pseudoknot often results in its folding into alternative structures (Wyatt *et al.*, 1990; Kolk *et al.*, 1998). This property probably allows the pseudoknot to function as a 'molecular switch' during minus-strand RNA synthesis (Deiman *et al.*, 1997).

b. Role in Viral RNA Replication

Conflicting reports exist about the possible role of tRNA-like structures in replication of viral RNAs so that the relationship that could possibly exist between tRNA functions and RNA replication is still not understood.

The tRNA-like structure is required for replication of TYMV RNA. It is a translational enhancer, which is a new role discovered for it (Matsuda and Dreher, 2004). Its other possible function could be to direct the replication complex to the correct 3`-end viral RNA and thus to prevent internal initiation of negative-strand RNA synthesis (Deiman *et al.*, 1997). Moreover, the valylated tRNA-like structure of TYMV functions as bait for host ribosomes, entraps them, directs them to second ORF of genomic RNA and so mediates synthesis of its polyprotein via a cap-independent process (Barends *et al.*, 2003). Elimination of tRNA-like structure from the native TYMV RNA completely abolishes polyprotein synthesis, whereas translation of the first ORF into MP and that of sgRNA into CP remain unchanged. The tRNA-linked valine gets incorporated into N terminus of polyprotein in a cap-independent and initiator-tRNA-independent manner. This initiation of TYMV polyprotein synthesis through mediation of tRNA-like structure is exceptional and circumvents the standard steps of translation initiation since normal eukaryotic translation initiation begins with initiation factor-dependent recognition of 5`-cap. However, the above scheme proposed by Barends *et al.* (2003) does not seem to be operative during the studies carried out by Matsuda and Dreher (2004).

Domain D2 of TMV tRNA-like structure and central core region are the most important elements for binding of TMV RdRp complex to viral 3`-terminal region (Osman *et al.*, 2000). The tRNA-like domain imparts improved stability and synergistic action to the 5`-capped terminal of viral RNA, contributes to messenger properties of viral genomic RNA, is important for RNA replication *in vivo* (Chandrika *et al.*, 2000), is important for negative-strand RNA synthesis *in vitro* (Osman and Buck, 2003) by regulating access of viral RNA to factors concerned with negative-strand RNA synthesis, and can substitute efficiently for 3`-poly(A) end of GUS or luciferase reporter genes and increase their translational efficiency (Gallie and Kobayashi, 1994). These properties of 3`-end UTR of TMV RNA are more due to the presence of a row of pseudoknots upstream of tRNA-like domain than due to the tRNA-like structure itself.

The BMV tRNA-like structure performs several functions during viral RNA replication: contains the signals and is required for initiation of minus-strand RNA synthesis for which replicase complex docks at the stem-loop structure C of tRNA-like structure (Sivakumaran *et al.*, 2003); is the site for capsid assembly and encapsidation of BMV RNAs to form virus particles (Choi *et al.*, 2002) and regulates translation of BMV RNAs (Barends *et al.*, 2004). The terminal triloop of tRNA-like structure is important for association with replicase. The central role of tRNA-like structure is the temporal regulation of the consecutive processes of RNA translation, RNA transcription and RNA encapsidation (Barends *et al.*, 2004). The tRNA-like structure, when present, is envisaged to perform identical roles during infection cycles of all plant viruses (Barends *et al.*, 2004).

AMV tRNA-like structure also contains the signals for and is required for initiation of minus-strand RNA synthesis. Minus-strand RNA promoter consists of a triloop (hpE) that is located in and attached to a 3`-tRNA-like structure. AMV tRNA-like structure also contains binding sites for viral CP. Binding of CP to tRNA-like structure favours formation of a linear conformation instead of the tRNA-like structure. In linear conformation, synthesis of minus-strand RNA (and hence of RNA replication) is inhibited while translation is stimulated (Neeleman *et al.*, 2001). Capsid protein thus appears to have a regulatory function and acts as a structural switch of 3`-UTR - from tRNA-like structure to a hairpin array - and thereby regulates the functional transition from replication to translation and vice versa (Olsthoorn *et al.*, 1999; Neeleman *et al.*, 2001). Accordingly, for efficient translation at early stage of virus infection, a few molecules of CP are required in inoculum. This explains the essential need for the presence of CP in inoculum for successful infection by AMV. Olsthoorn *et al.* (2004) show that minus-strand RNA promoter, upon detachment from tRNA-like structure, can function as a subgenomic promoter *in vitro* and can replace authentic subgenomic promoter in live virus. Accordingly, they concluded that AMV subgenomic and negative-strand promoters arc basically the same but, since the negative-strand promoter is linked to 3`-tRNA-like structure, it forces the polymerase to initiate negative-strand synthesis at 3`-end.

Efficient translation depends on 5`- to 3`-end communication of mRNA, established by a 3`-poly(A) stretch that communicates via poly(A)-binding protein (PABP) and eIF4G with eIF4E bound to capped 5`-end. This circular form is believed to assess mRNA integrity and to recycle ribosomes for multiple rounds of translation. Indeed, the 5`-cap-binding factors and PABP synergistically stimulate translation (Wilkie *et al.*, 2003). For TYMV RNA, this circularization appears to be established by

3`-tRNA-like structure communication with the start of ORF2 (Barends *et al.*, 2003). In AMV, either the CP (attached to 3`-end of viral RNA) or the exposed loop sequences of hairpin structures bind translation factors (attached to 5`-end of viral RNA) and establish 5`- to 3`-end communication (circularization) of viral RNA (Bol, 2003).

The tRNA-like structures at 3`-end of plus-strand RNA viruses are involved both in replication and translation of viral RNA while sequences and structures found at the 3`-end of minus-strand RNA copies of genomes are most probably involved in intensive replication initiation. The tRNA-like structure is also essential for replication initiation of CMV RNA3 (Boccard and Baulcombe, 1993).

c. Roles of Pseudoknot

Pseudoknots fulfill some significant structural and functional characteristics and roles and have been reviewed (Pleij, 1994; Deimen and Pleij, 1997). The role and presence of pseudoknot in tRNA-like structures of plant virus RNAs is a puzzle because of some of its contradictory attributes. Firstly, the pseudoknot, as a central part of tRNA-like structure, is essential for aminoacylation of bromoviruses and tymoviruses (Mans *et al.*, 1992) but aminoacylation is not always necessary for virus replication and accumulation in plants (Goodwin *et al.*, 1997). Secondly, the pseudoknot is not required for replication of viral RNA *in vitro* (Deiman *et al.*, 1998; Singh and Dreher, 1998; Chapman and Kao, 1999) but is essential for replication of these viruses *in vivo* and cannot be replaced by a genuine tRNA. However, the pseudoknot is essential for AMV RNA replication both *in vitro* and *in vivo*. It is also essential for RNA replication of BaMV (Cheng *et al.*, 2002). Pseudoknot possesses low stability so that it often folds into alternative structures (Wyatt *et al.*, 1990; Gluick *et al.*, 1997; Kolk *et al.*, 1998). This permits the pseudoknot to function as a 'molecular switch' between translation and replication of viral RNA genome (Deiman *et al.*, 2000). Pseudoknots occurring upstream of tRNA-like structures seem to have some function in RNA replication (Héricourt *et al.*, 1999). Out of the three pseudoknots in TMV, the one that is immediately upstream of tRNA-like structure is essential for TMV RNA replication. Similarly, the upstream pseudoknot of TYMV tRNA-like structure has a role in replication of this RNA (Deiman *et al.*, 1997) since it seems to act as an important, though nonspecific, spacer element ensuring proper presentation of a functional tRNA-like structure (Matsuda and Dreher, 2004). Pseudoknots permit the exposure of nucleotide determinants for protein recognition. The above data indicate a possible regulatory function for the pseudoknot in plant virus RNAs. Pseudoknot and the two stem regions in amino acid-acceptor arm of tRNA-like structure of BMV RNA are seemingly essential for recognition by aminoacyl-tRNA synthetase and for tyrosylation capacity of BMV RNA.

d. Other Functions

The tRNAs may protect viral RNA against degradation. Gallie and coworkers (Gallie and Walbot, 1990; Gallie *et al.*, 1991) found that if TMV tRNA-like structure is tagged to 3`-end of unrelated mRNAs, the resulting chimeric construct exhibits greater stability in host plants. The tRNA-like structure could help in possible encapsidation of viral RNA by CP molecules (Dreher, 1999). The tRNA may regulate translation of viral proteins since, in certain cases, 5`- and 3`-regions of viral RNAs can potentially base pair. In such situations, accessibility of 3`-end of viral RNA might be regulated by aminoacyl-tRNA synthetase.

Barends *et al.* (2004) suggested that tRNA-like structure plays a central role in temporal regulation of consecutive stages of RNA translation, RNA transcription, and RNA encapsidation during the life cycle of BMV. Parts of tRNA-like structure play crucial role in synthesis of virus enzymes for RNA replication upon entry of BMV virions into a host cell (start of infection). The tRNA-like structure performs dual functions during the intermediate stages of viral infection: tRNA-like structure binding to replicase leads to synthesis of minus-strand RNA synthesis from 3`- to 5`-direction while tRNA-like structure binding to ribosomes leads to translation of relevant proteins from 5`- to 3`-direction and there is no clash between the two opposing directions of these functions. Additional host factors may be involved in this dual regulation and act as docking sites for either the ribosome or the replicase during virus replication. At the end of infection, tRNA-like structure is a target for regulation as a nucleation site for capsid assembly as well as for encapsidation of viral RNA (Choi *et al.*, 2002) and thereby prevents alternative interactions with ribosome or replicase. It is possible that the above temporal regulation of viral life cycle by tRNA-like structure is also functional in other tRNA-like structure-containing viruses.

It seems that histidinylation of tRNA-like structure of TMV is required for interaction of 3`-UTR with a host protein, the elongation factor eEF1A (Zeenko *et al.*, 2002). The TYMV tRNA-like structure also binds eEF1A and this interaction enhanced viral RNA translation in plant cells (Matsuda and Dreher, 2004).

B. Poly(A) Tail

Poly(A) tail is present at 3`-ends of genomic RNAs of plant viruses of *Bymovirus, Capillovirus, Carlavirus, Comovirus, Nepovirus, Potexvirus, Potyvirus, Rymovirus, Trichovirus,* and *Waikavirus* genera. The 3`-poly(A) tail of *White clover mosaic potexvirus* (WClMV) genomic RNA has 100 to 300 adenylates while that of BaMV genomic RNA has 90 to 170 adenylates. The adenylate residues in CPMV ranged from about 25 to 170 for bottom RNA and 25 to 370 for middle RNA. Length of poly(A) tails varies with different RNAs in the same preparation.

However, intercistronic poly(A) tract exists in BMV and is required for BMV RNA replication (French and Ahlquist, 1988). Choi *et al.* (2004) found that the intercistronic poly(A) sequence binds the BMV replicase specifically, and that 16 nucleotides of the typically 22-nucleotide-long BMV poly(A) tract was sufficient to effectively bind the BMV replicase.

The exact biological role of poly(A) tail in plant viruses is not yet known but several functions have been attributed to it. It is suggested to protect virus RNAs from digestion and degradation, is regarded to perform important functions during viral RNA replication, performs several functions during translation of viral RNA, and acts as an essential *cis*-acting motif. A part of poly(A) tract of B RNA of CPMV can take up a hairpin structure that has a heptanucleotide sequence just upstream of poly(A) tract. Deletions from 3`-end, that prevent formation of hairpin, interfere with replication of RNA. Similarly, deletion of poly(A) tail from BNYVV RNA3 leads to a great decrease in ability of RNA to replicate. Significantly, in cases, where infection does take place with 3`-poly(A)-deficient RNA/transcripts, the poly(A) tail is restored in progeny RNA molecules in BNYVV, CPMV, PPV, and WClMV. Cheng *et al.* (2002) found that the synthesis of minus-strand RNA of BaMV initiates from multiple sites within the poly(A) tail. In contrast, TEV RNA molecules lacking poly(A) are as infectious as the polyadenylated molecules.

The poly(A) tract acts as the binding site for the PABP (Mangus *et al.*, 2003), which stabilizes the binding of eIF4F to a 5`-cap structure (Sachs, 2000). PABP is found in all eukaryotes and is involved in at least three major post-transcriptional processes: initiation of protein synthesis, mRNA biogenesis and mRNA turnover. The poly(A) tail also interacts with eIF4G and eIF4B and converts mRNA into a closed loop structure (after interacting with 5`-cap structure). Interaction of poly(A) tail with cap structure stabilizes the mRNA and enhances translation initiation. Thus, poly(A) tail of messengers enhances translation synergistically with cap structure by interaction with PABP and with cap-binding factors eIF4F and eIF4B (Gallie, 1991, 1998). Moreover, this interaction promotes recruitment of 40S ribosomal subunit by interaction with eIF4G and eIF3 and confers protection against 5`-exonucleolytic degradation of mRNA. Dissociation of PABP from mRNA disrupts interaction of PABP with initiation factors and renders the cap structure vulnerable to removal by decapping enzyme Dep1p. Decapping is quickly followed by degradation of mRNA by 5`- to 3`-RNA exoribonuclease.

C. Repair Mechanism of 3`-End of Viral RNA

Some RNA viruses can repair their missing 3`-end nucleotides. This 3`-end repair mechanism (analogous to cellular telomerases) is reported in some and proposed in some other single-stranded RNA plant viruses. Several repair mechanisms have been postulated or identified in plant viruses.

Tombusviruses do not possess a tRNA-like structure. The viral RNA repair has been extensively studied in tombusviruses and their satellite viruses. The genomic and satellite RNAs of *Cymbidium ringspot tombusvirus*, possessing a deletion of the 3`-terminal CCC area, could repair this deletion *in vivo* (Dalmay *et al.*, 1993; Dalmay and Rubino, 1995). The 3`-ends of viral TCV RNA genomes and their associated satellite (sat) RNAs can be repaired in several ways - by viral polymerase, by homologous and heterologous RNA recombination, by a host-terminal transferase including the poly(A) polymerase complex, by the use of abortive initiation products/by the abortive synthesis of 4 to 8 oligonucleotides coupled to the use of primers in the repair of satRNA ends, and by the synthesis of initiation products from mutated initiation sequence to prime the synthesis of the satRNA (Dalmay *et al.*, 1993; Dalmay and Rubino, 1995; Carpenter and Simon, 1996a, 1996b; Nagy *et al.*, 1997, 1999; Guan and Simon, 2000). Absence of tRNA-like structure perhaps necessitates more active participation of viral replicase in tombusvirus RNA-end repair.

Guan and Simon (2000) detected repair mechanism(s) for 3`-end of TCV (of family *Tombusviridae*). The 3`-ends of genomic TCV RNA and satellite sat-RNAs form a stable stem-loop structure followed by a short single-stranded motif CUGCCC-3`-. Deletion of up to 6 nucleotides [CUGCCC wild-type termination sequence bases at the 3`-end of Satellite (sat) RNA C (which is associated with TCV)] was regained/repaired both *in vivo* and *in vitro* by using the analogous motif of TCV. This 3`-end repair mechanism involves formation of 4 to 8 nucleotide oligoribonucleotides through abortive synthesis by viral replicase by employing 3`-end of viral genomic RNA as template. This novel viral replicase-mediated 3`-end repair model involves synthesis of a pool of oligoribonucleotides and their subsequent utilization to prime negative-strand synthesis, thereby correcting deletions at 3`-ends of TCV-associated RNAs. Thus, the 3`-end repair mechanism employs the 3`-end of positive-strand RNA of TCV as a template. The repair of satellite RNAs associated with the tripartite CMV also requires the presence of their viral RNA1 and RNA2 (Burgyan and Garcia-Arenal, 1998). Up to seven bases deleted from 3`-end of a CMV satRNA could be repaired *in vivo*.

The 3`-tRNA-like structure of BMV RNAs can be repaired by RNA-RNA recombination mediated by viral replicase, by other replication-associated activities of viral replicase such as the use of abortive products to prime RNA synthesis, or by cellular and/or viral enzymes [(ATP, CTP): tRNA nucleotidytransferase] that can directly repair the RNA (Bujarski and Kaesberg, 1986; Rao *et al.*, 1989; Rao and Hall, 1993; Boccard and Baulcombe, 1993; Nagy *et al.*, 1997).

Hema *et al.* (2005) found that the repair of nucleotide substitutions (but not short deletions) of the CCA sequence in BMV RNAs have different requirements. They found that presence of wild-type RNA1 is necessary for the repair of the other two BMV RNAs having nucleotide substitutions in their CC ends; that BMV RNA1 contributed to increased repair of nucleotide substitutions at the CCA end; that the 1a protein produced from RNA1 is likely the responsible component; that the repaired sequence is heterogeneous; and that there were different levels of repair for BMV RNA3 and RNA2. They suggested that the BMV replicase may be solely responsible for repair of CCA ends of BMV RNAs; that RNA recombination is not a major contributor to the repair as was also reported by Rao *et al.* (1989); and that either both cellular and viral replicase enzymes participate in reconstruction of the ends of BMV tRNA-like structure or the cellular enzymes act first to reconstruct the CCA sequence or to generate a sequence that the BMV replicase could subsequently use to direct minus-strand initiation. If cellular proteins do participate in BMV end repair, then the CCA tRNA nucleotidyltransferase (that repairs deletions of the CCA ends of tRNAs) is possibly involved in the 3`-end repair. This enzyme actually binds to the BMV tRNA-like structure sequence (Felden *et al.*, 1994) and was postulated to repair the tRNA-like structure of BMV RNAs (Rao *et al.*, 1989), and the tRNA-like structure of TYMV (Litvak *et al.*, 1970). The postulated dependence on cellular enzymes for repair of viral tRNA-like structure differs from observations on tombusviruses as mentioned above. Hema *et al.* (2005) could not assign any definite roles for the cellular or viral enzymes in end repair. It is possible that BMV replicase actively participates and/or is responsible for end repair.

IV. VIRAL GENES

The viral genes/motifs/proteins are involved in performing the whole gamut of functions essential for the existence and life of plant viruses. They are the genes associated with viral RNA replication and translation (RdRp, helicase, proteinase, methyl/guanylyltransferase), virion formation (CP), and cell-to-cell movement, and are shown in Table 2. However, there are several other genome-encoded proteins that are involved in various activities of the virus. Some of these are genome-linked protein, enzymes involved in viral RNA translation, proteinases that are concerned with maturation of polyproteins, proteins involved in host-specific long-distance virus movement and spread in plants, cytoplasmic and/or nuclear inclusion proteins, proteins determining pathogenicity and hypersensitivity, proteins determining virus-vector interactions, and still other proteins of as yet unknown functions. Out of these, only the proteins involved in plant virus replication and cell-to-cell movement are encoded by all viruses; out of the remaining proteins, only some or the other of them are encoded by different viruses. Some of positive-strand RNA viruses contain only three genes that control basic functions of replication, cell-to-cell movement, and encapsidation.

In positive-sense RNA viruses, the RdRp is the best-conserved region with RNA helicase holding the second position. In fact, RdRp is the only domain of positive-sense RNA viruses allowing an all-inclusive phylogenetic analysis (Koonin and Dolja, 1993). The coat protein gene in majority of plant viruses is located in vicinity of 3`-end of genomic RNA and in one of the RNA segments of a multipartite virus. This is true of genomes of AMV, BMV, CMV, TMV, and several others. However, CarMV, Norwich strain of BSMV, and CAM strain of TRV are exception to this generalization since their coat protein cistron is located proximal to 5`-end.

Héricourt *et al.* (1999) divide viral genes into those that perform coding functions and those that perform non-coding functions. Viral genes that perform coding functions are: polymerase gene, helicase-like activity gene, coat protein gene, cell-to-cell virus movement gene, systemic virus movement gene, proteinase gene, vector transmission of virus gene, and symptoms gene. The main functions of the products of these genes are self-evident from their nomenclature itself. But, almost every protein is a multifunctional entity and takes part in several viral functions. Description of these (subsidiary) functions, except where concerned with virus replication, is not the subject matter of this book. However, various details about the proteins, concerned with replication of plant viruses, are given in appropriate chapters of this book.

Viral RNAs also contain signals/motifs/elements/sequences/genes that perform non-coding functions. Such genes are essential for viral interaction with other viral components, are involved in viral life cycle and are essential for virus replication and translation processes and mostly function in *cis*, but can also function in *trans*. They are mentioned at appropriate places in this book.

A. Genes of Positive-Sense RNA Viruses

Héricourt *et al.* (1999) give general arrangement of genes in RNA genomes of different genera of positive-sense RNA plant viruses while Morozov and Solovyev (1999) discuss viral genes and genome maps/organisation in RNA viruses in great detail. Only a few examples are, therefore, given below in Table 2.

Closterovirus genome has 9 to 12 ORFs. Seven of these ORFs are conserved and contain two gene modules (in *Beet yellows virus*) consisting of ORFs 1a/1b and ORFs 2, 3, 4, 5 and 6. The 5`-proximal ORFs 1a/1b are expressed from genomic RNA itself while all other genes are expressed from subgenomic RNAs. Citrus tristeza virus RNA produces two polyproteins: ORF 1a encodes a 349-kDa polyprotein containing two papain-like protease domains and methyltransferase-like and helicase-like domains and ORF 1b encodes the second and the larger ~400 kDa polyprotein that has an additional RdRp domain that is thought to result from a +1 frameshift (Karasev *et al.*, 1995).

TABLE 2

Genes of positive-sense RNA genomes of plant viruses
(Modified from Héricourt *et al.*, 1999)

Genus	Virus	Protein motifs			Movement Protein	Capsid protein gene position
		Methyl-transferase	Helicase	RdRp		
Alfamovirus	AMV	126kDa (1a) Of RNA1	126kDa (1a) Of RNA1	90kDa (2a) of RNA2	32kDa (3a) of RNA3	3' on RNA3
Bromovirus	BMV	106kDa (1a) Of RNA1	106kDa (1a) Of RNA1	90kDa (2a) of RNA2	32kDa (3a) of RNA3	3' on RNA3
Bymovirus	BaYMV		270kDa (polyprot.) Of RNA1	270kDa (polyprot.) of RNA1	270kDa (polyprot.) of RNA1	3' on RNA1
Capillovirus	ASGV	241kDa (polyprot.)	241kDa (polyprot.)	241kDa (polyprot.)		3'
Carlavirus	PVM	223kDa (polyprot. in BISV)	223kDa (polyprot. In BISV)	223kDa (polyprot.)	Triple gene Block	Internal
Carmovirus	CarMV			86kDa (r/t)	7kDa (prot. 5)	3'
Closterovirus	BYV	295kDa (polyprot; 1a)	295kDa (polyprot; 1a)	348kDa (f/s polyprot; 1a/1b)	65kDa	Internal (duplicated)
Comovirus	CPMV		58kDa (of 202kDa polyprot.) of B-RNA	87kDa (of 202kDa polyprot.) Of B-RNA	48kDa (of 105/95kDa polyprot.) Of M-RNA	3' on M-RNA
Cucumovirus	CMV	109kDa (1a) Of RNA1	109kDa (1a) of RNA1	94kDa (2a) of RNA2	30kDa (3a) of RNA3; 15kDa (2b) of RNA2	3' on RNA3
Dianthovirus	CRSV			88kDa (f/s) of RNA1	33kDa prot. Of RNA2	3' on RNA1
Enamovirus	PEMV		84kDa (prot. 2) And 130kDa (f/s prot. 2/3) of RNA1	130kDa (f/s prot. 2/3) of RNA1; 65kDa of RNA2		Internal on RNA1
Furovirus	SBWMV	150kDa of RNA1	150kDa of RNA1	209kDa (r/t) of RNA1		5' on RNA2

	BNYVV	237kDa of RNA1	237kDa of RNA1	237kDa of RNA1	Triple gene Block	5' on RNA2
Hordeivirus	BSMV	130kDa of RNA1	130kDa of RNA1	87kDa of RNA3	Triple gene Block	5' on RNA2
Idaeovirus	RBDV	190kDa of RNA1	190kDa of RNA1	190kDa of RNA1		3' on RNA2
Illarvirus	TSV		123kDa (1a) of RNA1	91.6kDa (2a) of RNA2	32kDa RNA3 in PDV	3' on RNA3
Luteovirus	BYDV			99kDa (f/s)	17kDa in PLRV	Internal
Machlomovirus	MCMV	111kDa (r/t)				3'
Marafivirus	OBDV	227kDa	227kDa	227kDa		C-terminus of 227kDa; duplicated
Necrovirus	TNV			82kDa (r/t)	7.5kDa (7a) and 7.3kDa (7b) in TNV-D^H	3'
Nepovirus	TBRV		72kDA (of 254kDa polyprot.) of RNA1	92kDA (of 254kDa polyprot.) of RNA1	Postulated (on N-terminus of 150kDa polyprot.) of RNA2	3' on RNA3
Potexvirus	PVX	166kDa	166kDa	166kDa	Triple gene Block	3'
Potyvirus	PVY		(CI) of 336kDa polyprot.	(NIb) of 336kDa polyprot.		3'
Rymovirus	BrSMV		(CI) of 340kDa polyprot.	(NIb) of 340kDa polyprot.		3'
Sequivirus	PYFV		336kDa polyprot.	336kDa polyprot.		Internal
Sobemovirus	SBMV		105kDa	105kDa	17.8kDa (P1) in RYMV	3'
Tobamovirus	TMV	126kDa	126kDa	183kDa (r/t)	30kDa	3'
Tobravirus	TRV	134kDa	134kDa	194kDa (r/t)	29kDa (prot. 3)	Only gene on RNA2
Tombusvirus	TBSV			92kDa (r/t)	22kDa (and 19kDa)	Internal
Trichovirus	ACLSV	216kDa	216kDa	216kDa	36.5kDa in GVB	3'; internal in GVB
Tymovirus	TYMV	140kDa (of 206kDa polyprot.)	140kDa (of 206kDa polyprot.)	66kDa (of 206kDa polyprot.)	69kDa	3'
Umbravirus	GRV			94kDa (f/s prot. 1/2)	28kDa (prot. 3)	
Waikavirus	RTSV		390kDa (polyprot.)	390kDa (polyprot.)		Internal

The remaining 3`-ten ORFs of a closterovirus are expressed through a series of 3`-coterminal subgenomic RNAs. The 3`-most subgenomic RNA encodes a 23 kDa RNA-binding protein (López *et al.*, 2000); ORF2 encodes a small hydrophobic protein that is probably the membrane-binding protein; ORF3 encodes a protein homologous to HSP70 group of cellular heat shock proteins; ORF6 encodes CP while ORF5 encodes a diverged duplicate of CP that encapsulates a small portion of viral genomic RNA;

ORF7 encodes a unique viral protein while ORF8 encodes a protein that is an enhancer of viral replication.

GFLV RNA1 encodes polyprotein p1 of 253-kDa, which is processed by an embedded proteinase into five proteins required for virus replication: p1A of unknown function, p1B is the probable helicase, p1C is VPg, p1D is proteinase, and P1E is polymerase. These five proteins are the only proteins required for virus replication and they also function in *trans* to ensure replication of virus RNA2 (Ritzenthler *et al.*, 2002). GFLV RNA2 encodes 122-kDa p2 polyprotein that is processed in *trans* by 1D into three proteins (2A, 2B, and 2C). Protein 2A is essential for replication of RNA2, is associated with membranous structures and is recruited by RNA1-encoded replication machinery, and is hypothesized to mediate the transport of nascent p2-RNA2 complexes from their initiation site in cytosol to perinuclear replication sites where RNA2 replication and p2 cleavage occurs (Gaire *et al.*, 1999). Protein 2B is MP forming tubules through which virus particles travel to uninfected adjacent cells. Protein 2C is CP.

Tobamoviruses have capped and non-polyadenylated genomes with 3`-tRNA-like structure. There is a general agreement that TMV genome encodes four major proteins: the 126- and 183-kDa replication proteins, 30-kDa MP, and 17.5-kDa CP. The 5`-proximal ORF encodes the 126-kDa major protein that is terminated by a single amber stop codon UAG. Occasional translational readthrough of this 126-kDa ORF leaky termination codon results in the production of the 183-kDa protein. The N-terminal part of 126-kDa protein harbours methyltransferase/guanylyltransferase domain while its C-terminal part contains the helicase domain of superfamily 1. The C-terminal part of 183-kDa readthrough protein contains the RdRp of supergroup 3. Both these 5`-proximal genes (126-kDa and 183-kDa genes) are translated directly from genomic RNA. These two replication proteins, along with at least one host protein, form the active replication complex that is membrane-bound to endoplasmic reticulum of the infected cells. The 3`-proximal genes encode, through the respective subgenomic RNAs, a 30-kDa cell-to-cell MP and a 17.5-kDa CP; these two ORFs are located downstream of the 126/183-kDa ORF. The 30-kDa MP is formed at early stage of infection. TMV can potentially encode two more proteins (Carr, 2004). One is a 54-kDa protein; its ORF is contained in the third subgenomic RNA called I_1. However, this protein has never been successfully detected *in vivo*, including TMV-infected plants, transgenic plants engineered to constitutively express the 54-kDa ORF (Zaitlin, 1999), preparations of TMV replicase (Osman and Buck, 1997), or in TMV-infected tobacco protoplasts (Okada, 1999). Thus, it remains uncertain whether the 54-kDa protein really does occur during TMV infection. And even if it does exist, it seems clear that this protein may not be essential for virus replication (Okada, 1999). The second potential TMV gene product is p4 that was identified by Morozov *et al.* (1993) and later confirmed by Canto *et al.* (2004) who discovered this sixth ORF. The ORF 6 overlaps the 3`-end of the ORF for MP and the 5`-end of CP gene, exists in both TMV U1 and the crucifer strain (cr-TMV), and potentially encodes a 4.8-kDa protein. The biological significance ORF6 is now known - it is a determinant of viral pathogenicity in *Nicotiana benthamiana* (Canto *et al.*, 2004).

The two RNAs of bipartite genome of TRV are capped and non-polyadenylated and are separately encapsulated. The 3`-terminal region of both genomic RNAs is conserved and contains tRNA-like structure having a pseudoknot. TRV RNA1 encodes four proteins: 134-kDa protein that contains methyltransferase/guanylyltransferase domain and helicase domain of superfamily 1 at its N- and C-termini, respectively; translational readthrough of UGA terminator of 134-kDa protein ORF forms the 194-kDa protein that has the domain of RdRp of supergroup 3 at C-terminus; ORF 3 encodes MP through a subgenomic RNA; and a second short subgenomic RNA encodes the Cys-rich zinc-finger 16-kDa protein. The 16-kDa protein is distantly related to the hordeivirus Cys-rich γb-protein, which regulates virus RNA synthesis. RNA1 can replicate independently of replication of RNA2. The 5`-proximal gene of RNA2 is the CP gene, which is translated from a subgenomic RNA. The RNA2 of several TRV strains can encode two or three additional ORFs including the ORF that translates a protein essential for virus transmission by nematodes.

Umbravirus does not encode a CP, forms no virus particles and depends upon a helper virus. Its genome has four ORFs; ORF1 encodes a protein of unknown function, ORF2 encodes RdRp of supergroup 2, ORF3 encodes a protein that forms cytoplasmic and nuclear membrane-associated inclusions, and ORF4 encodes MP.

V. FUNCTIONS OF VIRAL RNA

The basic function of all nucleic acids is perpetuation and increase in population, by multiplication, of a living system. The RNA does so in viruses also. RNA is the genetic material in great majority of the plant viruses. It was established by the observation that viruses reconstituted from protein of one strain (or a distinct virus) and RNA of another strain (or virus) invariably exhibited characteristics of the strain/virus supplying RNA. When CCMV RNA was separately assembled with CP of CCMV, BMV, and BBMV, the infectivity of reconstituted hybrid virus in all cases was that of the virus providing RNA, i.e., CCMV. A hybrid TMV (TMV particles reconstituted from protein of one TMV strain and RNA of another strain) inoculated to susceptible hosts produces progeny, which has the coat of strain supplying RNA. These simple earlier experiments unequivocally established that the RNA is the genetic material in plant viruses, controls infectivity, and codes for the specific CP of virus. Virus RNA is thus typically regarded as a polycistronic messenger, which embodies information for 1,000 to 2,000 amino acids pertaining to a number of proteins. Second important property is the infectivity of viral RNAs and entire RNA molecule is required for this; any break destroys its infectivity.

Nucleic acid gives a definite length and breadth to elongated virus particles. TMV CP can reaggregate by itself to form rods with typical helical symmetry but such rods have no definite length. Nucleic acid also helps stabilize the shape and architectural structure of a virus, particularly in spherical viruses, where secondary and tertiary structures of RNA may be important parameters.

Coding genes, the genes that encode specific proteins, are generally about 3 to 6, but the total number of genes is rarely more than 12. Morozov and Solovyev (1999) have discussed in great detail the genes and genome organisation of RNA viruses. Some of the genes that are nearly common to all plant viruses, with few exceptions, are genes associated with viral RNA replication (polymerase gene, helicase gene, proteinase gene, CP gene, and a couple of other genes). Viral RNAs also contain signals that are necessary for interaction of RNA with other viral components. These signals can be involved in translation, replication, and encapsidation processes. All the functions performed by 5`- and 3`-end structures are basically the functions of viral RNA genome.

VI. REFERENCES

Ahola, T., and Ahlquist, P. 1999. Putative RNA capping activities encoded by *Brome mosaic virus*: Methylation and covalent bonding of guanylate by replicase protein 1a. *J. Virol.* 73: 10061-10069.

Ahola, T., Lampio, A., Auvinen. P., and Kääriäinen, L. 1999. Semliki Forest virus mRNA capping enzyme requires association with anionic membrane phospholipids for activity. *EMBO J.* 18: 3164-3172.

Ahola, T., de Boon, J. A., and Ahlquist, P. 2000. Helicase and capping enzyme active site mutations in brome mosaic virus protein 1a cause defects in template recruitmet, negative-strand RNA synthesis and viral RNA capping. *J. Virol.* 74: 8803-8811.

Anindya, R., Chittori, S., and Savithri, H. S. 2005. Tyrosine 66 of pepper vein banding virus genome-linked protein is uridylylated by RNA-dependent RNA polymerase. *Virology* 336: 154-162.

Aranda, M. A., and Maule, A. 1998. Virus-induced host gene shut-off in animals and plants. *Virology* 243: 261-267.

Atabekov, J. G., Radionova, N. P., Karpova, O. V., Kozlowsky, S. V., Novikov, V. K., and Arkhipenko, M. V. 2001. Translational activation of encapsidated potato virus X RNA by coat protein phosphorylation. *Virology* 286: 466-474.

Ayllón, M. A., Gowda, S., Satyanarayana, T., Karasev, A. V., Adkins, S., Mawassi, M., Guerri, J., Moreno, P., and Dawson, W. O. 2003. Effects of modification of the transcription initiation site context on citrus tristeza virus subgenomic RNA synthesis. *J. Virol.* 77: 9232-9243.

Bailey-Serres, J., Rochaix, J.-D., Wassenegger, M., and Fillipowicz, W. 1999. Plants, their organelles, viruses and transgenes reveal the mechanisms and relevance of post-transcriptional process. *EMBO J.* 18: 5153-5158.

Barends, S., Bink, H. H., van den Worm, S. H., Pleij, C. W. A., and Kraal, B. 2003. Entrapping ribosomes for viral translation: tRNA mimicry as a molecular Trojan horse. *Cell* 112: 123-129.

Barends, S., Rudinger-Thirion, J., Florentz, C., Giegé, R., Pleij, C. W. A., and Kraal, B. 2004. tRNA-like structure regulates translation of brome mosaic virus RNA. *J. Virol.* 78: 4003-4010.

Boccard, F., and Baulcombe, D. C. 1993. Mutational analysis of *cis*-acting sequences and gene function in RNA3 of *Cucumber mosaic virus. Virology* 193: 563-578.

Bol, J. F. 2003. *Alfalfa mosaic virus*: Coat protein-dependent initiation of infection. *Mol. Plant Pathol.* 4: 1-8.

Boonham, N., Henry, C. M., and Wood, K. R. 1998. The characterization of a subgenomic RNA and *in vitro* translation products of *Oat chlorotic stunt virus. Virus Genes* 16: 141-145.

Buck, K. W. 1996. Comparison of the replication of positive-stranded RNA viruses of plants and animals. *Adv. Virus Res.* 47: 159-251.

Bujarski, J. J., and Kaesberg, P. 1986. Genetic recombination between RNA components of a multipartite plant virus. *Nature* 321: 528-531.

Burgyan, J., and Garcia-Arenal, F. 1998. Template-independent repair of the 3`-end of cucumber mosaic virus satellite RNA controlled by RNAs 1 and 2 of helper virus. *J. Virol.* 72: 5061-5066.

Butsch, M., and Boris-Lawrie, K. 2002. Destiny of unspliced retroviral RNA: Ribosome and/or virion? *J. Virol.* 76: 3089-3094.

Canto, T., MacFarlane, S. A., and Palukaitis, P. 2004. ORF6 of *Tobacco mosaic virus* is a determinant of viral pathogenicity in *Nicotiana benthamiana. J. Gen. Virol.* 85: 3123-3133.

Carette, J. E., van Lent, J., MacFarlane, S. A., Wellink, J., and van Kammen, A. 2002. Cowpea mosaic virus 32- and 60-kilodalton replication proteins target and change the morphology of endoplasmic reticulum membranes. *J. Virol.* 76: 6293-6301.

Carpenter, C. D., and Simon, A. E. 1996a. *In vivo* restoration of biologically active 3`-ends of virus associated RNAs by nonhomologous RNA recombination and replacement of a terminal motif. *J. Virol.* 70: 478-486.

Carpenter, C. D., and Simon, A. E. 1996b. *In vivo* repair of 3`-end deletions in a TCV satellite RNA may involve two abortive synthesis and priming events. *Virology* 226: 153-160.

Carr, J. P. 2004. *Tobacco mosaic virus. Annu. Plant Rev.* 11: 27-67.

Chandrika, R., Ranbidran, S., Lewandowski, D. J., Manjunath, K. L., and Dawson, W. O. 2000. Full tobacco mosaic virus RNAs and defective RNAs have different 3`-replication signals. *Virology* 273: 198-203.

Chapman, M. R., and Kao, C. C. 1999. A minimal RNA promoter for minus-strand RNA synthesis by the brome mosaic virus polymerase complex. *J. Mol. Biol.* 286: 709-720.

Chen, J., Noueiry, A., and Ahlquist, P. 2001. Brome mosaic virus protein 1a recruits viral RNA2 to RNA replication through a 5`-proximal RNA2 signal. *J. Virol.* 75: 3207-3219.

Cheng, J.-H., Peng, C.-W., Hsu, Y.-H., and Tsai, C.-H. 2002. The synthesis of minus-strand RNA of *Bamboo mosaic potexvirus* initiates from multiple sites within the poly(A) tail. *J. Virol.* 76: 6114-6120.

Choi, S.-K., Hema, M., Gopinath, K., Santos, J., and Kao, C. C. 2004. Replicase-binding sites on plus- and minus-strand brome mosaic virus RNAs and their roles in RNA replication in plant cells. *J. Virol.* 78: 13420-13429.

Choi, Y. G., Dreher, T. W., and Rao, A. L. N. 2002. tRNA elements mediate the assembly of an icosahedral virus. *Proc. Natl. Acad. Sci. USA* 99: 655-660.

Dalmay, T., and Rubino, L. 1995. Replication of cymbidium ringspot virus satellite RNA mutants. *Virology* 206: 1092-1098.

Dalmay, T., Russo, M., and Burgyan, J. 1993. Repair *in vivo* of altered 3`-terminus of cymbidium ringspot tombusvirus RNA. *Virology* 192: 551-555.

Darós, J. A., Schaad, M. C., and Carrington, J. C. 1999. Functional analysis of the interaction between VPg-proteinase (NIa) and polymerase (NIb) of *Tobacco etch potyvirus*, using conditional and suppressor mutants. *J. Virol.* 73: 8732-8740.

Deimen, B. A. L. M., and Pleij, C. W. A. 1997. Pseudoknots: A vital feature in viral RNA. *Sem. Virol.* 8: 166-176.

Deiman B. A. L. M., Kortlever, R. M., and Pleij, C. W. A. 1997. The role of the pseudoknot end of turnip yellow mosaic virus RNA in minus-strand synthesis by the viral RNA-dependent RNA polymerase. *J. Virol.* 71: 5990-5996.

Deiman, B. A. L. M., Koenen, A. K., Verlaan, P. W. G., and Pleij, C. W. A. 1998. Minimal template requirements for initiation of minus-strand synthesis *in vitro* by the RNA-dependent RNA polymerase of *Turnip yellow mosaic virus. J. Virol.* 72: 3965-3972.

Deiman, B. A. L. M., Verlaan, P. W. G., and Pleij, C. W. A. 2000. *In vitro* transcription by the turnip yellow mosaic virus RNA polymerase: A comparison with alfalfa mosaic virus and brome mosaic virus replicases. *J. Virol.* 74: 264-271.

den Boon, J. A., Chen, J., and Ahlquist, P. 2001. Identification of sequences in brome mosaic virus replicase protein 1a that mediate association with endoplasmic reticulum membranes. *J. Virol.* 75: 12370-12381.

Dreher, T. W. 1999. Functions of 3`-untranslated regions of positive-stranded RNA viral genomes. *Annu. Rev. Phytopathol.* 37: 151-174.

Dreher, T. W., Tsai, C. H., and Skuzeski, J. M. 1996. Aminoacylation identity switch of turnip yellow mosaic virus RNA from valine to methionine results in an infectious virus. *Proc. Natl. Acad. Sci. USA* 93: 12212-12216.

Dumas, P., Moras, D., Florentz, C., Giegé, R., *et al.*, 1987. 3-D graphics modelling of the tRNA-like 3`-end of turnip yellow mosaic virus RNA: Structural and functional implications. *J. Biomol. Struct. Dyn.* 4: 707-728.

Fechter, P., Rudinger-Thirion, J., Florentz, C., and Giegé, R. 2001a. Novel features in the tRNA-like world of plant viral RNAs. *Cell. Mol. Life Sci.* 58: 1547-1561.

Fechter, P., Giegé, R., and Rudinger-Thirion, J. 2001b. Specific tyrosylation of the bulky tRNA-like structure of brome mosaic virus RNA relies solely on identity nucleotides present in its amino acid-accepting stem. *J. Mol. Biol.* 309: 387-399.

Felden, B., and Giegé, R. 1998. Resected RNA pseudoknots and their recognition by histidyl-tRNA synthetase. *Proc. Natl. Acad. Sci. USA* 95: 10431-10436.

Felden, B., Florentz, C., Giegé, R., and Westhof, E. 1994. Solution structure of the 3`-end of brome mosaic virus genomic RNAs: Conformational mimicry with canonical tRNAs. *J. Mol. Biol.* 235: 508-531.

Felden, B., Florentz, C., Giegé, R., and Westhof, E. 1996. A central pseudoknotted three-way junction imposes tRNA-like mimicry and the orientation of three 5`-upstream pseudoknots in the 3`-terminus of tobacco mosaic virus RNA. *RNA* 2: 201-212.

Fellers, J., Wan, J., Hong, Y., Collins, G. B., and Hunt, A. G. 1998. *In vitro* interaction between a potyvirus-encoded, genome-linked protein and RNA-dependent RNA polymerase. *J. Gen. Virol.* 79: 2043-2049.

Florentz, C., and Giegé, R. 1995. tRNA-like structures in plant viral RNAs. In: Soll, D., and RajBhandary, U. (Eds.). tRNA: Structure, Biosynthesis and Function. Am. Soc. Microbiol., Washington, USA. pp. 141-163.

French, R., and Ahlquist, P. 1988. Characterization and engineering of sequences controlling *in vitro* synthesis of brome mosaic virus subgenomic RNA. *J. Virol.* 62: 2411-2420.

Furuichi, Y., and Shatkin, A. J. 2000. Viral and cellular mRNA capping: Past and prospects. *Adv. Virus Res.* 55: 135-184.

Gaire, F., Schmitt, C., Stussi-Garaud, C., Pick, L., and Ritzenthaler, C. 1999. Protein 2A of *Grapevine fanleaf nepovirus* is implicated in RNA2 replication and colocalizes to the replication site. *Virology* 264: 25-36.

Gale, M., Jr., Tan, S.-L., and Katze, M. G. 2000. Translational control of viral gene expression in eukaryotes. *Microbiol. Mol. Biol. Rev.* 64: 239-280.

Gallie, D. R. 1991. The cap and poly(A) tail function synergistically to regulate mRNA translational efficiency. *Genes Dev.* 5: 2108-2116.

Gallie, D. R. 1996. Translational control of cellular and viral mRNAs. *Plant Mol. Biol.* 32: 145-158.

Gallie, D. R. 1998. A tail of two termini: A functional interaction between the termini of a mRNA is a prerequisite for efficient translation initiation. *Gene* 216: 1-11.

Gallie, D. R. 2001. Cap-independent translation conferred by the 5`-leader of *Tobacco etch virus* is eukaryotic initiation factor 4G dependent. *J. Virol.* 75: 12141-12152.

Gallie, D. R., and Kobayashi, M. 1994. The role of the 3`-untranslated region in non-polyadenylated plant viral mRNAs in regulating translational efficiency. *Gene* 142: 159-165.

Gallie, D. R., and Walbot, V. 1990. RNA pseudoknot domain of *Tobacco mosaic virus* can functionally substitute for a poly(A) tail in plant and animal cells. *Genes Dev.* 4: 1149-1157.

Gallie, D. R., Feder, J. N., Schimke, R. T., and Walbot, V. 1991. Post-transcriptional regulation in higher eukaryotes: The role of the reporter gene in controlling expression. *Mol. Gen. Genet.* 228: 258-264.

Gamarnik, A. V., and Andino, R. 1998. Switch from translation to RNA replication in a positive-stranded RNA virus. *Genes Dev.* 12: 2293-2304.

Giegé, R. 1996. Interplay of tRNA-like structures from plant viral RNAs with partners of the translation and replication machineries. *Proc. Natl. Acad. Sci. USA* 93: 12078-12081.

Giege, R., Frugier, M., and Rudinger, J. 1998a. tRNA mimics. *Curr. Opin. Struct. Biol.* 8: 286-293.

Giege, R., Sisler, M., and Florentz, C. 1998b. Universal rules and idiosyncratic features in tRNA identity. *Nucl. Acids Res.* 26: 5017-5035.

Gierer, A., and Schramm, G. 1956. Infectivity of ribonucleic acid from *Tobacco mosaic virus. Nature* (*London*) 177: 702-703.

Gluick, T. C., Wills, N. M., Gesteland, R., and Draper, D. E. 1997. Folding of an mRNA pseudoknot required for stop codon readthrough: Effects of mono- and divalent cations on stability. *Biochemistry* 36: 16173-16186.

Goodwin, J. B., and Dreher, T. W. 1998. Transfer RNA mimicry in a new group of positive-strand RNA plant viruses, the furoviruses: Differential aminoacylation between the RNA components of one genome. *Virology* 246: 170-178.

Goodwin, J. B., Skuzeski, J. M., and Dreher, T. W. 1997. Characterization of chimeric turnip yellow mosaic virus genomes that are infectious in the absence of aminoacylation. *Virology* 230: 113-124.

Guan, H. C., and Simon, A. E. 2000. Polymerization of non-template bases before transcription initiation at the 3`-ends of templates by an RNA-dependent RNA polymerase: An activity involved in 3`-end repair of viral RNAs. *Proc. Natl. Acad. Sci. USA* 97: 12451-12456.

Guo, D., Rajamäki, M., Saarma, M., and Valkonen, J. P. T. 2001.Towards a protein interaction map of potyviruses: Protein interaction mixtures of two potyviruses based on yeast two-hybrid system. *J. Gen. Virol.* 82: 935-939.

Guo, L., Allen, E. M., and Miller, W. A. 2001. Base-pairing between untranslated regions facilitates translation of uncapped, nonpolyadenylated virus RNA. *Mol. Cell* 7: 1103-1109.

Haenni, A.-L., and Chapeville, F. 1997. An enigma: The role of viral RNA aminoacylation. *Acta Biochem. Pol.* 44: 827-838.

Hämaläinen, J. H., Kekarainen, T., Gebhardt, C., Watanabe, K. N., and Valkonen, J. P. T. 2000. Recessive and dominant resistance interfere with the vascular transport of *Potato virus A* in diploid potatoes. *Mol. Plant-Microb. Interact.* 13: 402-412.

Han, S., and Sanfaçon, H. 2003. Tomato ringspot virus proteins containing the nucleoside triphosphate binding domain are transmembrane proteins that associate with the endoplasmic reticulum and cofractionate with replication complexes. *J. Virol.* 77: 523-534.

Hema, M., Gopinath, K., and Kao, C. C. 2005. Repair of the tRNA-like CCA sequence in a multipartite positive-strand RNA virus. *J. Virol.* 79: 1417-1427.

Héricourt, F., Jupin, I., and Haenni, A.-L. 1999. Genome of RNA viruses. In:. Mandahar, C. L. (Ed.). Molecular Biology of Plant Viruses. Kluwer Academic Publishers, Boston/Dordrecht/London. pp. 1-28.

Hershey, J. W. B., and Merrick, W. C. 2000. The pathway and mechanism of initiation of protein synthesis. In: Sonenberg, N., Hershey, J. W. B., and Mathews, M. B. (Eds.). Translational Control of Gene Expression. Cold Spring Harbor Laboratory Press, Cold Spring Harbor , New York, USA. pp. 33-88.

Hong, Y., Levay, K., Murphy, J. F., Klein, P. G., Shaw, J. G., and Hunt, A. G. 1995. A potyvirus polymerase interacts with the viral coat protein and VPg in yeast cells. *Virology* 214: 159-166.

Huang, Y.-L., Han, Y.-T., Chang, Y.-T., Hsu, Y.-H., and Meng, M. 2004. Critical residues for GTP methylation and formation of the covalent m^{7}GMP-enzyme intermediate in the capping enzyme domain of *Bamboo mosaic virus*. *J. Gen. Virol.* 78: 1271-1280.

Kao, C. C., and Sivakumaran, K. 2000. *Brome mosaic virus*, good for an RNA virologist's basic needs. *Mol. Plant Pathol.* 1: 91-97.

Karasev, A. V., Boykov, V. P., Gowda, S., Nikolaeva, O. V., Koonin, E. V., *et al.*, 1995. Complete sequence of the citrus tristeza virus RNA genome. *Virology* 208: 511-520.

Karasev, A. V., Hilf, M. E., Garnsey, S. M., and Dawson, W. O. 1997. Transcriptional strategy of closteroviruses: Mapping the 5'-termini of the citrus tristeza subgenomic RNAs. *J. Virol.* 71: 6233-6236.

Kawakami, S., Padgett, H. S., Hosokawa, D., Yoshimi, O., Beachy, R. N., and Watanabe, Y. 1999. Phosphorylation and/or presence of serine 37 in the protein of *Tomato mosaic tobamovirus* is essential for intracellular localization and stability *in vivo*. *J. Virol.* 73: 6831-6840.

Klug, A., and Caspar, D. L. D. 1960. The structure of small viruses. *Adv. Virus Res.* 7: 225-325.

Kolk, M. H., van der Graaff, M., Wijmenga, S. S., Pleij, C. W. A., Heus, H. A., and Hilbers, W. C. 1998. NMR structure of a classical pseudoknot: Interplay of single- and double-stranded RNA. *Science* 280: 434-438.

Kong, F., Sivakumaran, K., and Kao, C. C. 1999. The N-terminal half of the brome mosaic virus 1a protein has RNA capping-associated activities: Specificity for GTP and S-adenosylmethionine. *Virology* 259: 200-210.

Koonin, E. V., and Dolja, V. V. 1993. Evolution and taxonomy of positive-strand RNA viruses: Implications of comparative analysis of amino acid sequences. *Crit. Rev. Biochem. Mol. Biol.* 28: 375-430.

Kozak, M. 1978. How do eukaryotic ribosomes select initiation regions in messenger RNA? *Cell* 15: 1109-1123.

Kozak, M. 1980. Binding of wheat germ ribosomes to fragmented viral mRNA. *J. Virol.* 35: 748-756.

Kozak, M. 1989. The scanning model for translation: An update. *J. Cell Biol.* 108: 229-241.

Leathers, V., Tanguay, R., Kobayashi, M., and Gallie, D. R. 1993. A phylogenetically conserved sequence within viral 3'-untranslated RNA pseudoknots regulates translation. *Mol. Cell. Biol.* 13: 5331-5337.

Lellis, A. D., Kasschau, K. D., Witham, S. A., and Carrington, J. C. 2002. Loss-of-susceptibility mutants of *Arabidopsis thaliana* reveal an essential role for eIF(iso)4E during potyvirus infection. *Curr. Biol.* 12: 1046-1051.

Léonard, S., Plante, D., Wittmann, S., Daigneault, N., Fortin, M. G., and Laliberté, J.-F. 2000. Complex formation between potyvirus VPg and translation eukaryotic initiation factor 4E correlation with virus infectivity. *J. Virol.* 74: 7730-7737.

Léonard, S., Viel, C., Beauchemin, C., Daigneault, N., Fortin, M. G., and Laliberté, J.-F. 2004. Interaction of VPg-Pro of *Turnip mosaic virus* with the translation initiation factor 4E and the poly(A)-binding protein *in planta*. *J. Gen. Virol.* 85: 1055-1063.

Li, X. H., Valdez, P., Olvera, R. E., and Carrington, J. C. 1997. Functions of the tobacco etch potyvirus polymerase (NIb): Subcellular transport and protein-protein interactisn with VPg/proteinase (NIa). *J. Virol.* 71: 1598-1607.

Li, Y.-L., Chen, Y.-J., Hsu, Y.-H., and Meng, M. 2001a. Characterization of the AdoMet-dependent guanylyltransferase activity that is associated with the N-terminus of bamboo mosaic virus replicase. *J. Virol.* 75: 782-788.

Li, Y.-L., Shih, T. W., Hsu, Y.-H., Han, Y.-T., Huang, Y.-L., and Meng, M. 2001b. The helicase-like domain of potexvirus replicase participates in formation of RNA 5`-cap structure by exhibiting RNA 5`-triphosphatase activity. *J. Virol.* 75: 12114-12120.

Litvak, S., Carre, D. S., and Chapeville, F. 1970. TYMV RNA as a substrate of the tRNA nucleotidyltransferase. *FEBS Lett.* 11: 316-319.

López, C. J., Novas-Castillo, S., Govida, P., Moreno, R., and Flores, R. 2000. The 23-kDa protein coded by the 3`-terminal gene of *Citrus tristeza virus* is an RNA-binding protein. *Virology* 269: 462-470.

Maizels, N., and Weiner, A. M. 1993. The genomic tag hypothesis: Modern viruses as molecular fossils of ancient strategies for genomic replication. In: Gestaland, R. F., and Atkins, J. F. (Eds.). The RNA World. Cold Spring Harbor Laboratory Press, Plainview, NY., USA. pp. 577-602.

Mangus, D. A., Evans, M. C., and Jacobson, A. 2003. Poly(A)-binding proteins: Multiplication scaffolds for the post-transcriptional control of gene expression. *Genome Biol.* 4: 223.

Mans, R. M. W., Pleij, C. W. A., and Bosch, L. 1991. Transfer RNA-like structures: Structure, function and evolutionary significance. *Eur. J. Biochem.* 201: 303-324.

Mans, R. M. W., van Steeg, M. H., Verlaan, P. W. G., Pleij, C. W. A., and Bosch, L. 1992. Mutational analysis of the pseudoknot in the tRNA-like structure of turnip yellow mosaic virus RNA: Aminoacylation efficiency and pseudoknot stability. *J. Mol. Biol.* 223: 221-232.

Matsuda, D., and Dreher, T. W. 2004. The tRNA-like structure of turnip yellow mosaic virus RNA is a 3`-translatioinal enhancer. *Virology* 321: 36-46.

Matsuda, D., Yoshinari, S., and Dreher, T. W. 2004. eEF1A binding to aminoacylated viral RNA represses minus strand synthesis by TYMV RNA-dependent RNA polymerase. *Virology* 321: 47-56.

Merits, A., Guo, D., and Saarma, M. 1998. VPg, coat protein and five non-structural proteins of *Potato A potyvirus* bind RNA in a sequence-unspecific manner. *J. Gen. Virol.* 79: 3123-3127.

Merits. A., Kettunen, R., Mäkinen, K., Lampio, A., Auvinen, P., Kääriäinen, L., and Ahola, T. 1999. Virus-specific capping of tobacco mosaic virus RNA: Methylation of GTP prior to formation of covalent complex p126-m^{7}GMP. *FEBS Lett.* 455: 45-48.

Merits, A., Rajamäki, M. L., Lindholm, P., Runeberg-Roos, P., Kekarainen, T., *et al.*, 2002. Proteolytic processing of potyviral proteins and polyprotein processing intermediates in insect and plant cells. *J. Gen. Virol.* 83: 1211-1221.

Merrick, W. C., and Hershey, J. W. B. 1996. The pathway and mechanism of eukaryotic protein synthesis. In: Hershey, J. W. B., Mathews, M. B., and Sonenberg, N. (Eds.). Translational Control. Cold Spring Harbor Laboratory Press, Cold Spring Harbor, N. Y., USA. pp. 31-69.

Meulewaeter, F., van Montagu, M., and Cornelissen, M. 1998. Features of autonomous function of the translational enhancer domain of *Satellite tobacco necrosis virus*. *RNA* 4: 1347-1356.

Miller, E. D., Plante, C. A., Kim, K. H., Brown, J. W., and Hemenway, C. 1998. Stem-loop structure in the 5`-region of potato virus X genome required for plus-strand RNA accumulation. *J. Mol. Biol.* 284: 591-608.

Mizumoto, H., Hikichi, Y., and Okuno, T. 2002. The 3`-untranslated region of RNA1 as a primary determinant of temperature sensitivity of *Red clover necrotic mosaic virus* Canadian strain. *Virology* 293: 320-327.

Mizumoto, H., Tatsuta, M., Kaido, M., Mise, K., and Okuno, T. 2003. Cap-independent translational enhancement by 3`-untranslated region of red clover necrotic mosaic virus RNA1. *J. Virol.* 77: 12113-12121.

Morozov, S., and Solovyev, A. 1999. Genome organization in RNA viruses. In: Mandahar, C. L. (Ed.). Molecular Biology of Plant Viruses. Kluwer Academic Publ., Boston/Dordrecht/London. pp. 47-98.

Morozov, S. Y., Denisenko, O. N., Zelenina, D. A., Fedorkin, O. N., Solovyev, A. G., Maisse, E., Casper, R., and Atabekov, J. G. 1993. A novel open reading frame in tobacco mosaic virus genome coding for a putative small, positively charged protein. *Biochimie* 75: 659-665.

Murphy, J. F., Klein, P. G., Hunt, A. G., and Shaw, J. G. 1996. Replacement of the tyrosine residue that links a potyviral VPg to the viral RNA is lethal. *Virology* 220: 535-538.

Nagy, P. D., Carpenter, C. D., and Simon, A. E. 1997. A novel 3`-end repair mechanism in an RNA virus. *Proc. Natl. Acad. Sci. USA* 94: 1113-1118.

Nagy, P. D., Pogany, J., and Simon, A. E. 1999. RNA elements required for RNA recombination function as replication enhancers *in vitro* and *in vivo* in a plus-strand RNA virus. *EMBO J.* 18: 5653-5665.

Neeleman, L., Olsthoorn, R. C. L., Linthorst, H. J. M., and Bol, J. F. 2001. Translation of a nonpolyadenylated viral RNA is enhanced by binding of viral coat protein or polyadenylation of the RNA. *Proc. Natl. Acad. Sci., USA* 98: 14286-14291.

Nicolas, O., Dunnington, S. W., Gotow, L. F., Pirone, T. P., and Hellmann, M. 1997. Variations in the VPg protein allow a potyvirus to overcome *va* gene resistance in tobacco. *Virology* 237: 452-459.

Niepel, M., and Gallie, D. R. 1999. Identification and characterization of the functional elements within the tobacco etch virus 5`-leader required for cap-independent translation. *J. Virol.* 73: 9080-9088.

Ohsato, S., Miyanishi, M., and Shirako, Y. 2003. The optimal temperature for RNA replication in cells infected by *Soil-borne wheat mosaic virus* is 17°C. *J. Gen. Virol.* 84: 995-1000.

Okada, Y. 1999. Historical overview of research on the tobacco mosaic virus genome: Genome organization, infectivity and gene manipulation. *Phil. Trans. Roy. Soc. London* 354B: 569-582.

Olsthoorn, R. C. L., Martens, S., Brederode, F. T., and Bol, J. F. 1999. A conformational switch at the 3`-end of a plant virus RNA regulates virus replication. *EMBO J.* 18: 4856-4864.

Olsthoorn, R. C. L., Haasnoot, P. C. J., and Bol, J. F. 2004. Similarities and differences between the subgenomic and minus-strand promoters of an RNA virus. *J. Virol.* 78: 4048-4053.

Oruetxebarria, I., Guo, D., Merits, A., Makinen, K., Saarma, M., and Valkonen, J. P. T. 2001. Identification of the genome-linked protein in virions of *Potato virus A*, with comparison to other members in genus *Potyvirus*. *Virus Res.* 73: 103-112.

Osman, T. A. M., and Buck, K. W. 1997. The tobacco mosaic virus RNA polymerase complex contains a plant protein related to the RNA-binding subunit of yeast eIF3. *J. Virol.* 71: 6075-6082.

Osman, T. A. M., and Buck, K. W. 2003. Identification of a region of the tobacco mosaic virus 126- and 183-kilodalton replication proteins which binds specifically to the viral 3`-terminal tRNA-like structure. *J. Virol.* 77: 8669-8675.

Osman, T. A. M., Hemenway, C. L., and Buck, K. W. 2000. Role of the 3`-tRNA-like structure in tobacco mosaic virus minus-strand RNA synthesis by the viral RNA-dependent RNA polymerase. *J. Virol.* 74: 11671-11680.

Peremyslov, V. V., and Dolja, V. V. 2002. Identification of the subgenomic mRNAs that encode 6-kDa movement protein and Hsp 70 homolog of *Beet yellows virus*. *Virology* 295: 299-306.

Pleij, C. W. A. 1994. Pseudoknots. *Curr. Opin. Struct. Biol.* 4: 337-344.

Plochocka, D., Welnicki, M., Zielenkiewicz, P., and Ostoja-Zagorski, W. 1996. Three dimensional model of the potyviral genome-linked protein. *Proc. Natl. Acad. Sci. USA* 93: 12150-12154.

Puustinen, P., Rajamäki, M.-L., Ivanov, K. I., Valkonen, J. P. T., and Mäkinen, K. 2002. Detection of potyviral genome-linked protein VPg in virions and its phosphorylation by host kinases. *J. Virol.* 76: 12703-12711.

Rajamäki, M., and Volkenen, J. P. T. 1999. The 6K2 protein and the VPg of *Potato virus A* are determinants of systemic infection in *Nicandra physalloides*. *Mol. Plant-Microbe Interact.* 12: 1074-1081.

Rajamäki, M., and Volkenen, J. P. T. 2002. Viral genome-linked protein (VPg) controls accumulation and phloem-loading of a potyvirus in inoculated potato plants. *Mol. Plant-Microbe Interact.* 15: 138-149.

Rao, A. L. N., and Hall, T. C. 1993. Recombination and polymerase error facilitates restoration of infectivity in *Brome mosaic virus*. *J. Virol.* 67: 969-979.

Rao, A. L. N., Dreher, T. W., Marsh, L. E., and Hall, T. C. 1989. Telomeric function of the tRNA-like structure of brome mosaic virus RNA. *Proc. Natl. Acad. Sci. USA* 86: 5335-5339.

Ray, D., Na, H., and White, K. A. 2004. Structural properties of a multifunctional T-shaped RNA domain that mediate efficient tomato bushy stunt virus RNA replication. *J. Virol.* 78: 10490-10500.

Redondo, E., Krause-Sakate, R., Yang, S.-J., Lot, H., Le Gall, O., and Candresse, T. 2001. Lettuce mosaic virus pathogenicity determinants in susceptible and tolerant lettuce cultivars map to different regions of the viral genome. *Mol. Plant-Microbe Interact.* 4: 804-810.

Riechmann, J. L., Lain, S., and Garcia, J. A. 1989. The genome-linked protein and 5`-RNA sequence of *Plum pox potyvirus*. *J. Gen. Virol.* 70: 2785-2789.

Riechmann, J. L., Lain, S., and Garcia, J. A. 1992. Highlights and prospects of potyvirus molecular biology. *J. Gen. Virol.* 73: 1-16.

Rietveld, K., van Poelgeest, K., Pleij, C. W. A., van Boon, J. H., and Bosch, L. 1982. The tRNA-like structure at the 3`-terminus of turnip yellow mosaic virus RNA: Differences and similarities with canonical tRNA. *Nucl. Acids Res.* 10: 1929-1946.

Rietveld, K., Linschooten, K., Pleij, C. W. A., and Bosch, L. 1984. The three-dimensional folding of the tRNA-like structure of tobacco mosaic virus RNA. A new building principle applied twice. *EMBO J.* 3: 2613-2619.

Ritzenthler, C., Laporte, C., Gaire, F., Dunoyer, P., Schmitt. C., Duval, S., *et al.,* 2002. *Grapevine fanleaf virus* replication occurs on endoplasmic-derived membranes. *J. Virol.* 76: 8808-8819.

Rozanov, M. N., Koonin, E. V., and Gorbalenya, A. E. 1992. Conservation of the putative methyltransferase domain: A hallmark of the 'Sindbis-like' supergroup of positive-strand RNA viruses. *J. Gen. Virol.* 73: 2129-2134.

Ruffel, S., Dussault, M.-H., Palloix, A., Moury, B., Bendahmane, A., Robaglia, C., and Caranta, C. 2002. A natural recessive resistance gene against *Potato virus Y* in pepper corresponds to the eukaryotic initiation factor 4E (eIF4E). *Plant J.* 32: 1067-1075.

Sachs, A. B. 2000. Physical and functional interactions between the mRNA cap structure and poly(A) tail. In: Sonenberg, N., Hershey, J. W. B., and Mathews, M. B. (Eds.). Translational Control of Gene Expression. Cold Spring Harbor Laboratory Press, Cold Spring Harbor, N. Y., USA. p. 447-465.

Schaad, M. C., Lellis, A. D., and Carrington, J. C. 1997. VPg of *Tobacco etch potyvirus* is a host genotype-specific determinant for long-distance movement. *J. Virol.* 71: 8624-8631.

Schaad, M. C., Anderberg, R. J., and Carrington, J. C. 2000. Strain-specific interaction of the tobacco etch virus NIa protein with the translation initiation factor eIF4E in the yeast two-hybrid system. *Virology* 273: 300-306.

Schwartz, M., Chen, J., Janda, M., Sullivan, M., de Boon, J., and Ahlquist, P. 2002. A positive-strand RNA virus replication complex parallels form and function of retrovirus capsids. *Mol. Cell* 9: 505-514.

Shuman, S. 1995. Capping enzyme in eukaryotic mRNA synthesis. *Proc. Nucl. Acids Res. Mol. Biol.* 50: 101-129.

Shuman, S. 2002. What messenger RNA capping tells us about eukaryotic evolution. *Nat. Rev. Mol. Cell Biol.* 3: 619-625.

Singh, R. N., and Dreher, T. W. 1997. Turnip yellow mosaic virus RNA-dependent RNA polymerase: Initiation of minus-strand synthesis *in vitro. Virology* 233: 430-439.

Singh, R. N., and Dreher, T. W. 1998. Specific site selection in RNA resulting from the combination of non-specific secondary structure and -CCR- boxes: Initiation of minus-strand synthesis by turnip yellow mosaic virus RNA-dependent RNA polymerase. *RNA* 4: 1083-1095.

Sivakumaran, K., Hema, M., and Kao, C. C. 2003. Brome mosaic virus RNA synthesis *in vitro* and in barley protoplasts. *J. Virol.* 77: 5703-5711.

Skaf, J. S., Schutz, M. H., Hirata, H., and de Zoeten, G. A. 2000. Mutational evidence that the VPg is involved in the replication and not the movement of *Pea enation mosaic virus 1. J. Gen. Virol.* 81: 1103-1109.

Söll, D., and RajBhandary, U. L. (Eds.). 1995. tRNA: Structure, Biosynthesis and Function. ASM Press, Washington, D. C., USA.

ten Dam, E. B., Pleij, C. W. A., and Bosch, L. 1990. RNA pseudoknots: Translational frameshifting and readthrough on viral RNAs. *Virus Genes* 4: 121-136.

van der Heijden, M. W., Carette, J. E, Reinhoud, P. J., Haegi, A., and Bol, J. F. 2001. Alfalfa mosaic virus replicase proteins P1 and P2 interact and colocalize at the vacuolar membrane. *J. Virol.* 75: 1879-1887.

Vlot, A. C., and Bol, J. F. 2003. The 5`-untranslated region of alfalfa mosaic virus RNA1 is involved in negative-strand RNA synthesis. *J. Virol.* 77: 11284-11289.

Vlot, A. C., Menard, A., and Bol, J. F. 2002. Role of the alfalfa mosaic virus methyltransferase-like domain in negative-strand RNA synthesis. *J. Virol.* 76: 11321-11328.

Waigmann, E., Chen, M.-H., Bachmaier, R., Ghosroy, S., and Citovsky, V. 2000. Regulation of plasmodesmatal transport by phosphorylation of tobacco mosaic virus cell-to-cell movement protein. *EMBO J.* 19: 4875-4884.

Wang, A., Carrier, K., Chisholm, J., Wieczorek, A., Huguenot, C., and Sanfaçon, H. 1999. Proteolytic processing of tomato ringspot nepovirus 3C-like protease precursors: Definition of the domains for the VPg, protease and putative RNA-dependent RNA polymerase. *J. Gen. Virol.* 80: 799-809.

Wang, A., Han, S., and Sanfaçon, H. 2004. Topogenesis in membranes of the NTB-VPg protein of *Tomato ringspot nepovirus*: Definition of the C-terminal transmembrane domain. *J. Gen. Virol.* 85: 535-545.

Wang, S. P., Browning, K. S., and Miller, W. A. 1997. A viral sequence in the 3`-untranslated region mimics a 5`-cap in facilitating translation of uncapped mRNA. *EMBO J.* 16: 4107-4116.

Wells, D. R., Tanguay, R. L., Le, H., and Gallie, D. R. 1998. HSP101 functions as a specific translational regulatory protein whose activity is regulated by nutrient status. *Genes &Dev.* 12: 3236-3251.

Wilkie, G. S., Dickson, K. S., and Gray, N. K. 2003. Regulation of mRNA translation by 5`- and 3`-UTR-binding factors. *Trends Biochem. Sci.* 28: 182-188.

Wimmer, E. 1982. Genome-linked proteins of viruses. *Cell* 28: 199-201.

Wittmann, S., Chatel, H., Fortin, M. G., and Laliberti, J. F. 1997. Interaction of the viral protein genome-linked of *Turnip mosaic potyvirus* with the translational eukaryotic initiation factor (iso)4E of *Arabidopsis thaliana* using the yeast two-hybrid system. *Virology* 234: 84-92.

Wu, B., and White, K. A. 1999. A primary determinant of cap-independent translation is located in the 3`-proximal region of the tomato bushy stunt virus genome. *J. Virol.* 73: 8982-8988.

Wyatt, J. R., Puglist, J. D., and Tinoco, I., Jr. 1990. RNA pseudoknots: Stability and loop size requirements. *J. Mol. Biol.* 214: 455-470.

Yambao, Ma L. M., Masuta, C., Nakahara, K., and Uyeda, I. 2003. The central and C-terminal domains of VPg of *Clover yellow vein virus* are important for VPg-HCPro and VPg-VPg interactions. *J. Gen. Virol.* 84: 2861-2869.

Zaccomer, B., Haenni, A.-L., and Macaya, G. 1995. The remarkable variety of plant RNA virus genomes. *J. Gen. Virol.* 76: 231-247.

Zaitlin, M. 1999. Elucidation of genome organization of *Tobacco mosaic virus. Phil. Trans. Roy. Soc. London* 354B: 587-591.

Zeenko, V. V., Ryabova, V., Spirin, A. S., Rothnie, H. M., Hess, D., Browning, K. S., and Hohn, T. 2002. Eukaryotic elongation factor 1A interacts with the upstream pseudoknot domain in the 3`-untranslated region of tobacco mosaic virus RNA. *J. Virol.* 76: 5678-5691.

3 INFECTION BY AND UNCOATING OF VIRUS PARTICLES

I. INFECTION

Entry of invading virus particles into cytoplasm of a cell mostly takes place through the membrane system. But, plant viruses do not have any system by which they can enter the cell through barriers like cuticular layer, cell wall, and finally plasma membrane. Additionally, there are no specific virus receptors on the surface of plant cells. It is, thus, generally accepted that wounding of the plant cell is required for the virus particles to penetrate the host cell (Verduin, 1992). Virus entry into a cell is a complex process, which is still not fully understood.

Attachment of rod-shaped viruses PVX, TMV, and TRV is end-on on outer cell wall and also on cell walls that delimit intercellular spaces. Virus attachment is a random and nonspecific process, which indicates that attachment (infection) sites are not virus specific. Both the TMV and TRV particles injected into intercellular spaces of host plant (*Nicotiana tabacum* var. Xanthi-nc) and a non-host (*Zea mays*) plant, after 10 minutes of infiltration, had end-on association on cell walls bordering the intercellular spaces. Purified CPMV strain Sb or radioidonated purified virus (^{125}I-CPMV) were inoculated into protoplasts of immune, hypersensitive, and susceptible cowpea lines. There were no differences in the attachment of ^{125}I-CPMV suggesting that attachment of CPMV particles to protoplasts does not ensure virus infection and so is not a determining factor in susceptibility of cowpea protoplasts to CPMV infection. Binding of CPMV particles to resistant and fully susceptible cowpea protoplasts was also similar.

A. Entry of Virus Particle/Genome into a Cell

It is only through wounds, caused mechanically or by vectors, that viruses come in contact with cytoplasm or reach the interior of a cell. Wounding, which damages the cuticle and cell wall and/or breaks trichomes to generate infectible sites, is normally regarded an essential feature for viral infection. The former exposes cell membrane while the latter exposes plasmodesmata, which are abundant in cell walls of hairs and also between hair cells and underlying epidermal cells. In the ultimate analysis, therefore, plasmodesmata or protoplasmic membrane exposed in cell walls of broken hair or wounded epidermal cells may be the sites to which the invading virus particles get attached and through which virus gains entry into a cell.

Wounds can be caused by an abrasive, or by natural virus vectors like fungi, nematodes or insects and the virus particles are believed to be deposited directly into the cytoplasm. Immunological studies show that TMV particles are located predominantly within the cytoplasm of tobacco epidermal cells. The efficiency of this inoculation process is very slow, and several hundreds to thousands of particles must be inoculated to produce a single local lesion. Even the use of protoplasts has not improved this efficiency (Verduin, 1992). Three hypotheses exist about ingress of virus particles into host cells and protoplasts through a membrane system: entry through pinocytic uptake, through microlesions, and through electro-osmotic mechanism.

TMV particles, when in contact with plasmalemma of tobacco mesophyll protoplasts, are regarded to enter the protoplasts by pinocytosis or endocytosis – a process by which substances enter a cell from exterior and pass into cytoplasm within membrane-bound vesicles. This process reportedly occurs in all the mesophyll protoplast-virus combinations studied. Attachment of virus particles to plasmalemma induces its invagination at the point of attachment, followed by closure of neck of the invagination leading to imprisonment of the virus particle within an intracytoplasmic vesicle. The TMV particle is taken in by tobacco mesophyll protoplasts within 10 minutes or so. However, because of thermodynamic constraints, walled cells are thought unlikely to exhibit pinocytosis. Moreover, virus uptake by protoplasts is a passive process because of the electrostatic or electro-osmotic interactions between the two (Zhuravlev *et al.*, 1980; Watts and King, 1984).

Virus particles are thought to penetrate into the protoplasts through microlesions (transitory wounding) in plasmalemma caused by poly-L-ornithine (PLO) present in the suspending medium. However, this process does not explain viral infection of protoplasts in the absence of PLO, even if such reports are rare, and any demonstrable damage to plasmalemma of PVX-inoculated protoplasts was not revealed by electron microscope.

Electro-osmotic mechanism is based on the assumption that any one of the possible pre-existing cell mechanisms of nutrient uptake could account for virus uptake. This uptake is suggested to be selective, is thought to be the most suitable method for virus intake, and operates because of the interaction of surface charges of plasmalemma and the adsorbed substances. The forces responsible for transmembrane transport of assimilates/ions may also be responsible for simultaneous or parallel transmembrane transport of the virus particles. Electro-osmotic transport of ions may be responsible for penetration of adsorbed virus particles into protoplasts through plasmalemma (Zhuravlev *et al.*, 1980).

B. Cell-Infecting Unit

Each virus infection is regarded to be caused by a single virus particle because of the great dilution of virus preparation and also because of the existence of a linear relationship between inoculum concentration of WTV (that contains double-stranded RNA segments as genome and belongs to family *Reoviridae*) and *Potato yellow dwarf virus* (that contains negative-sense RNA as genome and belongs to family *Rhabdoviridae*) and infections produced on their respective vector cell monolayers. However, it is difficult to prove this experimentally because of the presence of very large number of virus particles as well as of a large number of inactive virions in an inoculum, difficulty in determining which of the virus particles actually take part in infection, the lack of knowledge of proper cultural conditions favouring optimal virus infection, due to the built-in limitations of the techniques and materials used for studying infection, and the infectivity assay. These limitations will stay until the minimal number of virus particles and more sensitive methods of virus detection are employed (Shaw, 1999). Where these limitations have been largely eliminated, the cell-infecting unit (CIU) has been determined to be as low as 1.7 in TMV, 10 to 30 in *Turnip yellow mosaic virus* (TYMV), and 250 to 410 in WTV.

The CIU is a certain minimum number of virus particles that must be present in inoculum for producing one infection. Widely differing estimates have been made about CIU even for the same virus. In TMV, the CIU has been calculated to be from as little as 20 to as high as 1,000,000 with 20,000 to 200,000, 100,000, 50,000 and 60,000 as some of the intermediate values. An average value of 1.7 TMV virions are required for infection of one tobacco protoplast at the nonsaturating concentration of inoculum but possibly a fairly large number of TMV particles are needed at the greatly saturated inoculum concentrations.

The estimated number of virus particles of some plant viruses, required for formation of one local lesion by mechanical inoculation on leaves of *Chenopodium amaranticolor*, is 10^7 for CMV and TEV, and 3×10^6 for CCMV. In contrast, only 10 to 30 TYMV particles are required to produce a single local lesion in mechanically inoculated Chinese cabbage leaves. The high CIU values are due to the difficulty in getting the virions into cells and not due to large proportions of inactive virions in inoculum, at least in WTV.

The number of virions adsorbed per infected (inoculated) protoplast has also been reported. About 600 to 8,000 virions of TMV, CCMV, and CPMV were adsorbed to tobacco/cowpea protoplasts; 30 long and 85 short TRV particles in phosphate buffer but 250 long and 700 short TRV virions in citrate buffer become attached to tobacco protoplasts at concentrations causing 50% infection. About 1400 to 1600 particles of BSMV were adsorbed per barley protoplast (Ben-Sin and Tien Po, 1982). However, it is doubtful if the number of virus particles adsorbed on a protoplast should be taken as the CIU.

C. Nucleo-Cytoplasmic Shuttling

Nearly all plant viruses multiply in cytoplasm where they uncoat and RNA genome moves into the nucleus for carrying out various functions in connection with its replication. The progeny RNA molecules later move out of the nucleus and into the cytoplasm where, in general, they undergo assembly into complete virus particles. Such regulation of bidirectional trafficking of macromolecules between nucleus and cytoplasm through nuclear pore complex is an essential feature of many fundamental cellular processes. This nucleo-cytoplasmic shuttling of proteins in plant cells is exhibited by transcription factors in nucleus and proteins of DNA viruses that replicate in nucleus. Some recent reports have established the existence of nucleo-cytoplasmic shuttling of plant virus proteins by which viral proteins also move from cytoplasm into nucleus and back.

Protein p25 encoded by RNA3 of BNYVV is present both in nuclei and cytoplasm of infected cells (Haeberle and Stussi-Garaud, 1995). Ryabov *et al.* (2004) found that the 27-kDa protein encoded by ORF3 of GRV is located in nucleus (particularly in nucleolus). Plant virus proteins like CMV 2b (Lucy *et al.*, 2000; Wang *et al.*, 2004) and TEV NIb (Li *et al.*, 1997) also move actively to nucleus of infected cells. This is made possible due to presence of a nuclear localization signal (NLS) in these viral proteins. Vetter *et al.* (2004) found that a NLS is present in the N-terminal part (amino acid residues 1-103) of a ~25-kDa protein (p25) and this signal is required for nuclear translocation of p25 from cytoplasm; that p25 also contains a nuclear-export sequence (NES) that involves a domain containing hydrophobic residues 164-196 and this sequence is required for export of p25 out of the nucleus to cytoplasm; that CRM1 pathway (Fornerod *et al.*, 1997) actively exports p25; and that p25 contains a highly basic N-terminal domain and an acidic C-terminal domain. Similarly, Ryabov *et al.* (2004) found amino acids 108 to 122 of 27-kDa protein of GRV contain an arginine-rich nuclear-localization signal, and a leucine-rich nuclear export signal is located at amino acids 148 to 156 and is essential for nuclear export of 27-kDa viral protein. The 27-kDa protein is required for movement of viral RNA through phloem and for viral RNA protection in the host plant. The NLS in plant potyviral NIa protein is a bipartite signal sequence, between amino acid residues 1 and 72 near the amino-terminal cleavage site, that mediates its nuclear translocation (Restrepo *et al.*, 1990; Carrington, *et al.*, 1991).

Wang *et al.* (2004) demonstrate that 2b protein of Fny-CMV localizes to host nucleus by using the karyopherin α-mediated system; show that 2b protein has two NLSs and either of the NLSs is sufficient to localize the 2b protein to nucleus; and provide evidence that both NLSs are capable of binding to AtKAPα, an *A. thaliana* karyopherin that is an essential nuclear membrane protein. Localization of 2b protein of Fny-CMV to nucleus by either of the two NLSs do not resemble a classical monopartite NLS as demonstrated in 2b protein of Q strain of CMV (Lucy *et al.*, 2000).

Plant viral proteins, as with other nuclear proteins, need to dock on the outside of nuclear pore complex for entering into nucleus as happens in case of structural proteins of DNA viruses like CaMV and *Tomato yellow leaf curl virus* and a negative-strand RNA virus like *Sonchus yellow net virus* (Wang *et al.*, 2004). Docking may occur via karyopherin α proteins, which are analogous to the importin α proteins of animals. Wang *et al.* (2004) show that either of the two NLSs of Fny-CMV 2b protein is capable of interacting with the importin-like docking and nuclear transport system, and that each NLS of the viral protein bound separately to karyopherin, although deletion of both prevented binding. Therefore, modification of either NLS of 2b protein did not prevent 2b nuclear transport through the nuclear pore complex.

The GRV ORF3-encoded 27-kDa protein (Ryabov *et al.*, 2004), BNYVV-encoded p25 (Vetter *et al.*, 2004), and Fny-CMV 2b protein (Wang *et al.*, 2004) are the first reports of virus-encoded proteins, which shuttle between the nucleus and cytoplasm (nucleo-cytoplasmic shuttling). It is significant that the ORF3 encoded 27-kDa GRV protein contains amino acid stretches, which are similar to several virus-encoded movement proteins, and that this umbravirus long-distance MP contains signals for both nuclear localization and nuclear export (Ryabov *et al.*, 2004). Presence of such signals in movement proteins can be very important for movement of viruses (from and to nucleus) in plants.

Kasamatsu and Nakanishi (1998) review the entry mechanisms of animal DNA viruses into the nucleus. Viral DNA enters the nucleus in at least three ways: DNA may be associated with nuclear localization signal-bearing structural proteins, which facilitate entry; NLS may be packaged into virions along with a virally encoded protease that disassembles the virion, exposing NLS that has been tightly or covalently bound to DNA and so facilitates entry; or packaged with cellular proteins that modify viral proteins to create functional NLS for DNA's nuclear entry. Nuclear pore complex (NPC) is the key structure in nucleo-cytoplasmic transport. The number of NPC per nucleus varies with the physiological condition of a cell; in general, actively dividing cells have more NPCs than non-dividing cells. NPC is generally 90 to 100 nm in external diameter, has cylindrical structure with an eight-fold rotational symmetry in the plane of nuclear envelope.

The two replicase proteins, p36 and its readthrough p95, of *Carnation Italian ringspot tombusvirus* are localized to outer membrane of mitochondria in yeast cells while p36 contains the signal for targeting the two replicases to mitochondria (Pantaleo *et al.*, 2003, 2004).

II. UNCOATING OF VIRUS PARTICLES

Virus disassembly/uncoating/destabilization/deproteinisation has been dealt with in nearly all the reviews concerned with assembly of plant viruses (Butler, 1999; Shaw, 1999; Stubbs, 1999; and Culver, 2002). Viral uncoating studies have been carried out both *in vitro* and *in vivo*; a wide range of denaturing chemicals or solution conditions induce viral disassembly under *in vitro* conditions. However, these chemical treatments do not accurately mimic the *in vivo* mechanism of uncoating. The uncoating process of TMV virions is powered mainly by two factors: by the effects of entry into the cell environment and by interaction between components of the virus and host factors.

The highly stable TMV virion, upon entry into a cell, has to be disassembled for initiation of virus translation and replication. These alterations in virion stability suggest the existence of a stability-switching mechanism that is responsive to changes in the surrounding environment. Such a switching mechanism is proposed to be due to the repulsive interaction of carboxyl-carboxylate groups from glutamate or aspartate residues located at the interface between adjacent capsid protein subunits. In the extracellular milieu, the repulsive interactions made by the negatively charged carboxylate groups would be stabilized by the presence of positively charged cations, such as Ca^{2+} ions or protons (Caspar and Namba, 1990). However, the lower pH and Ca^{2+} ion concentration found within plant cells compared to extracellular environment are thought to cause a loss of the hydrogen and calcium ions that stabilize CP-RNA interactions (Culver, 2002). The loss of stabilizing protons or cations permits the juxtaposed negatively charged carboxylate groups to interact. The repulsive forces derived from interacting carboxylate groups may then destabilize the virus particle to initiate uncoating. The Ω leader sequence contains no G residues and is thought to interact relatively loosely with CP subunits. Thus, the weakening of the CP-RNA interaction induced by the loss of hydrogen and calcium ions will be more pronounced in this region of TMV RNA (Shaw, 1999; Culver, 2002). This possibly could be the mechanism by which the 5`-terminal region of TMV RNA becomes uncovered and exposed to the translational machinery of the inoculated cell.

The intersubunit carboxylate interactions exist at three distinct locations within the virus (Namba and Stubbs, 1986). First interaction exists in the axial interface, that involves interaction of Glu50 from one subunit with Asp77 of the subunit below. The second interaction occurs laterally between Glu95 from one subunit and Glu106 of the adjacent subunit. This interaction is far more complex, with Glu106 capable of interacting with either Glu95, Glu97, or Asp109 of the adjacent subunit (Lu *et al.*, 1996). The third interaction is between Asp116 and phosphate from the RNA. The destabilizing forces in both the axial and lateral directions within the virion as well as between the RNA and protein are provided by the repulsive forces derived from all three of these carboxylate interactions. Mutational studies demonstrated that all three interactions affect TMV disassembly, the axial and lateral interactions contributing the maximum to disassembly (Culver *et al.*, 1995; Lu *et al.*, 1996). These workers also found, during *in vivo* studies, that carboxylate mutants exhibited increased abilities for self-assembling into virion-like rods that did not contain RNA. Thus, carboxylate interactions are essential for both assembly and disassembly (Culver, 2002).

No experimental evidence produced so far shows the in-coming virus particle to be attached to some receptor-type subcellular component, which causes some degree of particle instability that is essential for uncoating of the leader sequence.

A. Uncoating

It takes only minutes, after entry into a cell, for a virus particle to uncoat. TMV particles uncoat and disappear from view from pinocytic vesicles within 15 to 30 minutes after inoculation; about 15 to 20 per cent of RNA genome gets released from TMV particles within 7 minutes of inoculation. Similarly, about 60% of the TYMV and BMV virions are uncoated within 10 minutes of inoculation. Swelling of TYMV particles at 11.5 to 11.6 pH in M KCl takes place within 30 seconds and the release of RNA is completed within 3 to 10 minutes of inoculation. However, it could be a much more rapid process; many TYMV particles uncoat within 45 seconds, the whole process being nearly completed within 2 minutes. Degradation of TRV particles commences immediately after end-on attachment to cell walls of both host and nonhost plants.

Simple early experiments clearly suggested that uncoating does take place. Inactivation of TMV particles during irradiation experiments was one such approach. The general principle of these experiments is that TMV is more sensitive to UV irradiation and is inactivated soon if it exists as naked RNA than when its genome is sheltered in the protein coat. Thus, time taken for inactivation of viruses is taken as evidence indicating uncoating or not of viral RNA. TMV multiplication is detected some hours earlier when inoculation is done by RNA than if done by virus particles. Infection sites produced by TMV as inoculum become resistant to UV irradiation 2 to 5 hours after inoculation while those produced by TMV RNA inoculum become resistant immediately after inoculation. Disease symptoms appear several hours earlier if naked TMV RNA is used as inoculum rather than complete TMV particles.

B. Site(s) of Uncoating

1. Extracellular Sites

Solvents and detergents destabilize virus particles leading to the release of RNA. On this basis, Caspar (1963) suggested that lipid-containing plant structures and hydrophobic environment associated with cell wall or cell membrane may adversely effect the stabilizing interactions within virions resulting in uncoating of virus RNA. The cuticle and plasmalemma are rich in lipid contents so that both or either may be the site(s) of viral uncoating. This is supported by the suggestion that attachment of virus particles to membranes reduces their stability. No uncoating enzyme has been detected or is involved in the uncoating process.

Following mechanical inoculation, contact of plant viruses with cuticular plant surfaces destabilizes the virus particles so that uncoating occurs outside the cell, possibly on cuticle or cell wall of epidermis, and uncoated viral RNA then passes through cell wall and penetrates plasmalemma of an epidermal cell (de Zoeten, 1995). The released capsid protein molecules are thought to alter the cell membrane to permit entry of viral RNA genome. No lesions were produced on peeled cowpea half leaves when inoculated with TNV, TNV RNA, TMV, or TMV RNA. Virus activity could also not be detected. Local lesions were, however, produced in unpeeled cowpea leaves as also in the mesophyll if epidermis remained attached to it for a certain limited minimum period after inoculation. This period (which was at least 8 hour in cowpea-TMV, 6 hour in cowpea-TNV, 8.5 hour in Xanthi tobacco-TMV, and 6 hour in Xanthi tobacco-TNV systems) period could be decreased to at least 1 to 1.5 hour in cowpea and Xanthi tobacco leaves that were kept in complete darkness for 24 hour prior to inoculation. These experiments suggested epidermis to be necessary for infection by virions or RNA (Coutts, 1980; Wieringa-Brants, 1981). Intact leaf surface was also necessary for TYMV uncoating (Matthews and Witz, 1985).

Some other experimental evidence confirming that dissociation of virus particles occurs on the cell wall is: ultrastructural studies, altered morphology of the attached ends of virions, and by autoradiographic studies. During infection of host (tobacco) and nonhost (corn) leaves by TRV, about 80% of the virus particles showed some morphological change in their ends attached to the cell wall. In about 5% of these virions, the capsid protein at the attached end was very loosely associated with the particle-cuticle interface so that these attached virions simulated an inverted funnel.

This also accounted for shortening in length of TRV particles soon after attachment to both hosts. Moreover, uncoating of TRV is a nonspecific event since it can occur on nonhost cells as well. Similarly, uncoating of TYMV is not confined only to host plants, it uncoats with equal efficiency on its nonhosts (Matthews and Witz, 1985).

The suggestion has also been mooted that virus dissociation occurs extracellularly on plasma membrane. This may be valid for all viruses that come in direct contact with plasma membrane after mechanical inoculation and may have far-reaching consequences on the stability of virus particles *in situ* within the host tissues. This means that progeny virions formed intracellularly cannot be uncoated inside the cell but exist in a 'resting state' until they are placed again in a favourable environment for uncoating by mechanical inoculation or by a vector. The plasma membrane of pinocytic vesicles, by implication, may be the site of the uncoating of virus particles located in and attached to it. This, however, does not seem to be true since virus particles disappear from pinocytic vesicle a few minutes after their entry so that viral genome, rather than the complete virus particle, enters the cytosol.

2. Intracellular Sites

Virus particles are thought to uncoat during penetration of the cell membrane (see below).

C. Mechanism(s) of Uncoating

Rapid destabilization and uncoating of the invading virus particle, which takes place within minutes of mechanical inoculation, indicates that the uncoating process is physical in which no enzymes are involved; hence it is not a biochemical process. Studies with isolated membrane phospholipids also support this theory. Studies with TRV and TYMV, already mentioned above, indicate that virus degradation, like virus attachment, is also possibly a generalized non-specific event (Matthews and Witz, 1985) and ends in successful infection only on a wounded susceptible host plant. There is yet no complete agreement about the mechanisms and conditions necessary for deproteinisation.

1. Role of Ca^{2+} Divalent Cations

Possibly most, if not all, plant viruses harbour special cation-binding sites that behave like a switch controlling assembly/disassembly of a virus particle. TMV particle in its extracellular state contains bound Ca^{2+} ions which ensure its stability and inhibit TMV disassembly. The interior of cells contains a low concentration of intracellular Ca^{2+} so that virus particles, upon entry into a cell, release their Ca^{2+} ions. This Ca^{2+} ion dissociation results in a significant alteration in charge of a virus particle and triggers some structural and conformational changes in the particular virion causing release of RNA. The difference in the relative Ca^{2+} concentration in extracellular and intracellular environment sets in several actions: it helps the virus to distinguish between the two environments, ensures virion stability outside the cell, causes virions to uncoat only in low Ca^{2+} concentration in cell cytoplasm, and does indicate that uncoating is a nonspecific process.

The close proximity of carboxylate groups (with abnormal pK values) in virus particle is essential for this mechanism to be functional. Subsequent extensive research, that included X-ray diffraction analysis of capsid protein and virus particles and analysis of virus particles at 2.9 Å resolution, did identify three sites at which electrostatic repulsion could be operative between the adjacent carboxylate groups (Namba *et al.*, 1989). Two of these carboxyl-carboxylate interactions were functional between axillary or adjacent protein subunits while the third involved a carboxylate-phosphate interaction between capsid protein and viral RNA. Two of these three sites are considered as Ca^{2+}-binding sites.

In short, disassembly is suggested to occur because, virus particles during their passage/entry through cell membrane and cytoplasm, face a lowering of Ca^{2+} ion concentration and a raised pH so that Ca^{2+} and protons are removed from carboxyl-carboxylate and carboxyl-phosphate sites. This process causes electrostatic repulsion of proximal negative charges leading to destabilization of virus particles or that intersubunit repulsion causes particle disassembly (Namba *et al.*, 1989).

Two objections to the above theory have come up. The postulated role of Ca^{2+} in stability/disassembly of the virus particle does not appear to operate in AMV (Oostergetel *et al.*, 1981). Moreover, the above speculation rests on the belief that uncoating of a virus particle takes place within a cell – which, as already mentioned, is not an experimentally supported presumption. The above speculation has been modified now. Uncoating of a virus particle occurs outside a cell but the passage of RNA through plasmalemma for entering into cytosol is possibly mediated by the energy that is generated by differential Ca^{2+} concentration on two sides of a plasma membrane.

2. Role of pH

The pH of the *in vitro* solution can cause disassembly of plant viruses. As back as 1963, Caspar (1963) had hypothesized that disassembly of the TMV virus particles could occur/start because the virus particle (upon entry into a plant cell) encounters certain pH and ionic strength conditions in the cytoplasm of the invaded cell. This encounter, within cytoplasm in the interior of the cell, is suggested to exert the same effect as the particle pre-treatment (see below) of Mundry *et al.* (1991). The conditions required for *in vitro* RNA release from two tymoviruses, namely TYMV and *Belladonna mottle virus* (BDMV), are entirely different in the two viruses. The RNA of BDMV is released by merely increasing the pH to 7 or above while much harsher conditions (like high ionic strength and elevated temperature, denaturing agents, or very high alkaline pH of about 11.5 to 11.6) are needed for release of TYMV RNA (Virudachalam *et al.*, 1983). This is correlated with the conditions required for canceling the repulsive forces of the negatively charged phosphate groups of RNA so as to prevent swelling and change in capsid structure required for RNA release. Protein-RNA linkages in BDMV are sufficient for preventing RNA release below pH 6.8 as the capsid protein molecules cancel all the PO_4^- negative charges. At pH near 7.0 or above, the protein-RNA interactions are weakened resulting in repulsion of the negatively charged phosphate groups, which in turn swells the particle and causes RNA release. Thus, mere change in pH from 6.8 to 7.0 is enough to generate the driving force for RNA release (Virudachalam *et al.*, 1983). On the other hand, presence of polyamines, which bind to TYMV RNA, is necessary to effectively cancel repulsive forces of the phosphate groups. Much harsher conditions are needed for elimination of polyamine molecules and hence for TYMV RNA release.

In *in vitro* conditions, the AMV virus particles start to uncoat when pH is raised from 7 to 8 and a similar pH change could mediate its dissociation *in vivo*.

3. Co-translational and Co-replicational Mechanisms

It was commonly believed, prior to 1984, that viral RNA was first completely stripped off its protein coat followed by translation of the released RNA. Wilson and co-workers showed that it is not so. Wilson (1984, 1985a, 1985b) and Wilson and Watkins (1985) found that the encapsidated RNA in partially stripped but seemingly intact TMV rods, after brief pre-incubation in pH 8.0 buffer that released a small number of CP subunits from the 5`-end of viral RNA, directed synthesis of virus-specific proteins in cell-free systems. Mundry *et al.* (1991) also found that a subpopulation of virus particles, when the purified TMV particles are subjected to mild alkali or detergent treatment, rapidly uncoat partially by losing about 200 capsid protein subunits at 5` - end of viral RNA. This could occur because presumably the 68-nucleotide $\Omega$ 5`-leader sequence (which lack G residues) possesses weak interaction with capsid protein subunits in comparison to other parts of viral genome. Pre-treatment thus results in uncoating of leader sequence and also of the first AUG codon.

The exposed 5`-end of the RNA gets bound by ribosomes to begin translation initiation of 126/183-kDa replicase ORF. The ribosomes travel towards the 3`-end during translation and remove further coat protein subunits progressively to cause sequential disassembly of rod-shaped particles. Wilson (1984), thus, found that 40S ribosome subunits could start scanning and, simultaneously, could dislodge capsid protein subunits while the first ORF was being translated. He called this as the cotranslational disassembly of TMV RNA – that is, virus particles are uncoated *in vivo* during the translation process of the viral RNA and not earlier. Conceivably, active translation process generates the force essential for removing capsid protein subunits from the replicase ORF. Shaw *et al.* (1986) demonstrated this process *in vivo*. Co-translational disassembly can explain only the uncoating of TMV RNA from 5`-cap to the stop codon of 183-kDa protein gene. Since translation cannot be re-initiated internally on genomic RNA, co-translational disassembly could only permit the synthesis of 126- and 183-kDa proteins so that this process cannot expose the remaining viral sequences, including the ORFs for movement protein and capsid protein, in virus particle. Despite this, the viral RNA is entirely stripped of capsid within 20 minutes after inoculation (Wu *et al.*, 1994; Shaw, 1999). Wilson (1984), therefore, speculated that viral replicase could somehow participate in completing of striping process. This is the co-replicational hypothesis of Wilson (1984).

Studies of Wu and Shaw (1996, 1997) appear to support this; they found that in synchronously infected tobacco protoplasts the initiation of minus-strand synthesis, which is the first product of viral replicase activity, occurs simultaneously with the removal of capsid protein from the 3`-end of viral RNA. They therefore, proposed a mechanism for removal of the remaining capsid protein subunits from the 3`-end of the viral RNA. This mechanism involves interaction of the newly synthesized 126/183-kDa replicase proteins with 3`-end of viral RNA to begin synthesis of minus-strand RNA. The RNA transcription by replicase removes the remaining capsid protein subunits from viral RNA. This is the co-replicational disassembly mechanism. An early step in such a disassembly mechanism could be the formation of tRNA-like structure at 3`-end and its recognition by viral polymerase (Wilson, 1988). Wu and Shaw (1997) found that progeny negative-strand virus RNA begins to be synthesized at the same time as 3`- to 5`-disassembly is initiated – suggesting that disassembly and initiation of TMV RNA replication are possibly coupled processes. They also found that on inoculation of protoplasts with encapsidated RNA of a mutant that lacks functional replicase activity, neither uncoating of the 3`-end of these RNAs nor minus-strand synthesis was detected (Wu and Shaw, 1997).

The complexes that contained partially disassembled virus particles, ribosomes, and the presumed nascent polypeptide chains were given the name of 'striposomes' (Wilson, 1984). The uncoated viral RNA in these complexes was found associated with as many as 24 ribosomes. These translation complexes appear after 15 minutes of incubation in a cell-free translation system programmed with pH 8.0-washed TMV particles. The cotranslational disassembly and striposomes were also later observed *in vivo* during early stages (between 10 to 70 minutes) of TMV inoculation of tobacco epidermal cells (Shaw *et al.*, 1986). Later, striposomes were also identified in TMV-infected plant cells showing that co-translational disassembly does occur *in vivo* as well (Wu *et al.*, 1994). In case, co-translational disassembly also causes uncoating of viral RNA *in vivo*, the virus particles must undergo similar type of partial destabilization.

The above hypothesis envisages the following: that more or less complete virus particles enter plasmalemma; that during their passage through plasmalemma these particles encounter low ionic (Ca^{2+}) strength and Ca^{2+} ion gradients; that any one or more of the three factors (hydrophobic environment of plasmalemma or intracellular phospholipid membrane or subcellular low pH of compartments) causes the virions to swell or lose some protein subunits from the 5`-end leader sequence so that partially encapsidated virions enter the cytosol. There the ribosomes take over and result in co-translational and co-replicational disassembly in which two processes are initiated simultaneously – uncoating of viral RNA in 3`- to 5`-direction and synthesis of progeny negative-strand RNA molecules.

Two experimental evidences favouring co-translational/co-replicational disassembly exist. One, extracts collected a few minutes after inoculation with radio-active TMV-inoculated epidermal cells contained complexes that resembled *in vitro* generated stripo-somes in appearance in electron microscope and in sedimentation behaviour in isopycnic gradients (Shaw *et al.*, 1986). Second, about 75% of TMV RNA, that includes the leader sequence and most or whole of the 183-kDa ORF, was disassembled in the assumed 5`-to 3`-direction within the first two to three minutes after electroporation of protoplasts (Wu *et al.*, 1994). The capsid protein subunits were removed from the region close to origin-of-assembly nucleotide sequence several minutes later. No information about uncoating of 3`-end of viral RNA was obtained.

The above hypothesis (co-translational/co-replicational disassembly of TMV particles *in vivo*) of virion uncoating has several alluring features. Most significantly, it accounts for virus uncoating and RNA release at mild physiological conditions prevalent *in vivo* in contrast to the harsh conditions commonly required for *in vitro* virus disassembly. Deproteinisation of virus particles on external surfaces compels the naked and labile single-stranded RNA for undertaking a hazardous journey during which it is susceptible, even for the relatively short periods for which it may be in that state, to the ubiquitous RNAses. This means that complete or partial uncoating of viral genome must occur either just before or simultaneously with the association of RNA with polyribosomes. One more appealing feature is that the 3`-end of viral RNA remains protected until it starts functioning in initiation of synthesis of progeny viral RNA.

Thus, the salient features of the above uncoating mechanism are: establishment of infection causes the loss of a small number of capsid protein subunits to expose the 5`-UTR including the Ω leader sequence; translation is initiated on the exposed RNA during which host ribosomes synthesize the 126- and 183-kDa replication proteins and simultaneously strip the viral RNA up to the stop codon of the 183-kDa coding region; the 126- and 183-kDa replication proteins associate with host factors to form an active replicase complex; the replicase complex interacts with 3`-viral RNA sequences to initiate transcription of viral minus-sense RNA and to simultaneously remove the remaining capsid protein subunits from the positive-sense viral genomic RNA. From this point, viral RNA replication and gene expression is unimpeded.

Co-translational disassembly also seems to be functional in spherical viruses. The RNA encapsidated in pre-swollen particles of some spherical plant viruses (AMV, BMV, CCMV, and SBMV or of their compact particles after swelling) direct protein synthesis in wheat germ extract (Brisco *et al.*, 1986). It was suggested that pre-treatment of virus particles makes the RNA accessible to ribosomes which, during translation, must be extracting or releasing RNA from spherical capsid without disrupting the capsid completely. The TYMV RNA escapes through a hole formed in the capsid by removal of some capsid protein subunits.

D. Direction of Uncoating

Early workers suggested that TMV particles, upon *in vitro* treatment with alkali, urea, sodium dodecyl sulfate, and dimethyl sulfoxide, underwent uncoating predominantly, if not exclusively, from the 5`- to 3`-terminus (Wilson *et al.*, 1981; Blowers and Wilson, 1981). Thus, disassembly in general was considered as being polar from the 5`-end of the RNA. Disassembly of *Papaya mosaic virus* is also polar and starts from 3`-terminus towards 5`-end followed by sequential stripping of protein subunits exclusively from the 3`-end (Lok and AbouHaidar, 1981).

Occurrence of bidirectional stripping (that is, simultaneous uncoating from both the 5`- and 3`-ends) has also been suggested. Wilson *et al.* (1981) and Nicolaieff and Lebeurier (1979) suggested that majority of TMV particles uncoat rapidly and extensively from the 5`-terminus but a subpopulation of virus particles uncoat bidirectionally. The uncoating from 5`-end is rapid but that from 3`-terminus is slow and less extensive. Stripping from the 3`-end proceeds up to less than 500 nucleotides by which time the major 5`- to 3`-direction of uncoating acquires dominance (Wilson *et al.*, 1981). The directions of major and minor stripping, reported by Nicolaieff and Lebeurier (1979), are the reverse of the directions reported by Wilson *et al.* (1981). Using RT-PCR analysis, Wu and Shaw (1996) concluded that uncoating of TMV particles *in vivo* occurs bidirectionally, that a region at or near the origin-of-assembly sequence is the last to be uncoated, that disassembly at 3`-end of virus particles starts between 2 to 5 minutes after inoculation, that the removal of protein subunits was slower in initial stages of disassembly, and that disassembly primarily was in 3`- to 5`- direction.

This bidirectional disassembly model involves translation-mediated disassembly from the 5`-end of the virion followed by replicase-mediated disassembly from the 3`- end and is supported by PCR studies of Wu and Shaw (1996, 1997). A very important advantage of bidirectional disassembly is that viral RNA remains protected in cytoplasm till the initiation of active translation and replication processes.

III. REFERENCES

Ben-Sin, C., and Tien Po. 1982. Infection of barley protoplasts with *Barley stripe mosaic virus* detected and assayed by immunoperoxidase. *J. Gen. Virol.* 58: 323-327.

Blowers, L. E., and Wilson, T. M. A. 1981. The effect of urea on *Tobacco mosaic virus*: Polarity of disassembly. *J. Gen. Virol.* 61: 137-141.

Brisco, M. J., Hull, R., and Wilson, T. M. A. 1986. Swelling of isometric and of bacilliform plant virus nucleocapsids is required for virus-specific protein synthesis *in vitro. Virology* 148: 210-217.

Butler, P. J. G. 1999. Self-assembly of *Tobacco mosaic virus*: The role of an intermediate aggregate in generating both specificity and speed. *Phil. Trans. Roy. Soc. London, Ser. B.* 354: 537-550.

Carrington, J. C., Freed, D. D., and Leinicke, A. J. 1991. Bipartite signal sequence mediates nuclear translocation of the plant potyviral NIa protein. *Plant Cell* 3: 953-962.

Caspar, D. L. D. 1963. Assembly and stability of the tobacco mosaic virus particle. *Adv. Protein Chem.* 18: 37-121.

Caspar, D. L. D., and Namba, K. 1990. Switching in the self-assembly of *Tobacco mosaic virus. Adv. Biophys.* 26: 157-185.

Coutts, R. H. A. 1980. Virus induced local lesions on cowpea leaves: The role of the epidermis and the effects of kinetin. *Phytopath. Z.* 97: 307-316.

Culver, J. N. 2002. Tobacco mosaic virus assembly and disassembly determinants in pathogenicity and resistance. *Annu. Rev. Phytopath.* 40: 287-308.

Culver, J. N., Dawson, W. O., Plonk, K., and Stubbs, G. 1995. Site-directed mutagenesis confirms the involvement of carboxylate groups in the disassembly of *Tobacco mosaic virus. Virology* 206: 724-730.

de Zoeten, G. A. 1995. Plant virus infection: Another point of view. *Adv. Bot. Res.* 21: 105-124.

Fornerod, M., Ohno, M., Yoshida, M., and Mattaj, I. W. 1997. CRM1 is an export receptor for leucine-rich nuclear export signals. *Cell* 90: 1051-1060.

Haeberle, A. M., and Stussi-Garaud, C. 1995. *In situ* localization of the non-structural protein P25 encoded by beet necrotic yellow vein virus RNA3. *J. Gen. Virol.* 76: 643-650.

Kasamatsu, H., and Nakanishi, A. 1998. How do animal DNA viruses get to the nucleus? *Annu. Rev. Microbiol.* 52: 627-686.

Li, X.-H., Valdez, P., Olvera, R, E., and Carrington, J. C. 1997. Functions of the tobacco etch potyvirus polymerase (NIb): Subcellular transport and protein-protein interactions with VPg/proteinase (NIa). *J. Virol.* 71: 1598-1607.

Lok, S., and AbouHaidar, M. 1981. The polar alkaline disassembly of *Papaya mosaic virus*. *Virology* 113: 637-643.

Lu, B., Stubbs, G. J., and Culver, J. N. 1996. Carboxylate interactions involved in the disassembly of *Tobacco mosaic tobamovirus*. *Virology* 225: 11-20.

Lucy, A. P., Guo, H. S., Li, W. X., and Ding, S. W. 2000. Suppression of post-transcriptional gene silencing by a plant viral protein localized in the nucleus. *EMBO J.* 19: 1672-1680.

Matthews, R. E. F., and Witz, J. 1985. Uncoating of turnip yellow mosaic virus RNA *in vivo*. *Virology* 144: 318-327.

Mundry, K. W., Watkins, P. A. C., Ashfield, T., Plaskitt, K. A., Eisele-Walter, S., and Wilson, T. M. A. 1991. Complete uncoating of the 5`-leader sequence of tobacco mosaic virus RNA occurs rapidly and is required to initiate cotranslational virus disassembly *in vitro*. *J. Gen. Virol.* 72: 769-777.

Namba, K., and Stubbs, G. 1986. Studies of *Tobacco mosaic virus* at 3.6Å resolution: Implications for assembly. *Science* 231: 1401-1406.

Namba, K., Pattanayek, R., and Stubbs, G. 1989. Visualization of protein-nucleic acid interactions in a virus. Refinement of intact tobacco mosaic virus structure at 2.9Å resolution by fibre diffraction. *J. Mol. Biol.* 208: 307-325.

Nicolaieff, A., and Lebeurier, G. 1979. Polar uncoating of *Tobacco mosaic virus* (TMV) with dimethylsulphoxide (DMSO) and subsequent reassembly of partially stripped TMV. *Mol. Gen. Genet.* 171: 327-333.

Oostergetel, G. T., Krijgsman, P. C. J., Mellema, J. E., Cusack, S., and Miller, A. 1981. Evidence for the absence of swelling of *Alfalfa mosaic virus*. *Virology* 109: 206-210.

Pantaleo, V., Rubino, L., and Russo, M. 2003. Replication of carnation Italian ringspot virus defective interfering RNA in *Saccharomyces cerevisiae*. *J. Virol.* 77: 2116-2123.

Pantaleo, V., Rubino, L., and Russo, M. 2004. The p36 and p95 replicase proteins of *Carnation Italian ringspot virus* cooperate in stabilizing defective interfering RNA. *J. Gen. Virol.* 85: 2429-2433.

Restrepo, M. A., Freed, D. S., and Carrington, J. C. 1990. Nuclear transport of plant potyviral proteins. *Plant Cell* 2: 987-998.

Ryabov, E. V., Kim, S. H., and Taliansky, M. 2004. Identification of a nuclear localization signal and nuclear export signal of the umbraviral long-distance RNA movement protein. *J. Gen. Virol.* 85: 1329-1333.

Shaw, J. G. 1999. *Tobacco mosaic virus* and the study of early events in virus infections. *Phil. Trans. Roy. Soc. London* 354B: 603-611.

Shaw, J. G., Plaskitt, K. A., and Wilson, T. M. A. 1986. Evidence that tobacco mosaic virus particles disassemble cotranslationally *in vivo*. *Virology* 148: 326-336.

Stubbs, G. 1999. Tobacco mosaic virus particle structure and the initiation of disassembly. *Phil. Trans. Roy. Soc. London,*. 354B: 551-557.

Verduin, B. J. M. 1992. Early interactions between viruses and plants. *Semin. Virol.* 3: 423-431.

Vetter, G., Hily, J.-M., Klein, E., Schmidlin, L., Haas, M., Merkle, T., and Gilmer, D. 2004. Nucleo-cytoplasmic shuttling of the beet necrotic yellow vein virus RNA3-encoded p25 protein. *J. Gen. Virol.* 85: 2459-2469.

Virudachalam, R., Sitaraman, K., Heuss, K. L., Markley, J. L., and Argos, P. 1983. Effect of pH-induced release of RNA from *Belladonna mottle virus* and the stabilizing effect of polyamines and cations. *Virology* 130: 351-359.

Wang, Y., Tzfira, T., Gaba, V., Citovsky, V., Palukaitis, P., and Gal-On, A. 2004. Functional analysis of the *Cucumber mosaic virus* 2b protein: Pathogenicity and nuclear localization. *J. Gen. Virol.* 85: 3135-3147.

Watts, J. W., and King, J. M. 1984. The effect of charge on infection of tobacco protoplasts by bromoviruses. *J. Gen. Virol.* 65: 1709-1712.

Wieringa-Brants, D. H. 1981. The role of the epidermis in virus-induced local lesions on cowpea and tobacco leaves. *J. Gen. Virol.* 54: 209-212.

Wilson, T. M. A. 1984. Cotranslational disassembly increases the efficiency of expression of TMV RNA in wheat germ cell-free extracts. *Virology* 138: 353-356.

Wilson, T. M. A. 1985a. Nucleocapsid disassembly and early gene expression by positive-strand RNA viruses. *J. Gen. Virol.* 66: 1201-1207.

Wilson, T. M. A. 1985b. Does TMV uncoat cotranslationally *in vivo*? *Trends Biochem. Sci.* 10: 57-60.

Wilson, T. M. A. 1988. Structural interactions between plant RNA viruses and cells. *Oxf. Surv. Plant Mol. Cell. Biol.* 5: 89-144.

Wilson, T. M. A., and McNicol, J. W. 1995. A conserved precise RNA encapsidation pattern in tobamovirus particles. *Arch. Virol.* 140: 1677-1685.

Wilson, T. M. A., and Watkins, P. A. C. 1985. Cotranslational disassembly of a cowpea strain (Cc) of TMV: Evidence that viral RNA-protein interactions at the assembly origin block ribosome translation *in vitro*. *Virology* 145: 346-349.

Wilson, T. M. A., Lomonossoff, G. P., and Glover, J. F. 1981. Dimethylsulphoxide (DMSO) disassembles *Tobacco mosaic virus* predominantly from the 5`-end of the viral RNA. *J. Gen. Virol.* 53: 225-234.

Wu, X. J., and Shaw, J. G. 1996. Bidirectional uncoating of the genomic RNA of a helical virus. *Proc. Nat. Acad. Sci. USA* 93: 2981-2984.

Wu, X. J., and Shaw, J. G. 1997. Evidence that a viral replicase protein is involved in the disassembly of tobacco mosaic virus particles *in vivo*. *Virology* 239: 426-434.

Wu, X. J., Up, Z., and Shaw, J. G. 1994. Uncoating of tobacco mosaic virus RNA in protoplasts. *Virology* 200: 256-262.

Zhuravlev, Yu., Yudakova, Z. S., and Pisetskaya, N. F. 1980. Infection of tobacco protoplasts with TMV in the absence of poly-L-ornithine and electroosmotic mechanism of virus entry. *Phytopath. Z.* 98: 296-309.

4 REPLICATION OF PLUS-SENSE VIRAL RNA

I. INTRODUCTION

Viral RNA can not be categorized into any of the three RNA types (messenger RNA, transfer RNA, and ribosomal RNA) since it contains all the genetic information of a virus, can act as a template, can serve as a messenger and directs the synthesis of relevant virus-specific proteins in a host cell, and multiplies under its own direction. Viral RNA is thus a self-replicating polycistronic messenger, a messenger that contains genes for two or more proteins. Parental viral RNA, upon release from ribosomes, switches from translation mode (during RdRp synthesis) to replication mode, and so triggers its own replication in which several viral proteins are involved that are together called replication proteins. Many viral replication proteins have been identified and characterized.

Viral RNA replication (formation of progeny genomic RNA molecules identical to the original parental RNA molecule) always starts from 3`-end that must be unwound. Viral RNA replication begins with specific recognition of *cis*-acting RNA elements on the infecting viral positive-strand RNA by membrane-bound viral replicase and/or associated host factors (Lai, 1998; Restrepo-Hartwig and Ahlquist, 1996, 1999; Chen and Ahlquist, 2000; Boon *et al.*, 2001; Ahlquist *et al.*, 2003; Noueiry and Ahlquist, 2003; and Boguszewska-Chachulska and Haenni, 2005). Recognition of a promoter at 3`-end of positive-strand RNA genome/template of the infecting virus by RdRp starts the replication cycle of viral RNA genome. Specific *cis*- and *trans*-acting nucleotide sequences and RNA secondary structures within 5`-termini of virus genomic RNAs are central to virus RNA replication. The compulsion of unwinding of 3`-end of viral RNA is apparently correlated with the fact that several RdRps have a narrow template channel that can only accommodate single-stranded RNA or have mechanisms that discriminate against the use of double-stranded templates for *de novo* initiation (Lesburg *et al.*, 1999; Butcher *et al.*, 2001; Ranjith-Kumar *et al.*, 2003a).

Vlot *et al.* (2001, 2003), while working on replication/synthesis of AMV RNAs 1 and 2, arrived at three very significant conclusions. One, the virus possesses a mechanism that coordinates replication of RNAs 1 and 2 when present together in a cell. AMV p1 and p2 proteins may interact *in cis*; this interaction is between helicase domain of p1 and N-terminal region of p2 (van der Heijden *et al.*, 2001) and may play a role in coordinated replication of AMV RNAs 1 and 2, although other factors may also be involved in this process (Vlot *et al.*, 2001, 2003). Two, this mechanism also provides selectivity to the replicase in the use of particular viral template RNA. Replicase not only selects genome RNAs that are essential for production of progeny but also selects the fittest template from a population of variants of a given genome segment. Three, the same or some other identical mechanism permits AMV to optimize its replication strategy by replicating only those RNAs that are essential for producing progeny that is infectious to the type of cell in which it was produced.

Obviously, synthesis of progeny RNA molecules takes place in nuclei of the host cells so that plant viruses possess a signal for localization of viral RNA to nucleus. This nucleo-cytoplasmic shuttling has been dealt with in Chapter 3.

In increasing number of plant viruses, circularization of positive-strand genomic RNAs is involved in regulation of translation, replication and sgRNA synthesis. Circularization of RNAs occurs because of the long-distance RNA-RNA interactions between 5`- and 3`-terminal sequences. An interaction between the two termini [3`- and 5`-ends] of the RNA appears to take place in many cases during RNA replication so that 3`-end sequence often can regulate RNA synthesis or translation initiated from 5`-end of RNA. The 3`-end sequence of BMV as well as of TMV RNAs does affect synthesis of positive-strand that begins from the 5`-end. Synthesis of BMV negative-strand RNA, which starts from 3`-end of RNA, requires a *cis*-acting enhancer sequence upstream of the subgenomic RNA promoter. A similar situation obtains in AMV in which an identical long-distance interaction between various RNA regions is required for initiation of subgenomic RNA synthesis (van der Vossen *et al.*, 1995). Two short elements on negative strands of satellite RNA C of TCV, one located at 11 bases from 3`-end and the other located 41 bases from 5`-end, are important for plus-strand RNA synthesis (Guan *et al.*, 1997). Qiu and Scholthof (2000) found that 263 nucleotides of 3`-UTR plus 73 nucleotides upstream of capsid protein stop codon and the first 16 nucleotides in 5`-UTR are required for RNA amplification and/or systemic spread of *Satellite panicum mosaic virus* (SPMV). In CPMV, both 3`-terminal *cis*-acting elements and a stem-loop upstream of poly(A) are important for replication of its RNA. The RNAs of BYDV have neither a cap structure nor a poly(A) tail, but long-distance base pairing between a stem-loop structure in 5`-UTR and a translation enhancer in 3`-UTR have been proposed to ensure circularization of RNAs and transfer of initiation factors to 5`-end (Guo *et al.*, 2001). Possibly, 5`- and 3`-end interactions provide some advantage to viruses because this ensures that only intact viral RNAs are used as templates for RNA replication and transcription. Communication between 5`-cap and 3`-poly(A) tail, resulting in circularized mRNA, enhances translation of cellular mRNAs (Sachs, 2000). Moreover, this leads to circularization of virus genome that greatly facilitates viral RNA replication.

Host specificity of a virus implies that the virus is able to undergo replication in a specific host and to induce disease symptoms by cell-to-cell and systemic movement within it. Thus, ability of a virus to infect a particular host (host specificity) and induce disease symptoms (symptomatology) by cell-to-cell and systemic movement (virus transport/movement) in that host is a clear demonstration of the ability of the particular virus to replicate in the concerned host. These topics are extremely complicated in which replication, symptom induction, and virus movement in the specific host are determined by genes from both virus and host. This gene-for-gene interaction in plant viruses was reviewed by Dardick and Culver (1999). Only a couple of examples showing participation of virus genes/proteins in virus replication/host specificity/ symptomatology/virus transport are mentioned here.

Early work by genetic exchanges between BMV and CCMV demonstrated that BMV 1a protein controls some aspects of template specificity in RNA replication. The protein p19 of TBSV is involved in host-specific systemic invasion and symptom development (Scholthof, H. B. *et al.*, 1995a, 1995b; Chu *et al.*, 2000) and is also active as suppressor of gene silencing (Qu and Morris, 2002; Qiu *et al.*, 2002). The protein p23 of *Hibiscus chlorotic ringspot carmovirus* (HCRSV) is involved in host determination so that this protein determines the host-specific replicative function in kenaf (Liang *et al.*, 2002). There could be any one or more of the following three reasons: p23 interacts with specific host factors to regulate host replicational machinery involved in virus replication; p23 could act as a *cis-* or *trans-*activator to either directly or indirectly regulate expression of host genes involved in viral replication; and p23 may interact with viral polymerase complex and so regulate virus replication (Liang *et al.*, 2002).

Certain facts about replication of TMV RNA have been established experimentally. One, interaction between 126- and 183-kDa proteins has been demonstrated (Goregaoker *et al.*, 2001) and a purified TMV RdRp preparation containing a 1:1 dimer of 126- and 183-kDa proteins is able to synthesize negative-strand RNA templates *in vitro* (Watanabe *et al.*, 1999). Two, TMV-L (tomato strain) 126-kDa replication protein of RdRp complex and TMV-L RNA 3`-terminal region are actually bound and it is a region of 126- and 183-kDa replication proteins located downstream of the core methyltransferase domain that binds RdRp to RNA 3` -terminal region *in vitro* (Osman and Buck, 2003). Three, domain D2 and central core region of TLS are the most important elements for binding of TMV RdRp complex to viral 3`-terminal region (Osman *et al.*, 2000). Four, Osman and Buck (2003) show that two aromatic amino acids (at positions 409 and 416) in this region of 126- and 183-kDa replication proteins are essential for binding to RNA 3` -terminal region and for replication of TMV RNA in tomato protoplasts. These aromatic amino acids may either directly interact with nucleic acid bases of 3`-terminal region of RNA or be essential for maintaining the structure of P314-423 region of 126- or 183-kDa protein that binds to 3`-terminal region. It is likely that P314-423 region binds to core C and D2 domain of 3`-terminal region. The interacting sites of 126- and 183-kDa proteins are located within a 110-amino acid region just downstream of core methyltransferase domain and a region comprising of central core C and domain D2 in 3`-terminal region. Possibly, binding of this region of 126- and 183-kDa proteins to TMV-L 3`-terminal region of RNA could be essential for virus RNA replication.

Lewandowski and Dawson (2000) showed that TMV 126-kDa protein appears to function in *cis* whereas 183-kDa protein can function in *trans*; and suggested that 126-kDa protein binds to its mRNA and targets it for replication. The 183-kDa protein might then bind to 126-kDa protein and initiate replication. Subsequent binding of 183-kDa protein to already bound 126-kDa protein may position the catalytically active site of polymerase domain at 3`-terminus of template RNA to enable initiation of negative-strand synthesis. Possible, binding of polymerase to 3`-terminal CCA sequence may be relatively weak. Hence, the function of 126-kDa protein may be both to recruit RNA templates for replication and to subsequently bind 183-kDa protein to position the catalytically active site of polymerase domain close to 3`-end of template RNA.

II. MODELS OF VIRAL RNA REPLICATION

From the negative-strand RNA synthesis stage onwards, three models have been proposed (Buck, 1996) for replication of RNA. Conceivably, different mechanisms could operate in different viruses or virus groups. RNA genomes of many positive-strand plant viruses, like cellular mRNAs, contain a 5`-cap structure and a 3`-poly(A) tail but several exceptions exist (Chapter 2). Replication of these exceptional RNAs may be different.

In first model, the nascent negative-strand RNA remains base-paired to the positive-strand RNA but only in the area where RdRp is bound to the template positive-strand RNA and is involved in active synthesis of negative-strand RNA. The 5` tail of the nascent negative-strand RNA is not base-paired to the template so that a free negative-strand RNA product is produced with the result that the positive-strand RNA template gets released. The polymerase recognizes a promoter at 3`-end of the negative-strand-strand RNA, which is then employed as a template for synthesizing a positive-strand RNA resulting in the formation of replicative intermediate (RI). Here also, the nascent positive-strand RNA remains base-paired to the negative-strand RNA template only in the area at which polymerase is attached to template and is actively synthesizing RNA. Hence the RI is mostly single-stranded and is constituted by the full-length negative-strand RNA template to which several positive-strands of RNA are attached. The progeny RNA strands again are largely single-stranded. This model of RNA replication occurs in *Qβ virus* RNA and possesses certain characteristics: the replicase of Qβ RNA is a holoenzyme so that it catalyzes complete replication of viral RNA, the RIs are mainly single-stranded, and replicase is incapable of using Qβ dsRNA as template. This model presupposes the recognition of a single-stranded structure for initiating the synthesis of the positive-strand RNA.

However, the above model needs an explanation, which is not required for model 2 of RNA replication. How are the base pairs, formed at the time of RNA synthesis, almost immediately unwound by replicase complex? The possible explanation is - a helix-destabilizing protein, or a second molecule of helicase, suitably positioned in replication complex, could probably accomplish this. It is significant in this connection that the purified replication complexes of two plant viruses (BMV and CMV) seem to contain more of the respective 1a protein (which bears helicase-like domain) than 2a protein (which bears polymerase-like domain). Moreover, replication of BMV RNA is more sensitive to reductions in expression of 1a protein (that also has capping functions) than to reductions in expression of 2a protein. Hayes and Buck (1990) employed an *in vitro* system that was able to catalyze complete replication of CMV RNA and detected free positive and negative strands in the ratio of 7:1 as also some ds RNA. This observation suggests that replication occurs according to model 1. Additionally, they found that dsRNA did not act as a template for replicase complex;

however, this may also mean that replicase lacks an essential component required for initiating synthesis on a dsRNA template.

In second model, the negative-strand RNA synthesized stays base-paired with the positive-strand RNA template. This produces the replicative form RNAs (RF RNAs), which are initially partly double-stranded and partly single-stranded but finally become completely dsRNA. Thus, the RdRp recognizes a promoter at end of RF RNA containing 3`-end of negative-strand and 5`-end of positive-strand. Synthesis of progeny positive-strand RNA molecules starts by employing the negative-strand as template. This happens by a strand displacement method that displaces the negative-strand from the dsRF RNA. The negative strands, by acting as templates, produce the RIs constituted by dsRNA with one or, by repeated reinitiation, several single-stranded 5` tails of full-length positive-strands of RNA. Thus, multiple progeny positive-strands of RNA are synthe-sized by repeated reinitiation and are then released. The first full-length released positive-strand is the original template RNA strand. The above manner of synthesis of progeny positive strands from a RI RNA is analogous to the semi-conservative transcription of dsRNA by strand displacement mechanism. This mechanism is characteristic of the dsRNA viruses of families *Birnaviridae, Cystoviridae*, and *Partitiviridae*. This model presupposes that the unwinding of dsRF RNA occurs at 3`-end of the negative-strand prior to the beginning of RNA synthesis and that a ds structure has to be recognised for initiation of positive-strand RNA synthesis (and for synthesis of subgenomic RNA in viruses that utilize subgenomic promoters).

In third model, formation of dsRF RNA is exactly similar to that in model II. Synthesis of progeny positive-strand RNA occurs by employing the negative-strand of dsRNA as template. This synthesis takes place by transiently displacing the positive strand of dsRF RNA in the zone at which RNA is being synthesized. Synthesis of progeny positive-strands from a dsRF RNA is analogous to the conservative transcription of dsRNA that is characteristic of dsRNA viruses of families *Reoviridae* and *Togaviridae*. The RIs produced consist of double-stranded RNA that has one or several single-stranded tails. But unlike RIs in model 2, in which single-stranded tails are the displaced 5`-tails of full-length positive RNA strands, these tails are of the nascent, incomplete progeny positive RNA strands.

However, unequivocal distinction between the 'open' and 'closed' models (models I and II, respectively) of RNA replication for any eukaryotic positive-stranded RNA virus is not generally possible at present. Then, cellular and viral helicases cannot act on completely double-stranded structures. This means that unwinding of a completely dsRF RNA possibly requires another protein or helicases to bind to single-stranded regions close to the duplex region to be unwound and a 3`- to 5`-helicase (that is bound to the negative template of the unwound RF RNA) most probably unwinds the duplex subsequently. Viral helicase of PPV (Lain *et al.*, 1990) and RdRp complex of AMV (de Graaff *et al.*, 1995) have the required strand displacement property. Model 2 requires recognition of a double-stranded structure for initiation of positive-strand synthesis (and for subgenomic RNA synthesis for those viruses that utilize subgenomic promoters), whereas model 1 requires recognition of a single-stranded structure and thus model 2 appears to be more operative in plant viruses.

Several experiments support the participation of negative-strand RNA during replication of positive-strand RNA viruses. An isolated RdRp complex of AMV, BMV and CMV could use the negative-strand template to produce full-length and subgenomic RNA. Transgenic plants that express double-stranded RNA-specific ribonuclease of plant virus (*Tomato mosaic virus*, CMV, and PVY) exhibit resistance (although incomplete) to that particular virus (Watanabe *et al.*, 1995). This indicates the formation of double-stranded replication intermediate structures during replication of these plant viruses.

III. NEGATIVE-STRAND RNA SYNTHESIS

In general, RNA viruses have a specific structure at the 3`-end of the genome that is required for initiation of minus-strand RNA synthesis (Ahlquist, 2003), which is also the start of RNA replication process of positive-strand RNA viruses. Various structures and *cis* sequences located at 3`-end serve as promoters for minus-strand RNA synthesis (Buck, 1996). These include tRNA-like structures/sequences (TLSs) in family *Bromoviridae* and genera *Tobamovirus, Tymovirus,* and *Hordeivirus* (Dreher, 1999). Plant viruses lacking a TLS contain various other conserved elements that play a role in initiation of RNA synthesis. For example, the satellite RNA C (satC) of TCV contains a stable 3`-terminal stem-loop (SL) structure, which is indispensable for its synthesis (Song and Simon, 1995). TCV is associated with several dispensable noncoding RNAs including satC (356 bases) and satD (194 bases). The core promoter for synthesis of satC minus strands is identified as a 3` terminal SL flanked by the sequence (CCUGCCC-OH), which is also found at 3`-end of TCV genomic RNA and satD (Song and Simon, 1995; Stupina and Simon, 1997; Carpenter and Simon, 1998). Transcriptional repressors, leading to repression of minus-strand RNA synthesis, exist in some plant viruses (Zhang *et al.*, 2004a, 2004b; Sun *et al.*, 2005) and are discussed later.

The 5`-terminal bases of satC are involved in minus-strand initiation *in vitro.* Zhang *et al.* (2004a) proposed that the 5`-end interacts with sequences in or near 3` hairpin, stabilizing the structure of the promoter and permitting the RdRp to properly recognize 3`-end when derepressor relieves repression. This suggests involvement of 5`-guanylates in minus-strand synthesis. Work on a variety of viral genomic RNA and subviral RNA replicons also support a role for 5`-end in minus-strand synthesis, through protein-protein bridges or direct RNA-RNA interactions or proposed 5`-stabilization of 3`-elements (Wu *et al.,* 2001; Vlot and Bol, 2003).

Minus strands are synthesized early in infection. They are mostly single-stranded in tissues involved in viral RNA replication but are predominantly part of the duplex RF RNA on completion of replication. Level of minus strands stays constant during the phase of rapid TMV synthesis. The ratio of plus to minus strands at the end of this stage is highly variable depending upon the virus. The minus-strand RNA of *Pea seed-borne mosaic potyvirus* is most abundant in tissues found along the periphery of infected area but occurs in much smaller amounts within the infected area – indicating that the peripheral cells were most recently infected and so involved in active viral

RNA replication (Wang and Maule, 1995). Besides the viral template, three other inputs are essential for synthesis of negative-strand RNA: the TLS, promoter element, and RdRp. These are discussed below in some of the plant viruses.

Olsthoorn and Bol (2002) and Olsthoorn *et al.* (2004) found that the 3`-terminal 145 nucleotides (that are common between the three AMV RNAs) can adopt a TLS and act as a core promoter for initiation of negative-strand RNA synthesis at 3`-end of template. They identified a single stem-loop structure, the triloop hairpin E (hpE), as the negative-strand RNA promoter, which is the most important and essential element for negative-strand RNA synthesis *in vitro* while presence of B, C, and D hairpins is required for optimum negative-strand synthesis. The proposed major role of TLS is to enforce initiation of transcription by polymerase at the very end of genome. The structure of hpE is conserved in all AMV isolates sequenced so far and features a 10-base pair stem, a 4-nucleotide bulge and a UGG triloop. The UGG triloop sequence is not essential for *in vitro* transcription but may be playing some role *in vivo*. The negative-strand promoter hpE and subgenomic RNA promoter hairpin of AMV are equivalent in binding viral polymerase, share many common features and are essential for performing their respective assigned function. A similar stem-loop structure can be formed at the homologous position in 3`-UTR of RNAs of about all ilarviruses. In PDV RNA1 and RNA3, a triloop, that strongly resembles hpE, can be folded. In apple mosaic virus RNA3 and prunus necrotic ringspot virus RNA3, a pentaloop hairpin is present at this location in 3` -UTR.

Structures similar to AMV negative-strand promoter hpE have been identified in 3` -TLS of BMV (Chapman and Kao, 1999; Sivakumaran *et al.*, 2003) and CMV RNAs (Sivakumaran *et al.*, 2000). These are conserved structures and are required for their respective negative-strand synthesis. Stem-loop structure C (SLC) at 3`-end of BMV RNA consists of 11 base pairs, a triloop, a 4-nucleotide bulge and thus resembles the AMV hpE. The SLC of several isolates of CMV consists of 13 base pairs, a 5`-nucleotide bulge, and a triloop; but variants with a pentaloop do also exist (Sivakumaran *et al.*, 2000). The 5`-UTRs of BMV RNAs contain stem-loop structures with loop sequences resembling box B elements homologous to TΨC stem-loop of cellular tRNAs. These elements are required for negative-strand RNA synthesis (Chen *et al.*, 2001; Schwartz *et al.*, 2002). The top C-G base pair of SLC and A of the 5`-loop are the only conserved sequences between BMV and CMV stem-loop structures. BMV RdRp can specifically recognise the 5`-most A in AUA triloop that is involved in the so-called adenine motif. This feature is unlike AMV RdRp, which seems to be insensitive to loop sequence but can probably recognise specific base pairs in the stem. Thus, related viruses have evolved different strategies to recognise their negative-strand promoter hairpins.

The BMV 2a polymerase-like replication protein binds to BMV 1a helicase-like replication protein and initiates negative-strand RNA synthesis at 3`-terminus TLS (Sullivan and Ahlquist, 1997; Chen and Ahlquist, 2000). The 2a protein is directed to ER by the viral protein 1a, which also recruits RNAs 2 and 3 templates for replication (Sullivan and Ahlquist, 1997; Chen and Ahlquist, 2000).

The 3`-terminal region of TMV (strain L) RNA contains all *cis*-acting sequences needed for negative-strand synthesis, at least *in vitro*, and can be folded into a TLS and nearby upstream pseudoknots that are important for TMV RNA replication *in vivo* and negative-strand RNA synthesis *in vitro* (Osman *et al.*, 2000; Chandrika *et al.*, 2000). The 3`-terminal elements important for negative-strand synthesis *in vitro* include the CCA sequence domains D1 (equivalent to a tRNA acceptor arm), D2 (similar to a tRNA anticodon) and D3 (an upstream pseudoknotted region) and a central core region C that connects domains D1, D2, and D3.

Wang and Wong (2004) found that synthesis of HCRSV minus-strand RNA is initiated opposite the 3`-terminal two C residues at the 3`-end *in vitro* and *in vivo* and that the 3`-terminal CCC nucleotides have essential role in minus-strand RNA synthesis because minus-strand RNA initiation begins at 3`-terminal two Cs. This is similar to the results obtained from BMV (Chapman and Kao, 1999), TMV (Osman and Buck, 1996), and TYMV (Singh and Dreher, 1997), in which initiation of minus-strand RNA synthesis also starts opposite the two Cs while minus-strand RNA synthesis is reduced upon removal or substitution of terminal A in the CCA box. In TYMV RNA, the 3` TLS present the CCA-3` in a conformation that is easily accessible to the replicase (Dreher, 1999). The CCC-3` may also be involved in forming a conformation for replicase access both *in vitro* and *in vivo* in HCRSV (Wang and Wong, 2004). However, the presence of a CCC-3` terminal sequence alone is not sufficient for RNA synthesis as is evident from below.

In addition to the CCC-3` terminal sequence, two putative stem-loops located within the 3`-terminus 87 nucleotides of HCRSV plus-strand RNA are also essential for minus-strand RNA synthesis (Wang and Wong, 2004). The first SL is very similar to the single SL that is a minimum promoter for minus-strand RNA synthesis of TCV satellite RNA C (Song and Simon, 1995). This suggests that the two predicted SLs in HCRSV might also play a similar role. The U loop located on SL2 is an essential structure for RNA synthesis in HCRSV. The secondary structure of SLs in 3`-UTR of RNA viruses is required for protein binding (Lai, 1998). For example, the hpE loop of AMV is not essential for RNA synthesis, whereas the stem and base pairing of the lower triloop are essential (Olsthoorn and Bol, 2002). Similarly, failure to bind replication factors by HCRSV mutants, with disruption or deletion of SL1 or SL2, rendered them unable to initiate minus-strand synthesis. Disruption of certain SLs in BMV (Chapman and Kao, 1999), TYMV (Deiman *et al.*, 1997), AMV (Olsthoorn *et al.*, 1999), and CTV (Satyanarayana *et al.*, 2002a) resulted in reduction in minus-strand RNA synthesis, but not in complete inhibition of RNA synthesis as in HCRSV. In TCV 3`-UTR, elements located hundreds of nucleotides upstream of 3`-UTR, were needed for efficient replication (Carpenter *et al.*, 1995). Taken together, the specific sequence CCC at the 3`-terminus and the two SL structures located in the 3`-UTR are essential for efficient minus-strand RNA synthesis in HCRSV.

A. Model of Negative-Strand RNA Synthesis

Cheng *et al.* (2002) have propounded a working model for negative-strand RNA synthesis of BaMV. (a) The RdRp interacts with 3`-UTR [including the potexviral conserved hexamer motif (ACNUAA)] and about 20 nucleotides of poly(A) sequence immediately downstream of 3`-UTR (Huang *et al.*, 2001), which may be used to initiate minus-strand RNA synthesis. *In vivo* experiments support this and show that about 13 to 25 adenylates following 3`-UTR are involved in maintaining the integrity of pseudoknot structure and are required for efficient viral RNA replication in protoplasts (Cheng *et al.*, 2002). (b) After interaction of RdRp with 3`-UTR of BaMV RNA, the RdRp is suggested to initiate minus strand RNA synthesis opposite one of the adenylates within the pseudoknot. This is done perhaps selectively in the loop region of the pseudoknot, which is the most favoured position for initiation. (c) The negative strands are then used as templates to synthesize progeny RNAs. These progeny RNAs must be good templates for subsequent polyadenylation by up to about 90 to 170 adenylates, as in BaMV virion RNAs. This different-length short stretch of As would be at the very end of genomic RNA due to the variation of initiation sites of negative-strand RNA synthesis. Therefore, the enzyme functional in polyadenylation must recognise these short stretches of poly(A) tailed genomes and maintain structural integrity of virion RNA.

The above model has some notable features. First: Certain exceptional features exist during negative-strand RNA synthesis in some plant viruses. The minus-strand RNA synthesis was found to initiate from several positions within poly(A) tail of BaMV so that Cheng *et al.* (2002) deduced that initiation site for negative-strand RNA synthesis is not fixed at one position but resides opposite one of the 15 adenylates of poly(A) tail immediately downstream of 3`-UTR of genomic RNA. Thus, multiple putative initiation sites of BaMV negative-strand RNA synthesis exist and any one of the adenylates could be used to initiate negative-strand RNA.

Second: The poly(A)-tailed RNA of viruses is comprised of hundreds of adenylates so that it might be difficult for RdRp, bound to 3` UTR as in BaMV (Huang *et al.*, 2001), to initiate negative-strand RNA at the end of poly(A) tail hundreds of nucleotides downstream. It is, therefore, regarded that a positive-strand RNA virus with a poly(A)-tailed genome would use the poly(A) tail as a template to initiate negative-strand RNA synthesis. However, only a short stretch of poly(A) sequence connected to 3` UTR is necessary for initiation.

Third: Both BMV and TYMV initiate their negative-strand RNA synthesis opposite the penultimate cytidylate residue.

Fourth: The results of Osman and Buck (2003) are compatible with a model for initiation of TMV-L negative-strand RNA synthesis in which an internal region of TMV-L 126-kDa protein first binds to central core C and domain D2 region of TMV-L RNA 3`-terminal region and is then followed by binding of 183-kDa protein to this complex and positioning of catalytically active sites of polymerase domain close to 3`-terminal CCA initiation site.

Fifth: It is clear from above that both BMV and TMV appear to have evolved similar mechanisms for recruitment of RNA templates and initiation of negative-strand RNA synthesis. In membrane-bound BMV replication complexes, ratio of 1a to 2a proteins is about 25:1 and since 1a protein forms spherules, which bud into endoplasmic reticulum, it is suggested that 1a protein plays both structural and functional roles in assembling membrane-bound replication complexes and sequestering 2a polymerase and BMV RNA templates within them (Schwartz *et al.*, 2002). Similarly, TMV 126-kDa protein is present in much larger amounts than 183-kDa protein in insulated membrane-bound replication complexes (Osman and Buck, 1996; Watanabe *et al.*, 1999) and may play a role in assembling TMV replication complexes and recruiting 183-kDa protein and RNA template that is similar to the role of BMV 1a protein in assembling BMV replication complexes. Oligomerization of helicase domain of TMV 126- and 183-kDa proteins is reported. A difference in mechanisms of sequestering RNA templates between the two viruses is that, unlike BMV, there is no recognizable tRNA-like TΨC stem-loop structure at 3` terminus or internally in TMV positive-strand RNA. Hence, TMV 126-kDa protein binds directly to 3`-terminal TLS.

IV. DOUBLE-STRANDED FORMS OF REPLICATIVE RNAs

Two diametrically opposite views have been expressed about the existence, significance and importance of dsRNAs: that they exist and participate in viral RNA replication and that they do not exist, are only artifacts and consequently do not perform any role in RNA replication. Thus, from the first review on dsRNAs (Ralph, 1969) till date, their nature and importance in viral RNA replication is still not finally settled (Buck, 1996). Satyanarayana *et al.* (2002a) state that a long-standing conundrum of virology has been the question of the *in vivo* relationship of complementary positive and negative RNA strands: whether they exist in a double-stranded helix in the cell or only become double-stranded during extraction. This problem still exists but virologists have decided to let it be and have moved on.

Early studies involving pulse chase experiments and kinetics of labeling (on BMV, CPMV, TMV, and several other plant viruses) showed that the label could be chased from RF and RI RNAs into full-length genomic sized single-stranded RNA molecules so that RF and RI RNAs act as progenitors for the synthesis of complementary and progeny RNA strands, respectively, of the virus concerned. The *in vitro* ^{32}P incorporation in TMV RF and RI RNAs showed that their synthesis and synthesis of progeny RNA stops at about the same time. Double-stranded RNA of TomRSV and of certain other plant viruses has been isolated in high yields only from leaves in which virus was increasing due to its multiplication. The switch-over from RF synthesis to RI synthesis (that is, from symmetric positive strand and minus strand synthesis to asymmetric plus strand progeny RNA synthesis) is controlled by genomic AMV RNA3. Although the nature and significance of dsRNAs in viral RNA replication is still debatable, yet their detection in virus infected cells continues to be reported - by Ritzenthaler *et al.* (2002) in GFLV-infected cells, by Dunoyer *et al.* (2002) in modified vesiculated ER in cells infected by PCV, and by still others. Transgenic plants expressing double-stranded RNA-specific ribonuclease showed resistance (but incomplete) to *Tomato mosaic virus*, CMV and PVY (Watanabe *et al.*, 1995) – indicating the formation of double-stranded RNA as the replication intermediate that does play some part during RNA replication.

Each genomic RNA of a multipartite virus replicates through its own dsRNA. A regulatory mechanism appears to govern the rates at which each species of genomic RNA of a multicomponent virus is synthesized relative to others. In wild type CCMV, even as the rate of RNA synthesis changes greatly during infection period, relative rates of synthesis of the three genomic RNAs remain constant as also that of the corresponding dsRNAs. This ratio is approximately two RNA3 molecules produced to one molecule each of RNA1 and RNA2. Thus, more RNA3 is synthesized than can be encapsidated so that it accumulates in infected cells to a higher level than other viral RNAs. Similarly, CMV RNA3 and its corresponding RF RNA occur predominantly in CMV strain Y-infected protoplasts. Moreover, a dsRNA species, which corresponds in size to subgenomic RNA of RCNMV, is found in infected tissue (Sit *et al.*, 1998).

A. Replicative Form RNA

The RF RNAs are full-length double-stranded structures composed entirely of base pairs, are never infectious unless denatured to release the viral genome strand, and are usually found in nucleic acid preparations from cells infected with positive-strand RNA viruses or in extracted products of *in vitro* RNA synthesis conducted in presence of crude or partially purified polymerase preparations. The 3` termini of minus strands of RF RNAs are exactly complementary to the corresponding 5` sequences of the RNA and so form a perfect duplex with 5`-ends of virion-sense RNA. This is even true of both the middle and bottom component RNAs of CPMV so that a duplex (with 5`-end of virion-sense RNA of each of the two components) is formed. Moreover, terminal sequences of both CPMV RF RNAs are identical; this may have some relevance in replication of this multicomponent virus.

The RF RNA in TomRSV-infected leaf tissue is positively correlated with the concentration of virus specific RdRp and the concentration of virus particles. Moreover, amounts of RF RNA and RdRp actively increase just prior to and during the period of rapid virus synthesis. Polymerase activity falls rapidly with cessation of virus synthesis followed by a similar decline in the amount of RF RNAs. The RF RNA accumulates with the increase in concentrations of viral RNA in the infected cells as in *Qβ virus* in which model 1 of RNA replication has been well established.

B. Replicative Intermediate RNA

The nature and significance of RIs in the replication of plus-stranded RNA is less clear (Buck, 1996). The RIs in infected cells are branched chains and are partly double-stranded. Each RI is usually composed of one complete full-length negative RNA strand to which are bound several partially synthesized positive strands that are being elongated and are in different stages of formation. RI RNAs are formed because of the simultaneous production of many complementary plus strands, which are displaced one after another from the same negative-strand RNA template molecule. The isolated RI RNA of *Poliovirus* is constituted by a full-length negative-strand RNA template to which 6-8 nascent positive RNA strands (in different stages of elongation/formation) are attached. Similar RI RNAs have also been detected in virus-infected plants. Tracer techniques have established that RI is the direct precursor of viral RNA. Presence of RIs is accounted for by the semi-conservative mechanism of RNA replication.

C. Double-Stranded RNAs as Artifacts

The RI RNAs are suggested to be mainly single-stranded structures like the RIs of TYMV detected *in vivo*. The *in vivo* double-stranded RNAs could be the dead-end products formed by annealing of positive and negative strands of the infecting RNA. Annealing of the two RNAs may also occur during extraction of viral nucleic acids from infected cells during deproteinization by phenol and are regarded as isolation artifacts formed during RNA extraction (Garnier *et al.*, 1980). The dsRNA extracted from *in vitro* RNA synthesis systems may also be produced in the same manner or may form as a result of inadequacies in *in vitro* systems. However, dsRNA may be formed *in vivo* in late stages of infection.

V. SYNTHESIS OF PROGENY POSITIVE-STRAND RNA

The minus RNA strands, released from the double-stranded RF RNA, serve as templates for synthesis of complementary plus strands of viral RNA (progeny RNA) from the 3`-end. The total number of minus-strand templates available for synthesis of the progeny plus RNA strands is limited so that the former have to behave as templates for synthesis of plus strands over and over again. This could occur in two ways: the semiconservative and conservative methods.

In the asymmetric semiconservative mechanism, the minus strand remains constant while the first plus strand of double-stranded RNA is elbowed out by the second plus strand, the second by the third and so on. Thus, there is multiple simultaneous formation of progeny plus RNA strands, which are at different stages of completion and appear to hang as tails. Up to five partially completed plus strands may be associated with RI. In conservative mechanism, the double-stranded structure remains conserved but its secondary structure is disrupted at the replication point to permit the copying to proceed along the minus strand. Experimental evidence favours the synthesis of progeny positive RNA strands through semiconservative asymmetric mechanism. Large amounts of plus strands are synthesized as compared to the minus strands; thus, synthesis of the two strands is asymmetric and proceeds at different rates. Correct positive-strand BMV RNA initiation requires 5`-terminal sequences and a non-template guanylate added to 3`-end of negative-strand RNA *in vitro* (Sivakumaran *et al.*, 1999).

RNA templates for plus-strand and minus-strand RNA synthesis most likely possess different sequences near initiation site. These RNAs initiate with a pyrimidine (but a cytidylate in BMV). Bromoviral templates for positive-strand RNA synthesis are rich in A or U nucleotides in contrast to the templates for minus-strand RNA synthesis (Chapman and Kao, 1999; Sivakumaran *et al.*, 1999). This is significant because plus-strand RNA synthesis by BMV polymerase is more efficient in case the template contains an A/U-rich sequence near initiation site (Sivakumaran *et al.*, 1999). Hema and Kao (2004) found that mutations at positions adjacent to initiation cytidylate in templates for genomic and subgenomic plus-strand RNA synthesis significantly decreased RNA accumulation and that different requirements exist for template sequence near initiation nucleotide for BMV RNA accumulation in plant cells and that an A/U-rich sequence is preferred for accumulation of subgenomic RNA.

On this basis, Hema and Kao (2004) hypothesized that an A/U-rich template sequence regulates level of RNA accumulation. The nucleotide position of the first C or G in template controls the relative amount of the four BMV RNAs: RNA4 (10) > RNA3 (7) > RNA2 (4) > RNA1 (2). The second and third nucleotides of genomic plus-strand and subgenomic RNAs of all bromoviruses have A and U, respectively (Adkins *et al.*, 1998). They derived three conclusions pertaining to BMV RNA replication: template sequence does not significantly affect the minus-strand RNA synthesis; specific identities of +2A and +3U are required for plus-strand RNA synthesis; and an A/U-rich sequence is needed for efficient subgenomic RNA synthesis. BMV genomic plus-strand and subgenomic RNA formation needs highly specific nucleotides at three positions: +1C, +2A, and +3U in the template-sense RNA.

VI. ASYMMETRY IN NEGATIVE-STRAND AND POSITIVE-STRAND PROGENY RNA SYNTHESIS

The replication process is usually asymmetric so that 10- to 1000-fold more of positive RNA strands over negative RNA strands are produced. Plus-strand viral RNAs contain sequences and structural elements that permit cognate RdRp to correctly initiate and transcribe asymmetric levels of plus and minus strands during RNA replication. This asymmetry is characteristically found in all positive-strand RNA viruses investigated. Molar ratio for positive-strand:negative-strand progeny RNA production of about one is achieved, during transfection of barley protoplasts with BMV RNAs 1and 2 (but in the absence of RNA3); however, >100 progeny plus-strands of progeny RNA molecules are synthesized for every negative-strand progeny RNA in the presence of RNA3. The presence of RNA3 increases the total plus-strand RNA production by over 200-fold and that of RNAs 1 and 2 by about 30-fold. The ratio of production of positive- to negative-strand genomic RNAs is 10:1 for flaviviruses (animal viruses); 40 to 50:1 for CTV (Satyanarayana *et al.*, 1999, 2002b); 50 to 100:1 for coronaviruses (animal viruses); 100:1 for BMV and TMV; and 1000:1 for AMV. Moreover, the ratios of positive- to negative-strand sgRNAs in wild-type CTV is estimated to be 10 to 20:1 (Satyanarayana *et al.*, 2002b). Asynchronous accumulation of RNAs 1 and 2 of *Lettuce infectious yellows virus* (LIYV) also occurs (Yeh *et al.*, 2000).

Mechanism that controls ratio of positive- to negative-strands of genomic RNAs, leading to strand asymmetry due to asymmetric virus RNA replication, has been investigated. In BMV, RNA3 controls this process. The subgenomic RNA promoter and upstream inter-cistronic sequences present in central region of RNA3 ensure switch-over to asymmetric RNA replication. These two types of switches act in *trans*. The intercistronic region is adjacent to $ICR2_{1100}$ (ICR2-like sequence present at nucleotide 1100), exerts *trans*-acting effect on BMV replication, and is essential for positive-strand RNA3 amplification as well as for asymmetry of BMV replication. Possibly, host factors bind to the intercistronic sequence and/or the subgenomic promoter to profoundly affect the replicase activities and thereby affect strand asymmetry. In AMV, a frameshift in capsid protein gene results in a 100-fold reduction in positive-strand RNA accumulation but a 3- to 10-fold increase in negative-strand accumulation indicating that capsid protein is involved in strand asymmetry in favour of positive RNA strands.

Satyanarayana *et al.* (2002b) found that protein 23 (p23) gene of CTV regulated asymmetric accumulation of positive- and negative-stranded subgenomic RNAs; that this protein mainly down-regulated the accumulation of negative-stranded sgRNAs and caused only a modest increase in accumulation of positive-stranded sgRNA; that amino acid residues 46 to 180, which contained RNA-binding and zinc finger domains, were indispensable for asymmetric RNA accumulation while N-terminal 5 to 45 and C-terminal 181 to 209 amino acids residues were not required; that zinc finger is possibly involved in asymmetric RNA accumulation; and that the excess negative-stranded sgRNA reduces the availability of corresponding positive-strand sgRNA as a messenger. Thus, p23 protein serves as a switch to convert replication from symmetrical to asymmetrical production of positive- and negative–stranded RNA of both genomic and sgRNAs by down regulating negative-strand RNA accumulation. López *et al.* (2000) demonstrated that CTV p23 gene product binds RNA *in vitro* and RNA-binding domain mapped between amino acid residues 50 to 68, which include the putative zinc finger domain.

Other regulatory factors favouring asymmetric synthesis of positive-strand RNA are: the ability of an RNA genome to compete for the limited polymerase, characteristic features of RNA synthesis like the processivity of polymerase, frequency of initiation by polymerase, capability of the polymerase to transition out of initiation mode, and frequency of template switch (Hema and Kao, 2004).

VII. TIME COURSE OF VIRAL RNA, VIRAL PROTEIN, AND VIRUS PARTICLE SYNTHESIS

Rates of TMV RNA synthesis, TMV protein synthesis, replicase synthesis, and formation of TMV particles have been studied in infected tobacco leaves, inoculated tissue culture and inoculated protoplasts and generally closely parallel each other.

Synthesis of RNA increases almost linearly during the first 60 hr period but declines subsequently. In fact, RNA synthesis in infected protoplasts begins in less than 4 hr post-inoculation (hpi), becomes exponential and remains so until 8 hpi, drops sharply and becomes linear at about 10 hpi and then continues to increase linearly. Synthesis of single-stranded and double-stranded RNAs (RF and RI RNAs) also nearly follows the same curve: they are first detected at 6-8 hpi, increase exponentially till about 20 hpi after which their increase becomes linear. Maximum synthesis of double-stranded RNA occurs at 18-20 hpi and that of single-stranded RNA at 24-26 hpi. About 20 minutes are required for synthesis of a TMV RNA molecule during exponential stages of replication. The synthesized viral RNA is quickly incorporated into viral protein to form virions. First progeny TMV appears in cytoplasm of mesophyll protoplasts 6 hr after inoculation. TMV particles are formed exponentially during early stages of virus replication but linearly during the late stages. TMV virion formation is exponential during 50-80 hpi but declines to become linear subsequently. Large number of virus particles is produced. Similarly, replication of *Pea seed-borne mosaic potyvirus* is rapid (Wang and Maule, 1995). So is of TEV, which infects approximately one new cell every two hours (Dolja *et al.*, 1992).

Yeh *et al.* (2000) report that simultaneous inoculation of bipartite LIYV resulted in asynchronous accumulation of progeny LIYV RNAs. LIYV RNA1 progeny genomic (both positive- and negative-sense RNAs) and subgenomic RNAs (including subgenomic RNA of ORF2-encoded protein p32) were detected in protoplasts 12 hpi and accumulate to maximum high levels by 24 hpi. The ORF2 subgenomic RNA was most abundant of all the LIYV RNAs. In contrast, RNA2 progeny (positive- and negative-sense RNA2) was detected only between 24-36 hpi or at about 36 hpi. The genomic and subgenomic RNA of HCRSV could be detected in infected kenaf protoplasts as early as 6 hpi (Liang *et al.*, 2002). In all plant viruses, appearance of virus particles is closely related with synthesis of capsid protein.

In pulse-chase experiments, TEV polyprotein appeared 5 minutes after initiation of chase, processed NIa protein appeared 10 minutes after initiation of chase while proteolysis of polyprotein occurred with a half time of about 20 to 30 minute after the chase (Restrepo-Hartwig and Carrington, 1992).

VIII. REPLICATION PROMOTERS, ENHANCERS AND REPRESSORS

The terms replication 'promoter' (that promotes or initiates replication) and 'enhancer' (that enhances or increases replication) are often employed in the literal sense. Therefore, replication promoters and replication enhancers are similar in their ability to bind RdRp but are different in their abilities to support initiation of RNA synthesis. Replication promoters are present in several plant viruses like CNV (Panavas *et al.*, 2002a, 2002b, 2003) and other plant viruses mentioned in the text. The secondary structure of stem-loop structure (called SL1-III or hairpin) of replication promoter region III are important for recognition and/or binding by tombusvirus (TBSV) RdRp. The box B consensus sequence element in 5`-UTRs of BMV RNAs 1 and 2, and also in cucumovirus RNAs, is GGUUCAANNCC where N is any possible nucleotide (Chen *et al.*, 2001). These stem-loop structures have loop sequences that resemble box B elements homologous to TΨC stem-loop of cellular tRNAs and act as promoters of negative-strand synthesis (Schwartz *et al.*, 2002; Chen *et al.*, 2001). Negative-strand RNA synthesis from BMV RNA3 templates requires 1a and 2a replication proteins and is driven by a promoter in the 3` TLS of RNA3 (Ishikawa *et al.*, 1997; Chapman and Kao, 1999). In turn, negative-stand RNA3 acts as a template for positive-strand RNA3 synthesis and sgRNA transcription and is driven by promoters adjacent to their respective initiation sites (Sivakumaran *et al.*, 1999; Sivakumaran and Kao, 1999).

BMV RNA3 acts as a template for RNA replication as well as for sgRNA transcription, which is initiated internally on negative-strand RNA3 templates. Both these processes (positive-strand RNA3 synthesis/replication and sgRNA transcription from RNA3) depend upon common viral replication factors and RNA templates. Then how are they coordinated? It is noteworthy that GTP is the priming nucleotide for synthesis of negative-strand RNA3, positive-strand RNA3, and sgRNA. Synthesis of all these three types of RNAs requires stem-loop RNA secondary structures for recognition of corresponding promoters (Haasnoot *et al.*, 2002; Kao, 2002). The promoters for replication and sgRNA transcription have different nucleotide sequences (Kao, 2002; Ranjith-Kumar *et al.*, 2003b), which suggest that recognition of these varied promoters may involve distinct domains of viral and/or host factors. An alternative or additional theory is the 'induced fit' mechanism, which anticipates that viral replicase may adjust to different promoters (Stawicki and Kao, 1999; Williamson, 2000).

Replication enhancers generally occur in viral RNA minus-strands, may not be proximal to the core promoter, contain sequence and/or structural features of core promoters, and can promote transcription in the absence of sequences resembling the transcription site (Nagy *et al.*, 1999; Panavas and Nagy, 2003; Ray and White, 2003). Replication enhancers have been detected in several plant viruses like AMV (van Rossum *et al.*, 1997), TCV (Nagy *et al.*, 1999), TBSV (Ray and White, 2003; Panavas and Nagy, 2003), and must be existing but not yet detected in many other plant viruses. Ray and White (2003) detected *in vivo* a replication enhancer in TBSV and found that the 5`-proximal segment of region III is a modular RNA replication element that functions mainly by forming an RNA hairpin structure (a stem-loop structure designated SL1-III) in negative strand. The two stem-loop structures [SL1-III(-) and SL2-III(-)] in region III play interchangeable roles in enhancement of tombusvirus RNA synthesis. No additive stimulatory effect comes from combination of the two stem-loops. Region SL1-III contains a replication enhancer that functions as a strong enhancer in negative-stranded RNA (that is, acts as a strong replication enhancer for synthesis of positive-strand *in vitro*) and a weak enhancer in positive-strands (for formation of negative strands of RNA) in tombusviruses (Panavas and Nagy, 2003). Panavas and Nagy (2003) propose that the putative binding of RdRp to replication enhancer of TBSV may facilitate correct positioning of RdRp over 3`-end of template including CCU initiation site. In the absence of proper initiation site, replication enhancer cannot support RNA synthesis.

TCV and its related carmoviruses and its associated satellite C RNA contain replication enhancer elements (Nagy *et al.*, 1999). The motif 1 replication enhancer RNA element of TCV is believed to function in negative strand synthesis by recruiting viral RdRp (recruiting role), thereby facilitating its initiation at positive strand promoter – similar to the possible TBSV SL1-III(-) function in an analogous capacity. A recruiting role in negative strand would require subsequent delivery of RdRp to positive strand promoter at 3`-end of template. Such long-distance interactions could be mediated by either a protein bridge or RNA-RNA interactions; the latter have been demonstrated for activation of subgenomic RNA transcription in TBSV (Zhang *et al.*, 1999; Choi and White, 2002). In contrast to TBSV, replication enhancer in genomic TCV RNA shows an additive effect on RNA synthesis. A single-stranded region between SL1-III(-) and SL2-III(-) hairpins is required for full enhancer activity.

Replication enhancers have also been reported in several multipartite plant viruses: γRNA-encoded γβ protein of BSMV RNA; CPMV M RNA-encoded 58-kDa protein; BNYVV-encoded protein p14; and PCV RNA1-encoded p15. It is significant that BSMV γB, BNYVV p14, and PCV p15 proteins belong to a group of cysteine-rich proteins, and BNYVV p14 shares statistically significant similarity with other nucleic acid-binding proteins. These cysteine-rich proteins enhance or influence replication of all genomic components of their respective viruses. In contrast, CPMV 58-kDa protein is a template-selective replication enhancer and is needed for replication of M RNA but not for B RNA; so it is a *cis* replication enhancer. RCNMV genomic RNA2 is a transcriptional enhancer and functions in *trans* for synthesis of RCNMV RNA1 subgenomic RNA (Sit *et al.*, 1998).

Carla-, furo-, hordei-, and tobraviruses contain 3`-proximal genes for small cysteine-rich proteins, some of which have RNA binding activity. One of the functions of these proteins is to regulate synthesis of capsid protein, which in these viruses is encoded by 5`-proximal genes. These cysteine-rich proteins are not absolute requirements for RNA replication but do affect replication in some cases.

Replication/transcriptional repressors (also called transcriptional silencers) for negative-strand RNA synthesis have been detected in members of family *Tombusviridae* in the last couple of years. They are located on plus-strands just upstream from the core promoter. The presence of transcriptional (replicational) enhancers and repressors on opposite strands (on minus- and plus-RNA strands, respectively) suggests that these elements regulate asymmetric levels of plus- and minus-strand RNA synthesis (Pogany *et al.*, 2003).

A repressor performs negative regulation of minus-strand synthesis and has been identified in TCV and its satellite RNA satC (Zhang *et al.*, 2004a, 2004b; Zhang and Simon, 2005), TBSV (Pogany *et al.*, 2003) and predicted for BYDV (Koev *et al.*, 2002). Zhang *et al.* (2004a, 2004b) and Zhang and Simon (2005) worked extensively on the repressor of TCV and satC. Members of genus *Carmovirus*, family *Tombusviridae*, contain a structurally conserved 3`-proximal hairpin (H5) with a large 14-base internal symmetrical loop (LSL). The H5 is proximal to the core promoter and functions as a repressor of minus-strand synthesis *in vitro* through an interaction between LSL and 3` terminal bases in TCV satellite RNA satC (which has partial sequence similarity with its helper virus *Turnip crinkle virus*). Of the 14-base satC H5 LSL, specific sequences in the middle and upper regions on both sides of the LSL are necessary for robust TCV as well as of satC accumulation in plants and protoplasts. LSL and lower stem of H5 were of greater importance for satC accumulation (i.e., multiplication) than the upper stem (Zhang and Simon, 2005).

Repression of minus-strand synthesis by H5 was due to sequestration of the 3`-end from RdRp through interactions by base pairing between the 3`-terminus and LSL; in fact between four of the seven bases on the 3` side of the LSL (5`GGGC) and the satC 3` terminal bases (GCCC-OH). A second sequence, located in a single-stranded region upstream of H5, is predicted to function as a derepressor by disrupting the 3` end-H5 interaction (Zhang *et al.*, 2004b). Possibly, the LSL might be involved in other processes (like satC fitness and permitting robust replication) in addition to repression of minus-strand synthesis. The minus-strand repressor in TBSV (SL3) contains an internal loop with both similarities and striking differences with carmovirus H5. In contrast to the symmetrical or nearly symmetrical H5 large internal loops of all carmoviruses, the TBSV SL3 is asymmetrical, with only a single adenylate occupying the 5`-side (Pogany *et al.*, 2003). However, the 3`-side of SL3 and H5 internal loops are similar, with five of eight SL3 bases (GGGCU) identical to their carmoviral counterpart.

IX. TEMPLATE SELECTION BY COGNATE VIRAL REPLICASES

The mechanism of template selection by cognate replicases for viral RNA replication in plus-strand RNA viruses is poorly understood, so is the contribution of viral and cellular proteins to RNA template recognition. Two types of experimental evidence have been published in this connection. The first type shows that, in infected cells, viral replicases specifically recognize and selectively replicate their cognate viral RNAs from a heterogeneous pool of cellular RNAs. Thus, virus-encoded proteins facilitate selective template recruitment to viral replicase complex *in vivo* as is true of the 1a protein of BMV and the 126-kDa protein of *Tomato mosaic virus* (Sullivan and Ahlquist, 1999; Chen *et al.*, 2001; Osman and Buck, 2003). In contrast, many purified viral replicase complexes effectively utilize heterologous promoter or initiation elements during *in vitro* studies (Yoshinari *et al.*, 2000; Kao, *et al.*, 2001; Rajendran *et al.*, 2002). These conflicting *in vivo* and *in vitro* results were explained on the basis of the suggestion that host factors are involved in selective recognition of viral templates by the relevant RdRp (Buck, 1996). Even in the two above-mentioned cases of specific template selection by viral RdRps, it is uncertain whether the respective viral RNAs are recognised directly by these replicase proteins or some assistance is provided by the host proteins (Diez *et al.*, 2000).

This issue appears to be settled at least in TBSV by Pogany *et al.* (2005), on which this section is mainly based. The TBSV p33 and p92 replication proteins are essential for RNA replication; are part of viral replicase complex; accumulate *in vivo* in 20:1 ratio, respectively (Scholthof, K. B.-G *et al.*, 1995; Panaviene *et al.*, 2004); and less plentiful p92 functions as viral RdRp, while role of the more abundant p33 was not known (White and Nagy, 2004) although it was regarded to play some unknown essential but auxiliary role. Pogany *et al.* (2005) demonstrated that recombinant p33 replicase protein binds specifically *in vitro* to a conserved internal replication element (IRE) located within the p92 RdRp coding region of viral genome; specific recognition and binding of p33 to IRE RNA element *in vitro* depended on the presence of a C . C mismatch within a conserved RNA helix; binding of p33 to RNA depended on the presence of the conserved RII(+)-SL hairpin, which serves as an internal recognition element; and strong correlation existed between p33:IRE complex formation *in vitro* and viral replication *in vivo*, so that mutations in IRE that disrupted selective p33 binding *in vitro* also abolished TBSV RNA replication both in plant and in *Saccharomyces cerevisiae* cells. They proposed that p33:IRE interaction provides a mechanism to selectively recruit viral RNAs into cognate viral replicase complexes, specifically binds the cognate viral RNA template *in vitro*, directs viral template recruitment into replication mode; that the major RNA element recognized by p33 is a C_{99} . C_{143} mismatch in an internal loop within the RII(+)-SL; and proposed that one of the functions of the C_{99} . C_{143} mismatch is to open up the central helix in RII(+)-SL to facilitate binding or positioning of the arginine-proline-arginine-motif (RPR-motif) of p33.

Pogany *et al.* (2005) proposed a model, on the bases of their *in vitro* and *in vivo* studies and the results of Monkewich *et al.* (2005), for the central role for the RII(+)-SL:p33 interaction in *Tombusviridae* RNA replication. They proposed that the RII(+)-SL, p33, and possibly p92 act together at an early step in replication to facilitate template selection and co-recruitment of replication factors and proposed a multistep model for explaining the occurrence of this perceived process. The p33 and p92 proteins are translated from the genome soon after infection; the RII(+)-SL is possibly formed in the genome transiently (Monkewich *et al.*, 2005); p33 interacts productively with RII(+)-SL when a required p33 threshold concentration is reached; and recognition of the C · C mismatch is critical for specific binding of p33 to the RNA template; is dependent on protein dimerization or oligomerization (which may also involve p92); and requirement of oligomerization suggests that the interface of the protein subunits could form an RNA-binding pocket that in turn specifically recognizes the C · C mismatch. The RII(+)-SL:p33 complex formed is then transported to membranes via the membrane targeting signals present in the N terminus of p33 and p92 (Rubino and Russo, 1998). Recruitment to membranes could also down-regulate the translation process [as in *Brome mosaic virus* (Janda and Ahlquist, 1998)]. In contrast, subgenomic mRNAs transcribed during infection would not be recruited to replication complexes since they lack RII(+)-SL, and so would remain dedicated to translation. In this model, the selective binding of p33 (and/or p92) to RII(+)-SL-containing RNAs is proposed to be the primary factor *in vivo* for the observed specificity of *Tombusvirus* RNA replication. In short, the essential role of the p33:p33/p92 interaction domain in selective RNA binding suggests that intermolecular interaction between two or more p33s (and/or possibly p92) proteins results in the formation of an RNA-binding pocket that has high specificity for the C · C mismatch within RII(+)-SL.

All genera in *Tombusviridae* encode comparable replicase proteins and all the sequenced members of the genus *Tombusvirus* are predicted to form RII(+)-SL-like structures with the C · C mismatch. Thus, the above mechanism and conclusions may be relevant to other members of *Tombusviridae* so that perhaps the ability of p33 replicase protein to specifically recognize a C · C mismatch is conserved in all tombusviruses (Pogany *et al.*, 2005). Other viral proteins with arginine-rich RNA-binding domains also use somewhat similar mechanisms to recognize cognate RNAs. It may be mentioned that RII(+)-SL is not the only essential RNA element for tombusvirus RNA replication; other elements also required for tombusvirus RNA replication are: a replication silencer element (Pogany *et al.*, 2003) and the minus-strand initiation promoter (Panavas *et al.*, 2002a; Fabian *et al.*, 2003).

X. CAPSID PROTEIN AND VIRAL RNA REPLICATION

Viral capsid protein does not have any role in replication of majority of plant viruses (including alpha-, bromo-, and tombusviruses) but has a definite role during the replication of AMV, ilarviruses and in some other plant viruses. Probably, capsid protein of AMV and ilarviruses confers a competitive advantage to viral RNAs over polyadenylated cellular mRNAs. The role of capsid protein in AMV is the best investigated and reviewed (Jaspars, 1999; Bol, 1999, 2003).

The capsid protein is associated with RNAs of AMV and ilarviruses and is invariably present in purified RdRp preparations from AMV-infected plants. The capsid protein peptides and the 3`-terminal nucleotides of AMV RNAs, which interact with each other, are the N-terminal amino acid numbers 25, 26 or 38 of the capsid protein and the 39 nucleotides that constitute the minimal coat protein-binding site at the 3`-end of viral RNAs (Baer *et al.,* 1994; Houser-Scott *et al.,* 1994a, 1994b). The 3`-terminal 145 nucleotides of the AMV RNAs (genomic RNAs 1, 2, and 3, and subgenomic RNA4) are homologous and can adopt two mutually exclusive and alternative conformations: TLS or a linear array of hairpins (Olsthoorn *et al.,* 1999; Bol, 1999, 2003). This is a unique property of 3`-UTRs of AMV and ilarviruses.

The linear array of hairpins consists of a series of hairpins A to E with flanking AUGC motifs 1 to 4, which possess high-affinity capsid protein binding site. Conceivably, AUGC repeats are sequence-specific determinants for capsid protein/RNA interactions so that AUGC motifs in viral RNAs are involved in capsid protein binding. Hairpin D contains a sequence UCCU in its loop. This sequence can potentially base pair with sequence AGGA located in loop and stem of hairpin A to form a pseudoknot that creates a structure which strongly resembles TLS of bromoviruses and cucumoviruses. This AMV TLS is formed by a pseudoknot interaction between nucleotides 5 to 8 and 90 to 93 from 3`-terminus so that genomic RNAs, under physiological conditions, are postulated to be mainly in pseudoknotted configuration (Olsthroon *et al.,* 1999). The TLS cannot be aminoacylated but is required for and acts as promoter for minus-strand RNA synthesis. Thus, the 3`-end of RNAs of AMV and ilarviruses has a dual role – formation of minus-strand RNA and binding of capsid protein.

The 3`-UTR of AMV RNA3 bears at least two high-affinity capsid protein-binding sites. Site 1 is situated in 3`-end 39 nucleotides and is made up of hairpins A and B and AUGC motifs 1, 2, and 3. The AMV capsid protein amino acids required for binding to site 1 are the N-terminal sequence of 25 amino acids; arginine at position 17 is particularly critical for binding. capsid protein-binding site 2 exists outside the homologous region of 145 nucleotides and is constituted by hairpins F and G and AUGC motifs 4 and 5. However, relevance of capsid protein binding site 2 in RNA3 is not clear. Major capsid protein-binding sites also exist in internal positions of the three AMV RNAs and they can also form stable stem-loop structures. Thus, nucleotides U844, C846, and A877 are important for capsid protein binding (Ansel-McKinney and Gehrke, 1998; Rocheleau *et al.,* 2004). These sites can be involved in virus assembly. Different CP domains are possibly involved in binding of capsid protein to internal and 3`-terminal sites of AMV RNAs (Neeleman *et al.,* 2004).

A. Functions of Capsid Protein

The capsid protein is involved in several functions in plant viruses - encapsidation of virus RNA, cell-to-cell virus movement, long-distance virus movement, virus transmission by vectors, and in still other functions. Different amino acids of capsid protein are involved in these multifarious functions. This ensures that the CP motifs involved in these functions are independent of other sequences so that any one of these functions can be mutated without influencing other functions. However, RNA-associated CP found in AMV and ilarviruses is an exceptional case because of its intimate involvement in virus RNA replication. Besides, CP is involved in some replication step of TMV; it regulates formation of replication complexes (Asurmendi *et al.*, 2004). Moreover, genome-associated CP may also prevent collision between translating ribosomes and replicase molecules synthesizing negative-strands.

1. *Alfalfa mosaic alfamovirus*

CP taking part in AMV life cycle can be categorized into two types - CP that is originally present in viral inoculum since it is associated with viral RNAs; and CP that is synthesized from viral subgenomic RNA4 and appears later in viral life cycle. Possibly these CPs have different functions during viral life cycle so that many functions have been proposed for CP (Bol, 1999, 2003; Jaspars, 1999).

a. Functions of Capsid Protein Bound to Inoculum Viral RNAs

i. Genome Activation / Role in Translation of Viral RNAs

The CP, originally attached to viral RNAs in inoculum, induces genome activation for initiating AMV infection and results in synthesis of genomic and subgenomic RNAs followed by translation of these RNAs (Jaspars, 1999; Bol, 1999, 2003; Choi *et al.*, 2003). The role of CP in translation of inoculum RNAs is supported by recent data and consists of following sequence of events according to Bol (2003).

First step: After uncoating of virus particles of inoculum, a few molecules of CP remain bound to the high-affinity binding site 1 at 3`-end of viral RNAs. Binding of CP to site 1 is essential for efficient translation of AMV RNAs *in vivo*.

Capsid protein bound to termini of AMV RNAs enhances translation of viral RNAs by acting as a functional analogue of poly(A)-bound protein (PABP), by promoting recruitment of 40S ribosomal subunits and/or by enhancing stability of viral RNAs and by interacting with host translation initiation factors (eIF4G and other factors) bound to 5`-cap structure and stimulates translation of RNAs 1 and 2 leading to formation of replicase proteins p1 and p2 (Neeleman *et al.*, 2001; Bol, 2003). Krab *et al.* (2005) found that AMV CP interacts specifically with eIF4G and eIFiso4G subunits from wheat eIF4F and eIFiso4G, respectively, so that their results support the hypothesis that the role of CP in translation of viral RNAs mimics the role of PABP in translation of cellular mRNAs and converts viral RNAs into closed-loop structures. Simultaneously, CP could stop formation of RNA structure that acts as negative-strand promoter for preventing collision between translating ribosomes and replicase molecules. By analogy to BMV, AMV P1 could bind to ICR2 motifs present in AMV

RNAs and are essential for replication of AMV. Efficient translation of AMV RNA4 in plant cells is dependent on the ability of CP to bind to 3`-end. In fact, both CP gene and 3`-UTR are necessary for *in vivo* translation of AMV RNA4 (Neeleman *et al.*, 2001).

Third step: Protein P1 is proposed to recruit viral RNAs from translation machinery and also targets a complex of P1, P2 and viral RNAs to membrane structures where replication complexes are formed – similar to the targeting mechanism employed by BMV (den Boon *et al.*, 2001). The BMV replication complexes are located within endoplasmic reticulum-derived vesicles while AMV replication complexes are located within vesicles derived from tonoplast.

In vesicles, CP possibly dissociates from 3`-terminus of inoculum RNAs to permit the formation of TLS and of negative-strand RNA promoter and binding to hairpin E in negative-strand RNA promoter. Presence of TLS (more specifically, formation of pseudoknot structure) at 3`-end permits initiation of AMV negative-strand RNA synthesis, and is essential for viral replication *in vivo* and *in vitro*. Thus, CP has no role in formation of AMV negative-strand RNA.

Fourth step: The AMV progeny RNA positive strands are synthesized. The sequences, in 5` -UTR of AMV RNAs, which are required for RNA replication *in vivo* or for synthesis of positive-strand RNA *in vitro*, have been identified. The natural negative-strand RNAs, acting as templates for synthesis of positive-strand RNA molecules *in vivo*, contain a single non-coded G residue at 3`-end (Houwing *et al.*, 2001). The strong stimulation to synthesis of positive-strand RNA molecules by CP could be more due to protection of viral RNA from degradation than to the actual stimulation of RNA synthesis (Vlot *et al.*, 2001). Thus, at present no proof exists about the role of CP in synthesis of positive-strand viral RNAs.

Fifth step: The final step is the assembly of complete virus particles. The 3` -UTRs of AMV RNAs contain high-affinity CP-binding sites but are not required for encapsidation. Thus, the origins of assembly in AMV RNAs still need to be identified and CP bound to 3`-termini of AMV RNAs plays no role in assembly of virus particles (Vlot *et al.*, 2001). The CP also binds to several internal sites in AMV RNAs and these sites may have some as yet unknown role in virus assembly.

Whole of the above process can be briefly summarised as below. The CP binds to each of three genomic RNAs in inoculum; the RNAs are then targeted to membrane-bound replication complexes where each RNA is transcribed into the corresponding negative strand. Capsid protein is not involved in this transcription process but, after complete synthesis of the negative strand, CP associates with replicase to make the enzyme complex competent for synthesis of the progeny positive-strand RNAs. The CP bound to AMV and ilarvirus RNAs most likely gives them some competitive advantage over the cellular mRNAs since it behaves like PABP, interacts with host translation factors bound to 5`-cap structure of viral genome, and stimulates translation of viral RNAs. The AUGC motif is suggested to be part of a translational determinant involved in genome activation process.

ii. Other Functions

CP, originally attached to AMV viral RNAs in inoculum, protects 3`-end of viral positive-strand inoculum (mRNAs) from exonucleolytic degradation (Neeleman and

Bol, 1999; Vlot *et al.*, 2001). The bound CP molecules do not allow the loss of 3`-terminal nucleotides of positive-strand viral RNAs during their repeated translation cycles. This is similar to the function of the host proteins, bound to TLS or poly(A) tails located at 3`-termini of RNA genome of other viruses, for protecting the 3`-end against exonucleolytic degradation during translation process of viral RNAs.

CP is proposed to target genomic RNAs to membrane-bound replication complexes (Neeleman *et al.*, 2001; and Bol, 2003). AMV CP may stabilize the replication complex, consisting of viral RNAs/viral replication proteins P1 and P2 and the host translation initiation factors, and also promote the recruitment of 40S ribosome subunits by forming a closed-loop configuration. Dimer formation is possibly essential for the putative interaction of CP with translation initiation factors (Neeleman *et al.*, 2004).

AMV CP may promote circularization of viral RNAs analogous to the function of PABP in circularization of cellular RNAs. A protein-protein bridge between initiation factors (bound to cap structure) and CP (bound to 3`-end) of AMV RNAs is proposed to convert these RNAs into a closed-loop structure that is essential for translation of RNAs (Neeleman *et al.*, 2001). However, the 3`-terminal sequences in AMV RNAs, that could potentially base pair with 5` sequences involved in negative-strand RNA synthesis, have not yet been identified (Vlot and Bol, 2003).

AMV CP in inoculum is not involved in negative-strand RNA synthesis (de Graff *et al.*, 1995; Neeleman and Bol, 1999; Vlot *et al.*, 2001) on the infecting positive-strand viral RNAs both *in vivo* and *in vitro*; on the contrary, CP binding to 3`-end of AMV RNA disrupts a conformation of 3` UTR that is required for minus-strand RNA promoter activity *in vitro* leading to inhibition of negative-strand RNA production due to interference with pseudoknot formation (Olsthoorn *et al.*, 1999; Neeleman *et al.*, 2001). This indicates that AMV CP is required in a step prior to viral negative-strand synthesis, possibly during translation of inoculum RNAs (Neeleman and Bol, 1999; Olsthoorn *et al.*, 1999).

The CP, derived from parental virions or obtained from translation of viral RNA4 in inoculum, could possibly have a dual role – in promoting translation of viral RNAs and in preventing premature initiation of minus-strand RNA synthesis by newly synthesized replication proteins. Later, in infection cycle, the *de novo* synthesized CP will result in increased CP levels that may in turn cause encapsidation of viral RNAs into the progeny virus particles (Neeleman *et al.*, 2001).

Presence of AMV CP in inoculum causes an approximately 100-fold increase of positive-strand RNA accumulation in infected protoplasts (var der Kuyl *et al.*, 1991). This strong stimulation to synthesis of positive-strand RNA molecules by CP could be more due to protection of viral RNA from degradation than to the actual stimulation of RNA synthesis (Vlot *et al.*, 2001). Thus, at present no proof exists about the role of CP in synthesis of positive-strand viral RNAs.

The 3`UTR and CP bound to 3`-termini of AMV RNAs is not required for encapsidation (Vlot *et al.*, 2001). But CP also binds to several internal sites in AMV RNAs and these sites may act as the origin of assembly of virus particles. This is supported by the fact that a coding sequence of BMV RNA 1 is required for encapsidation whereas 3`-UTRs of RNAs are not needed (Duggal and Hall, 1993).

AUGC$_{865-868}$ sequences are conserved among RNAs of AMV and ilarviruses. Rocheleau *et al.* (2004) suggest that CP binding to AUGC sequences determines the orientation of the 3` hairpin relative to one another, while local structural features within these hairpins are also critical determinants of their functional activity; that activation of AMV replication is dependent on both the initial fold of the CP-free RNA and the final RNA fold established through viral CP binding so that the mechanism of AMV genome activation centers on CP-mediated RNA conformational changes that organize 3` terminus of RNA for replication function.

b. Functions of Capsid Protein Expressed from Viral RNAs 3 and 4

All functions (like genome encapsidation, virus transmission, etc.), normally performed by CP of a virus, are also performed by AMV CP. In addition, this CP performs several other functions. AMV CP is required for viral RNA positive-strand synthesis *in vivo* as well as *in vitro*; is proposed to release viral positive-strand RNAs from the membrane-bound replication complexes that contain the negative-strand template (Houwing *et al.*, 1998); is implicated during synthesis of positive-strands of RNAs 3 and 4; enhances translation of viral mRNAs like binding of CP to 3`-UTRs of RNAs 3 and 4 strongly enhanced translation of these RNAs (Neeleman *et al.*, 2001); and binding of newly synthesized CP to 3`-end of plus-strand RNAs most likely shuts off minus-strand synthesis triggering a switch to asymmetric synthesis of plus-strand viral RNAs (Olsthoorn *et al.*, 1999; Neeleman *et al.*, 2001). The CP acts in *trans* for synthesis of positive-strand RNAs 1 and 2 but acts in *cis* for synthesis of the positive-strand RNA3.

c. Switch

Olsthroon *et al.* (1999) postulated that presence of TLS or a linear array of hairpins, the mutually exclusive and alternative conformations, at 3`-termini of RNAs of AMV and ilarviruses may be the switch that enables viral RNA to shift from translation to replication mode and vice versa. The linear array of stem-loop structures contains several high-affinity CP-binding sites while tRNA-like conformation is specifically recognised by RdRp. It is proposed that 3`-UTR acts as a molecular switch that regulates the transition from translation to replication of parental RNAs. In this model, CP bound to inoculum RNAs would force the 3`-UTR into the CP-binding conformation to enhance translation and/or to prevent premature initiation of negative-strand RNA synthesis. Subsequently, CP dissociates from parental RNA to allow TLS formation and initiation of negative-strand RNA synthesis.

Later, targeting parental AMV RNAs to chloroplasts membranes (where replication complexes are assembled) could result in dissociation of 3` terminally attached CP molecules. This collapses the pseudoknot with consequent shut-off of minus-strand RNA synthesis and start of progeny positive-strand RNA synthesis. This is the trigger that starts asymmetric synthesis of positive-strand viral RNAs. In this way, CP has a regulatory function during synthesis of progeny positive-strand RNAs.

In the later step of replication, *de novo* synthesized CP could shut-off negative-strand synthesis by binding to 3`-UTR of progeny RNAs.

2. Other Plant Viruses

Recently, CP has also been found to have important roles in replication of other plant viruses. Asurmendi *et al.* (2004) found that TMV CP enhances production of movement protein; enhances formation of virus replication complexes and so has a role in their formation. It is well known that TMV CP is not needed for TMV RNA replication. But Asurmendi *et al.* (2004) consider it likely that TMV CP regulates production of subgenomic RNAs that encode movement protein [and perhaps CP (Bendahmane *et al.*, 2002)] or translation of mRNAs; they suggest a regulatory role of CP in establishing replication complexes and in regulating transcription. Bendahmane *et al.* (2002) proposed a positive regulatory effect on production of movement protein by wild-type CP. Interaction between viral RNA and CP is also considered a regulating element of TCV RNA life cycle.

XI. REFERENCES

Adkins, S., Stawicki, S. S., Faurote, G., Siegel, R. W., and Kao, C. C. 1998. Mechanistic analysis of RNA synthesis by RNA-dependent RNA polymerase from two promoters reveals similarities to DNA-dependent RNA polymerase. *RNA* 4: 455-470.

Ahlquist, P., Noueiry, A. O., Lee, W.-M., Kushner, D. B., and Dye, B. T. 2003. Host factors in positive-strand RNA virus genome replication. *J. Virol.* 77: 8181-8186.

Ansel-McKinney, P., and Gehrke, L. 1998. RNA determinants of a specific RNA-coat protein peptide interaction in *Alfalfa mosaic virus:* Conservation of homologous features in ilarvirus RNAs. *J. Mol. Biol.* 278: 767-785.

Asurmendi, S., Berg, R. H., Koo, J. C., and Beachy, R. N. 2004. Coat protein regulates formation of replication complexes during TMV infection. *Proc. Nat. Acad. Sci. USA* 101: 1415-1420.

Baer, M. L., Houser, F., Loesch-Fries, L. S., and Gehrke, L. 1994. Specific RNA binding by amino-terminal peptides of alfalfa mosaic virus coat protein. *EMBO J.* 13: 727-735.

Bendahmane, A., Szécsi, J., Chen, I., Berg, R. H., and Beachy, R. N. 2002. Characterization of mutant tobacco mosaic virus coat protein that interferes with virus cell-to-cell movement. *Proc. Nat. Acad. Sci. USA* 99: 3645-3650.

Boguszewska-Chachulska, A. M., and Haenni, A.-L. 2005. RNA viruses redirect host factors to better amplify their genomes. *Adv. Virus Res.* In Press

Bol, J. F. 1999. *Alfalfa mosaic virus* and ilarviruses: Involvement of coat protein in multiple steps of the replication cycle. *J. Gen. Virol.* 80: 1089-1102.

Bol, J. F. 2003. *Alfalfa mosaic virus*: Coat protein-dependent initiation of infection. *Mol. Plant Pathol.* 4: 1-8.

Boon, J. A., Chen, J., and Ahlquist, P. 2001. Identification in brome mosaic virus replicase protein 1a that mediates association with endoplasmic reticulum membrane. *J. Virol.* 75: 12370-12381.

Brigneti, G., Voinnet, O., Li, W.-X., Ji, L.-H., Ding, S.- W., and Baulcombe, D. C. 1998. Viral pathogenicity determinants are suppressors of transgene silencing in *Nicotiana benthamiana. EMBO J.* 17: 6739-6746.

Buck, K. W. 1996. Comparison of the replication of positive-stranded RNA viruses of plants and animals. *Adv. Virus Res.* 47: 159-251.

Butcher, S. J., Grimes, J. M., Makeyev, E. V., Bamford, D. H., and Stuart, D. I. 2001. A mechanism for initiating RNA-dependent RNA polymerization. *Nature* 410: 235-240.

Carpenter, C. D., and Simon, A. E. 1998. Analysis of sequences and putative structures required for viral satellite RNA accumulation by *in vivo* genetic selection. *Nucl. Acids Res.* 26: 2426-2432.

Carpenter, C. D., Oh, J. W., Zhang, C., and Simon, A.E. 1995. Involvement of a stem-loop structure in the location of junction sites in viral RNA recombination. *J. Mol. Biol.* 245: 608-622.

Chandrika, R., Ranbidran, S., Lewandowski, D. J., Manjunath, K. L., and Dawson, W. O. 2000. Full tobacco mosaic virus RNAs and defective RNAs have different 3` replication signals. *Virology* 273: 198-203.

Chapman, M. R., and Kao, C. C. 1999. A minimal promoter for minus-strand RNA synthesis by the brome mosaic virus polymerase complex. *J. Mol. Biol.* 286: 709-720.

Chen, J., and Ahlquist, P. 2000. Brome mosaic virus polymerase-like 2a is directed to the endoplasmic reticulum by helicase-like viral protein 1a. *J. Virol.* 74: 4310-4318.

Chen, J., Noueiry, A., and Ahlquist, P. 2001. Brome mosaic virus protein 1a recruits viral RNA2 to RNA replication through a 5` proximal RNA2 signal. *J. Virol.* 75: 3207-3219. .

Cheng, J.-H., Peng, C.-W., Hsu, Y.-H., and Tsai, C.-H. 2002. The synthesis of minus-strand RNA of *Bamboo mosaic potexvirus* initiates from multiple sites within the poly(A) tail. *J. Virol.* 76: 6114-6120.

Choi, I. R., and White, K. A. 2002. An RNA activator of subgenomic mRNA1 transcription in *Tomato bushy stunt virus. J. Biol. Chem.* 277: 3760-3766.

Choi, J., Kim, B. S., Zhao, X., and Loesch-Fries, S. 2003. The importance of alfalfa mosaic virus coat protein dimers in the initiation of replication. *Virology* 305: 44-49.

Chu, M., Desvoyes. B., Turina, M., Noad, R., and Scholthof, H. B. 2000. Genetic dissection of tomato bushy stunt virus p19 protein-mediated host-dependent symptom induction and systemic invasion. *Virology* 266: 79-87.

Dardick, C. D., and Culver, J. N. 1999. Gene-for-gene interactions: In. Mandahar, C. L (Ed.). Molecular Biology of Plant Viruses. Kluwer Academic Publishers, Boston / Dordrecht / London. p. 211-224.

de Graaff, M., Thorburn, C., and Jaspars, E. M. J. 1995. Interaction between RNA-dependent RNA polymerase of *Alfalfa mosaic virus* and its template: Oxidation of vicinal hydroxyl groups blocks *in vitro* RNA synthesis. *Virology* 213: 650-654.

Deiman, B. A. L. M., Kortlever, R. M., and Pleij, C. W. A. 1997. The role of the pseudoknot at the 3` end of turnip yellow mosaic virus RNA in minus strand synthesis by the viral RNA-dependent RNA polymerase. *J. Virol.* 71: 5990-5996.

den Boon, J. A., Chen, J., and Ahlquist, P. 2001. Identification of sequences in brome mosaic virus replicase protein 1a that mediate association with endoplasmic reticulum membranes. *J. Virol.* 75: 12370-12381.

Diez, J., Ishikawa, M., Kaido, M., and Ahlquist, P. 2000. Identification and characterization of a host protein required for efficient template selection in viral RNA replication. *Proc. Nat. Acad. Sci. USA* 97: 3913-3918.

Dolja, V. V., McBride, H. J., and Carrington, J. C. 1992. Tagging of plant potyvirus replication and movement by insertion of β-glucuronidase into the viral polyprotein. *Proc. Natl. Acad. Sci. USA* 89: 10208-10212.

Dreher, T. W. 1999. Functions of 3`-untranslated regions of positive stranded RNA viral genomes. *Annu. Rev. Phytopath.* 37: 151-174.

Duggal, R., and Hall, T. C. 1993. Identification of domains in brome mosaic virus RNA1 and coat protein necessary for specific interaction and encapsidation. *J. Virol.* 67: 6406-6412

Dunoyer, P., Ritzenthaler, C., Hemmer, O., Michler, P., and Fritsch, C. 2002. Intracellular localization of the peanut clump virus replication complex in tobacco BY-2 protoplasts containing green fluorescent protein-labeled endoplasmic reticulum on Golgi apparatus. *J. Virol.* 76: 865-874.

Fabian, M. R., Na, H., Ray, D., and White, K. A. 2003. 3`-Terminal RNA secondary structures are important for accumulation of tomato bushy stunt virus DI RNAs. *Virology* 313: 567-580.

Felden, B., Florentz, C., Giegé, R., and Westhof, E. 1996. A central pseudoknotted three-way junction imposes tRNA-like mimicry and the orientation of three 5` upstream pseudoknots in the 3` terminus of tobacco mosaic virus RNA. *RNA* 2: 201-212.

Garnier, M., Mamoun, R., and Bové, J. M. 1980. TYMV RNA replication *in vivo*: Replicative intermediate is mainly single stranded. Virology 104: 357-374.

Goregaoker, S. P., Lewandowski, D. J., and Culver, J. N. 2001. Identification of functional analysis of an interaction between domains of the 126/183-kDa replicase-associated proteins of *Tobacco mosaic virus. Virology* 282: 320-328.

Guan, H. C., Song, C., and Simon, A. E. 1997. RNA promoters located on (-)-strands of a subviral RNA associated with *Turnip crinkle virus. RNA* 3: 1401-1412.

Guo, L., Allen, E. M., and Miller, W. A. 2001. Base-pairing between untranslated regions facilitates translation of uncapped, nonpolyadenylated viral RNA. *Mol. Cell* 7: 1103-1109.

Haasnoot, P. C., Brederode, F. T., Olsthoorn, R. C., and Bol, J. F. 2000. A conserved hairpin structure in alfamovirus and bromovirus subgenomic promoters is required for efficient RNA synthesis *in vitro*. *RNA* 6: 708-716.

Hayes, R. J., and Buck, K. W. 1990. Complete replication of a eukaryotic virus RNA *in vitro* by a purified RNA-dependent RNA polymerase. *Cell* 63: 363-368.

Hema, M., and Kao, C. C. 2004. Template sequence near the initiation nucleotide can modulate brome mosaic virus RNA accumulation in plant protoplasts. *J. Virol.* 78: 1169-1180.

Herzog, E., Hemmer, O., Hauser, S., Meyer, G., Bouzoubaa, S., and Fritsch, C. 1998. Identification of genes involved in replication and movement of *Peanut clump virus*. *Virology* 248: 312-322.

Houser-Scott, F., Baer, M. L., Liem, K. F., Jr., Cai, J.-M., and Gehrke, L. 1994a. Nucleotide sequence and structural determinants of specific binding of coat protein or coat protein peptides to the 3` -untranslated region of alfalfa mosaic virus RNA4. *J. Virol.* 68: 2194-2205.

Houser-Scott, F., Ansel-McKinney, P. A., Cai, J.-M., and Gehrke, L. 1994b. *In vitro* genetic selection analysis of alfalfa mosaic virus coat protein binding to 3`-terminal AUGC repeats. *J. Virol.* 71: 2310-2319.

Houwing, C. J., van de Putte, P., and Jaspars, E. M. J. 1998. Regulation of single strand RNA synthesis of *Alfalfa mosaic virus* in nontransgenic cowpea protoplasts by the viral coat protein. *Arch. Virol.* 143: 489-500.

Houwing, C. J., Veld, M. R. Man In'T., Zuidema, D., de Graaff, M. , and Jaspars, E. M. J. 2001. Natural minus-strand RNAs of *Alfalfa mosaic virus* as *in vitro* templates for viral RNA polymerase. 3`-Terminal non-coded guanosine and coat protein are insufficient factors for full-size plus-strand synthesis. *Arch. Virol.* 146: 571-588.

Huang, C. Y., Huang, Y. L., Meng, M., Hsu, Y. H., and Tsai, C. H. 2001. Sequences at the 3` -untranslated region of bamboo mosaic potexvirus RNA interact with the viral RNA-dependent RNA polymerase. *J. Virol.* 75: 2818-2824.

Ishikawa, M., Diez, J., Restrepo-Hartwig, M., and Ahlquist, P. 1997. Yeast mutations in multiple complementation groups inhibit brome mosaic virus RNA replication and transcription and perturb regulated expression of the viral polymerase-like gene. *Proc. Natl. Acad. Sci. USA* 94: 13810-13815.

Janda, M., and Ahlquist, P. 1998. Brome mosaic virus RNA replication protein 1a dramatically increases *in vivo* stability but not translation of viral genomic RNA3. *Proc. Natl. Acad. Sci. USA* 95: 2227-2232.

Jaspars, E. M. J. 1999. Genome activation in alfamo- and ilarviruses. *Arch. Virol.* 144: 843-863.

Kao, C. C. 2002. Lessons learned from the core RNA promoters of *Brome mosaic virus* and *Cucumber mosaic virus*. *Mol. Plant Pathol.* 3: 53-59.

Kao, C. C., Singh, P., and Ecker, D. J. 2001. *De novo* initiation of viral RNA-dependent RNA synthesis. *Virology* 287: 251-260.

Kasschau, K. D., and Carrington, J. C. 2001. Long-distance movement and replication maintenance functions correlate with silencing suppression activity of potyviral HC-Pro. *Virology* 285: 71-81.

Koev, G., Liu, S., Beckett, R., and Miller, W. A. 2002. The 3`-terminal structure required for replication of barley yellow dwarf virus RNA contains an embedded 3` -end. *Virology* 292: 114-126.

Krab, I. M., Caldwell, C., Gallie, D. R., and Bol, J. F. 2005. Coat protein enhances translational efficiency of alfalfa mosaic virus RNAs and interacts with the eIF4G component of initiation factor eIF4F. *J. Gen. Virol.* 86: 1841-1849.

Lesburg, C. A., Cable, M. B., Ferrari, E., Hong, Z., Mannarino, A. F., and Weber, P. C. 1999. Crystal structure of the RNA-dependent RNA polymerase from *Hepatitis C virus* reveals a fully encircled active site. *Nat. Struct. Biol.* 6: 937-943.

Lai, M. M. C. 1998. Cellular factors in the transcription and replication of viral RNA genomes: A parallel to DNA-dependet RNA transcription. *Virology* 244: 1-12.

Laín, S., Reichmann, J. L., and Garcia, J. A. 1990. RNA helicase: A novel activity associated with a protein encoded by a positive-strand RNA virus. *Nucl. Acids Res.* 18: 7003-7006.

Lewandowski, D. J., and Dawson, W. O. 2000. Functions of the 126- and 183-kDa proteins of *Tobacco mosaic virus*. *Virology* 271: 90-98.

Li, X.-H., Valdez, P., Olvera, R. E., and Carrington, J. C. 1997. Functions of the tobacco etch virus RNA polymerase (NIb): Subcellular transport and protein-protein interaction with VPg/proteinase (NIa). *J. Virol.* 71: 1598-1607.

Liang, X.-Z., Lucy, A. P., Ding, S.-W., and Wong, S. M. 2002. The p23 protein of *Hibiscus chlorotic ringspot virus* is indispensable for host-specific replication . *J. Virol.* 76: 12312-12319.

López, C. J., Novas-Castillo, S., Govida, P., Moreno, R., and Flores, R. 2000. The 23-kDa protein coded by the 3`-terminal gene of *Citrus tristeza virus* is an RNA-binding protein. *Virology* 269: 462-470.

Monkewich, S., Lin, H.-X., Fabian, M. R., Xu, W., Na, H., Ray, D., Chernysheva, O. A., Nagy, P. D., and White, K. A. 2005. The p92 polymerase coding region contains an internal RNA element required at an early step in tombusvirus genome replication. *J. Virol.* 79: 4848-4858.

Nagy, P. D., Pogany, J., and Simon, A. E. 1999. RNA elements required for RNA recombination function as replication enhancers *in vitro* and *in vivo* in a plus-strand RNA virus. *EMBO J.* 18: 5653-5665.

Neeleman, L., and Bol, J. F. 1999. *cis*-Acting functions of alfalfa mosaic virus proteins involved in replication and encapsidation of viral RNA. *Virology* 254: 324-333.

Neeleman, L., Olsthoorn, C. L., Linthorst, H. J. M., and Bol, J. F. 2001. Translation of a nonpolyadenylated viral RNA is enhanced by binding of viral coat protein or polyadenylation of the RNA. *Proc. Nat. Acad. Sci., USA* 98: 14286-14291.

Neeleman, L., Linthorst, H. J. M., and Bol, J. F. 2004. Efficient translation of alfamovirus RNAs requires the binding coat protein dimers to the 3` -termini of the viral RNAs. *J. Gen. Virol.* 85: 231-240.

Noueiry, A. O., and Ahlquist, P. 2003. Brome mosaic virus RNA replication: Revealing the role of the host in RNA-virus replication. *Annu. Rev. Phytopath.* 41: 77-98.

Olsthoorn, R. C. L., and Bol, J. F. 2002. Role of an essential triloop hairpin and flanking structures in the 3` -untranslated region of alfalfa mosaic virus RNA in *in vitro* transcription. *J. Virol.* 76: 8747-8756.

Olsthoorn, R. C. L., Martens, S., Brederode, F. T., and Bol, J. F. 1999. A conformational switch at the 3`-end of a plant virus RNA regulates virus replication. *EMBO J.* 18: 4856-4864.

Olsthoorn, R. C. L., Haasnoot, P. C. J., and Bol, J. F. 2004. Similarities and differences between the subgenomic and minus-strand promoters of an RNA virus. *J. Virol.* 78: 4048-4053.

Osman, T. A. M., and K. W. Buck. 1996. Complete replication *in vitro* of tobacco mosaic virus RNA by a template-dependent membrane-bound RNA polymerase. *J. Virol.* 70: 6227-6234.

Osman, T. A. M., and K. W. Buck. 1997. The tobacco mosaic virus RNA polymerase complex contains a plant protein related to RNA-binding subunit of yeast elF-3. *J. Virol.* 71: 6075-6082.

Osman, T. A. M., and K. W. Buck. 2003. Identification of a region of the tobacco mosaic virus 126- and 183-kilodalton replication proteins which binds specifically to the viral 3`-terminal tRNA-like structure. *J. Virol.* 77: 8669-8675.

Osman, T. A. M., Hemenway, C. L., and Buck, K. W. 2000. Role of the 3` tRNA-like structure in tobacco mosaic virus minus-strand RNA synthesis by the viral RNA-dependent RNA polymerase. *J. Virol.* 74: 11671-11680.

Panavas, T., and Nagy, P. D. 2003. The RNA replication enhancer element of tombusviruses contains two interchangeable hairpins that are functional during plus-strand synthesis. *J. Virol.* 77: 258-269.

Panavas, T., Pogany, J., and Nagy, P. D. 2002a. Analysis of minimal promoter sequences for plus-strand synthesis by the cucumber necrosis virus RNA-dependent RNA polymerase. *Virology* 296: 263-274.

Panavas, T., Pogany, J., and Nagy, P. D. 2002b. Internal initiation by the cucumber necrosis virus RNA-dependent RNA polymerase is facilitated by promoter-like sequences. *Virology* 296: 275-287.

Panavas, T., Panaviene, Z., Pogany, J., and Nagy, P. D. 2003. Enhancement of RNA synthesis by promoter duplication in tombusviruses. *Virology* 310: 118-129.

Panaviene, Z., Panavas, T., Serva, S., and Nagy, P. D. 2004. Purification of the cucumber necrosis virus replicase from yeast cells: Role of coexpressed viral RNA in stimulation of replicase activity. *J. Virol.* 78: 8254-8263.

Pogany, J., Fabian, M. R., White, K. A., and Nagy, P. D. 2003. A replication silencer element in a plus-strand RNA virus. *EMBO J.* 22: 5602-5611.

Pogany, J., White, K. A., and Nagy, P. D. 2005. Specific binding of tombusvirus replication protein p33 to an internal replication element in the viral RNA is essential for replication. *J. Virol.* 79: 4859-4869.

Pogue, G. P., and Hall, T. C. 1992. The requirement for a 5` stem-loop structure in brome mosaic virus replication supports a new model for viral positive-strand RNA initiation. *J. Virol.* 66: 674-684.

Qiu, W. P., and Scholthof, K.-B. G. 2000. *In vitro-* and *in vivo*-generated defective RNAs of *Satellite panicum mosaic virus* define *cis*-acting RNA elements required for replication and movement. *J. Virol.* 74: 2247-2254.

Qiu, W. P., Park, J. W., and Scholthof, H. B. 2002. Tombusvirus p19-mediated suppression of virus-induced gene silencing is controlled by genetic and dosage features that influence pathogenicity. *Mol. Plant-Microbe Interact.* 15: 269-280.

Qu, F., and Morris, T. J. 2002. Efficient infection of *Nicotiana benthamiana* by *Tomato bushy stunt virus* is facilitated by the coat protein and maintained by p19 through suppression of gene silencing. *Mol. Plant-Microbe Interact.* 15: 193-202.

Rajendran, K. S., Pogany, J., and Nagy, P. D. 2002. Comparison of turnip crinkle virus RNA-dependent RNA polymerase preparations expressed in *Escherichia coli* or derived from infected plants. *J. Virol.* 76: 1707-1717.

Ralph, R. K. 1969. Double-stranded viral RNA. *Adv. Virus Res.* 15: 61-158.

Ranjith-Kumar, C. T., Gutshall, L., Sarisky, R. T., and Kao, C. C. 2003a. Multiple interactions within the hepatitis C virus RNA polymerase repress primer-dependent RNA synthesis. *J. Mol. Biol.* 330: 675-685.

Ranjith-Kumar, C. T., Zhang, X., and Kao, C. C. 2003b. Enhancer-like activity of a brome mosaic virus RNA promoter. *J. Virol.* 77: 1830-1839.

Ray, D., and White, K. A. 2003. An internally located RNA hairpin enhances replication of tomato bushy stunt virus RNAs. *J. Virol.* 77: 245-257.

Restrepo-Hartwig, M. A., and P. Ahlquist. 1996. Brome mosaic virus helicase- and polymerase-like proteins colocalize on the endoplasmic reticulum at sites of viral RNA synthesis. *J. Virol.* 70: 8908-8916.

Restrepo-Hartwig, M. A., and Ahlquist, P. 1999. Brome mosaic virus RNA replication proteins 1a and 2a colocalize and 1a independently localizes on the yeast endoplasmic reticulum *J. Virol.* 73: 10303-10309.

Restrepo-Hartwig, M. A., and Carrington, J. C. 1992. Regulation of nuclear transport of a plant potyvirus protein by autoproteolysis. *J. Virol.* 66: 5662-5666.

Ritzenthaler, C., Laporte, C., Gaire, F., Dunoyer, P., Schmitt, C., *et al.*, 2002. Grapevine fanleaf virus replication occurs on endoplasmic-derived membrane. *J. Virol.* 76: 8808-8819.

Rocheleau, G., Petrillo, J., Guogas, L., and Gehrke, L. 2004. Degenerate *in vitro* genetic selection reveals mutations that diminish alfalfa mosaic virus RNA replication without affecting coat protein binding. *J. Virol.* 78: 8036-8046.

Rubino, L., and Russo, M. 1998. Membrane targeting sequences in tombusvirus infections. *Virology* 252: 431-437.

Sachs, A. B. 2000. Physical and functional interactions between the mRNA cap structure and poly(A) tail. In: Sonenberg, N., Hershey, J. W. B., and Mathews , M. B. (Eds.). Translational Control of Gene Expression. Cold Spring Harbor Lab. Press, Cold Spring Harbor, N. Y., USA. p. 447-465.

Satyanarayana, T., Gowda, S., Boyko, V. P., Albiach-Marti, M. R., Mawassi, M., Navas-Castillo, J., *et al.*, 1999. An engineered closterovirus RNA replicon and analysis of heterologous terminal sequences for replication. *Proc. Nat. Acad. Sci. USA* 96: 7433-7438.

Satyanarayana, T., Gowda, S., Ayllón, M. A., Albiach-Marti, M. R., and Dawson, W. O. 2002a. Mutational analysis of the replication signals in the 3`-nontranslated region of *Citrus tristeza virus*. *Virology* 300: 140-152.

Satyanarayana, T., Gowda, S., Ayllón, M. A., Albiach-Marti, M. R., Rabindran, S., and Dawson, W. O. 2002b. The p23 protein of *Citrus tristeza virus* controls asymmetrical RNA accumulation. *J. Virol.* 76: 473-483.

Scholthof, H. B., Scholthof, K.-B. G., Kikkert, M., and Jackson, A. O. 1995a. Tomato bushy stunt virus spread is regulated by two nested genes that function in   cell-to-cell movement and host-dependent systemic invasion. *Virology* 213: 425-438.

Scholthof, H. B., Scholthof, K.-B. G., and Jackson, A. O. 1995b. Identification of tomato bushy stunt virus host-specific symptom determinants by expression of individual genes from a potato virus X vector. *Plant Cell* 7: 1157-1172.

Scholthof, K.-B. G., Scholthof, H. B., and Jackson, A. O. 1995. The tomato bushy stunt virus replicase proteins are coordinately expressed and membrane-associated. *Virology* 208: 365-369.

Schwartz, M., Chen, J., Janda, M., Sullivan, M., den Boon, J., and Ahlquist, P. 2002. A positive-strand RNA virus replication complex parallels form and function of retrovirus capsids. *Mol. Cell* 9: 505-514.

Singh, R. N., and T. W. Dreher. 1997. Turnip yellow mosaic virus RNA-dependent RNA polymerase: Initiation of minus-strand synthesis *in vitro*. *Virology* 233: 430-439.

Sit, T. L., Vaewhongs, A. A., and Lommel, S. A. 1998. RNA-mediated transactivation of transcription from a viral RNA. *Science* 281: 829-832.

Sivakumaran, K., and Kao, C. C. 1999. Initiation of genomic plus-strand RNA synthesis from DNA and RNA templates by a viral RNA-dependent RNA polymerase. *J. Virol.* 73: 6415-6423.

Sivakumaran, K., Kim, C. H., Tayon, R., Jr., and Kao, C. C. 1999. RNA sequence and secondary structural determinants in minimal viral promoter that directs replicase recognition and initiation of genomic plus-strand RNA synthesis. *J. Mol. Biol.* 294: 667-682.

Sivakumaran, K., Bao, Y., Roossinck, M. J., and Kao, C. C. 2000. Recognition of the core RNA promoter for minus-strand RNA synthesis by the replicase of *Brome mosaic virus* and *Cucumber mosaic virus.* *J. Virol.* 74: 10323-10331.

Sivakumaran, K., Hema, M., and Kao, C. C. 2003. Brome mosaic virus RNA synthesis *in vitro* and in barley protoplasts. *J. Virol.* 77: 5703-5711.

Song, C., and Simon, A. E. 1995. Requirement for a 3`-terminal stem-loop for *in vitro* transcription by an RNA-dependent RNA polymerase. *J.. Mol. Biol.* 254: 6-14.

Stawicki, S., and Kao, C. C. 1999. Spatial perturbations within an RNA promoter specifically recognized by a viral RNA-dependent RNA polymerase (RdRp) reveal that RdRp can adjust its promoter binding sites. *J. Virol.* 73: 189-204.

Stupina, V., and Simon, A. E. 1997. Analysis *in vivo* of turnip crinkle virus satellite RNA C variants with mutations in the 3`-terminal minus-strand promoter. *Virology* 238: 470-477.

Sullivan, M. L., and Ahlquist, P. 1997. *cis*-Acting signals in bromovirus RNA replication and gene expression: Networking with viral proteins and host factors. *Semin. Virol.* 8: 221-230.

Sullivan, M. L., and Ahlquist, P. 1999. A brome mosaic virus intergenic RNA3 replication signal functions with viral replication protein 1a to dramatically stabilize RNA *in vivo. J. Virol.* 73: 2622-2632.

Sun, X., Zhang, G., and Simon, A. E. 2005. Short internal sequences involved in replication and virion accumulation in a subviral RNA of *Turnip crinkle virus. J. Virol.* 79: 512-524.

van Rossum, C. M. A., Reusken, C. B. E. M., Brederode, F. Th., and Bol. J. F. 1997. The 3` untranslated region of alfalfa mosaic virus RNA3 contains a core promoter for minus-strand RNA synthesis and an enhancer element. *J. Gen. Virol.* 78: 3045-3049.

van der Heijden, M. W., Carette, J. E., Reinhoud, P. J., Haegi, A., and Bol, J. F. 2001. Alfalfa mosaic virus replicase proteins P1 and P2 interact and colocalize at the vacuolar membrane. *J. Virol.* 75: 1879-1887.

van der Kuyl, A. C., Neeleman, L., and Bol, J. F. 1991. Role of alfalfa mosaic virus coat protein in regulation of the balance between plus and minus RNA synthesis. *Virology* 185: 496-499.

van der Vossen, E. A. G., Notenboom, T., and Bol, J. F. 1995. Characterization of sequences controlling the synthesis of alfalfa mosaic virus subgenomic RNA *in vivo. Virology* 212: 663-672.

Vlot, A. C., and Bol, J. F. 2003. The 5` untranslated region of alfalfa mosaic virus RNA1 is involved in negative-strand RNA synthesis. *J. Virol.* 77: 11284-11289.

Vlot, A. C., Neeleman, L., Linthorst, H. J. M., and Bol, J. F. 2001. Role of the 3`-untranslated regions of alfalfa mosaic virus RNAs in the formation of a transiently expressed replicase in plants and in the assembly of virions. *J. Virol.* 75: 6440-6449.

Vlot, A. C., Laros, S. M., and Bol, J. F. 2003. Coordinate replication of alfalfa mosaic virus RNAs 1 and 2 involves *cis*- and *trans*-acting functions of the encoded helicase-like and polymerase-like domains. *J. Virol.* 77: 10790-10798.

Wang, D., and Maule, A. J. 1995. Inhibition of host gene expression associated with plant virus replication. *Science* 267: 229-231.

Wang, H.-H., and Wong, S.-M. 2004. Significance of the 3`-terminal region in minus-strand RNA synthesis of *Hibiscus chlorotic ringspot virus. J. Gen. Virol.* 85: 1763-1776.

Watanabe, Y., Ogawa, T., Takahashi, H., Ishida, I., Takeuchi, Y., Yamamoto, M., and Okada, Y. 1995. Resistance against multiple plant viruses in plants mediated by a double-stranded RNA-specific ribonuclease. *FEBS Lett.* 372: 165-168.

Watanabe, T., Honda, A., Iwata, A., Ueda, S., Hibi, T., and Ishihama, A. 1999. Isolation from tobacco mosaic virus-infected tobacco of a solubilized template-specific RNA-dependent RNA polymerase containing a 126/183K protein heterodimer. *J. Virol.* 73: 2633-2640.

White, K. A., and Nagy, P. D. 2004. Advances in the molecular biology of tombusviruses: Gene expression, genome replication, and recombination. *Prog. Nucl. Acid Res. Mol. Biol.* 78: 187-226.

Williamson, J. R. 2000. Induced fit in RNA-protein recognition. *Nat. Struct. Biol.* 7: 834-837.

Wu, B., Vanti, W. B., and White, K. A. 2001. An RNA domain within the 5`-untranslated region of the tomato bushy stunt virus genome modulates viral RNA replication. *J. Mol. Biol.* 305: 741-756.

Yeh, H.-H., Tian, T., Rubio, L., Crawford, B., and Falk, B. W. 2000. Asynchronous accumulation of lettuce infectious yellows virus RNAs 1 and 2 and identification of an RNA1 *trans* enhancer of RNA2 accumulation. *J. Virol.* 74: 5762-5768.

Yoshinari, S., Nagy, P. D., Simon, A. E., and Dreher, T. W. 2000. CCA initiation boxes without unique promoter elements support *in vitro* transcription by three viral RNA-dependent RNA polymerases. *RNA* 6: 698-707.

Zhang, G., Slowinski, V., and White, K. A. 1999. Subgenomic mRNA regulation by a distal RNA element in a (+)-strand RNA virus. *RNA* 5: 550-561.

Zhang, G., Zhang, J., and Simon, A. E. 2004a. Repression and derepression of minus-strand synthesis in a plus-strand RNA virus replicon. *J. Virol.* 78: 7619-7633.

Zhang, J., and Simon, A. E. 2005. Importance of sequence and structural elements within a viral replication repressor. *Virology* 333: 301-315.

Zhang, J., Stuntz, R. M., and Simon, A. E. 2004b. Analysis of a viral replication repressor: Sequence requirements for a large symmetrical internal loop. *Virology* 326: 90-102.

5 RNA-DEPENDENT RNA POLYMERASES AND REPLICASES

I. INTRODUCTION

Some important reviews and papers on virus-encoded RNA-dependent RNA polymerases are: Kamer and Argos, 1984; Koonin, 1991; Koonin and Dolja, 1993; Ishihama and Barbier, 1994; Buck, 1996, 1999; Quadt and Jaspars, 1989; Kao *et al.*, 2001; Ahlquist, 2002; van der Heijden and Bol, 2002; and van Dijk *et al.*, 2004.

RNA replicases are specific to the virus and faithfully produce full-length negative-strand, positive-strand and subgenomic RNA chains. The CNV replicase, present in the enriched membrane fractions obtained from yeast and *Nicotiana benthamiana* protoplasts, synthesized both plus- and minus-stranded RNAs *in vitro* (Panaviene *et al.*, 2004). In contrast, RNA-dependent RNA polymerase (RdRp or polymerase) produces full-length negative-polarity RNA from the infecting positive-sense viral RNA template but not the full-length progeny positive-strand RNA chains from the complementary negative-sense RNA produced earlier. All RdRps isolated to-date do so. However, this clear distinction between a replicase and a polymerase is usually not followed by researchers so that in published literature the two terms are often used indiscriminately and interchangeably. In most cases, it is not the replicase that has been described but rather an RdRp under the name of replicase.

Two hypotheses have been enunciated to explain the inability of RdRp to synthesize positive-strand RNA progeny from the negative-strand template: the polymerase might be used only once as suggested for CPMV RdRp since, having produced a negative strand, the RdRp could be inactivated; second, some defect occurs in post-transcriptional modification of the negative strands produced *in vitro* and this defect, possibly in polymerase complex itself, may render the RdRp unable to synthesize positive strands.

No pre-existing cellular machinery for replication of RNA viral genomes exists in host cells so that presence of virus-encoded replicase is essential for viral RNA replication. Therefore, RNA genomes of all replication-competent plus-strand RNA viruses, that have been sequenced, have a specific RdRp sequence and encode an RNA polymerase. Thus, the RNA polymerases/replicases are universal and well-conserved enzymes/motifs present in all positive-strand RNA viruses but viral replicase systems have been developed only for a limited number of plus-stranded RNA viruses. In a few cases, the RdRp activity of this gene product has been demonstrated biochemically [*Tobacco vein mottling virus* (TVMV) (Hong and Hunt, 1996), TMV (Osman and Buck, 1996); BaMV (Li *et al.*, 1998), CNV (Panaviene *et al.*,

2004)]. However, in most cases, the putative RdRps/genes have been identified on the bases of two types of observations: from conserved amino acid sequences, including a GDD   triplet   (Argos, 1988; Koonin and Dolja, 1993) in polypeptides whose sequences are deduced from ORFs in nucleotide sequences of viral RNA, and requirement for these ORFs (containing those motifs) in virus replication as established by mutation studies. Mutations in RdRp domains, that disable virus replication, have been shown in BMV (Kroner *et al.*, 1989; Traynor *et al.*, 1991), PVX (Longstaff *et al.*, 1993; Davenport and Baulcombe, 1997), TMV (Ogawa *et al.*, 1991) and TYMV (Weiland and Dreher 1993). Rajendran *et al.* (2002) suggest that RdRp molecules are reused sequentially several times during viral RNA replication and RdRp preparations are stable during prolonged incubation.

The RdRps can only use single-stranded RNA as template either because the template channel of RdRps is narrow and can only accommodate single-stranded RNA or because RdRps have mechanisms that discriminate against the use of double-stranded templates for *de novo* initiation (Lesburg *et al.*, 1999; Butcher *et al.*, 2001; Ranjith-Kumar *et al.*, 2003a). The template-dependent RdRp preparations are the enzyme preparations which are able to copy exogenously added template *in vitro* while the template-independent RdRps are the enzyme preparations which copy RNAs endogenous to the enzyme preparation – that is, copy the RNAs that are originally present in the preparations. It is more difficult to obtain template-dependent RdRp preparations than the template-independent RdRp preparations. Specific template-dependent enzymes have been obtained from infected plants as well as from heterologous systems. Viral replicases can be simple products and are composed of a single protein that contains both the RdRp sequence and any other essential element (as in carla-, hordei-, potex-, and trichoviruses) or they are complex products and are composed of two or more different proteins.

Plant viruses adopt different strategies for formation of virus polymerases (Table 1). The polymerases can be produced by proteolytic processing (as in tymoviruses), or are encoded by two distinct ORFs (as in *Bromoviridae*) or are formed by frameshifting or by translational readthrough (as in furo-, necro-, tobamo, tobra-, and tombusviruses) (Buck, 1996; Drugeon *et al.*, 1999; Morozov and Solovyev, 1999). Thus, the RdRps of RNA plant viruses are synthesized by a variety of different strategies. In superfamily of alpha-like plant viruses, replicase is encoded by 5`-proximal genomic ORF(s). Closteroviral RdRp domain is expressed via the +1 frameshift mechanism that is not found in other positive-strand RNA viruses. The expression of RdRp of BYV likely involves a usual +1 frameshift mechanism but it is proposed that replicase domains are proteolytically separated (Erokhina *et al.*, 2000). Two ORFs at 5`-end of RNA of GRV produce the possible RdRp by -1 frameshift (Ryabov *et al.*, 2004).

The RdRp of several viruses, including BaMV, TEV, and TCV are active when expressed without other virus-coded auxiliary proteins. On the other hand, RdRps of AMV, BMV, and of several other plant viruses require the presence of a viral auxiliary protein in order to be functional *in vitro* (Quadt *et al.*, 1995; Vlot *et al.*, 2001). The two overlapping replicase genes encoding p33 and p92 replica-tion proteins of tombusviruses (TBSV and CNV) are essential for replication of genomic RNA in plant cells (Scholthof *et al.*, 1995; Oster *et al.*, 1998; Panaviene *et al.*, 2003, 2004); p92 has the RdRp signature motifs in its C terminus, function

TABLE 1

Classification, lineage, and mode of expression of RNA-dependent RNA
polymerases of families or genera of plus-sense RNA plant viruses
(Based on Table II of Buck, 1996)

RdRp Supergroup[a]	RdRp Lineage[a]	Virus Family Or Genus	Virus Examples	Expression mode of RdRp[b]
1	Picorna	*Sequiviridae*	*Parsnip yellow fleck virus, Rice tungro spherical virus*	PP
1	Picorna	*Comoviridae*	*Cowpea mosaic virus, Tobacco ringspot virus*	PP
1	Poty	*Potyviridae*	*Potato Y virus, Tobacco etch virus, Plum pox virus, Barley yellow mosaic Virus*	PP
1	Sobemo	*Sobemovirus*	*Southern bean mosaic virus, Cocksfoot mottle virus*	PP
1	Sobemo	*Luteovirus* subgroup II	*Potato leaf roll virus, Beet western yellows virus, Barley yellow dwarf virus*	FS, PP?
1	Sobemo	*Enamovirus*	Pea enation mosaic RNA1	FS, PP?
2	Carmo	*Tombusviridae*	*Tomato bushy stunt virus, Cucumber Necrosis virus, Cymbidium ringspot Virus, Carnation mottle virus, Turnip Crinkle virus*	RT
2	Carmo	*Machlomovirus*	*Maize chlorotic mottle virus*	RT
2	Carmo	*Necrovirus*	*Tobacco necrosis virus*	RT
2	Carmo	*Dianthovirus*	*Red clover necrotic mosaic virus*	FS
2	Carmo	*Luteovirus* subgroup I	*Barley yellow dwarf virus (MAV, PAV, SGV)*	FS
2	Carmo	*Enamovirus*	Pea enation mosaic RNA2	FS
2	Carmo	*Umbravirus*	*Carrot mottle virus, Groundnut rosette virus*	?
2	Carmo	*Unclassified*	Beet western yellows ST9- Associated RNA	FS
3	Tymo	*Capillovirus*	*Apple stem grooving virus*	D
3	Tymo	*Carlavirus*	*Carnation latent virus, Potato M virus, Blueberry scorch virus*	

3	Tymo	*Trichovirus*	*Apple chlorotic leaf spot virus, Potato T virus*	D
3	Tymo	*Tymovirus*	*Turnip yellow mosaic virus, Eggplant mosaic virus*	PP
3	Tymo	*Potexvirus*	*Potato X virus, Cymbidium mosaic Virus, Foxtail mosaic virus*	D
3	Tobamo	*Tobamovirus*	*Tobacco mosaic virus, Pepper mild mottle virus*	RT
3	Tobamo	*Tobravirus*	*Tobacco rattle virus, Pea early browning virus, Pepper ringspot virus*	RT
3	Tobamo	*Hordeivirus*	*Barley stripe mosaic virus*	D
3	Tobamo	*Furovirus*	*Soil-borne wheat mosaic virus*	RT
3	Tobabo	*Idaeovirus*	*Raspberry bushy dwarf virus*	D
3	Tobamo	*Bromoviridae*	*Brome mosaic virus, Cucumber mosaic virus, Alfalfa mosaic virus, Tobacco streak virus*	D
3	Tobamo	*Closterovirus*	*Beet yellows virus, Citrus tristeza virus*	FS, PP
3	Tobamo	*Closterovirus (tentative)*	*Lettuce infectious yellows virus*	FS, PP
3	Rubi	*Furovirus*	*Beet necrotic yellow vein virus*	D

a = RdRp supergroups and lineages are from Koonin and Dolja, 1993.
b = PP - polyprotein processing; RT - readthrough; FS - frameshift; D - direct translation

of p33 is presently unknown although it has RNA-binding site that is an arginine-proline-rich RPR motif (Rajendran and Nagy, 2003). Thus, the functional replicase complex of CNV in yeast contains both p33 and p92 replicase proteins. The CNV replicase, once formed in cells, is a stable complex (Panaviene *et al.*, 2004).

The RdRp of tombusviruses is able to function *in trans* (Rubino and Russo, 1995). This was confirmed by Boonrod *et al.* (2005) since the ability of TBSV KB1 [a TBSV clone carrying mutations within the RPR motif (see below) and hence cannot replicate by itself] to replicate was restored by expression of the viral RdRp *in trans*. They found that a mutant carrying three arginine substitutions in this motif rendered the virus unable to replicate in plants and protoplasts of *Nicotiana benthamiana*. When the replicase function was provided *in trans*, by expressing the TBSV RdRp in *N. benthamiana* plants, an infectious variant could be isolated.

There is indication that strong selection pressure is active for maintaining necessary sequence of the viral RdRp. The glycine residue in the GDD motif of TBSV RdRp in position 620 was substituted by alanine. The resulting viral mutant was tested for infectivity on mechanically inoculated *N. benthamiana* plants; these plants did not show any symptoms and no systemic infection 5 days post-inoculation but about half of these plants established systemic infection 14 days post-inoculation. This suggests a reversion of the mutant; in fact, the substituted alanine residue had reverted to the original glycine (G_{620}) in all three TBSV clones tested (Boonrod *et al.*, 2005).

Replicases isolated early in BMV infection process preferentially used the promoter for minus-strand RNA synthesis, while the replicases isolated later increased the ability for genomic and subgenomic plus-strand RNA synthesis. These results parallel the production of the different classes of RNAs produced *in vivo* (Sivakumaran *et al.*, 2003). Thus, the BMV replicases extracted from plants at different times after infection have different levels of recognition of the minimal promoters (minimal length RNAs) for plus- and minus-strand RNA synthesis.

An RdRp is present in uninfected host plants (host RdRp). This enzyme can utilize single-stranded RNA and single-stranded DNA as template, possesses no marked sequence specificity, synthesizes comparatively short and heterogeneous mixture of RNA products, has been purified from several plants, and is constituted by single polypeptide chains having molecular weights 100- to 140-kDa. Its function in the plants is still not definitely known. Various types of traumatizations enhance activity of host RdRp, including infection by plant viruses such as AMV, CCMV, CMV, CPMV, and by many other viruses. It is unequivocally established that host RdRp is distinct from viral RdRp.

A. Time Course of Replicase Production

Viral RNA, soon after partial or complete uncoating of virions after their entry into a cell, directs synthesis of some early virus-specific protein(s); one of these proteins is the virus-encoded replicase. The replicase is produced within 1-10 min after infection and is usually synthesized in sufficient amount after 20-30 min of infection (Mandahar, 1989). Translation of replicase has to be an early event in life cycle of positive-sense RNA viruses because virus particles are not accompanied by their specific RdRps.

Replicases formed in early stages of infection process preferentially use the promoter for minus-strand RNA synthesis to produce the minus-strand RNAs while replicases isolated in late stage of infection possessed the enhanced ability for synthesis of genomic and subgenomic plus-strand RNAs. These shifts in recognizing the minimal promoters parallel the formation of different classes of BMV RNAs produced *in vivo* (Sivakumaran *et al.*, 2003). Similarly, the CNV replicase synthesized more plus-stranded products at the late time of infection than at the early time of infection (Panaviene *et al.*, 2004). It is also proposed that CNV replicase is involved in minus-strand synthesis during the entire replication process and there was no shutdown of minus-strand RNA synthesis in cells of both infection-independent yeast system and the protoplast-based infection-based *N. benthamiana* system. The amounts of plus- and minus-stranded RNAs, of both genomic and DI RNAs, of TBSV (Ray and White, 2003) and CNV (Panaviene *et al.*, 2004) increase continuously over 24 to 30 hour of infection in cucumber and *N. benthamiana* protoplasts. The amount of p33, and possibly of p92 also, of CNV increases up to 24 hour in plant protoplasts. This is likely to favour synthesis of plus-strand RNA at the late time of infection, as has also been suggested for the unrelated TMV.

B. Isolation and Purification of Viral Replicases

The RdRps/replicase complexes of positive-strand RNA viruses are always associated with membranes of virus-infected cells (Restrepo-Hartwig and Ahlquist, 1996; Schaad *et al.*, 1997; Noueiry and Ahlquist, 2003). The membrane localization domains in replicase proteins are present within the N-terminal overlapping domain of the two replicase proteins of TBSV (Rubino and Russo, 1998). The attachment of RdRp complexes to membranes is responsible for their appearance in particulate fraction of crude extracts in most of the cases. Asurmendi *et al.* (2004) found that most or all the TMV replication complexes assembled on, and eventually surround, strands or bodies of ER; contain movement protein; and that replicase is located within the movement protein bodies. The MP is an integral membrane protein (Brill *et al.*, 2000), which promotes formation of aggregates of ER and facilitates formation of TMV replication complexes. Plant cells infected with BMV also contain electron-dense cytoplasmic inclusion bodies, which are similar to that of the TMV replication complexes, and also contain 1a and 2a replicase proteins along with MP 3a (Dohi *et al.*, 2001) that is also associated with membrane.

Solubilization of the membrane-located enzyme complex is a major step during preparation/purification and isolation of RdRps. It is difficult to obtain viral RdRps in the active form in large quantities. Different procedures have been employed for releasing the RdRp from membranes. These procedures have generally been only partially effective except in the following cases: isolation of CMV replicase in which non-ionic Nonidet P-40 served as detergent; isolation of BMV replicase in which non-ionic detergents such as P-40 or Tween-80 were used; and use of the some additives for isolation of RdRps of some viruses such as Lubrol W for TYMV, CHAPS for TMV, and Triton-X-100 for CPMV. These treatments ultimately solubilized RdRp activity but they consistently decreased template specificity.

Functional replicases have been isolated and purified to various extents from virus-infected plant hosts or from heterologous systems. Purified replicase enzymes obtained from plant hosts are constituted by single polypeptide chains of molecular weights ranging from 10- to 140-kDa. Purified/partially purified template-specific viral RdRp preparations have been obtained from virus-infected plants in case of AMV (Quadt *et al.*, 1991), BaMV (Cheng *et al.*, 2001), BMV (Miller and Hall, 1983; Quadt and Jaspars, 1990; Sivakumaran *et al.*, 2003), CMV (Hayes and Buck, 1990; Gal-On *et al.*, 2000), CNV (Nagy and Pogany, 2000; Panaviene *et al.*, 2004), CCMV, *Foxtail mosaic virus* (Rouleau *et al.*, 1993), *Hibiscus chlorotic ringspot carmovirus* (Wang and Wong, 2004), PVX (Doronin and Hemenway, 1996; Plante *et al.*, 2000; Cheng *et al.*, 2001), RCNMV (Bates *et al.*, 1995), TMV (Osman and Buck, 1996; Watanabe *et al.*, 1999), TBSV (Nagy and Pogany, 2000), TCV (Song and Simons, 1994), and TYMV (Mouches *et al.*, 1974; Deiman *et al.*, 1997; Singh and Dreher, 1997). These RdRp preparations initiated complementary RNA synthesis either with short primers or in the absence of primers (*de novo* synthesis). As already mentioned, the first true replicase to be extracted from a eukaryotic plus-strand RNA plant virus-infected tissue was obtained from CMV-infected tissue and it completely and specifically replicated CMV genome *in vitro* (Hayes and Buck, 1990).

The heterologous systems, with the names of plant viruses whose replicases have been purified/expressed from the respective systems being shown in parenthesis, include *Escherichia coli* [TVMV (Hong and Hunt, 1996); TCV (Rajendran *et al.*, 2002); and the animal virus *GB virus-B* (Ranjith-Kumar *et al.*, 2003b)], yeast *Saccharomyces cerevisiae* [BMV (Quadt *et al.*, 1995); CNV (Panavas and Nagy, 2003a; Panaviene *et al.*, 2004; Stork *et al.*, 2005); *Carnation Italian ringspot tombusvirus* (Pantaleo *et al.*, 2004)], and insect cell system [TYMV (Héricourt *et al.*, 2000)]. Besides these, mammalian cells have been used for replicase purification of animal viruses like *Poliovirus* and *Hepatitis C virus* and *Xenopus* for purification of poliovirus replicase.

There is one major drawback in obtaining purified replicase preparations from infected plants - production of the replicase proteins entirely depends on virus replication. This makes it difficult to study the effect of detrimental mutations within the viral RNA or the replicase proteins on replication in the plant-derived *in vitro* viral replicase assay. On the other hand, expression of the replicase proteins can be achieved in heterologous systems without dependence on virus replication. This facilitates mutational analysis of the replicase genes. Moreover, viral RdRps purified from infected plant cells are multi-subunit enzymes while those obtained from heterologous systems are single-subunit RdRps.

Biochemical features of RdRps of some positive-strand RNA plant viruses [TVMV (Hong and Hunt, 1996) and BaMV (Li *et al.*, 1998)] have been examined in some detail by using purified RdRp preparations obtained from heterologous expression systems. In contrast, RdRp of TYMV, expressed in insect cells, had no replicase activity *in vitro* (Héricourt *et al.*, 2000). Both BMV RdRp (Quadt *et al.*, 1995) and RdRp of CNV (Panaviene *et al.*, 2004), isolated from recombinant *S. cerevisiae*, were infective. The template-dependent replicase complex of CNV, obtained from *S. cerevisiae*, is highly active (Panaviene *et al.*, 2004).

The use and development of heterologous systems for expression and purification of highly active recombinant viral replicases has several advantages (Panaviene *et al.*, 2004). Purified recombinant viral replicases (obtained from heterologous systems) are useful for studying the mechanism of viral RNA replication *in vitro*, for understanding the role of replication proteins and RNA templates in replication and such studies complement the studies conducted by plant-based and protoplast-based CNV replicase assays.

II. PLANT VIRAL POLYMERASES

Table 2 lists RdRps of some plant viruses. Purified CMV replicase, the first true replicase extracted from a plus-strand RNA virus (CMV)-infected plants that completely replicated CMV genome *in vitro*, contains at least three proteins: 111- kDa 1a protein (encoded by CMV RNA1), 94-kDa 2a protein (encoded by CMV RNA2), and a plant host protein of about 50-kDa (Hayes and Buck, 1990). The 1a protein contains putative methyltransferase-like and helicase-like domains and 2a protein contains the domains specifying polymerase activity. Since then, RdRp complexes of a number of other positive-strand RNA viruses have been isolated from infected plants (Plante *et al.*, 2000).

TABLE 2
RNA-dependent RNA polymerases of some plus-sense RNA plant viruses
(Based on Morozov and Solovyev, 1999, with additions)

Alfalfa mosaic alfamovirus: RdRp is the 2a protein of 90-kDa encoded by RNA2

Alpha-like Supergroup: The viral genomes encode RdRp of supergroup 3.

Apple stem grooving capillovirus: RdRp is part of 241-kDa polyprotein, which is encoded by ORF1; RdRp of supergroup 3 and released by protease cleavage.

Bamboo mosaic potexvirus: RdRp is the C-terminal 473 amino acid residues of 155-kDa polyprotein

Barley stripe mosaic hordeivirus: 87-kDa protein, encoded by RNA3 or RNA γ, contains the RdRp domain of supergroup 3.

Barley yellow dwarf luteovirus: ORF2-encoded protein expressed only as ORF1/ORF2 frameshift fusion and includes 60-kDa RdRp domain of supergroup 2.

Barley yellow mosaic bymovirus: RdRp encoded by RNA1 and released from polyprotein by protease cleavage

Beet necrotic yellow vein benyvirus: RdRp of supergroup 3, released by proteolytic cleavage of RNA1-encoded 237-kDa polyprotein

Beet soil-borne pomovirus: RNA1 encodes 145-kDa protein and its readthrough product of 204-kDa contains the RdRp of supergroup 3.

Beet yellows closterovirus: ORF1b codes for RdRp of supergroup 3 (tobamo lineage) by translational frameshift resulting in ORF1a/1b fusion protein

Brome mosaic bromovirus: RdRp is 2a protein of 90-kDa encoded by RNA2

Carnation Italian ringspot tombusvirus: ORF2 encodes a 95-kDa protein that contains RdRp motif and is translated by readthrough of 5`-proximal ORF1 that encodes a 36-kDa protein (Weber-Lotfi *et al.*, 2002).

Carmo-like supergroup: Viral genomes encode RdRp of supergroup 2.

Carrot mottle mimic umbravirus: ORF2 encodes RdRp of supergroup 2, which is expressed as ORF1/ORF2 frameshifting fusion and is of 64-kDa of supergroup 2.

Citrus tristeza closterovirus: ORF1b codes for RdRp of supergroup 3 (tobamo lineage) by a translational frameshift resulting in ORF1a/1b fusion protein

Closteroviridae Family: ORF1b codes for RdRp of supergroup 3 (tobamo lineage) by translational frameshift resulting in ORF1a/1b fusion protein

Cocksfoot mottle sobemovirus: RdRp of 61-kDa, of supergroup 1, formed by ORF2-ORF3 fusion protein due to frameshifting

Cucumber mosaic cucumovirus: 2a protein of 94-kDa encoded by RNA2

Pea enation mosaic virus RNA1: 67-kDa RdRp of supergroup 1 produced as ORF1/ORF2 frameshift product

Pepper vein banding potyvirus: The NIb protein is the viral polymerase and it could uridylylate the amino acid Tyr 66 of VPg in a template independent manner (Anindya *et al.*, 2005).

Potato leaf roll polerovirus: 70-kDa RdRp domain of supergroup 1 translated through a −1 frameshift as ORF1/ORF2 polyprotein

Potato virus Y: NIb protein is RdRp of supergroup 1, cleaved from a polyprotein by proteolytic cleavage

Raspberry bushy dwarf idaeovirus: RdRp of supergroup 3 (tobamo lineage) located in 2a protein as C-terminal part of 190-kDa polyprotein encoded by RNA1

Soil-borne wheat mosaic furovirus: 209-kDa protein encoded by RNA1 contains RdRp motif of supergroup 3 and synthesized by translational readthrough of 150-kDa protein gene

Southern bean mosaic sobemovirus: RdRp of supergroup 1, expressed as C-terminal part of ORF2-encoded polyprotein, released by proteolytic cleavage

Tobacco mosaic tobamovirus: 183-kDa protein synthesized by translational readthrough of 126-kDa ORF leaky termination codon, C-terminal region of 183-kDa protein contains domain of RdRp of supergroup 3.

Replicase is located within the MP bodies (Asurmendi *et al.*, 2004). The 126-kDa protein is present in infected cells in 10-fold excess over 183-kDa protein but both proteins co-immunoprecipitate in a 1:1 ratio (Watanabe *et al.*, 1999).

Tobacco rattle virus: Translational readthrough of 134-kDa ORF UGA terminator in RNA1 gives rise to 194-kDa protein having the domain of RdRp of supergroup 3 at its C-terminus.

Tomato black ring nepovirus: RNA1 encodes 87-kDa RdRp as part of 254-kDa polyprotein, RdRp released by proteolytic cleavage

Tomato bushy stunt tombusvirus: ORF2 codes for 92-kDa RdRp of supergroup 2, produced by readthrough of 33-kDa protein stop codon

Turnip crinkle carmovirus: p88 contains RdRp motif (O'Reilly and Kao, 1998; Rajendran *et al.*, 2002)

Turnip yellow mosaic tymovirus: 66-kDa protein; the 206-kDa polyprotein undergoes autocatalytic cleavage, giving products of 141-kDa and 66-kDa; the 66-kDa protein contains RdRp domain.

Umbravirus genus: ORF2 encodes 64-kDa RdRp of supergroup 2, expressed as ORF1/ORF2 frameshifting fusion protein

An isolated BMV/AMV RdRp can use negative-strand RNA template to produce subgenomic BMV RNA (Dreher and Hall, 1988) /full-length AMV genomic RNA (de Graaf *et al.*, 1995a, 1995b). BMV RdRp is highly template-specific and template-dependent, has been isolated from BMV-infected barley leaves, and polymerase-recognition site of BMV RNAs is located within the 134 nucleotides at 3`-end.

Replication of CNV needs two overlapping replication proteins, the p33 replication cofactor (or auxiliary protein) and the p92 RdRp. TBSV also codes for p33 and p92 replicase proteins; both are proposed to be part of replication complex and p33 is two-fold more abundant than p92 replicase protein in infected cells (Scholthof *et al.*, 1995). The p92 contains RdRp motif in its non-overlapping C-terminal part, and is expressed from genomic RNA via a readthrough mechanism of p33 termination codon so that its N-terminal part overlaps with p33, which is a single-stranded nucleic acid-binding protein. This ability of p33 replicase protein to bind to single-stranded RNA is similar to other plant viral proteins such as viral MPs (Citovsky *et al.*, 1990), βb protein of BSMV (Donald *et al.*, 1997), protein p20 encoded by bamboo mosaic virus satellite RNA (Tsai *et al.*, 1999), and a protein encoded by sequences at 3`-UTR of BaMV RNA (Huang *et al.*, 2001). RCNMV encodes putative replicase components - p88 that contains the RdRp motif and p27 that is presumably the accessory protein of the replicase complex.

The BaMV ORF1 encodes a 155-kDa polypeptide postulated to be involved in replication of viral genome and formation of cap structure at 5`-ends of viral transcripts (Huang *et al.*, 2004). It is the C-terminal 472 amino acids that specifically recognise 3`-UTR of this virus and can perform RNA-dependent RNA synthesis and so acts as RdRp (Huang *et al.*, 2001).

Two genes (of p28 and p88) of TCV are required for replication; p88 overlaps p28 and contains signature RdRp motif (O`Reilly and Kao, 1998) in the unique C-terminal portion. Role of p28 is not known. Rajendran *et al.* (2002) expressed and purified TCV replicase proteins from *E. coli* and found that p88 alone possessed polymerase activity that is specific to TCV. The N-terminally truncated p88 (designated p88c) lacks p28 overlapping domain, is an active RdRp, is approximately 60-kDa and has ten-fold higher activity than the full-length p88. The p88c is one of the smallest RdRps, similar in size to polymerase 3D of *Poliovirus*. Both p88 and p88c are single unit RdRps while TCV RdRps are multisubunit.

Erokhina *et al.* (2000) found that the 5`-terminal ORF of BYV codes for a 295-kDa polyprotein having papain-like cysteine proteinase, methyltransferase and helicase domains while ORF 1b encodes RdRp and that the 63-kDa and 100-kDa BYV products *in planta* are the methyltransferase-like and helicase-like proteins, respectively. Several features distinguish BYV replicase from those of other members of this supergroup: expression of BYV RdRp likely involves a usual +1 frameshift mechanism; a unique domain of unknown function is present between methyltransferase and RNA helicase domains; and it is proposed that replicase domains are proteolytically separated. Work of Dolja and coworkers (Dolja, 2003) implies that BYV replicase is able to recognise seven distinct subgenomic promoters or that replicase recognises some still undetected common features of these promoters, for example, a unique secondary structure element.

AMV RdRp has been partially purified from AMV-infected tobacco plants and transgenic P12 plants. *In vitro*, these RdRp preparations supported negative-strand synthesis on positive-strand AMV templates as well as subgenomic positive-strand synthesis on a negative-strand RNA3 template. In *Carnation Italian ringspot tombusvirus* (CIRV), the 5`-proximal ORF1 encodes a 36-kDa protein (p36) and ORF2 encodes a 95-kDa protein (p95) by readthrough of p36 stop codon. The p95 contains the conserved RdRp motif (Weber-Lotfi *et al.*, 2002). Later, Pantaleo *et al.* (2003, 2004) confirmed that these two proteins are the replicase proteins and are required for replication of tombusviruses. Both these replicase proteins are targeted and membrane-anchored at mitochondria in plant and yeast cells (Rubino *et al.*, 2001; Weber-Lotfi *et al.*, 2002) and this targeting is independent of each other (Pantaleo *et al.*, 2004). CIRV replicase proteins are directly involved in stabilization and recruitment of template RNA, similar to proteins 1a and 2a of BMV.

BMV, other alphaviruses and many other positive-strand RNA viruses produce partially double-stranded replicative intermediate RNAs bearing multiple nascent strands. This shows that several elongating polymerases can simultaneously traverse a single RNA and can simultaneously copy a single viral RNA template (Buck, 1996). In BMV-infected cells, 1a replication protein is more limiting than 2a polymerase for viral RNA Synthesis (Grdzelishvili *et al.*, 2005). The number of 2a polymerases within individual replication complexes is limited to 10 to 15 copies (Schwartz *et al.*, 2002) but still polymerase is not a limiting factor in the replication of BMV RNAs (Grdzelishvili *et al.*, 2005). This shows that although genomic and subgenomic promoters may compete locally for 2a BMV polymerases within individual replication complexes, ample polymerase in all probability is available for both genomic and sgRNA syntheses. This suggests that genomic and subgenomic RNA syntheses might occur simultaneously on a single negative strand RNA template without interference.

Another significant property of RdRps is that RdRp can adjust its promoter binding sites indicating that an 'induced fit' mechanism may adjust the viral replicase to different promoters (Stawicki and Kao, 1999; Williamson, 2000).

III. SPECIFICITY

The purified viral RdRps possess variable specificity in template selection and stringency of promoter recognition (Buck, 1996, 1999; Deiman *et al.*, 2000). In fact, two contrasting observations have been made about the specificity of virus-encoded viral RdRps: that RdRps are specific to the virus and faithfully produce full-length negative-strand (and positive-strand) RNA chains, and that the core viral polymerase can conduct only the basic RdRp function of RNA-dependent RNA synthesis but fails to determine their template specificity (Lai, 1998).

A. Specificity of Viral Polymerase Action

The TYMV RdRp preparation was TYMV RNA-specific (Deiman *et al.*, 1997). Polymerase binds to a specific sequence or structure in genome followed directly by initiation of transcription (Deiman *et al.*, 2000). TBSV encodes p33 and p92 replicase proteins, both of which bind to viral RNA *in vitro*. Binding of p33 to single-stranded RNA is stronger than binding to double-stranded RNA in *vitro*, and p33 binds to a TBSV-derived sequence with high affinity (Rajendran and Nagy, 2003). The functional RdRp complex from BaMV-infected plants specifically uses short transcript templates for both minus- and plus-strand RNA synthesis *in vitro* (Cheng *et al.*, 2001). All the TCV RdRps [p88 and p88c (single unit RdRp) and plant TCV RdRp (multisubunit RdRp)] are recognised efficiently and specifically by the 3` -proximal minimal plus-strand initiation promoter of satC (Rajendran *et al.*, 2002). Replication of *in vitro* tobravirus recombinants showed that specificity of template recognition was determined by 5` -non-coding and not by 3` -non-coding sequence (Mueller *et al.*, 1997). Thus, isolated replication complexes containing virus-encoded RdRp, but free from host RdRp, are able to synthesize full-length viral RNA in several plant viruses - like BMV, CMV, CNV, CPMV, TCV, TBSV, TYMV and other plant viruses. The RdRps in about all these cases replicated and transcribed viral RNAs specifically.

1. Host Factors and Viral Polymerase Specificity

According to Lai (1998), it is the cellular proteins/factors in the polymerase-host protein replication complex that ensure specificity and that are essential for template-specific RNA synthesis (Lai, 1998). Compelling evidence exists in its support. Partially purified RdRp complexes of all well-studied eukaryotic plus-strand RNA plant viruses like BMV, CMV, TMV, and TYMV contain copurified host membranes and host proteins (Mouches *et al.*, 1984; Buck, 1996; Lai, 1998; van der Heijden and Bol, 2002; Noueiry and Ahlquist, 2003; Ahlquist *et al.*, 2003). These specific host proteins constitute an intrinsic part of viral RdRp and have been identified so far in only a few plant viruses. The eukaryotic translation initiation factor (elF3) binds to RdRp of BMV and TMV. The elF3 binds directly to 2a protein (viral polymerase protein) of BMV but its mode of action in TMV replicase has not yet been worked out.

It is possible that mechanistic and functional roles of elF3 translation factor are different in these two plant viruses.

An RdRp preparation can replicate any viral RNA without exhibiting any viral RNA template-specificity in the absence of the participating cellular factors. Many RdRps do not bind to viral RNA specifically or at all and also mostly fail to bind directly to *cis*-acting regulatory or promoter sequences on viral RNA. Binding of viral polymerase to template RNA is mediated by cellular proteins that dock specifically to viral RNAs. Cellular factors are thus necessary for template-specific RNA-dependent RNA synthesis (Lai, 1998; Strauss and Strauss, 1999, Ahlquist *et al.*, 2003). The ability of RdRps for initiation of RNA synthesis at specific sites also appears to depend upon their interaction with cellular proteins that bind directly to viral templates. It is generally assumed now that polymerases/replicases of all positive-strand RNAs are composed of both virus- and host-coded proteins. In fact, Lai (1998) regards cellular proteins as being components of the replicase. The interactions between viral RdRp and cellular factors may form an RdRp complex even in the absence of RNA template in situations when cellular factor is already a part of RdRp as in BMV and TMV. In other cases, cellular factors first dock with viral RNA template, as in TYMV, followed by interaction of this RNA template with viral RdRp. Thus, viral RdRps and cellular factors together form the transcription or replication complexes on viral RNA. Role of host proteins during replication of plant viruses is discussed in greater detail in a separate Chapter.

B. Absence of Specificity of Viral Polymerase Action

Following reports suggest lack of specificity of viral RdRp action. The promoter for CMV negative-strand RNA synthesis is recognized by heterologous viral polymerase(s): pseudorecombinants and recombinants formed between CMV and *Tomato aspermy virus* undergo normal replication in protoplasts as well as in plants in presence of CMV RdRp. Similarly, the pseudorecombinant obtained by replacing 3`-terminus of BMV RNA3 by that of the CMV was replicated by BMV replicase (Rao and Grantham, 1994) - suggesting that BMV RdRp could recognize the 3`-terminus promoter in RNA of the distantly related CMV. Likewise, a TMV recombinant carrying 3`-terminus of BMV RNA could be replicated by TMV replicase (Ishikawa *et al.*, 1991).

The transgene mRNA that contained 3`-UTR of *Lettuce mosaic virus* (LMV), upon expression in transgenic tobacco plants, acted as template for synthesis of complementary negative-strand RNA of *Pepper mottle virus* (PMV), TEV, and TVMV but not of CMV when the transgenic plants were individually infected by these viruses (Teycheney *et al.*, 2000). Synthesis of negative-strand RNA was abolished upon deletion of 3`-UTR from the transgene. Teycheney *et al.* (2000) derived following conclusions: viral replicase of several potyviruses and TEV recognized heterologous 3`-UTR upon inclusion in transgene mRNAs and used them as transcription promoters; polymerases of PMV, TEV, and TVMV could initiate transcription of the negative-strand RNA from LMV 3`-UTR incorporated in cellular mRNAs as transgene; and RdRp of potyviruses could act *in trans* on a viral promoter located internally on a cellular mRNA. This is the first report of *in trans* action of a viral RdRp on a viral promoter located on a cellular mRNA.

Similarly, the common sequence elements, such as CCA repeats, can be recognised by RdRps of several plant viruses (Deiman *et al.*, 2000; Yoshinari *et al.*, 2000) - suggesting significant similarities among these viral RdRps. Thus, lack of specificity in polymerase action could be more widespread in plant viruses than thought earlier (Teycheney *et al.*, 2000).

IV. STRUCTURE

A. Conserved Domains

Kamer and Argos (1984) identified several similar motifs between known poliovirus RdRp and putative RdRps of several other positive-strand RNA animal and plant viruses and concluded that all polymerases possess a set of conserved sequence motifs. This sequence similarity was most pronounced in some short motifs scattered over a region of about 300 amino acids in which several conserved RdRp motifs were later identified. Koonin (1991) and Koonin and Dolja (1993) found eight such conserved motifs out of which only three 'core' motifs (IV, V, and VI) exhibited unequivocal conservation throughout the positive-strand RNA viruses and had six invariant amino acid residues. Despite this, the overall sequence similarity among RdRps of these viruses is very poor. Koonin (1991) found that some of the conserved motifs were of more general occurrence, with modified versions existing in the RdRps of double-stranded RNA viruses, negative-stranded RNA viruses and, even in, reverse transcriptases and DNA-dependent DNA and RNA polymerases. Most of these conserved domains are in the palm domain, with the motifs A, B and C being most prominent.

One of these conserved domains (motif VI) is GDD (glycine-aspartic acid – aspartic acid residues) motif and the other two conserved motifs (motifs IV and V) are located upstream. The GDD motif contains an invariant central amino acid sequence of Gly-Asp-Asp tripeptide as part of consensus sequence. This sequence/domain is found in one of the nonstructural proteins of all positive-strand RNA viruses and is also referred to as the polymerase domain. The nonstructural protein (polymerase) is assumed to be part of viral replicase.

The above consensus sequence is flanked by pentapeptides constituted mainly by hydrophobic amino acids. This suggests a β-hairpin structure composed of two hydrogen-bonded antiparallel β-strands connected to a short exposed loop containing GDD amino acids. *In vitro* mutations in GDD box of several animal and plant viruses greatly reduced or abolished RNA replication as in PVX and TYMV. For example, when GDD box of 166-kDa protein of PVX was mutated to GED, ADD or GAD, virus infectivity was abolished and RNA replication in protoplasts was greatly reduced (Longstaff *et al.*, 1993). Similarly, TYMV RNA replication was abolished by substitution of G to R in GDD box of 66-kDa TYMV protein (Weiland and Dreher, 1993). Mutations were incorporated in the central polymerase-like motif of BMV 2a protein that encompasses motif IV and the flanking sequences. This produced BMV mutants in which replication became *ts* or was completely abolished.

On the basis of structural similarities and the conserved motifs, it has been proposed that all polymerases make use of a common two-metal mechanism of catalysis. This involves two conserved aspartic acid residues from A and C motifs, respectively, and two divalent metal ions for the formation of the phosphodiester bonds. The carboxylates anchor a pair of divalent metal ions that perform the major role in catalysis. One divalent ion (Mg^{2+} ion) promotes the deproteination of the 3`-hydroxyl of the nascent strand, while the second (Mg^{2+} ion) facilitates the formation of the pentacovalent transition state at $^{7-}$phosphate of dNTP and the exit of the inorganic pyrophosphate group (PP_i).

B. Shape and Structure

RNA-dependent RNA polymerases, DNA-dependent RNA polymerases, DNA-dependent DNA polymerases and reverse transcriptase have a similar structure, including the crystal structure (Hansen *et al.*, 1997; O`Reilly and Kao, 1998; O`Reilly *et al.*, 1998; Lesburg *et al.*, 1999; Ahlquist, 2002). The RdRp structure, given below, was arrived at after studying a large number of RdRps of animal viruses and bacteriophages. The structures of RdRps of six animal viruses have been determined to date. These RdRps are $3D^{Pol}$ of *Poliovirus* (PV); NS5B of *Hepatitis C virus* BK strain; NS5B of *Hepatitis C virus* 14 strain; bacteriophage φ6 protein P2; $3D^{Pol}$ of *Rabbit haemorrhagic disease virus* (RHDV); and reovirus protein λ3 (van Dijk *et al.*, 2004). All studies and structural data show that all nucleic acid polymerases have similar structure and mechanism of catalysis and that the determined structural characteristics of all viral RdRps closely resemble each other. Structural elements that ensure *de novo* initiation have also been identified in the above six RdRps of the animal viruses.

The RdRps in all the above viruses possess a basic right hand-like structure with 'palm', 'thumb', and 'finger' domains and with a cleft existing between fingers, palm, and thumb subdomains. It is the palm domain structure that is especially conserved and has four sequence motifs that are conserved in polymerases of all RNA and DNA viruses. However, the finger domains differ significantly among viral RNA polymerases. This may be due to specific adaptations to structurally diverse substrates. The basic polymerase right hand shape provides the correct geometrical arrangement of substrate molecules and metal ions at the active site for catalysis.

The location of template and primer was in cleft in HIV reverse transcriptase. The location of sequence motifs equivalent to IV (β-strand) and VI (β-strand-loop-β-strand) was close together on floor of the cleft within the palm subdomain. The three conserved GDD residues in these motifs were lying close to 3`-OH of primer and NTP-binding site in the polymerase catalytically active site. Function of the three GDD residues could be to bind Mg^{2+} ions that are required for polymerase activity. Poliovirus RdRp also seems to have a 'hand' structure. The NTP-binding site in this case is located in an area that spans motifs II and III.

The template binds to the catalytically active site of enzyme in RdRps of CMV and RCNMV (Hayes *et al.*, 1994; Bates *et al.*, 1995).

Despite the above resemblances in structure, RdRps have specific features that distinguish them from other polymerases. One of these structural attributes is their 'closed hand' conformation, in contrast to the 'open hand' shape of other known polymerases. The closed conformation is achieved by interconnection of the fingers and thumb domains with several loops (fingertips) that protrude from the fingers. The closed structure creates a well-defined template channel that possibly may regulate recognition of the initiation site. The function of the template channel and additional structures near the active site is to ensure that initiation of minus-strand RNA synthesis occurs at or near the end of 3`-termini of *Hepatitis C virus* and bacteriophage φ6 RNAs (van Dijk *et al.*, 2004). The template channel of RdRps is narrow and can only accommodate single-stranded RNA templates (Lesburg *et al.*, 1999; Ranjith-Kumar *et al.*, 2003a).

Another common structural RdRp element is the NTP (substrate) tunnel. A hypothesis envisages that negatively charged incoming NTPs interact sequentially with positively charged amino acids in the tunnel in order to reach the active site. Substrate-binding site has been mapped while the binding of Mg^{2+} and/or Mn^{2+} have mapped the active site of the enzyme. Thus, there is structural similarity and the conservation of secondary and tertiary structure elements in the palm and thumb domains of polymerases of the families *Cysto-, Flavi-, Picorna-,* and *Retroviridae* (van Dijk *et al.*, 2004). According to Poch *et al.* (1989), the 'palm domain' containing motifs A, B, C, and D is found in many polymerases whereas the E motif is unique to RdRps and reverse transcriptases, while motif C comprises the highly conserved GDD motif of RdRps.

Dohi *et al.* (2002) investigated the gross structure of the active BMV RdRp complex. In the solubilized RdRp complex, the intermediate area lying between the N-terminal methyltransferase-like domain and the C-terminal helicase-like domain of 1a protein as well as the N terminus region of 2a protein are exposed on surface of the solubilized complex. On the other hand, in membrane-bound RdRp complex, the intermediate region (between the methyltransferase-like and the helicase-like domains of 1a protein) is located at the border of the region that is buried within a membrane structure or with membrane-associated material. Thus, BMV RdRp complex is comprised of the RdRp itself and, in nearly all cases, of an NTP-binding helicase motif located on the same or another protein (Kadarè and Haenni, 1997).

It is thought that replicase complexes synthesizing positive-strands, negative-strands and subgenomic RNAs are conceivably different from each other. However, it looks more probable that the replication complex synthesizing one type of RNA strand is modified to synthesize another type of RNA strand.

C. Miscellaneous

The structures of six viral RdRps have been determined and van Dijk *et al.* (2004) have tabulated the relevant information, part of which is given below. The six viral RdRps are – 3D^{pol} of *Poliovirus* (*Picornaviridae*), NS5B of *Hepatitis C virus* (BK strain) (*Flavivirdae*), NS5B of *Hepatitis C virus* (J4 strain) (*Flavivirdae*), P2 of *Bacteiophage Φ6* (*Cystoviridae*), 3D^{pol} of *Rabbit haemorrhagic disease virus* (*Caliciviridae*), and λ3 of *Reovius* (type 3; strain Dearing) (*Reoviridae*). Their structures have been resolved at 1.85Å to 3.0Å, depending upon the virus. Complete RdRp protein has been crystallized in *Poliovuirus* (461 residues crystallized), *Hepatitis C virus* BK strain (531 to 578 residues crystallized and protein lacks 21- to 55 C-terminal amino acids as reported by different workers), *Hepatitis C virus* J4 strain (570 residues crystallized, protein lacks 21 C-terminal amino acid residues, and has a C-terminal His tag), *Bacteiophage Φ6* (complete 664-residue protein crystallized), *Rabbit haemorrhagic disease virus* (complete 516 residue-protein crystallized) and *Reovius* (type 3; strain Dearing) (complete 1267 residue-protein crystallized).

RdRp possesses certain structural features that facilitate and ensure *de novo* initiation of RNA synthesis. Such features have been identified in the above-mentioned polymerases and van Dijk *et al.* (2004) discuss these structural features of individual polymerases of these viruses in detail but are stated here briefly. In *Bacteiophage Φ6*, the two initiation nucleotides stack against each other as well as against a specialized priming platform followed by repositioning of RdRp for permitting egress of the product. This mechanism may also apply to RdRps of *Hepatitis C virus* and *Reovirus* and could be the initiation paradigm in many RdRps. Structural features of reovirus RdRp that facilitate initiation of RNA synthesis include a cube-like cage structure, tunnels and a priming loop. *Rabbit haemorrhagic disease virus* and *Poliovirus* have virus-encoded VPg covalently linked to 5`-terminus of their genome for minus-strand synthesis. *In vitro,* the rabbit haemorrhagic disease virus RdRp uses back priming from 3`-hydroxyl end of genome for minus-strand synthesis. In *Poliovirus*, VPg facilitates protein-primed RNA synthesis and Paul *et al.* (1998) predicted that this mechanism should function in most plus-strand RNA viruses with protein-linked genome.

The finger domains differ significantly among viral RNA polymerases, possibly due to specific adaptations to structurally diverse substrates. The finger domain of HCV polymerase has a long binding groove, which constitutes a template tunnel; a loop (the A or A1-loop) connects the fingers with the thumb; and a ß-hairpin protrudes from the thumb domain toward the active site at the base of palm domain. The A1-loop may be responsible for closed conformation of polymerase while Leu-30 is a critical element for polymerase activity. The ß-hairpin seems to be important for positioning the 3`-terminus of viral genome for correct initiation of replication since it allows only the single-stranded 3`-terminus of an RNA template to bind productively to the active site and may prevent the 3`-terminus of template RNA from slipping through the active site in order to ensure terminal initiation of replication. The thumb domain of RHDV RdRp seems to play an important role in RNA synthesis.

V. FUNCTIONS

Viral RdRps are involved in transcription and replication of viral RNA, in recombination and evolution of viruses, and in several other viral functions (Buck,

1996. 1999; Lai, 1998; Kao *et al.*, 2001; Ahlquist, 2002). The viral RdRps, thus, play multiple and crucial roles in viral life cycle and are basic to survival of a virus. It is mystifying as to how a single RdRp is able to recognize diverse types of linear and hairpin promoter structures. This could involve different binding sites in RdRp or different host factors. BMV polymerase recognizes the subgenomic RNA core promoter *in vitro* via a sequence-specific mechanism (Siegel *et al.*, 1997, 1998).

Out of the two overlapping TBSV replicase genes (p33 and p92) (Oster *et al.*, 1998; Panaviene *et al.*, 2003), the p92 bears the RdRp signature motifs in its C-terminus while the function(s) of p33 is at present unknown. Mutagenesis of the RNA-binding site [the arginine-proline-rich RPR motif (Rajendran and Nagy, 2003)] in p33 affected genomic RNA replication (Panaviene *et al.*, 2003), subgenomic RNA synthesis and RNA recombination (Panaviene and Nagy, 2003). This suggests that p33 is a multifunctional protein. Other RdRp molecules may also be multifunctional proteins though unequivocal data are not yet available.

The importance of each amino acid in the GDD motif has been worked out: the first aspartate residue of the GDD motif is strictly required for polymerase function (Molinari *et al.*, 1998) so that any changes at this position are not tolerated for *in vivo* virus replication and/or *in vitro* RNA synthesis (Longstaff *et al.*, 1993; Lohmann *et al.*, 1997). However, requirement for glycine is flexible and not that strict: its replacement does not completely abolish the functioning of RdRps (Hong and Hunt, 1996; Lohmann *et al.*, 1997); similarly, replacement of G_{620} of TBSV RdRp did not completely abolish virus replication (Boonrod *et al.*, 2005).

A. Replication of Viral RNA

The RdRps can utilize both single-stranded RNA and single-stranded DNA as templates with no marked specificity and cause synthesis of relatively short and heterogeneous RNA products that are usually of < 500 nucleotides. The development of template-dependent RdRp preparations from infected tissues or cells has greatly helped in studying replication of viral RNAs and its various stages. Viral RdRps mediate two steps during RNA replication: synthesis of complementary RNA strand from the invading plus-strand RNA template and serving of the new complementary negative-strand as template for producing large amounts of progeny positive-sense RNAs in an asymmetric manner. RdRps are involved in several substeps of these two steps: differentiation of genomic RNA replication from mRNA transcription, recruitment of viral RNA to site of replication, selection of template RNAs, selection of initiation sites for RNA synthesis, initiation and maintaining elongation of RNA synthesis, mRNA synthesis, modification of progeny RNAs with addition of 5`-cap or 3`-poly-adenylate. Moreover, the RdRp preparations have facilitated analyses of host and viral protein components and factors contained in RdRp complexes and several of the RdRp complexes have been useful for determining template requirements for initiation of RNA synthesis.

RNA synthesis can be initiated through only two different mechanisms: *de novo* RNA initiation and primer-dependent RNA initiation (Kao *et al.*, 2001). A characteristic feature of most of the viral RdRps is their ability to initiate RNA synthesis *de novo* i.e., without the need for an RNA primer for initiating RNA synthesis and is hence also called primer-independent initiation. This means that transcription promoters must have at least two important attributes: to be able to recruit to RdRp and to promote complementary RNA synthesis from the initiating nucleotide. *de novo* initiation requires interactions of at least four components: RdRp, RNA template with virus-specific initiation nucleotide, initiation nucleoside triphosphate (NTP), and a second NTP. The first phosphodiester bond is formed between initiation NTP and the second NTP.

The advantage of *de novo* RNA synthesis is that no information is lost during viral RNA replication and no additional enzymes are required for generating the primer or to cleave the region between template and newly synthesized RNA. In most cases, the productive *de novo* initiation event is immediately followed by elongation. *de novo* RNA initiation is employed by viruses possessing positive-, negative- (like *Vesicular stomatitis virus*), double-stranded- (like *Cystoviridae* and rotavirus) and ambisense RNA genomes (Kao *et al.*, 2001). *de novo* initiation is widely used by positive-strand RNA viruses like plant alphavirus-like viruses (Goldbach *et al.*, 1991; Strauss and Strauss, 1994), members of family *Flaviridae* namely *Hepatitis C virus* and *Dengu 2 virus* and bacteriophage Qβ.

The initiation nucleotide seems to affect initiation efficiency by the affinity of polymerase for initiation nucleotide of template and initiation NTPs. GTP is the preferred nucleotide for RdRps of BMV, *Bovine diarrhea virus*, *Hepatitis C virus*, and some others. Viruses have distinct initiation preferences even if they all use *de novo* initiation mechanism as in case of RdRps of flaviviruses, *Hepatitis C virus*, *Bovine diarrhea virus* and GBV-C. The RdRp of *Bovine diarrhea pestivirus* prefers to initiate from the 3`-termianl cytidylate, but can also use a penultimate cytidylate. On the other hand, the BMV, TYMV, and phage Qβ RdRps use penultimate cytidylate for initiating RNA synthesis (Singh and Dreher, 1998; Sivakumaran *et al.*, 1999). RdRp of BMV and CNV have distinct modes of initiation site recognition for replication initiation at the 3`-terminus and internal initiation. Enhancer-like activity of a viral RNA promoter is apparently important for this differentiation (Panavas *et al.*, 2002a, b; Ranjith-Kumar *et al.*, 2003c).

Wang *et al.* (2004) found that a single amino acid change of Phe50 to Ser50 in the 126/183 kDa replication proteins of *Odontoglossum ringspot tobamovirus* abolished virus replication and that corresponding mutations in TMV showed identical results. No minus-strand synthesis occurred in such mutants - indicating that Phe50 plays some crucial role in functioning of viral RdRps of tobamoviruses.

Asurmendi *et al.* (2004) suggested that TMV may possess a mechanism for regulating the amount of replicase that participates in its replication and thereby limits the amount of virus replication taking place. This conclusion was based on several types of observations: a significant proportion of replicase was not in close proximity with the viral replication complexes throughout the infection (Asurmendi *et al.*, 2004); replicase can accumulate in sites that lack viral RNA and possibly does not participate in virus replication (Más and Beachy, 1999); and a pool of replicase does not have any replicase activity (Hagiwara *et al.*, 2003). Restrepo *et al.* (1990) had also earlier suggested that some mechanism might be controlling the participation of NIa and NIb proteins in potyvirus replication. RNA genome of TCV contains two regions that specifically interact with viral CP. One of these regions consists of three RNA motifs that cover about 300 nucleotides in the gene that encodes the putative RdRp and spans a suppressible termination codon. Binding of CP subunits to this region may regulate the level of readthrough and hence of RdRp formation. This could be one such regulating mechanism.

1. Binding of Replicase and Role of *cis*-Acting Elements

Binding of replicase proteins to viral RNA (RNA-binding domains of RdRps) and recognition of *cis*-acting sequences is important during many steps of viral infection cycle. Thus, TBSV RdRp and RdRp of the closely related tombusvirus CNV binds to TBSV/CNV RNA *in vitro* and recognises *cis*-acting sequences such as the genomic and complementary promoters (Panavas *et al.*, 2002a, 2002b) and a replication enhancer (Panavas and Nagy, 2003b), which leads to RNA transcription *in vitro*. Rajendran and Nagy (2003) demonstrated that both replication proteins of TBSV bind to TBSV-derived RNA sequences (viral RNA) *in vitro*; that p33 and the overlapping domain of p92 contain an arginine- and proline-rich RNA-binding motif (termed RPR motif), which has the sequence RPRRRP; that presence of RPR motif is essential for tombusvirus replication; that this motif is highly conserved in tombusviruses and related carmoviruses and is critical for efficient RNA binding; that the non-overlapping C-terminal domain of p92 contains additional RNA-binding regions; that binding of p33 to single-stranded RNA is stronger than binding to double-stranded RNA *in vitro*; and that p33 binds to a TBSV-derived sequence with high affinity. Protein analysis predicts that RPR motif in p33 constitutes a hydrophobic pocket and is exposed Rajendran and Nagy (2003) propose that both p33 and p92 bind to RNA cooperatively. The functional significance of this cooperative binding by p33 and p92 is currently not known. It is possible that p33 can coat viral single-stranded RNA in infected cells, which may be beneficial during template recruitment and/or replication. The p33-coated RNAs may be more resistant to nucleases and less accessible to host-mediated gene silencing than are free viral RNAs (Waterhouse *et al.*, 2001; Vance and Vaucheret, 2001). The RNA-binding RPR region of p33 is essential for RNA binding and replica-tion of a tombusvirus *in vivo* as well (Panaviene *et al.*, 2003; Rajendran and Nagy, 2003). Boonrod *et al.* (2005) investigated which amino acid in the RPR motif play an essential role in RNA binding and found that the third arginine within the RPR motif is important for RNA binding. This confirms the earlier finding of Panaviene *et al.* (2003) who showed that replacing separately the second and third arginine residue by lysine residues decreased CNV replication by 95 to 98%. Panaviene *et al.* (2003) also demonstrated that mutations within the RNA-binding domains of the CNV replicase proteins affected the frequency of recombination by delaying the formation of recombinants; this could be true in case of TBSV as well. Thus, RPR motif plays a central role in viral RNA synthesis and metabolism and is also highly conserved among replicase proteins of tombusvirus-related TCV.

Lewandowski and Dawson (2000) showed that TMV 126-kDa replication protein appears to function in *cis* while 183-kDa protein can function in *trans*. They suggested that 126-kDa protein binds to its mRNA and targets it for replication. The 183-kDa protein might then bind to 126-kDa protein and initiate replication. The binding of polymerase domain occurs to 3`-terminal CCCA sequence of RNA. Osman and Buck (2003) propose that later binding of the 183-kDa protein to the already bound 126 kDa protein possibly orients the catalytically active site of RdRp domain close to 3`-terminus of template RNA to enable initiation of minus-strand RNA synthesis. Interaction of 126- and 183-kDa proteins has been demonstrated (Goregaoker *et al.*, 2001) and purified RdRp preparation containing a 1:1 dimer of 126- and 183-kDa proteins is able to synthesize negative-strand from a positive-strand RNA template *in vitro* (Watanabe *et al.*, 1999).

BMV replicase binds to the 3`-regions of all plus- and minus-strand RNAs; internal binding occurs only with RNA3. Choi *et al.* (2004) identified replicase-binding sites in all the BMV RNAs. Only one replicase-binding site was present in RNA1 and RNA2 near the initiation site for RNA synthesis, while both plus- and minus-strand RNA3 had two effective replicase binding sites. Stem-loop C within the tRNA-like structure is a major binding site in each of the plus-strand BMV RNAs. Subgenomic core promoter confers binding to BMV replicase. The cB box in minus-strand RNA1 and RNA2 is required for efficient BMV replicase binding and is more important for replicase binding than the B box. The intercistronic B box in RNA3 binds to 1a protein and forms the membrane-associated structure where BMV replicase is located in yeast and the B box of plus-strand RNA2 also binds 1a. Both the B and cB boxes seem to possess independent and necessary roles in BMV RNA replication; the B box binds 1a and regulates translation and replicase assembly while the cB box binds the replicase for initiation of genomic plus-strand RNA synthesis.

B. Transcription of Viral RNA

RdRps initiate as well as terminate transcription of viral RNA and are responsible for formation of subgenomic RNAs by transcription.

C. Recombination of Viral RNA

RNA recombination, leading to formation of viable recombinant RNAs, requires RNA replication (Lai, 1992; Simon and Bujarski, 1994; Bujarski, 1999) – indicating that viral replicase is involved in RNA recombination (Bujarski and Nagy, 1996). 'Copy choice (template switching)' is the favoured mechanism. This mechanism requires the polymerase, along with the nascent strand RNA, to switch from one template (called donor) to another template (called recipient). The switching occurs either after the polymerase-nascent RNA complex dissociates from the donor template or without this dissociation from donor. The former is the non-processive model and the latter the processive model. Recombination can take place during negative-strand RNA synthesis as in BMV (Bujarski *et al.*, 1994) or during positive-strand RNA synthesis as in TCV (Carpenter *et al.*, 1995). Thus, BMV replicase proteins do participate in RNA recombination. It is supported by the observation that alterations in replicase proteins may affect the occurrence of crossovers. For example, a single amino acid mutation within the core domain of BMV polymerase, protein 2a, inhibited the frequency of nonhomologous recombination below the level of detection (Figlerowicz *et al.*, 1997, 1998). Figlerowicz *et al.* (1998) reveal that N terminal domain of BMV 2a protein (the putative RdRp) participates in different ways in homologous and nonhomologous BMV RNA recombination. They map specific locations within N terminus that are involved in 1a-2a interaction and in recombination. The role of replicase proteins in RNA recombination in other viruses is still to be established.

Strand switching during RdRp copying is a means of viral RNA recombination (Cheng and Nagy, 2003). This, in turn, results in various favourable effects: it allows RNA viruses to repair deleterious mutations, to rearrange genes, and to acquire genes from other viruses or even hosts (Lai, 1992). Comparison of genomes of various viruses brings out resemblances in their genomes, which is a clear proof that virus recombination between viral genomes is the major force operating during evolution of RNA viruses. Thus, in addition to template-directed complementary RNA synthesis during normal RNA replication, many viral RdRps are capable of template switching which leads to formation of recombinant RNA molecules (Lai, 1992; Nagy and Simon, 1997). This property could also be responsible for the presence of defective interfering (DI) RNAs in carmo-, luteo-, and tombusviruses (Cheng *et al.*, 2002; Fabian *et al.*, 2003).

D. Evolution of Plant Viruses

Plant viruses, which depend upon RdRp for replication of their genomic and subgenomic RNAs, often show high inherent misincorporation of nucleotides during RNA copying by RdRps because RNA polymerases apparently lack proof-reading activities leading to error rates of 10^{-3} to 10^{-5} (Domingo and Holland, 1994; Domingo *et al.*, 1995, 2001). This leads to two very significant outcomes. One, it makes RNA copying an intrinsically erroneous process. This, coupled with high rates of viral RNA replication and large viral populations, enables the viruses to have intrinsically high mutation rates. This leads to a heterogeneous mixture of related species (Holland *et al.*, 1982) so that RNA viruses have a potential for much genetic variability (Roossinck, 1997; Drakke and Holland, 1999; Domingo *et al.*, 2001; Malpica *et al.*, 2002) and do actually show wide variability in many virus populations. This results in rapid virus evolution under all types of adverse conditions (for the virus) like improved resistance of the host, drug treatments, etc. so that RNA viruses sooner or later adapt to these adverse conditions – particularly since a number of potentially advantageous mutations are already present at the onset of selective pressure. Many RNA viruses, therefore, are very difficult to control. The success of RNA viruses is most probably a result of their high mutation rates.

Second, this gave rise to the quasispecies model for viruses. Such RNA viruses are thought to have evolved methods for correcting high error rates by combining mutant genomes with wild-type genomes – which maintains functional integrity of genomes. This also led to the formulation of concept of viral RNA 'quasi' species (Eigen and Biebricher, 1988) consisting of a population of closely related sequences. This model envisages populations of RNA replicons as a mixture of distinct but clearly related genotypes (Domingo *et al.*, 1995, 2001; Eigen, 1996).

Thus, high mutation rates and quasispecies dynamics (Eigen ad Biebricher, 1988; Drake and Holland, 1999; Domingo *et al.*, 2001) confer great adaptability to RNA viruses and represent a major obstacle for prevention and control of RNA diseases (Domingo *et al.*, 2001). Indeed, many RNA viruses, including *Human immunodeficiency virus* and *Hepatitis C virus*, efficiently escape host immune responses and medical treatment by promptly accumulating resistant mutants.

However, for any replication system there exists a theoretical maximum error rate that is compatible with maintenance of the information encoded in replicating genome (Eigen and Biebricher, 1988). Therefore, the larger the complexity of the genome, the higher the copying fidelity needed for maintaining the encoded information.

E. Miscellaneous Functions

Viral replicase systems, although developed only for a limited number of plus-strand RNA viruses, have proved very useful for studying the protein (*trans*-acting) and RNA (*cis*-acting) factors controlling viral replication.

The replicase preparations from BMV have been used as a model system for in-depth analysis of *cis*-acting elements that direct RNA synthesis and the mechanism of RNA synthesis (Dreher and Hall, 1988; Sun *et al.*, 1996; Siegel *et al.*, 1997, 1998; Adkins *et al.*, 1997, 1998; Chapman and Kao, 1998; Sivakumaran *et al.*, 1999) The results of BMV RNA synthesis *in vitro* have correlated with RNA replication in plant cells (Dreher and Hall, 1988; Chapman and Kao, 1999; Kim *et al.*, 2000). Moreover, being a tripartite positive-strand RNA virus, the replication proteins of BMV are encoded by RNA1 and RNA2 while proteins for encapsidation and cell-to-cell spread are encoded by RNA3. Hence, the replicase has to locate the specificity elements in RNA3 in *trans* (Ahlquist, 1992; Kao and Sivakumaran, 2000) leading to the study of *trans*-factors affecting viral RNA replication. Similarly, the plant-derived TBSV and CNV replicase preparations provide *in vitro* data that complement the *in vivo* data on genus *Tombusvirus* (Park *et al.*, 2002; Ray and White, 1999, 2003; Fabian *et al.*, 2003; Ray *et al.*, 2003).

TBSV and CNV (both tombusviruses) replicases obtained from infected plants have also been used extensively to characterize viral RNA elements that affect plus- and minus-strand RNA synthesis, including promoter (initiation) elements (Panavas *et al.*, 2002a, 2002b), replication enhancers (Panavas and Nagy, 2003a; Panavas *et al.*, 2003), a replication silencer element (Pogany *et al.*, 2003), and template-switching *in vitro* (Cheng and Nagy, 2003; Cheng *et al.*, 2002). These elements and processes are important for replication and recombination (including DI formation) of tombusviruses (Cheng *et al.*, 2002; Fabian *et al.*, 2003). Tombusviruses require the viral-coded p33 replication cofactor for template selection and recruitment into replication in infected cells.

Alterations in different regions of TMV RdRp may alter synthesis of subgenomic RNA (Watanabe *et al.*, 1987), symptom expression (Shintaku *et al.*, 1996; Goregaoker *et al.*, 2001), cell-to-cell movement (Hirashima and Watanabe, 2001), and host resistance responses (Hamamoto *et al.*, 1997). The coding regions of protease co-factor and C-terminal half of the putative helicase of *Bean pod mottle comovirus* are determinants of symptom severity (Gu and Ghabrial, 2005). The polymerase protein has been implicated in movement of TMV and CMV in infected plants (Deom *et al.*, 1997; Hirashima and Watanabe, 2001, 2003; Choi *et al.*, 2005). Both the CMV 3a MP and the CMV 2a polymerase protein affect the rate of CMV movement and timing of symptom development in zucchini squash and the two proteins function independently of each other in their interactions with the host (Choi *et al.*, 2005). Two alterations in CMV polymerase protein influenced the elicitation of a hypersensitive response and restriction of the virus to local lesions in cowpea (Kim and Palukaitis, 1997).

Virus replicase has been implicated in disease development (Bao *et al.*, 1996; Padmanabhan *et al.*, 2005). Padmanabhan *et al.* (2005) established during *in vitro* interaction assays that the helicase domain of TMV 126- or 183-kDa replicase protein(s) interacts with auxin/indole-3-acetic acid (Aux/IAA) protein called phytochrome-associated protein 1 (PAP1, also named IAA26) of *Arabidopsis thaliana*. The Aux/IAA protein PAP1 is a putative regulator of auxin response genes involved in plant development. Thus, interaction of TMV replicase with PAP1 interferes with and disrupts the plant's auxin response system and auxin signaling pathway and may play significant roles in inducing specific disease symptoms and determining disease severity. They hypothesized that TMV replicase protein disrupts PAP1 function by either destabilizing PAP1 through a ubiquitin-mediated process or by inappropriate sequestration of PAP1 by disrupting its nuclear localization. But precise mechanism by which TMV disrupts PAP1 function remains to be determined.

VI. CLASSIFICATION

The RdRps of positive-strand RNA viruses are classified into three supergroups (1, 2, and 3) on the basis of sequence similarities between more than 300 amino acids. After comparing amino acid sequences of the (putative) RdRps of twenty four plant viruses, apart from that of many animal viruses and some bacteriophages, Koonin (1991) formulated these three supergroups of viral RdRps because they shared a set of conserved sequence motifs. This was particularly so for the 'core' RdRp motifs IV, V, and VI. Each RdRp supergroup contains a number of different lineages and is also split into well-defined lineages compatible with viral groups (Koonin, 1991; Koonin and Dolja, 1993). The phylogenetic analysis of RdRps (Koonin and Dolja, 1993) revealed that each of the three supergroups and even the smaller groups comprising the supergroups contained plant viruses, animal viruses and bacteriophages with very different genome size, genome organisation, gene expression strategy and host specificities. However, only plant viruses are included in the classification of RdRps given below after Koonin (1991) and phylogenetic analysis by Koonin and Dolja (1993).

RdRp Supergroup 1: It contains plant viruses of picorna-like lineage (picornaviruses, comoviruses, nepoviruses, RTSV, PYFV), viruses of the poty lineage (potyviruses and bymoviruses), and viruses of the lineage with small genomes (sobemoviruses and luteoviruses).

RdRp Supergroup 2: It contains viruses of the lineage comprising of plant viruses with small genomes (BYDV – PAV, dianthoviruses, carmoviruses, tombusviruses, and necroviruses).

RdRp Supergroup 3: It contains viruses of tymo-like lineage (tymoviruses, carlaviruses, potexviruses, and capilloviruses), viruses of rubi-like lineage (*Beet necrotic yellow vein virus* and possibly alphaviruses), and viruses of tobamo-like lineage (tobamoviruses, trichoviruses, RBDV, hordeiviruses, tobraviruses, and closteroviruses)

All RNA polymerases belonging to supergroup 3 occur in viruses that possess superfamily 1 helicase-like protein. Additionally, complete correspondence exists between the helicase-like and polymerase lineages of these viruses. Such a relationship between polymerase and helicase-like genes occurs both in animal and plant viruses belonging to the alphavirus-like supergroup. Both animal and plant viruses of the picornavirus supergroup possess supergroup 1 RNA polymerases as well as superfamily 3 helicase-like proteins. The above two types of associations cannot be a mere coincidence but could suggest coevolution and a long–time association of helicase-like and polymerase-like genes of these two groups of viruses (Goldbach and de Haan, 1994). However, plant viruses of the potyvirus group contain supergroup 1 RNA polymerases and superfamily 2 helicase-like proteins indicating evolution of new combination of these genes in these viruses.

VII. REFERENCES

Adkins, S., Siegel, R. W., Sun, J. H., and Kao, C. C. 1997. Minimal templates directing accurate initiation of subgenomic RNA synthesis *in vitro* by the brome mosaic virus RNA-dependent RNA polymerase. *RNA* 3: 634-647.

Adkins, S., Seasick, S. S., Faurote, G., Siegel, R. W., and Kao, C. C. 1998. Mechanistic analysis of RNA synthesis by RNA-dependent RNA polymerase from two promoters reveals similarities to DNA-dependent RNA polymerase. *RNA* 4: 455-470.

Ahlquist, P. 1992. Bromovirus RNA replication and transcription. *Cur. Opin. Gen. Dev.* 2: 71-76

Ahlquist, P. 2002. RNA-dependent RNA polymerases, viruses and gene silencing. *Science* 296: 1270-1273.

Ahlquist, P., Noueiry, A. O., Lee, W.-M., Kushner, D. B., and Dye, B. T. 2003. Host factors in positive-strand RNA virus genome replication. *J. Virol.* 77: 8181- 8186.

Anindya, R., Chittori, S., and Savithri, H. S. 2005. Tyrosine 66 of pepper vein banding virus genome-linked protein is uridylylated by RNA-dependent RNA polymerase. *Virology* 336: 154-162.

Argos, P. 1988. A sequence motif in many polymerases. *Nucl. Acids Res.* 16: 9909-9916.

Asurmendi, S., Berg, R. H., Kao, J. C., and Beachy, R. N. 2004. Coat protein regulates formation of replication complexes during TMV infection. *Proc. Natl. Acad. Sci. USA* 101: 1415-1420.

Bao, Y., Carter, S. A., and Nelson, R. S. 1996. The 126- and 183-kilodalton proteins of *Tobacco mosaic virus*, and not their common nucleotide sequence, control mosaic symptom formation in tobacco. *J. Virol.* 70: 6378-6383.

Bates, H. J., Farjah, M., Osman, T. A. M., and Buck, K. W. 1995. Isolation and characterization of an RNA-dependent RNA polymerase from *Nicotiana clevelandii* plants infected with *Red clover necrotic mosaic dianthovirus*. *J. Gen. Virol.* 76: 1483-1491.

Boonrod, K., Chotewutmontri, S., Galetzka, D., and Krczal, G. 2005. Analysis of tombusvirus revertants to identify essential amino acid residues within RNA-dependent RNA polymerase motifs. *J. Gen. Virol.* 86: 823-826.

Brill, L. M., Nunn, R. S., Kahn, T. W., Yeager, M., and Beachy, R. N. 2000. Recombinant tobacco mosaic virus movement protein is an RNA-binding α-helical membrane protein. *Proc. Natl. Acad. Sci. USA* 97: 7112-7117.

Buck, K. W. 1996. Comparison of the replication of positive-stranded RNA viruses of plants and animals. *Adv. Virus Res.* 47: 159-251.

Buck, K. W. 1999. Replication of tobacco mosaic virus RNA. *Phil. Trans. Roy. Soc., London* 354: 613-627.

Bujarsky, J. J. 1999. Molecular basis of genetic variability in RNA viruses. In: Mandahar, C. L. (Ed.). Molecular Biology of Plant Viruses. Kluwer Academic Publishers, Boston / Dordrecht / London. p. 121-141.

Bujarski, J. J., and Nagy, P. D. 1996. Different mechanisms of homologous and nonhomologous recombination in *Brome mosaic virus*: Role of RNA sequences and replicase proteins. *Semin. Virol.* 7: 363-372.

Bujarski, J. J., Nagy, P. D., and Flasinki, S. 1994. Molecular studies of genetic RNA-RNA recombination in *Brome mosaic virus. Adv. Virus Res.* 43: 275-302.

Butcher, S. J., Grimes, J. M., Makeyev, E. V., Bamford, D. H., and Stuart, D. I. 2001. A mechanism for initiating RNA-dependent RNA polymerization. *Nature* 410: 235-240.

Carpenter, C. D., Oh, J. W., Zhang, C., and Simon, A. E. 1995. Involvement of a stem-loop structure in the location of junction sites in viral RNA recombination. *J. Mol. Biol.* 245: 608-622.

Chapman, M. R., and Kao, C. C. 1999. A minimal RNA promoter for minus-strand RNA synthesis by the brome mosaic virus polymerase. *J. Mol. Biol.* 286: 709-720.

Cheng, C. P., and Nagy, P. D. 2003. Mechanism of RNA recombination in carmo- and tombusviruses: Evidence for template switching by the RNA-dependent RNA polymerase. *J. Virol.* 77: 12033-12047.

Cheng, C. P., Pogany, J., and Nagy, P. D. 2002. Mechanism of DI RNA formation in tombusviruses: Dissecting the requirement for primer extension by the tombusvirus RNA-dependent RNA polymerase *in vitro. Virology* 304: 460-473.

Cheng, J.-H., Ding, M.-P., Hsu, Y.-H., and Tsai, C.-H. 2001. The partial purified RNA-dependent RNA polymerase from bamboo mosaic potexvirus- and potato virus X-infected plants containing the template dependent activities. *Virus Res.* 80: 41-52.

Choi, S.-K., Hema, M., Gopinath, K., Santos, J., and Kao, C. C. 2004. Replicase-binding sites on plus- and minus-strand brome mosaic virus RNAs and their roles in RNA replication in plant cells. *J. Virol.* 78: 13420-13429.

Choi, S. K., Palukaitis, P., Min, B. E., Lee, M. Y., Choi, J. K., and Ryu, K. H. 2005. Cucumber mosaic virus 2a polymerase and 3a movement proteins independently affect both virus movement and the timing of symptom development in zucchini squash. *J. Gen. Virol.* 86: 1213-1222.

Citovsky, V., Knorr, D., Schuster, G., and Zambryski, P. 1990. The p30 movement protein of *Tobacco mosaic virus* is a single strand nucleic acid-binding protein. *Cell* 60: 637-647.

Davenport, G. F., and Baulcombe, D. C. 1997. Mutation of the GKS motif of the RNA-dependent RNA polymerase from *Potato virus X* disables or eliminates virus replication. *J, Gen. Virol.* 78:1247-1251.

de Graaff, M., Thorburn, C., and Jaspars, E. M. J. 1995a. Interaction between RNA-dependent RNA polymerase of *Alfalfa mosaic virus* and its template: Oxidation of vicinal hydroxyl groups blocks *in vitro* RNA synthesis. *Virology* 213: 650-654.

de Graaff, M., Man, M. R., In'T Veld, and Jaspars, E. M. J. 1995b. *In vitro* evidence that the coat protein of *Alfalfa mosaic virus* plays a direct role in the regulation of plus and minus RNA synthesis: Implications for the life cycle of *Alfalfa mosaic virus. Virology* 208: 583-589.

Deiman, B. A. L. M., Séron, K., Jaspars, E. M. J., and Pleij, C. W. A. 1997. Efficient transcription of the tRNA-like structure of *Turnip yellow mosaic virus* by a template-dependent and specific RNA polymerase obtained by a new procedure. *J. Virol. Methods* 64: 181-195.

Deiman, B. A. L. M., Verlaan, P. W. G., and Pleij, C. W. A. 2000. *In vitro* transcription by the turnip yellow mosaic virus RNA polymerase: A comparison with alfalfa mosaic virus and brome mosaic virus replicases. *J. Virol.* 74: 264-271.

Deom, C. M., Quan, S., and He, X. Z. 1997. Replicase proteins as determinants of phloem-dependent long-distance movement tobamoviruses in tobacco. *Protoplasma* 199: 1-8.

Dohi, K., Mori, M., Furusawa, I., Mise, K., and Okuno, T. 2001. Brome mosaic virus replicase proteins localize with the movement protein at infection-specific cytoplasmic inclusions in infected barley leaf cells. *Arch. Virol.* 146: 1607-1615.

Dohi, K., Mise, K., Furusawa, I., and Okuno, T. 2002. RNA-dependent RNA polymerase complex of *Brome mosaic virus*: Analysis of the molecular structure with monoclonal antibodies. *J. Gen. Virol.* 83: 2879-2890.

Dolja, V. V. 2003. *Beet yellows virus:* The importance of being different. *Mol. Plant Pathol.* 4: 91-98.

Domingo, E., and J. J. Holland. 1994. Mutation rates and rapid evolution of RNA viruses. In: Morse, S. S. (Ed.). The Evolutionary Biology of Viruses. Raven Press, New York. p. 161-184.

Domingo, E., Holland, J. J., Biebricher, C., and Eigen, M. 1995. Quasi-species: The concept and the word. In: Gibbs, A., Calisher, C. H., and Garcia-Arenal, F. (Eds.). Molecular Basis of Virus Evolution. Cambridge Univ. Press, Cambridge. p. 181-191.

Domingo, E., Biebricher, K., Eigen, M., and Holland, J. 2001. Quasispecies and RNA Virus Evolution: Principles and Consequences. Landes Bioscience, Georgetown, Texas.

Donald, R. G. K., Lawrence, D. M., and Jackson, A. O. 1997. The barley stripe mosaic virus 58-kilodalton ßb protein is a multifunctional RNA binding protein. *J. Virol.* 71:1538-1546.

Doronin, S. V., and Hemenway. C. 1996. Synthesis of potato virus X RNAs by membrane-containing extracts. *J. Virol.* 70: 4795-4799.

Drake, J., and Holland, J. 1999. Mutation rates among RNA viruses. *Proc. Natl. Acad. Sci. USA* 96: 13910-13913.

Dreher, T. W., and Hall, T. C. 1988. Replication of BMV and related viruses. In: Holland, J., Domingo, E., and Ahlquist, P. (Eds.). RNA Genetics. CRC Press, Boca Raton, Fl., USA. p. 91-113.

Drugeon, G., Urcuqui-Inchima, S., Milner, M., Kadaré, G., Valle, R. P. C., Voyatzakis, A., Haenni, A.-L., and Schirawski, J. 1999. The strategies of plant virus gene expression: Models of economy. *Plant Sci.* 148: 77-88.

Eigen, M. 1996. On the nature of virus quasispecies. *Trends Microbiol.* 4: 216-218.

Eigen, M., and Biebricher, C. K. 1988. Sequence space and quasispecies distribution. In: Domingo, E., Holland, J. J., and Ahlquist, P. (Eds.). RNA Genetics, Volume 3. Variability of RNA Genomes. CRC Press Inc., Boca Raton, FL. p. 211-245.

Erokhina, T. N., Zinovkin, R. A., Vitushkina, M. V., Jelkmann, W., and Agranovsky, A. A. 2000. Detection of beet yellows closterovirus methyltransferase-like and helicase-like proteins *in vivo* using monoclonal antibodies. *J. Gen. Virol.* 81: 597-603.

Fabian, M. R., Na, H., Ray, D., and White, K. A. 2003. 3`-Terminal RNA secondary structures are important for accumulation of tomato bushy stunt virus DI RNAs. *Virology* 313: 567-580.

Figlerowicz, M., Nagy, P. D., and Bujarski, J. J. 1997. A mutation in the putative RNA polymerase gene inhibits nonhomologous, but not homologous, genetic recombination in an RNA virus. *Proc. Natl. Acad. Sci. USA* 94: 2073-2078.

Figlerowicz, M., Nagy, P. D., Tang, N., Kao, C. C., and Bujarski, J. J. 1998. Mutations in the N terminus of the brome mosaic virus polymerase effect genetic RNA-RNA recombination. *J. Virol.* 72: 9192-9200.

Gal-On, A., Canto, T., and Palukaitis, P. 2000. Characterization of genetically modified *Cucumber mosaic virus* expressing histidine-tagged 1a and 2a proteins. *Arch. Virol.* 145: 37-50.

Goldbach, R., and de Haan, P. 1994. RNA viral supergroups and the evolution of RNA viruses. In: Morse, S. S. (Ed.). The Evolutionary Biology of Viruses. Raven Press, New York. p. 105-120.

Goldbach, R., Le Gall, O., and Wellink, J. W. 1991. Alpha-like viruses in plants. *Semin. Virol.* 2: 19-25.

Goregaoker, S. P., Lewandowski, D. J., and Culver, J. N. 2001. Identification and functional analysis of an interaction between domains of the 126/183-kDa replicase-associated proteins of *Tobacco mosaic virus*. *Virology* 282: 320-328.

Grdzelishvili, V. Z., Garcia-Ruiz, H., Watanabe, T., and Ahlquist, P. 2005. Mutual interference between genomic RNA replication and subgenomic mRNA transcription in *Brome mosaic virus*. *J. Virol.* 79: 1438-1451.

Gu, H., and Ghabial, S. A. 2005. Bean pod mottle virus proteinase cofactor and putative helicase are symptom severity determinants. *Virology* 333: 271-283.

Hagiwara, Y., Komoda, K., Yamanaka, T., Tamai, A., Meshi, T., Funada, R., Tsuchiya, T., Naito, S., and Ishikawa, M. 2003. Subcellular localization of host and viral proteins associated with tobamovirus RNA replication. *EMBO J.* 22: 344-353.

Hamamoto, H., Watanabe, Y., Kamada, H., and Okada, Y. 1997. Amino acid changes in the putative replicase of *Tomato mosaic tobamovirus* that overcome resistance in *Tm-1* tomato. *J. Gen. Virol.* 78: 461-464.

Hansen, J. L., Long, A. M., and Schultz, S. C. 1997. Structure of the RNA-dependent RNA polymerase of *Poliovirus. Structure* 5: 1109-1122.

Hayes, R. J., and Buck, K. W. 1990. Complete replication of a eukaryotic virus RNA *in vitro* by a purified RNA-dependent RNA polymerase. *Cell* 63: 363-368.

Hayes, R. J., Pereira, V. C. A., McQuillin, A., and Buck, K. W. 1994. Localization of functional regions of cucumber mosaic virus RNA replicase using monoclonal and polyclonal antibodies. *J. Gen. Virol.* 75: 3177-3184.

Héricourt, F., Blanck, S., Redeker, V., and Jupin. I. 2000. Evidence for phosphorylation and ubiquitinylation of the turnip yellow mosaic virus RNA-dependent RNA polymerase domain expressed in a baculovirus-insect cell system. *Biochem. J.* 349: 417-425.

Hirashima, K., and Watanabe, Y. 2001. Tobamovirus replicase coding region is involved in cell-to-cell movement. *J. Virol.* 75: 8831-8836.

Hirashima, K., and Watanabe, Y. 2003. RNA helicase domain of tobamovirus replicase executes cell-to-cell movement possibly through collaboration with its nonconserved region. *J. Virol.* 77: 12357-12362.

Holland, J., Spindler, K., Horodyski, F., Grabau, E., Nichols, S., and van de Pol, S. 1982. Rapid evolution of RNA genomes. *Science* 215: 1577-1585.

Hong, Y., and Hunt, A. G. 1996. RNA polymerase activity by a potyvirus-encoded RNA-dependent RNA polymerase. *Virology* 226: 146-151.

Huang, C. Y., Huang, Y. L., Meng, M., Hsu, Y. H., and Tsai, C. H. 2001. Sequences at the 3`-untranslated region of bamboo mosaic potexvirus RNA interact with the viral RNA-dependent RNA polymerase. *J. Virol.* 75: 2818-2824.

Huang, Y.-L., Han, Y.-T., Chang, Y.-T., Hsu, Y.-H., and Meng, M. 2004. Critical residues for GTP methylation and formation of the covalent m^{7}GMP-enzyme intermediate in the capping enzyme domain of *Bamboo mosaic virus. J. Gen. Virol.* 78: 1271-1280.

Ishihama, A., and Barbier, P. 1994. Molecular anatomy of viral RNA-directed RNA polymerases. *Arch. Virol.* 134: 235-258.

Ishikawa, M., Kroner, P., Ahlquist, P., and Meshi, T. 1991. Biological activities of hybrid RNAs generated by 3`-end exchanges between tobacco mosaic and brome mosaic viruses. *J. Virol.* 65: 3451-3459.

Kadaré, G., and Haeni, A.-L. 1997. Virus-encoded helicases. *J. Virol.* 71: 2583-2590.

Kamer, G., and Argos, P. 1984. Primary structural comparisons of RNA-dependent polymerases from plant, animal and bacterial viruses. *Nucl. Acids Res.* 12: 7269-7282.

Kao, C. C, and P. Ahlquist. 1992. Identification of the domains required for direct interaction of the helicase-like and polymerase-like RNA replication proteins of *Brome mosaic virus. J. Virol.* 66: 7293-7302.

Kao, C. C., and Sivakumaran, K. 2000. *Brome mosaic virus,* good for an RNA virologist's basic needs. *Mol. Plant Pathol.* 1: 91-97.

Kao, C. C., Quadt, R., Hershberger, R. P., and Ahlquist, P. 1992. Brome mosaic virus RNA replication proteins 1a and 2a form a complex *in vitro. J. Virol.* 66: 6322-6329.

Kao, C. C., Singh, P., and Ecker, D. J. 2001. *De novo* initiation of viral RNA-dependent RNA synthesis. *Virology* 287: 251-260.

Kim, C. H., and Palukaitis, P. 1997. The plant defence response to *Cucumber mosaic virus* in cowpea is elicited by viral polymerase gene and affects virus accumulation in single cells. *EMBO J.* 16: 4060-4068.

Kim, M., Zhong, W., Hong, Z., and Kao, C. C. 2000. Template nucleotide moieties required for *de novo* initiation of RNA synthesis by a recombinant viral RNA-dependent RNA polymerase *J. Virol.* 74: 10312-10322.

Kim, S. H., Palukaitis, P., and Park, Y. I. 2002. Phosphorylation of cucumber mosaic virus RNA polymerase 2a protein inhibits formation of replicase complex. *EMBO J.* 21: 2292-2300

Koonin, E. V. 1991. The phylogeny of RNA-dependent RNA polymerases of positive-strand RNA viruses. *J. Gen. Virol.* 72: 2197-2206.

Koonin, E. V., and Dolja, V. V. 1993. Evolution and taxonomy of positive-strand RNA viruses: Implications of comparative analysis of amino acid sequences. *Crit. Rev. Biochem. Mol. Biol.* 28: 375-430.

Kroner, P. A., Richards, D., Traynor, P., and Ahlquist, P. 1989. Defined mutations in a small region of brome mosaic virus 2a gene cause diverse temperature-sensitive RNA replication phenotypes. *J. Virol.* 63: 5302-5309.

Lai, M. M. C. 1992. RNA recombination in animal and plant viruses. *Microbiol. Rev.* 56: 61-79.

Lai, M. M. C. 1998. Cellular factors in the transcription and replication of viral RNA genomes: A parallel to DNA-dependent RNA transcription. *Virology* 244: 1-12.

Lesburg, C. A., Cable, M. B., Ferrari, E., Hong, Z., Mannarino, A. F., and Weber, P. C. 1999. Crystal structure of the RNA-dependent RNA polymerase from *Hepatitis C virus* reveals a fully encircled active site. *Nat. Struct. Biol.* 6: 937-943.

Lewandowski, D. J., and Dawson, W. O. 2000. Functions of the 126- and 183-kDa proteins of *Tobacco mosaic virus. Virology* 271: 90-98.

Li, Y.-L., Cheng, Y.-M., Huang, Y.-L., Tsai, C.-H., Hsu, Y.-H., and Meng, M. 1998. Identification and characterization of the *Escherichia coli*-expressed RNA-dependent RNA polymerase of *Bamboo mosaic virus. J. Virol.* 72: 10093-10099.

Lohmann, V., Korner, F., Herian, U., and Bartenschlager, R. 1997. Biochemical properties of hepatitis C virus NS5B RNA-dependent RNA polymerase and identification of amino acid sequence motifs essential for enzyme activity. *J. Virol.* 71: 8416-8428.

Longstaff, M., Brigneti, G., Boccard, F., Chapman, S. N., and Baulcombe, D. C. 1993. Extreme resistance to potato virus X infection in plants expressing a modified component of the putative viral replicase. *EMBO J.* 12: 379-386.

Malpica, J., Fraile, A., Moreno, I., Obies, C., Drake, J., and Garcia-Arenal, F. 2002.The rate and character of spontaneous mutation in an RNA virus. *Genetics* 162: 1505-1511.

Mandahar, C. L. 1989. Multiplication of plus-sense RNA viruses. In: Mandahar, C. L. (Ed.). Plant Viruses. Volume 1. Structure and Replication. CRC Press, Boca Raton, FL., USA. p. 125-171.

Más, P., and Beachy, R. N. 1999. Replication of *Tobacco mosaic virus* on endoplasmic reticulum and role of the cytoskeleton and virus movement protein in intracellular distribution of viral RNA. *J. Cell. Biol.* 147: 945-958.

Miller, W. A., and Hall, T. C. 1983. Use of micrococcal nuclease in the purification of highly template-dependent RNA-dependent RNA polymerase from *Brome mosaic virus*-infected barley. *Virology* 125: 236-241.

Molnari, P., Marusic, C., Lucioli, A., Tavazza, R., and Tavazza, M. 1998. Identification of artichoke mottled crinkle virus (AMCV) proteins required for virus replication complementation of AMCV p33 and p92 replication-defective mutants. *J. Gen. Virol.* 79: 639-647.

Morozov, S., and Solovyev, A. 1999. Genome organisation in RNA viruses. In: Mandahar, C. L. (Ed.). Molecular Biology of Plant Viruses. Kluwer Academic Publ., Boston / Dordrecht / London. p. 47-98.

Mouches, C., Bové, C., and Bové, J. M. 1974. Turnip yellow mosaic virus RNA replicase: Partial purification of the enzyme from the solubilized enzyme-template complex. *Virology* 58: 409-423.

Mouches, C., Candresse, T., and Bové, J. M. 1984. Turnip yellow mosaic virus RNA-replicase contains host and virus-encoded subunits. *Virology* 134: 78-90.

Mueller, A.-M., Mooney, A. L., and MacFarlane, S. A. 1997. Replication of *in vitro* tobravirus recombinants shows that the specificity of template recognition is determined by 5`-non-coding but not 3`-non-coding sequences. *J. Gen. Virol.* 78: 2085-2088.

Nagy, P. D., and Pogany, J. 2000. Partial purification and characterization of cucumber necrosis virus and tomato bushy stunt virus RNA-dependent RNA polymerases: Similarities and differences in template usage between tombusvirus and carmovirus RNA-dependent RNA polymerases. *Virology* 276: 279-288.

Nagy, P. D., and Simon, A. E. 1997. New insights into the mechanisms of RNA recombination. *Virology* 235: 1-9.

Noueiry, A. O., and Ahlquist, A. 2003. Brome mosaic virus RNA replication: Revealing the role of host in RNA-virus replication. *Annu. Rev. Phytopathol.* 41: 77-98.

Ogawa, T., Watanabe, Y., Meshi, T., and Okada, Y. 1991. *Trans* complementation of virus-encoded replicase components of *Tobacco mosaic virus. Virology* 185: 580-584.

O'Reilly, E. K., and Kao, C. C. 1998. Analysis of RNA-dependent RNA polymerase structure and function as guided by known polymerase structures and computer prediction of secondary structures. *Virology* 252: 287-303.

O'Reilly, E. K., Wang, Z., French, R., and Kao, C. C. 1998. Interactions between the structural domains of the RNA replication proteins of plant-infecting RNA viruses. *J. Virol.* 72: 7160-7169.

Osman, T. A. M., and Buck, K. W. 1996. Complete replication *in vitro* of tobacco mosaic virus RNA by a template-dependent membrane-bound RNA polymerase. *J. Virol.* 70: 6227-6234.

Osman, T. A. M., and Buck, K. W. 1997. The tobacco mosaic virus RNA polymerase complex contains a plant protein related to RNA-binding subunit of yeast elF-3. *J. Virol.* 71: 6075-6082.

Osman, T. A. M., and Buck, K. W. 2003. Identification of a region of the tobacco mosaic virus 126- and 183-kilodalton replication proteins which binds specifically to the viral 3`-terminal tRNA-like structure. *J. Virol.* 77: 8669-8675.

Oster, S. K., Wu, B., and White, K. A. 1998. Uncoupled expression of p33 and p92 permits amplification of tomato bushy stunt virus RNAs. *J. Virol.* 72: 5845-5851.

Padmanabhan, M. S., Goregaoker, S. P., Golem, S., Shiferaw, H., and Culver, J. N. 2005. Interaction of the tobacco mosaic virus replicase proteins with the Aux/IAA protein PAP1/IAA26 is associated with disease development. *J. Virol.* 79: 2549-2558.

Panavas, T., and Nagy, P. D. 2003a. Yeast as a model host to study replication and recombination of defective interfering RNA of *Tomato bushy stunt virus. Virology* 314: 315-325.

Panavas, T., and Nagy, P. D. 2003b. The RNA replication enhancer element of tombusviruses contains two interchangeable hairpins that are functional during plus-strand synthesis. *J. Virol.* 77: 258-269.

Panavas, T., Pogany, J., and Nagy, P. D. 2002a. Analysis of minimal promoter sequences for plus-strand synthesis by the Cucumber necrosis virus RNA-dependent RNA polymerase. *Virology* 296: 263-274.

Panavas, T., Pogany, J., and Nagy, P. D. 2002b. Internal initiation by the cucumber necrosis virus RNA-dependent RNA polymerase is facilitated by promoter-like sequences. *Virology* 296: 275-287.

Panavas, T., Panaviene, Z., Pogany, J., and Nagy, P. D. 2003. Enhancement of RNA synthesis by promoter duplication in tombusviruses. *Virology* 310: 118-129.

Panaviene, Z., and Nagy, P. D. 2003. Mutations in the RNA-binding domains of tombusvirus replicase proteins affect RNA recombination *in vivo. Virology* 317: 359-372.

Panaviene, Z., Baker, J. M., and Nagy, P. D. 2003. The overlapping RNA-binding domains of p33 and p92 replicase proteins are essential for tombusvirus replication. *Virology* 308: 191-205.

Panaviene, Z., Panavas, T., Serva, S., and Nagy, P. D. 2004. Purification of the cucumber necrosis virus replicase from yeast cells: Role of coexpressed viral RNA in stimulation of replicase activity. *J. Virol.* 78: 8254-8263.

Pantaleo, V., Rubino, L., and Russo, M. 2003. Replication of carnation Italian ringspot virus defective interfering RNA in *Saccharomyces cerevisiae. J. Virol.* 77: 2116-2123.

Pantaleo, V., Rubino, L., and Russo, M. 2004. The p36 ad p95 replicase proteins of *Carnation Italian ringspot virus* cooperate in stabilizing defective interfering RNA. *J. Gen. Virol.* 85: 2429-2433.

Park, J.-W., Desvoyes, B., and Scholthof, H. B. 2002. Tomato bushy stunt virus genome RNA accumulation is regulated by inter-dependent *cis*-acting elements within the movement protein open reading frame. *J. Virol.* 76: 12747-12757.

Paul, A. V., van Boon, J. H., Filippov, D., and Wimmer, E. 1998. Protein-primed RNA synthesis by purified poliovirus RNA polymerase. *Nature* 393: 280-284.

Plante, C. A., Kim, K. H., Pillai-Nair, N., Osman, T. A. M., Buck, K. W., and Hemenway, C. L. 2000. Soluble, template-dependent extracts from *Nicotiana benthamiana* plants infected with *Potato virus X* transcribe both plus- and minus-strand RNA templates. *Virology* 275: 444-451.

Poch, O., Sauvaget, I., Delarue, M., and Tordo, N. 1989. Identification of four conserved motifs among RNA-dependent polymerase encoding elements. *EMBO J.* 8: 3867-3874.

Pogany, J., Fabian, M. R., White, K. A., and Nagy, P. D. 2003. A replication silencer element in a plus-strand RNA virus. *EMBO J.* 22: 5602-5611.

Quadt, R., and Jaspars, E. M. J. 1989. RNA polymerases of plus-strand RNA viruses of plants. *Mol. Plant-Microbe Interact.* 2: 219-223.

Quadt, R., and Jaspars, E. M. J. 1990. Purification and characterization of brome mosaic virus RNA-dependent RNA polymerase. *Virology* 178: 189-194.

Quadt, R., Rosdorff, H. J. M., Hunt, T. W., and Jaspars, E. M. J. 1991. Analysis of the protein composition of alfalfa mosaic virus RNA-dependent RNA polymerase. *Virology* 182: 309-315.

Quadt, R., Kao, C. C., Browning, K. S., Hershberger, R., and Ahlquist, P. A. 1993. Characterization of a host protein associated with brome mosaic virus RNA-dependent RNA polymerase. *Proc. Natl. Acad. Sci. USA* 90: 1498-1502.

Quadt, R., Ishikawa, M., Janda, M., and Ahlquist, P. 1995. Formation of brome mosaic virus RNA-dependent RNA polymerase in yeast requires coexpression of viral proteins and viral RNA. *Proc. Natl. Acad. Sci. USA* 92: 4892-4896.

Rajendran, K. S., and Nagy, P. D. 2003. Characterization of the RNA-binding domains in the replicase proteins of *Tomato bushy stunt virus. J. Virol.* 77: 9244-9258.

Rajendran, K. S., Pogany, J., and Nagy, P. D. 2002. Comparison of turnip crinkle virus RNA-dependent RNA polymerase preparations expressed in *Escherichia coli* or derived from infected plants. *J. Virol.* 76: 1707-1717.

Ranjith-Kumar, C. T., Gutshall, L., Sarisky, R. T., and Kao, C. C. 2003a. Multiple interactions within the hepatitis C virus RNA polymerase repress primer-dependent RNA synthesis. *J. Mol. Biol.* 330: 675-685.

Ranjith-Kumar, C. T., Santos, J. L., Gutshall, L. L., Johnston, V. K., Lin-Goerke, J., Kim, M. J., Porter, D. J., Maley, D., Greenwood, C., Earnshaw, D. L., Baker, A., Gu, B., Silverman, C., Sarisky, R. T., and Kao, C. C. 2003b. Enzymatic activities of the GB virus-B RNA-dependent RNA polymerase. *Virology* 312: 270-280.

Ranjith-Kumar, C. T., Zhang, X., and Kao, C. C. 2003c. Enhancer-like activity of a brome mosaic virus RNA promoter. *J. Virol.* 77: 1830-1839.

Rao, A. L. N., and Grantham, G. L. 1994. Amplification *in vivo* of brome mosaic virus RNAs bearing 3`-noncoding region from *Cucumber mosaic virus*. *Virology* 204: 478-481.

Rao, A. L. N., and Hall, T. C. 1993. Recombination and polymerase error facilitates restoration of infectivity in *Brome mosaic virus*. *J. Virol.* 67: 969-979.

Ray, D., and White, K. A. 1999. Enhancer-like properties of an RNA element that modulates tombusvirus RNA accumulation. *Virology* 256: 161-171.

Ray, D., and White, K. A. 2003. An internally located RNA hairpin enhances replication of tomato bushy stunt virus RNAs. *J. Virol.* 77: 245-257.

Ray, D., Wu, B., and White, K. A. 2003. A second functional RNA domain in the 5`-UTR of the tomato bushy stunt virus genome: Intra- and interdomain interactions mediate viral RNA replication. *RNA* 9: 1232-1245.

Restrepo, M. A., Freed, D. D., and Carrington, J. C. 1990. Nuclear transport of plant potyviral proteins. *Plant Cell* 2: 987-998.

Restrepo-Hartwig, M. A., and Ahlquist, P. 1996. Brome mosaic virus helicase- and polymerase-like proteins colocalize on the endoplasmic reticulum at sites of viral RNA synthesis. *J. Virol.* 70: 8908-9816.

Roossinck, M. J. 1997. Mechanisms of plant virus evolution. *Annu. Rev. Phytopathol.* 35: 191-209.

Rouleau, M., Bancroft, J. B., and Mackie, G. A. 1993. Partial purification and characterization of foxtail mosaic potexvirus RNA-dependent RNA polymerase. *Virology* 197: 695-703.

Rubino, L., and Russo, M. 1995. Characterization of resistance to *Cymbidium ringspot virus* in transgenic plants expressing a full-length viral replicase gene. *Virology* 212: 240-243.

Rubino, L., and Russo, M. 1998. Membrane targeting sequences in tombusvirus infections. *Virology* 252: 431-437.

Rubino, L., Weber-Lotfi, F., Dietrich, A., Stussi-Garaud, C., and Russo, M. 2001. The open reading frame 1-encoded ('36K') protein of *Carnation Italian ringspot virus* localizes to mitochondria. *J. Gen. Virol.* 82: 29-34.

Ryabov, E. V., Kim, S. H., and Taliansky, M. 2004. Identification of a nuclear localization signal and nuclear export signal of the umbraviral long-distance movement protein. *J. Gen. Virol.* 85: 1329-1333.

Schaad, M. C., Jensen, P. E., and Carrington, J. C. 1997. Formation of plant RNA virus replication complexes on membranes: Role of an endoplasmic reticulum-targeted viral protein. *EMBO J.* 16: 4049-4059.

Scholthof, K.-B. G., Scholthof, H. B., and Jackson, A. O. 1995. The tomato bushy stunt virus replicase proteins are coordinately expressed and membrane-associated. *Virology* 208: 365-369.

Schwartz, M., Chen, J., Janda, M., Sullivan, M., de Boon, J., and Ahlquist, P. 2002. A positive-strand RNA virus replication complex parallels form and function of retrovirus capsids. *Mol. Cell* 9: 505-514.

Shintaku, M. H., Carter, S. A., Bao, Y., and Nelson, R. S. 1996. Mapping nucleotides in the 126-kDa protein gene that control the differential symptoms induced by two strains of *Tobacco mosaic virus*. *Virology* 221: 218-225.

Siegel, R. W., Adkins, S., and Kao, C. C. 1997. Sequence-specific recognition of a subgenomic RNA promoter by a viral RNA polymerase. *Proc. Natl. Acad. Sci. USA* 94: 11238-11243.

Siegel, R. W., Bellon, L., Beigelman, L., and Kao, C. C. 1998. Moieties in an RNA promoter specifically recognised by a viral RNA-dependent RNA polymerase. *Proc. Natl. Acad. Sci. USA* 95:11613-11618.

Simon, A. E., and Bujarski, J. J. 1994. RNA-RNA recombination and evolution in virus-infected plants. *Annu. Rev. Phytopathol.* 32: 337-362.

Singh, R. N., and Dreher, T. W. 1997. Turnip yellow mosaic virus RNA-dependent RNA polymerase: Initiation of minus-strand synthesis *in vitro*. *Virology* 233: 430-439.

Singh, R. N., and Dreher, T. W. 1998. Specific site selection in RNA resulting from the combination of nonspecific secondary structure and –CCR-boxes: Initiation of minus-strand synthesis by turnip yellow mosaic virus RNA-dependent RNA polymerase. *RNA* 4: 1083-1095.

Sivakumaran, K., Kim, C. H., Tayon, R., Jr., and Kao, C. C. 1999. RNA sequence and secondary structural determinants in minimal viral promoter that directs replicase recognition and initiation of genomic plus-strand RNA synthesis. *J. Mol. Biol.* 294: 667-682.

Sivakumaran, K., Hema, M., and Kao, C. C. 2003. Brome mosaic virus RNA synthesis *in vitro* and in barley protoplasts. *J. Virol.* 77: 5703-5711.

Song, C., and Simon, A. E. 1994. RNA-dependent RNA polymerase from plants infected with *Turnip crinkle virus* can transcribe (+) and (-) strands of virus-associated RNAs. *Proc. Natl. Acad. Sci. USA* 91: 8792-8796.

Stawicki, S., and Kao, C. C. 1999. Spatial perturbations within an RNA promoter specifically recognized by a viral RNA-dependent RNA polymerase (RdRp) reveal that RdRp can adjust its promoter binding sites. *J. Virol.* 73: 189-204.

Stork, J., Panaviene, Z., and Nagy, P. D. 2005. Inhibition of *in vitro* RNA binding replicase activity by phosphorylation of the p33 replication protein of *Cucumber necrosis tombusvirus. Virology*, in press.

Strauss, J. H., and Strauss, E. G. 1994. The alphaviruses: Gene expression, replication and evolution. *Microbiol. Rev.* 58: 491-562.

Strauss, J. H., and Strauss, E. G. 1999. With a little help from the host. *Science* 283: 802-804.

Sun, J.-H., Adkins, S., Faurote, G., and Kao, C. C. 1996. Initiation of (-)-strand RNA synthesis catalyzed by the BMV RNA-dependent RNA polymerase: Synthesis of oligoribonucleotides. *Virology* 226: 1-12.

Teycheney, P.-Y., Aaziz, R., Dinant, S., Salánki, K., Tourneur, C., Balázs, E., Jacquemond, M., and Tepfer, M. 2000. Synthesis of (-)-strand RNA from the 3`-untranslated region of plant viral genomes expressed in transgenic plants upon infection with related viruses. *J. Gen. Virol.* 81:1121-1126.

Traynor, P., Young, B. M., and Ahlquist, P. 1991. Deletion analysis of brome mosaic virus 2a protein: Effects of RNA replication and systemic spread. *J. Virol.* 65: 2807-2815.

Tsai, M. S., Hsu, Y. H., and Lin, N. S. 1999. Bamboo mosaic potexvirus satellite RNA (sat BaMV RNA)-encoded P20 protein preferentially binds to satBaMV RNA. *J. Virol.* 73: 3032-3039.

Tsujimoto, Y., Numaga, T., Ohshima, K., Yano, M. A., Ohsawa, R., Gotu, D., Naito, S., and Ishikawa, M. 2003. *Arabidopsis* tobamovirus multiplication (TOM) 2 locus encodes a transmembrane protein that interacts with TOM1. *EMBO J.* 22: 335-343.

van Dijk, A. A., Makeyev, E. V., and Bamford, D. H. 2004. Initiation of viral RNA-dependent RNA polymerization. *J. Gen. Virol.* 85: 1077-1093.

van der Heijden, M. W., and Bol, J. F. 2002. Composition of alphavirus-like replication complexes: Involvement of virus- and host-encoded proteins. *Arch. Virol.* 147: 875-898.

Vance, V., and Vaucheret, H. 2001. RNA silencing in plant-defence and counterdefence. *Science* 292: 2277-2280.

Vlot, A. Neeleman, C., L., Linthorst, H. J. M., and Bol, J. F. 2001. Role of the 3`-untranslated regions of alfalfa mosaic virus RNAs in the formation of a transiently expressed replicase in plants and in the assembly of virions. *J. Virol.* 75: 6440-6449.

Vlot, A. C., Laros, S. M., and Bol, J. F. 2003. Coordinate replication of alfalfa mosaic virus RNAs 1 and 2 involves *cis*-acting and *trans*-acting functions of the encoded helicase-like and polymerase-like domains. *J. Virol.* 77: 10790-10798.

Wang, H.-H., and Wong, S.-M. 2004. Significance of the 3`-terminal region in minus-strand RNA synthesis of *Hibiscus chlorotic ringspot virus. J. Gen. Virol.* 85: 1763-1776.

Wang, H.-H., Yu, H.-H., and Wong, S.-K. 2004. Mutation of Phe50 to Ser50 in the 126/183 kDa proteins of *Odontoglossum ringspot virus* abolishes virus replication but can be complemented and restored by exact reversion. *J. Gen. Virol.* 85: 2447-2457.

Watanabe, T., Honda, A., Iwata, A., Ueda, S., Hibi, T., and Ishihama, A. 1999. Isolation from tobacco mosaic virus-infected tobacco of a solubilized template-specific RNA-dependent RNA polymerase containing a 126/183K protein heterodimer. *J. Virol.* 73: 2633-2640.

Watanabe, Y., Meshi, T., and Okada, T. 1987. Infection of protoplasts with *in vitro* transcribed *Tobacco mosaic virus* RNA. *FEBS Lett.* 219: 65-69.

Waterhouse, P. M., Wang, M. B., and Lough, T. 2001. Gene silencing as an adaptive defence against viruses. *Nature* 411: 834-842.

Weber-Lotfi, F., Dietrich, A., Russo, M., and Rubino, L. 2002. Mitochondrial targeting and membrane anchoring of a viral replicase in plant and yeast cells. *J. Virol.* 76: 10485-10496.

Weiland, J. J., and Dreher, T. W. 1993. *cis-* Preferential replication of the turnip yellow mosaic virus RNA genome. *Proc. Natl. Acad. Sci. USA* 90: 6095-6099.

Williamson, J. R. 2000. Induced fit in RNA-protein recognition. *Nat. Struct. Biol.* 7: 834-837.

Yoshinari, S., Nagy, P. D., Simon, A. E., and Dreher, T. W. 2000. CCA initiation boxes without unique promoter elements support *in vitro* transcription by three viral RNA-dependent RNA polymerases. *RNA* 6: 698-707.

6 HELICASES

I. INTRODUCTION

Helicases have been reviewed and/or dealt with in several important papers (Gorbalenya and Koonin, 1989; Gorbalenya *et al.*, 1989a; Koonin, 1991; Koonin and Dolja, 1993; Gorbalenya and Koonin, 1993; Buck, 1996; Kadaré and Haenni, 1997; Caruthers and McKay, 2002). They are a well-established class of enzymes that unwind double-stranded DNA (dsDNA), double-stranded DNA/RNA hybrids, or double-stranded RNA (dsRNA) structures during replication and/or transcription of cellular and viral genomes by disrupting the hydrogen bonds (that keep the two DNA/RNA strands together) by a reaction that is coupled with hydrolysis of an NTP (Kadaré and Haenni, 1997). They contain nucleoside 5`-triphosphate (NTP)-binding motif (Gorbalenya and Koonin, 1989; Gorbalenya *et al.*, 1989b, 1990) and catalyze separation of duplex oligonucleotides into single strands in an ATP-dependent process by which the energy required for driving the uncoupling process is obtained from hydrolysis of the bound ATP (Caruthers and McKay, 2002). Thus, helicase proteins contain the NTP-binding motif, separate the duplex oligonucleotides into single strands in an ATP-dependent reaction, generate energy, and are encoded by most of the positive-strand RNA viruses. This means that all helicases possess nucleoside triphosphatase (NTPase) activity so that all helicases have a NTP–dependent RNA helicase activity. Thus, helicases employ some molecular mechanism that changes the chemical energy generated by ATPase activity into a force that causes separation and displacement of strands of an oligonucleotide. A functional helicase has two characteristics: a conformational rearrangement of the protein that is driven by ATP and can generate energy, and at least two nucleic acid-binding sites between which the generated energy could exert itself for promoting unwinding/strand displacement. However, cylindrical inclusion (CI) RNA helicase of PPV possesses only one RNA-binding site, which is located in the carboxy-terminal part of protein (Fernándes *et al.*, 1995) and exhibits nucleic acid-stimulated ATPase activity (Lain *et al.*, 1991).

By 1997, the helicase enzyme had been definitely detected, identified, and established in only in two RNA plant viruses [PPV and *Tamarillo mosaic potyvirus* (TaMV)] and three RNA animal viruses (including that of *Hepatitis C virus* - (HCV) (Kadaré and Haenni, 1997). In all other cases, the helicase motif has been detected in the genome sequence. These are the putative helicases since they are determined only on the basis of sequence similarities/homologies with definitive helicases but possess no supporting biochemical data. It is possible that many of these putative helicases have helicase activity that is dependent upon ATP hydrolysis. Stimulation of ATPase activity by single-stranded DNA or RNA is the general circumstantial evidence suggesting that a particular protein is likely to be a helicase. The best-studied viral helicase is that of HCV. Helicases and putative helicases are placed together in a 'helicase-like family'.

A. Occurrence

The DNA/RNA helicases (definitive and putative) are found in bacteria, yeast, humans and viruses. Helicases may be involved in the replication of all viruses, whatever be the nature of their genome, because partially double-stranded polynucleotide structures are found during replication of all viruses studied so far. Presence of helicases in viruses having dsDNA or dsRNA genomes is understandable since they are needed for disrupting the stable duplex genome structures during replication and/or transcription. In fact, a putative helicase exists in genomes of all dsDNA viruses for which complete nucleotide sequence is known. The dsRNA viruses with large genomes, like reoviruses, also seem to possess helicase motifs while viruses with small dsRNA genomes, like birnaviruses, lack helicase motifs – indicating a relationship between genome size and existence of putative helicases (Gorbalenya and Koonin, 1989; Koonin and Dolja, 1993). The situation in positive-strand RNA viruses is discussed below. The retroviruses, negative-stranded RNA viruses, and ssDNA viruses do not seem to contain a putative helicase (Kadaré and Haenni, 1997).

The single-stranded positive-sense RNA viruses present a curious mixture. During their RNA replication, the genomes can be partially double-stranded and the RNA templates must be repeatedly available for a new cycle of replication process. The replication machinery may require an RNA unwinding activity both to break intramolecular base pairing in template RNA as well as to prevent formation of extensive base pairing between template RNA and the nascent complementary strand. In fact, scanning of positive-sense RNA genome for translation needs two essential factors; the canonical translation initiation factor eIF4A and a DEAD (Asp-Glu-Ala-Asp)- box RNA helicase for unwinding secondary structure in 5` -region (McCarthy, 1998). Only RNA helicases are discussed here.

RNA helicases appear to be universally present because modulation of RNA structure is essential for many basic metabolic processes like RNA synthesis, RNA splicing, RNA replication and RNA translation. Thus, most of the positive-strand RNA viruses possess a helicase/helicase-like structure in their genomes, which contain the relevant ORFs that encode or have the potential to encode helicases. In fact, about 80% of the positive-strand RNA viruses, whose nucleotide sequences are known, have the potential to code for at least one potential helicase. All positive-strand RNA viruses containing genomes larger than 5.8 kb generally encode a protein or protein domain having homologies with known helicases (Gorbalenya and Koonin, 1989; Kadaré and Haenni, 1997). Only the positive-strand RNA viruses having a small genome (< 5.8kb) seem not to encode a helicase possibly, because, unwinding is not required for replication of this small genome. Thus, the relationship between genome size and existence of putative helicases also operates in single-stranded positive-sense RNA viruses (Gorbalenya *et al.*, 1989; Koonin, 1991). Around 80% of such viruses, whose genome sequence is known, possess at least one potential helicase except those having genome smaller than < 5.8 kb. The above observations suggest that viruses with a small double-stranded or single-stranded RNA genome may not need RNA unwinding for its replication.

A need for viral and/or cellular helicases can be thought to exist at several steps of viral RNA synthesis. The RNA helicases might therefore play multiple roles in translation and in other steps of plant viral RNA replication. Some of these roles have been established and well characterized. Thus, the role of helicase enzyme has been definitely established in two RNA plant viruses (PPV and TaMV potyviruses) and three RNA animal viruses (Kadaré and Haenni, 1997). The roles of helicases in great majority of viruses have only been suggested and lack proper experimental proof. It is so because the helicase motif in all these cases has merely been detected in the genome sequence (Kadaré and Haenni, 1997). This in turn makes the efforts to assign any definite functions to them during viral life cycle largely speculative. There is thus a lack of definite information about viral helicases and the role played by them in majority of cases.

The PPV CI protein was the first helicase (of a positive-strand RNA virus) of which RNA helicase activity was actually detected experimentally (Lain *et al.*, 1990, 1991). This protein was purified from infected leaves of tobacco plants, and was later expressed and purified, in the form of a fusion protein, from bacteria. Presence of NTP and Mg^{2+} and hydrolysis of NTP is absolutely essential for helicase activity of PPV CI protein and for separation of the two RNA strands of the duplex. The PPV CI protein efficiently binds poly(A), its helicase activity is specific for PPV RNA duplex, direction of unwinding is from 3`- to 5`-end, contains an ATP-dependent RNA-unwinding activity and amino acid sequence similarity with known helicases (Lain *et al.*, 1990, 1991). Helicase activity of PPV CI protein needs a double-strand substrate with a 3`-single-strand terminal. Similarly, TaMV CI protein was purified from infected tobacco leaves, was found to be an RNA helicase, possessed essentially the same features as the PPV helicase and functions in a 3`- to 5`-polarity (Eagles *et al.*, 1994). Both PPV CI protein and TaMV CI protein show RNA-stimulated ATPase, RNA helicase, and RNA binding activities and belong to helicase superfamily 2.

The carboxy-terminal of 1a protein of both BMV (Kroner *et al.*, 1990) and CMV (Hayes *et al.*, 1994) has a helicase-like domain; in these two plant viruses as well as in many others such ORFs are essential for RNA replication. Peters *et al.* (1994) demonstrated the importance of the putative helicases of CPMV (58-kDa protein region of B polyprotein precursor). Purified RdRp complex, isolated from AMV-infected plants, could unwind double-stranded viral RNA in 3`- to 5`-direction (de Graaff *et al.*, 1995). Helicase-like p1 protein was present in these RdRp preparations (Quadt *et al.*, 1991) but it was not conclusively demonstrated if unwinding was accomplished by p1 protein, p2 protein, or by some other component of AMV complex.

Some positive-strand RNA plant viruses (carla-, furo-, hordei-, and potexviruses) contain TGB, which contains three homologous genes concerned with the transport of those viruses within the host plants. The first protein of TGB is a helicase (NTP-binding helicase motif) so that these plant viruses have two helicase-like motifs – one is embedded in replicative protein and is possibly involved in RNA replication while the second is the helicase of TGB. The second helicase-like protein is needed for cell-to-cell virus transport in infected plants but is not concerned with viral RNA replication (Petty *et al.*, 1990; Beck *et al.*, 1991; Gilmer *et al.*, 1992; Mushegian and

Koonin, 1993). In BNYVV, it binds viral nucleic acids (Bleykasten *et al.*, 1996) but its precise role in cell-to-cell virus movement is not yet known. The second helicase-like protein is regarded to originate from the replicative helicase-like protein by a gene duplication event, followed by allotment of a different function to these second proteins (Koonin and Dolja, 1993). Viral movement in infected plants is an active process and so in all likelihood involves an energy-dependent step and the second helicase presumably meets this need. It is supported by the fact that in *Foxtail mosaic potexvirus* the TGB-helicase-like protein exhibits both the ATPase and RNA-binding activities (Rouleau *et al.*, 1994). These properties could be central to the tagging and/or sequestering viral RNA molecules during their transport to adjacent cells.

Most RNA helicases, including cellular and viral helicases and most of the DNA helicases, bind to single-stranded regions adjacent to the duplex to be unwound, and do not act on completely double-stranded structures. Unwinding of a completely double-stranded replication form (RF) RNA could therefore involve another protein. In case of positive-stranded RNA replication, a 3`- to 5`-helicase bound to negative-template strand of the unwound origin (unwinding promoter) could perform subsequent duplex unwinding. The PPV helicase activity (Lain *et al.*, 1990) possesses the required displacement activity, as does the AMV RdRp complex (de Graaff *et al.*, 1995).

B. Characteristics of Viral RNA Helicases

All definite viral helicases show NTPase activity, which depends upon the existence of an NTP and a divalent cation, usually Mg^{++}. The products of NTP hydrolysis are NDP and P*i,* accompanied with the release of energy (due to ATP hydrolysis). All viral RNA helicases conduct unidirectional unwinding of the RNA duplex in 3`- to 5`-direction; in contrast, the DNA helicases do so either in a 5`- to 3`- or 3`- to 5`-direction. This unidirectionality may be an intrinsic characteristic of RNA helicases. Viral helicases have low substrate/template specificity so that no specific RNA sequence is required for activity of viral RNA helicases *in vitro*. Thus, viral RNA helicases hydrolyze ATP in presence of double-stranded and single-stranded RNAs, synthetic homopolymers, and double-stranded and single-stranded DNA. Helicases bind both single-stranded RNA as well as single-stranded DNA but with varying affinities.

Nature of substrate is an important factor during the strand displacement. The NTPase activity is generally stimulated by the presence of single-stranded nucleic acids or single-stranded regions of a double-stranded RNA. Completely double-stranded regions cannot be unwound; presence of unpaired regions, to which RNA helicase binds, is a pre-requisite for initiation of unwinding. In fact, viral RNA helicases must react with a 3`-single-stranded region (of the substrate) to which they first bind, then move toward the duplex region of substrate followed by unwinding.

It is possible that multiple binding sites could occur in a single protein molecule. Presence of multiple binding sites appears to be essential for unwinding to occur – because it makes it possible for a helicase to bind either single-stranded and duplex areas of a protein simultaneously or to bind two complementary single strands of an unwinding fork. In fact, majority of DNA virus helicases do seem to form oligomeric structures. This may be true of RNA helicases as well so that oligomerization most likely enables helicases to acquire multiple nucleic acid-binding sites. For instance, it is only when eIF4A (a helicase), which possesses a single RNA-binding site, is dimerized with eIF4B that the former functions as an RNA helicase. Possibly, eIF4B stabilizes the eIF4A/mRNA complex by providing its two RNA-binding domains. The CI RNA helicase of PPV has only one RNA-binding site that is located in its C-terminal part. But this protein always purifies as a large complex that in all probability is a multimer (Lain *et al.*, 1991).

To sum up, many positive-stranded RNA viruses encode helicases (or putative helicases) for duplex unwinding during replication; the RNA helicase activity needs a 3`-single-stranded RNA extension for initiating unwinding in a 3`- to 5`-direction; the known helicases are associated with NTPase activity; and unwinding process is coupled with NTP hydrolysis since energy released during NTP hydrolysis presumably powers the unwinding process (Kadaré and Haenni, 1997).

C. Nucleoside Triphosphate Connection

All the most important biochemical phenomena are coupled to hydrolysis of NTP along with release of energy in the form of ATP. Some of these processes could be DNA and RNA replication and transcription (especially during unwinding of DNA and RNA duplex, mRNA translation, DNA packaging and generation of dNTP), recombination, repair, protein synthesis, membrane transport, signal transduction and still other processes (Gorbalenya and Koonin, 1989). The NTP-binding pattern is one of the best-conserved sequences in viral proteins. All double-stranded DNA viruses, most groups of positive-strand RNA and of single-stranded DNA viruses have such proteins while negative-strand RNA viruses and retroid viruses appear to lack NTP proteins (Gorbalenya and Koonin, 1989).

Gorbalenya and Koonin (1989), after an extensive systemic survey, arrived at the following conclusions: The NTP-binding pattern is non-randomly distributed among various virus classes and is specifically typical of dsDNA viruses and positive-strand RNA viruses; most groups of positive-strand RNA viruses contain NTP-binding pattern; the helicases of all the three virus supergroups possess the NTP-binding motif which is made up of sites A and B; all of the positive-strand RNA viruses whose genome size is over about 5.8 kb contain NTP binding fingerprint while viruses with smaller genomes do not contain NTP proteins so that a strong correlation exists between genome size and existence of NTP proteins in these viruses and possibly also in double-stranded RNA viruses. This correlation seems to exist in the double-stranded DNA viruses also so that some groups of double-stranded DNA viruses have the NTP proteins while others do not. All NTP proteins bind ATP and carry the classical Walker 'A' (phosphate-binding loop or 'P-loop') and 'B' (Mg^{2+}-binding aspartic acid) motif.

The presence of NTP-binding motif in all the three superfamilies of helicases is the best common denominator between them. Thus, all helicases encoded by positive-strand RNA viruses, that have been studied so far (PPV CI, TaMV CI, and three helicases of animal viruses), show high levels of NTPase activity – showing that thus far a NTP-dependent RNA helicase activity has been reported in only a few helicases. The NTP-binding motif is constituted by A and B sites. The site A is made up of a stretch of hydrophobic amino acid residues followed by the conserved sequence GxxxGKS/T, x is any amino acid. This site directly binds ß and γ phosphates of NTP. The B site is made up of an Asp preceded by a stretch of hydrophobic amino acids and serves to chelate Mg^{2+} of Mg-NTP complex.

Most, but not all, of the nucleotide triphosphatases contain a common sequence pattern consisting of two separate motifs – the N-terminal 'A' site and the C-terminal 'B' site and both contain hydrophobic stretches. The phosphoryl moiety of ATP or GTP is bound to the central flexible Gly-rich loop of 'A' site while the invariant D of 'B' site chelated Mg^{2+} of Mg-NTP. The site A directly binds ß and γ phosphates of NTP while site B chelates Mg^{2+} of Mg-NTP complex. The NTPase activity depends upon the presence of an NTP and a divalent cation, which mostly is Mg^{2+}, and the products of this hydrolysis are DNP and P*i* for all helicases studied. This hydrolysis is an indispensable condition for the unwinding activity to occur. The NTPase function of helicases is commonly stimulated by the existence of single-stranded nucleic acid.

The NTP-binding elements A (I) and B (II) are apparently absent in several families of the plus-strand RNA viruses. This is true of some viruses possessing supergroup I polymerases of sobemo lineage (in plant virus genera *Sobemovirus* and *Luteovirus* subgroup II) and of some viruses having supergroup II polymerases as in plant virus genera *Dianthovirus*, *Luteovirus* subgroup I, *Machlomavirus*, *Necrovirus*, and family *Tombusviridae*. The genomes of these viruses are smaller than 6 kb. Characteristic NTP-binding sites are also not present in single-stranded negative-strand RNA viruses of family *Rhabdoviridae*, in some dsRNA viruses of families *Birnaviridae* and *Retroviridae*, and the DNA virus genus *Caulimovirus* of retrovirus/ plant pararetrovirus group (Gorbalenya and Koonin, 1989).

Four different explanations have been put forward for explaining the unwinding that can occur even in the absence of NTPase activity. First, NTP-binding sequences are possibly present but, having diverged so extensively, that they are not easily recognizable by primary sequence comparisons. Possible variants of various motifs of the NTP-binding site have been recognized in plant viruses: motif A (I) variants exist in SBMV and SbDV; motif B (II) variants reported in SbDV; motif IV variants are found in BYDV, BWYV, CarMV, CNV, MCMV, SbDV, and PLRV; and motif VI variants are found in BWYV, BYDV-strain PAV, CMoV, CNV, MCMV, PLRV, SBMV, and SbDV (Habili and Symons, 1989). The significance of this variation is not yet known. Second, viral RdRps may also possess strand-displacement (RNA unwinding) property as demonstrated for polymerase of *Poliovirus*. Third, a helix-destabilizing protein could be responsible for melting of a duplex in the absence of NTP hydrolysis. Energy of stoichiometric binding to single-strand nucleic acid may be made use of for accomplishing unwinding of the RNA duplex. Fourth, a cellular helicase may be co-opted by the virus for unwinding of the RNA duplex.

The RNA helicases of PPV CI and TaMV CI possess high NTPase activity in the absence of nucleic acid. This is unlike the RNA helicases of most cellular and DNA viruses; their helicases exhibit no NTPase activity in absence of RNA as in eIF4A,

SV40T antigen, and other such cases. This led Kadaré and Haenni (1997) to suggest that high basal NTPase activity may be an in-borne characteristic of helicases of plus-strand RNA viruses.

Residues 514 to 892 of 155-kDa polypeptide, encoded by 4.1 kb ORF1 of BaMV, contain nucleoside triphosphate-binding and helicase-like motifs and so has nucleoside triphosphatase activity and an RNA 5`-triphosphatase activity that specifically cleaves the γ-phosphate off from 5`-end of nascent RNA (Li *et al.*, 2001). The NTPase activities or NTP-binding motifs of helicase domains of 126-kDa TMV protein (Lewandowski and Dawson, 2000; Goregaoker *et al.*, 2001; Goregaokar and Culver, 2003), of TYMV protein p150 (Weiland and Dreher, 1993; Kadaré *et al.*, 1996), and of PVX polymerase (Davenport and Baulcombe, 1997) are required for virus and/or RNA replication.

II. CLASSIFICATION

On the basis of computer-assisted similarities obtained during comparisons of nucleotide sequences as well as during identification of NTPase/helicase motifs in such nucleotide sequences, cellular and viral RNA and DNA helicases as well as the putative helicases were placed into five superfamilies (SFs) and a number of smaller groups (Gorbalenya *et al.*, 1988, 1989a, 1989b, 1990; Gorbalenya and Koonin, 1989) but later all RNA helicases and putative helicases of plus-strand RNA viruses were placed into three superfamilies (Table 1), each of which also contains cellular and DNA viral helicases (Gorbalenya and Koonin, 1993; Koonin and Dolja, 1993; Buck, 1996; Kadaré and Haenni, 1997). Kadré and Haenni (1997) have listed the plant viruses suggested to contain the particular replicative proteins that exhibit virus-encoded NTP-helicase motif in each of the proposed superfamilies. However, the actual helicase activity has been established in only one plant virus genus, namely *Potyvirus*, while in all other cases the helicase domain has been detected on the basis of the proper nucleotide sequence. Out of the DNA plant viruses, only genus *Geminivirus* contains a protein (AL1) showing the proposed helicase-like domain (Eagle *et al.*, 1994).

The tree generated by Koonin and Dolja (1993) consists of five distinct lineages: arteri-like virus lineage; lineage composed of helicases of tobamo-like viruses and alphaviruses; rubi-like lineage; tymo-like lineage; and lineage of 'secondary' or 'accessory' helicases that are typical of a subset of tymoviruses, hordeiviruses, and BNYVV and are encoded by genes that exist outside the block of genes mediating RNA replication. The 'accessory' helicase-like proteins of hordeiviruses, potexviruses and BNYVV are involved in cell-to-cell movement of viruses in infected plants.

Viruses of helicase superfamily 1 (SF1) contain the helicase-like proteins of alpha-like plant viruses. Following positive-strand RNA plant virus genera contain the proposed SF1-type helicase - the particular replicative virus protein (p) that contains the helicase domain is shown for each virus within parenthesis: *Alfamovirus* (p1a), *Bromovirus* (p1a), *Capillovirus* (p241), *Carlavirus* (p26), *Closterovirus* (p295), *Cucumovirus* (p1a), *Furovirus* (p237), *Hordeivirus* (p130), *Idaeovirus* (p190),

Potexvirus (p180), *Tobamovirus* (p126), *Tobravirus* (p134), and *Tymovirus* (p206). The p206 of TYMV is a non-structural polyprotein and possesses NTPase activity (Kadré *et al.*, 1996). Some plant viruses with superfamily 1 putative helicases (hordeiviruses, potexviruses, and BNYVV) encode a second protein with helicase-like motif, which is not required for RNA replication but is involved in cell-to-cell virus translocation.

TABLE 1

Plant viruses belonging to different helicase superfamilies and lineages
(Modified from and based on Tables 5A to 5D of Koonin and Dolja, 1993)

Superfamily 1

Tobamovirus lineage: *Alfalfa mosaic virus, Barley stripe mosaic virus, Beet yellows virus, Brome mosaic virus, Cucumber mosaic virus, Raspberry bushy dwarf virus* (RBDV), *Soil-borne wheat mosaic virus, Tobacco mosaic virus,* and *Tobacco rattle virus*

Rubi lineage: *Beet necrotic yellow vein virus* 1

Tymo lineage: *Apple chlorotic leafspot virus, Apple stem grooving virus,* K*ennedya yellow mosaic virus, Narcissus mosaic virus* 1*, *Papaya mosaic virus* 1, *Potato virus M* 1, *Potato virus X* 1, *Shallot virus X* 1, *Strawberry mild yellow edge-associated virus* 1, *Turnip yellow mosaic virus,* and *White clover mosaic virus* 1

'Accessory helicase' lineage: *Barley stripe mosaic virus* 2, *Beet necrotic yellow vein virus* 2, *Narcissus mosaic virus* 2, *Potato virus M* 2, *Potato virus X* 2, *Shallot virus X* 2, *Strawberry mild yellow edge-associated virus* 2, and *White clover mosaic virus* 2

Superfamily 2

Poty lineage: *Barley yellow mosaic virus, Pea seed-borne mosaic virus, Plum pox virus,* and *Tobacco vein mottling virus*

Superfamily 3

Como lineage: *Cowpea mosaic virus, Hungarian grapevine chrome mosaic virus, Rice tungro spherical virus,* and *Tomato black ring virus*

---

*Certain viruses encode two putative helicases; the one containing a domain of a large protein and is involved in genome replication is designated '1' while the "accessory" helicase encoded by a stand-alone gene is designated '2'.

Note: In the original figure by Koonin and Dolja (1993), viruses (both animal and plant) are organised as per their relationships in the form of a 'tree' but the names of only plant viruses are given in alphabetical order in this table.

Viruses having SF1 helicase-like proteins also have supergroup 3 RNA polymerases (Buck, 1996). Complete correspondence exists between the helicase and polymerase lineage of these viruses. This association between these two replication proteins shows their co-evolution and a long-standing relationship between their respective genes in the concerned viruses. These viruses have both animal and plant hosts and belong to the alpha-like virus supergroup.

Duplex unwinding activities of several viral SF1 helicases have been characterized (Vlot *et al.*, 2003). These are the helicases of several animal viruses but only one plant virus [TMV (Goregaoker and Culver, 2003)]. Purified RdRp complex obtained from AMV-infected plants unwound the RNA duplex in a 3` to 5`-direction. This could be due to helicase-like P1 protein present in the purified RdRp complex. But these observations need confirmation.

SF1 helicases share seven conserved sequence motifs called I, Ia, and II to VI (Gorbalenya and Koonin, 1993; Kadaré and Haenni, 1997; Caruthers and McKay, 2002). Helicases of SFs 1 and 2 share motifs, which align with ATP-binding and ATP-hydrolyzing Walker A and B motifs, respectively (Gorbalenya ad Koonin, 1993; Caruthers and McKay, 2002). Lysine residue of motif I (K844 in AMV P1) is essential for ATP binding; its mutation inhibits ATP hydrolysis and unwinding of duplex (Kadaré and Haenni, 1997; Caruthers and McKay, 2002). SF1 helicase motif VI is thought to be interacting with ATP and functioning in transition of energy from ATP hydrolysis to duplex unwinding (Caruthers and McKay, 2002).

Helicase superfamily 2 (SF2) contains potyvirus-flavivirus-pestivirus helicase-like proteins (Gorbalenya *et al.*, 1989a, 1989b; Lain *et al.*, 1989; Kadaré and Haenni, 1997). Positive-strand RNA plant virus genera containing (putative) superfamily 2-type helicases are: *Bymovirus* (p270) and *Potyvirus* (CI protein). The genome organisation of potyviruses is similar to that of the picorna-like viruses but their helicase proteins are closely related to the corresponding flavivirus- and pestivirus-encoded helicase proteins. The motif-based predictions of enzymatic activities of helicases have been substantiated in the members of SF2, particularly in viruses of genus *Potyvirus*. Thus, RNA-binding NTPase activity and RNA helicase activity have been demonstrated for CI proteins of PPV (Lain *et al.*, 1991) and TaMV (Eagles *et al.*, 1994). Superfamily 2 helicases of poty-like viruses contain supergroup 1 RNA polymerases of poty lineage. Thus, as compared to superfamily 1 helicase proteins, possibly new combinations may have developed between the helicase and polymerase genes of these viruses during the course of evolution. Viral RNA helicases of superfamilies 2 and 3 function in 3`- to 5`-direction.

Helicase superfamily 3 (SF3) includes small putative helicase domains of ~100 amino acid residues found in DNA and RNA viruses, has only three conserved motifs and includes two classical ATP-binding motifs. Helicases of superfamily 3 are exclusively associated with supergroup 1 polymerases. This superfamily contains picornavirus-like helicase-like proteins (Gorbalenya *et al.*, 1989a, 1989b; Lain *et al.*, 1989). The positive-strand RNA plant virus genera containing superfamily 3-type helicases (picornavirus-like supergroup containing members of SF3) are: *Comovirus* (p58), *Nepovirus* (p27), *Sequivirus* (p336), and *Waikavirus* (ORF 1). Mutations in the NTP-binding motif of CPMV polyprotein B rendered the virus unable to replicate in protoplasts (Peters *et al.*, 1994). The simian virus 40 large tumor antigen (SV40T antigen) is the only protein of SF3 with a well-characterized DNA and RNA helicase activity. Poliovirus 2C protein shows NTPase activity but its helicase activity could not be detected. Thus, superfamily 3 includes proteins implicated in genome replication of positive-strand RNA viruses, and single-stranded DNA-containing small DNA viruses.

A. Relationships between Helicase Superfamilies

The helicase SFs 1 and 2 are distantly related to each other but are distinct from helicase SF3. Helicase SFs 1 and 2 each include proteins that are involved in replication of genome of positive-strand RNA viruses and of large DNA viruses.

Helicases of SFs 1 and 2 possess seven conserved amino acid segments while superfamily 3 helicases contain only three conserved segments. These motifs are called I, Ia, II, III, IV, V and VI for superfamilies 1and 2 but are designated as A, B and C for superfamily 3 (Gorbalenya and Koonin, 1993). Out of these motifs, two motifs (motifs I and II of superfamilies 1 and 2 and motifs A and B of superfamily 3) are common to all the superfamilies and are variations of the ATP-binding sequences. The seven conserved motifs are organised in a collinear manner. Ia, III, and IV are conserved to the minimum and no potential function is as yet ascribed to them. Segments I, II, V, and VI are best conserved; motif VI has an abundance of basic amino acid residues, is suggested to bind nucleic acids and is essential and sufficient by itself to enable CI protein of PPV to bind RNA. Mutations in the corresponding area of eIF-4A eliminated its RNA binding capacity and also decreased its ATPase activity. Thus, segment VI is involved both in ATPase activity and in interaction with nucleic acids. Variants of motif A are found in a large group of ATP- and GTP-binding proteins. Sequence A (I) is located in the NTP-binding site situated within a phosphate-binding loop (P-loop) between a ß-strand and γ-helix. The B (II) element may provide D residues needed for chelating Mg^{2+} ions which bind to terminal phosphates and enhance hydrolysis of NTP.

The SFs 1 and 2 encompass a large number of DNA and RNA helicases from eubacteria, eukaryotes and viruses; they include representatives that unwind duplexes in a 3`- to 5`-direction, as well as examples that unwind 5`- to 3`-direction; hence the motif-based segregation into SF-1 and SF-2 does not correlate with any particular substrate (DNA or RNA), phylogenetic source or feature of activity, such as direction of unwinding. The (putative) helicases of superfamily 2 of positive-stranded RNA viruses form a subset of proteins that contain variants of the DEAD box proteins.

Several viruses encode two putative helicases of SF 1 (Koonin and Dolja, 1993). It should be borne in mind that many proteins have been assigned to the different helicase families on the sole basis of presence of the defining sequence fingerprints. The question whether they actually catalyze a *bona fide* duplex unwinding event has been left for tackling by later workers. It is not known yet whether the positive-strand RNA viruses having small genomes do not need helicase activity at all or do they recruit cellular helicases. The relationship between the occurrence of helicase gene and genome size has been correlated with ensuring the fidelity of complementary nucleotide incorporation that is necessary for replication of relatively large genomes' (Koonin and Dolja, 1993).

III. STRUCTURE

The Per A protein of *Bacillus stearothermophilus* was the first helicase of which structure was elucidated (Subramanya *et al.*, 1996). It is a DNA helicase of superfamily 1; possesses four structural motifs – two parallel α-ß domains encompassing the canonical helicase sequence motifs and two more domains that have predominantly α-helical secondary structure. Each of these two domains is encoded as a single insertion with the polypeptide sequence of the two α-ß domains. The ATP is bound to the amino-proximal α-ß domain, where the Walker A and B domains occur.

Structure of the first helicase of SF2 to be worked out was of the helicase of *Hepatitis C virus* (Yao *et al.*, 1997). This helicase has three structural domains encoded sequentially in the polypeptide chain – two α-ß domains followed by an all α domain. Caruthers and McKay (2002) give details of these structures. Each helicase of SF1 and SF2 has two parallel α-ß domains out of which the amino-proximal domain binds ATP. Both α-ß domains possess a core that is regarded to include five contiguous strands in a parallel β sheet, plus associated helicases. The classical helicase sequence motifs are located within the cores of the two α-ß domains and at identical sites in each of the helicase families.

It seems that conservation of certain characters among helicase structure is dictated by their essential function(s) that are common to, or required by, all the helicases of SFs 1 and 2. This implies that the structural elements diagnostic of each family of helicases or individual helicases are possibly controlled by protein-specific functions, like DNA or RNA target recognition or interaction with accessory proteins. Other features presumably common to these helicases could be the single-stranded and/or double-stranded oligonucleotide binding. Caruthers and McKay (2002) discuss the relationship of structure to mechanism of action of some helicases (PerA, Rep and DnaB, eIF-4A, and RuvB) of some animal viruses.

IV. FUNCTIONS

Helicases possess nucleotide binding and hydrolysis activities and so have the capacity to interact with nucleic acids. This unwinding of duplexes generated during RNA replication is essential for permitting the strands to act as templates for continued RNA replication. Helicases also enzymatically unwind duplex RNA structures formed between template and newly synthesized strands during RNA replication. Helicases also generally unwind DNA duplex during its replication.

The helicases are also important for unwinding secondary structure of the RNA template. This is necessary for initiation of negative-strand synthesis on templates having extensive secondary structure at 3`-end as in templates having 3`- tRNA-like termini. The unwinding also increases the functions of polymerases on RNA regions possessing internal secondary structure. Similarly, duplex unwinding and removal of secondary structure from the resultant single-stranded templates is performed by helicases. Removal of secondary structure from RNA templates has, therefore, important bearing for RNA replication. The unwinding is done by disrupting RNA-protein or protein-protein interactions by breaking hydrogen bonds that keep together the two strands of RNA duplex. Fernándes *et al.* (1997) found that motif V of PPV CI RNA helicase is involved in NTP hydrolysis, is essential for virus RNA replication, and suggested the coupling of ATP hydrolysis and RNA unwinding activity of the helicase. Energy generated by ATP hydrolysis is apparently essential for disrupting hydrogen-bonded base pairs of RNA duplex.

Other functions performed by helicases are: a DEAD box RNA helicase of BMV RNA2 can selectively activate translation of a specific mRNA (Noueiry *et al.*, 2000); a correlation exists between virus viability and NTPase or duplex-unwinding activities of viral helicase proteins of PPV (Fernándes *et al.*, 1997), *Semliki Forest virus*, and other animal viruses; helicase domains of BMV 1a (Ahola *et al.*, 2000) and AMV p1 (Vlot *et al.*, 2003) are required for negative-strand synthesis; AMV proteins p1 and p2 may interact in *cis* and this interaction occurs between helicase domain of p1 with N-terminal region of p2 (van der Heijden *et al.*, 2001) and may be the mechanism that plays a role in coordinating replication of AMV RNAs 1 and 2, provide selectivity to the replicase in use of viral template RNA but other factors may also be involved (Vlot *et al.*, 2003); replication of AMV RNA1 requires helicase function in *cis* located in AMV p1 protein (Vlot *et al.*, 2003) like the *cis* requirement of TMV 126-kDa protein (Lewandowski and Dawson, 2000); BMV helicase domain of 1a is involved in recruitment of viral RNA template to replication complexes (Ahola *et al.*, 2000) and helicase domain of AMV p1 may also be involved in functions preceding initiation of negative-strand synthesis (Vlot *et al.*, 2003); mutations corresponding to K844A (of AMV p1) in helicase or helicase-like domains of a number of other plant viruses interfere with replication of virus or of virus RNA [BMV (Ahola *et al.*, 2000); PVX (Davenport and Baulcombe, 1997); TMV (Lewandowski and Dawson, 2000; Goregaoker *et al.*, 2001); TYMV (Weiland and Dreher, 1993)]; helicases of BMV (Ahola *et al.*, 2000) and TMV (Goregaoker *et al.*, 2001; Goregaoker and Culver, 2003) are believed to act as oligomers so that helicase domains may be involved in oligomerization of replicase proteins. It is clear from the above well-characterized helicase functions that helicases are involved in mRNA splicing, export, and translation.

Apart from the above established functions, helicases are also suggested to be involved in following roles in many of the plant viruses: helicases might ensure fidelity of replication because of a proof-reading energy-dependent mechanism and so reduce the error rate of polymerase (Koonin, 1991), which might be the reason as to why viruses with large genomes need a helicase to replicate efficiently; helicases might have some role during initiation of translation since eIF4A (a helicase), in combination with eIF4B, possibly disrupts secondary structure in mRNA upstream of the initiation codon, thereby facilitating attachment of 40S ribosome; helicases are involved in rRNA processing; helicases are reported to be involved in synthesis of all classes of RNA. The BMV 1a-helicase-like region (i.e., carboxy-terminal part) of 1a protein appears to be required for synthesis of all classes of RNA (positive-strand, negative-strand, and subgenomic RNA) in bromoviruses (Kroner *et al.*, 1990). Helicases also appear to influence RNA recombination in BMV (Nagy *et al.*, 1995).

The best-investigated helicase-like/NTPase domain is that of the BMV 1a. Its functions are mentioned separately in order to bring out the overall importance of helicases in plant viruses. The BMV 1a reportedly facilitates RNA synthesis in several ways. Firstly, it possibly recruits RNA from translation to replication mode by one or more presumed ways: it could be involved in RNA recognition; could modify RNA structure in order to facilitate other recognition events; could inhibit translation; and could control some type of RNA transport process which may be an essential requirement for formation of replication complexes on endoplasmic reticulum. Secondly, BMV 1a protein unwinds secondary structure at 3`-ends of viral RNAs. This in turn may provide RdRp-like (BMV 2a) protein access to RNA initiation sites. The unwinding of secondary structure of template, ahead of the elongating RdRp, may help elongation of RNA synthesis. Thirdly, it separates the template and product strands after RNA synthesis. Fourthly, it is essential for synthesis of BMV negative-strand RNA, positive-strand RNA, and subgenomic RNA. Fifthly, it may be involved in converting the capacity of replication complexes from synthesis of negative-strand to positive-strand synthesis. Sixthly, it may be involved in stabilization of BMV RNA3 (Ahola *et al.*, 2000). Seventhly, this protein domain also appears to influence BMV RNA recombination (Nagy *et al.*, 1995). Moreover, the BMV helicase and polymerase domains form a complex (O'Reilly *et al.*, 1997, 1998) and colocalize on endoplasmic reticulum (Restrepo-Hartwig and Ahlquist, 1996).

Chen *et al.* (2003) have discovered a box B motif-independent pathway for BMV 1a-mediated recruitment of BMV RNA2 to replication complex. This is a new pathway of template recruitment and requires the previously documented interaction of 1a with translating mRNA via 1a interaction with N-terminal region of nascent 2a polypeptide. Such interaction with nascent 2a protein could also be involved in recruitment of 2a polymerase to membranes by 1a. Thus, the helicase/NTPase-like BMV 1a protein may have distinct functions at different stages of RNA replication (Ahola and Ahlquist, 1999). In fact, BMV 1a protein is suggested to have distinct conformational states during negative-strand RNA synthesis, positive-strand RNA synthesis, subgenomic RNA synthesis and RNA3 stabilization (Ahola *et al.*, 2000).

The coding regions of protease co-factor together with C-terminal half of the putative helicase of *Bean pod mottle comovirus* are determinants of symptom severity (Gu and Ghabrial, 2005).

V. REFERENCES

Ahola, T., and Ahlquist, P. 1999. Putative RNA capping activities encoded by *Brome mosaic virus*: Methylation and covalent bonding of guanylate by replicase protein 1a. *J. Virol.* 73: 10061-10069.

Ahola, T., den Boon, J. A., and Ahlquist, P. 2000. Helicase and capping enzyme active site mutations in brome mosaic virus protein 1a cause defects in template recruitment, negative-strand RNA synthesis and viral RNA capping. *J. Virol.* 74: 8803-8811.

Beck, D. L., Guilford, P. J., Voot, D. M., Anderson, M. T., and Forster, R. L. S. 1991. Triple gene block proteins of *White clover mosaic potexvirus* are required for transport. *Virology* 183: 695-702.

Bleykasten, C., Gilmer, D., Guilley, H., Richards, K. E., and Jonard, G. 1996. Beet necrotic yellow vein virus 42-kDa triple gene block protein binds nucleic acid *in vitro*. *J. Gen. Virol.* 77: 889-897.

Buck, K. W. 1996. Comparison of the replication of positive-stranded RNA viruses of plants and animals. *Adv. Virus Res.* 47: 159-251.

Caruthers, J. M., and McKay, D. B. 2002. Helicase structure and mechanism. *Curr. Opin. Struct. Biol.* 12: 123-133.

Chen, J., Noueiry, A., and Ahlquist, P. 2003. An alternative pathway for recruiting template RNA to the brome mosaic virus replication complex. *J. Virol.* 77: 2568-2577.

Davenport, G. F., and Baulcombe, D. C. 1997. Mutation of the GKS motif of the RNA-dependent RNA polymerase from *Potato virus X* disables or eliminates virus replication. *J. Gen. Virol.* 78: 1247-1251.

de Graaff, M., Houwing, C. J., Lukacs, N., and Jaspars, E. M. J. 1995. RNA duplex unwinding activity of alfalfa mosaic virus RNA-dependent RNA polymerase. *FEBS Lett.* 371: 219-222.

Eagle, P. A., Orozco, B. M., and Hanley-Bowdoin, L. 1994. A DNA sequence required for geminivirus replication also mediates transcriptional regulation. *Plant Cell* 6: 1157-1170.

Eagles, R. M., Balmori-Melian, E., Beck, D. L., Gardner, R. C., and Forster, R. L. 1994. Characterization of NTPase, RNA-binding, and RNA-helicase activities of the cytoplasmic inclusion protein of *Tamarillo mosaic potyvirus*. *Eur. J. Biochem.* 224: 677-684.

Fernández, A., Lain, S., and Garcia, J. A. 1995. RNA helicase activity of the plum pox potyvirus CI protein expressed in *Escherichia coli*: Mapping of an RNA-binding domain. *Nucl. Acids Res.* 23: 1327-1232.

Fernández, A., Guo, H. S., Saenz, P., Simon-Buela, L., de Cedrón, M. G., and Garcia, J. A. 1997. The motif V of plum pox potyvirus CI RNA helicase is involved in NTP hydrolysis and is essential for virus RNA replication. *Nucl. Acids Res.* 25: 4474-4480.

Gilmer, D., Bouzoubaa, S., Hehn, A., Guilley, H., Richards, K., and Jonard, G. 1992. Efficient cell-to-cell movement of *Beet necrotic yellow vein virus* requires 3`-proximal gene located on RNA2. *Virology* 189: 40-47.

Gorbalenya, A. E., and Koonin, E. V. 1989. Viral proteins containing the purine NTP-binding sequence pattern. *Nucl. Acids Res.* 17: 8413-8440.

Gorbalenya, A. E., and Koonin, E. V. 1993. Helicases: Amino acid sequence comparisons and structure-function relationships. *Curr. Opin. Cell Biol.* 3: 419-429.

Gorbalenya, A. E., Koonin, E. V., Donchenko, A. P., and Blinov, V. M. 1988. A novel superfamily of nucleotide triphosphate-binding motif containing proteins which are probably involved in duplex unwinding in DNA and RNA replication and recombination. *FEBS Lett.* 235: 16-24.

Gorbalenya, A. E., Koonin, E. V., Donchenko, A. P., and Blinov, V. M. 1989a. Two related superfamilies of putative helicases involved in replication, recombination, repair and expression of DNA and RNA genomes. *Nucl. Acids Res.* 17: 4713-4730.

Gorbalenya, A. E., Blinov, V. M., Donchenko, A. P., and Koonin, E. V. 1989b. An NTP-binding motif is the most conserved sequence in a highly diverged monophyletic group of proteins involved in positive-strand RNA viral replication. *J. Mol. Evol.* 28: 256-268.

Gorbalenya, A. E., Koonin, E. V., and Wolf, Y. I. 1990. A new superfamily of putative NTP-binding domains encoded by genomes of small DNA and RNA viruses. *FEBS Lett.* 262: 145-148.

Goregaoker, S. P., and Culver, J. N. 2003. Oligomerization and activity of the helicase domain of the tobacco mosaic virus 126- and 183-kilodalton replicase proteins. *J. Virol.* 77: 3549-3556.

Goregaoker, S. P., Lewandowski, D. J., and Culver, J. N. 2001. Identification and functional analysis of an interaction between domains of the 126/183-kDa replicase-associated proteins of *Tobacco mosaic virus*. *Virology* 282: 320-328.

Gu, H., and Ghabial, S. A. 2005. Bean pod mottle virus proteinase cofactor and putative helicase are symptom severity determinants. *Virology* 333: 271-283.

Habili, N., and Symons, R. H. 1989. Evolutionary relationship between luteoviruses and other RNA plant viruses based on sequence motifs in their putative RNA polymerases and nucleic acid helicases. *Nucl. Acids Res.* 17: 9543-9555.

Hayes, R. J., Pereira, V. C. A., McQuillin, A., and Buck, K. W. 1994. Localization of functional regions of cucumber mosaic virus RNA replicase using monoclonal and polyclonal antibodies. *J. Gen. Virol.* 75: 3177-3184.

Kadaré, G., and Haenni, A.-L. 1997. Virus-encoded helicases. *J. Virol.* 71: 2583-2590.

Kadaré, G., David, C., and Haenni, A.-L. 1996. ATPase, GTPase, and RNA binding activities associated with the 206-kilodalton protein of *Turnip yellow mosaic virus*. *J. Virol.* 70: 8169-8174.

Koonin, E. V. 1991. Similarities in RNA helicases. *Nature* 352: 290.

Koonin, E. V., and Dolja, V. V. 1993. Evolution and taxonomy of positive-strand RNA viruses: Implications of comparative analysis of amino acid sequences. *Crit. Rev. Biochem. Mol. Biol.* 28: 375-430.

Kroner, P. A. D., Young, B. M., and Ahlquist, P. 1990. Analysis of the role of brome mosaic virus 1a protein domains in RNA replication, using linker insertion mutagenesis. *J. Virol.* 64: 6110-6120.

Lain, S., Reichman, J. L., Martin, M. T., and Garcia, J. A. 1989. Homologous potyvirus and flavivirus proteins belonging to a superfamily of helicase-like proteins. *Gene* 82: 357-362.

Lain, S., Reichman, J. L., and Garcia, J. A. 1990. RNA helicase: A novel activity associated with a protein encoded by a positive-strand RNA virus. *Nucl. Acids Res.* 18: 7003-7006.

Lain, S., Martin, M. T., Reichman, J. L., and Garcia, J. A. 1991. Novel catalytic activity associated with positive-strand RNA virus infection; Nucleic acid-stimulated ATPase activity of the plum pox potyvirus helicase-like protein. *J. Virol.* 65: 1-6.

Lewandowski, D. J., and Dawson, W. O. 2000. Functions of the 126- and 183-kDa proteins of *Tobacco mosaic virus*. *Virology* 271: 90-98.

Li, Y.-L., Shih, T. W., Hsu, Y.-H., Han, Y.-T., Huang, Y.-L., and Meng, M. 2001. The helicase-like domain of potexvirus replicase participates in formation of RNA 5`-cap structure by exhibiting RNA 5`-triphosphatase activity. *J. Virol.* 75: 12114-12120.

McCarthy, J. 1998. Posttranscriptional control of gene expression in yeast. *Microbiol. Mol. Biol. Rev.* 62: 1492-1553.

Mushegian, A. R., and Koonin, E. V. 1993. Cell-to-cell movement of plant viruses: Insights from amino acid sequence comparisons of movement proteins and from analysis with cellular transport systems. *Arch. Virol.* 133: 239-257.

Nagy, P. D., Dzianott, A., Ahlquist, P., and Bujarski, J. J. 1995. Mutations in the helicase-like domain of protein 1a alter the sites of RNA-RNA recombination in *Brome mosaic virus*. *J. Virol.* 69: 2547-2556.

Noueiry, A. O., Chen, J., and Ahlquist, P. 2000. A mutant allele of essential general translation initiation factor *DED1* selectively inhibits translation of a viral mRNA. *Proc. Natl. Aacd. Sci. USA* 97: 12985-12990.

O'Reilly, E. K., Paul, J. D., and Kao. C. C. 1997. Analysis of the interaction of viral RNA replication protein by using the yeast two-hybrid assay. *J. Virol.* 71: 7526-7532.

O'Reilly, E. K., Wang, Z., French, R., and Kao. C. C. 1998. Interactions between the structural domains of the RNA replication proteins of plant-infecting RNA viruses. *J. Virol.* 72: 7160-7169.

Peters, S. A., Verver, J., Nollen, E. A. A., van Lent, J. W. M., Wellink, J., and van Kammen, A. 1994. The NTP-binding motif in cowpea mosaic virus B polyprotein is essential for viral replication. *J. Gen. Virol.* 75: 3167-3176.

Petty, I. T. D., French, R., Jones, R. W., and Jackson, A. O. 1990. Identification of barley stripe mosaic virus genes involved in viral RNA replication and systemic movement. *EMBO J.* 9: 3453-3457.

Quadt, R., Rosdorff, H. J. M., Hunt, T. W., and Jaspars E. M. J. 1991. Analysis of the protein composition of alfalfa mosaic virus RNA-dependent RNA polymerase. *Virology* 182: 309-315.

Restrepo-Hartwig, M. A., and P. Ahlquist. 1996. Brome mosaic virus helicase- and polymerase-like proteins colocalize on the endoplasmic reticulum at sites of viral RNA synthesis. *J. Virol.* 70: 8908-8916.

Rouleau, M., Smith, R. J., Bancroft, J. B., and Mackie, G. A. 1994. Purification, properties and subcellular localization of foxtail mosaic potexvirus 26-Da protein. *Virology* 204: 254-265.

Subramanya, H. S., Bird, L. E., Brannigan, J. A., and Wigley, D. B. 1996. Crystal structure of a DExx DNA helicase. *Nature* 384: 379-383.

van der Heijden, M. W., Carette, J. E., Reinhoud, P. J., Haegi, A., and Bol, J. F. 2001. Alfalfa mosaic virus replicase proteins P1 and P2 interact and colocalize at the vacuolar membrane. *J. Virol.* 75: 1879-1887.

Vlot, A. C., Neeleman, L., Linthorst, H. J. M., and Bol, J. F. 2001. Role of the 3`-untranslated regions of alfalfa mosaic virus RNAs in the formation of a transiently expressed replicase in plants and in the assembly of virions. *J. Virol.* 75: 6440-6449.

Vlot, A. C., Laros, S. M., and Bol, J. F. 2003. Coordinate replication of alfalfa mosaic virus RNAs 1 and 2 involves *cis-* and *trans*-acting functions of the encoded helicase-like and polymerase-like domains. *J. Virol.* 77: 10790-10798.

Weiland, J. J., and Dreher, T. W. 1993. *cis*-Preferential replication of the turnip yellow mosaic virus RNA genome. *Proc. Natl.. Acad. Sci. USA* 90: 6095-6099.

Yao, N., Hesson, T., Cable, M., Hong, Z., Kwong, A. D., Le, H. V., and Weber, P. C. 1997. Structure of the hepatitis virus C RNA helicase domain. *Nat. Struct. Biol.* 4: 463-467.

7 PROTEINASES

I. INTRODUCTION

Proteinases (proteases) have been reviewed (Strauss, 1990; Gorbalenya *et al.*, 1991; Koonin and Dolja, 1993; Dougherty and Semler, 1993; Maia *et al.*, 1996a; Agranovsky and Morozov, 1999; Garcia *et al.*, 1999).

Proteolytic processing of viral genome-encoded polyproteins (post-translational proteolytic cleavage of polyproteins) is a very important translation strategy employed by many plus-strand RNA plant and animal viruses during genome expression. Plant virus-encoded large precursor polyproteins (produced by como-, nepo-, and potyviruses of picorna-like superfamily) undergo post-translational processing by one, two or even three virus-encoded proteinases to produce mature structural and non-structural virus proteins. Alpha-like, carmo-like, and sobemo-like viruses mostly express their genes through different combination of strategies that operate at the level of transcription (synthesis of subgenomic RNA) and/or translation (alternative translation initiation sites, readthrough mechanism due to the existence of suppressed termination codons, and frameshifting). However, some plant viruses of these three groups also express their genes by post-translational proteolysis by virus-encoded proteinases. Plant pararetroviruses (dsDNA plant viruses that replicate through RNA intermediates) also express their genes through proteolytic processing (Hohn and Fütterer, 1997).

Proteolytic cleavage has been detected in the following plant genera with the name of the plant virus in which it has been reported, apart from the other viruses of that particular genus, being mentioned in parenthesis: *Carlavirus* (BISV), *Closterovirus* (BYV), *Comovirus* (CPMV), *Benyvirus* (BNYVV), *Marafivirus* (OBDV), *Nepovirus* (TBRV), *Potyvirus* (PYV), *Tymovirus* (TYMV), and *Waikavirus* (RTSV). Proteolytic cleavage has been proposed in the following plant genera with the name of the plant virus in which it was proposed being mentioned in parenthesis: *Bymovirus* (BaYMV), *Capillovirus* (ASGV). *Enamovirus* (PEMV), *Luteovirus* (BYDV), *Rymovirus* (BrSMV), *Sequivirus* (PYFV), *Sobemovirus* (SBMV), and *Trichovirus* (ACLSV and GVB). Names of other plant viruses in which proteolytic cleavage has been reported or proposed are mentioned in the text and the tables.

In general, host proteinases process virion envelope proteins while viral proteinases process nonstructural viral proteins and also CPs in many viruses. Viral proteinases become functional in host cytoplasm in early stages of infection; and catalytically hydrolyze peptide bonds between amino acid residues while cleaving viral polyproteins for generating intermediate precursors and end products. Proteolytic polyprotein cleavage can be of *cis* and/or *trans* type. In *cis* cleavage type, proteinase and cleavage site are located in the same precursor and so it is monomolecular. The *cis* cleavage sites may exist on flanks of the proteinase gene and could be used for releasing the proteinase from a precursor or could be distal when their hydrolysis releases the proteinase in an intermediate polyprotein form. The *cis* cleavage events are very fast, occur cotranslationally and are also called autocatalytic. Proteinase and cleavage site are located in different proteins in *trans* type of cleavage and so it is bimolecular, i.e., a protein possessing the proteolytic property cleaves a second protein possessing the substrate cleavage site.

The structure and function of plant and animal virus proteinases appear very similar; moreover, these proteinases are related in sequence and activity to known cellular proteinases so that RNA plant viruses synthesize/are regarded to synthesize chymotrypsin-like serine or cysteine and papain-like cysteine (thiol) proteinases. Viruses producing proteinases show no correlation with viral capsid complexity, presence or absence of lipid envelope, and nature of genome. Thus, proteinases occur in non-enveloped single-stranded RNA viruses, enveloped single-stranded RNA viruses (flaviviruses, retroviruses), non-enveloped dsDNA viruses (adenoviruses and caulimoviruses) and enveloped dsDNA viruses (herpesviruses).

The cleavage sites of polyproteins are generally characterized by two factors: presence of a conserved amino acid sequence at cleavage sites and a favourable secondary structure of proteins around cleavage sites so that the substrate cleavage sites are mostly highly specific. The proteolytic processing site at the N-terminus of the subgroup II luteovirus VPg is predicted to be glutamic acid↓serine/threonine. Actual or predicted proteolytic processing sites/specific cleavage sites in polyproteins of various other plant viruses (Garcia *et al.*, 1999) are mentioned later at appropriate places.

Each proteinase is distinguished from all other proteinases of viral or cellular origin by its substrate-binding zone. This pocket has a complex three-dimensional structure, which is quite different amongst the members of a particular class of proteinases. This pocket may show preference for only a single amino acid or may possess an absolute requirement for a specific substrate even if it is up to seven amino acids in length in some cases. The complexity of the binding pocket determines the substrate specificity.

A. Classification

Proteinases are of two types on the basis of their site of action: exoproteinases remove amino acids from amino- or carboxy-termini of proteins and in general continue to do so in a progressive manner while endoproteinases break specific peptide bonds that exist between two amino acids located in internal parts of a protein (Dougherty and Semler, 1993).

Proteinases are also classified on the basis of amino acids constituting the catalytic site. The catalytic site is conserved so that amino acids constituting the active site provide a better and a more popular basis for classification of proteinases, which are divided into four different classes on this basis: serine and serine-like proteinases, cysteine (thiol) proteinases, aspartic (or acidic) proteinases, and metalloproteinases (Dougherty and Semler, 1993). Only the cysteine, serine and serine-like proteinases are involved in proteolytic processing of polyproteins encoded by positive-strand RNA plant viruses while aspartic proteinases and metalloproteinases do not do so (Maia *et al.*, 1996b; Garcia *et al.*, 1999) (Tables 1 and 2). Serine and serine-like proteinases have a catalytic triad of histidine (His), aspartic acid (Asp), and serine (Ser) amino acids while cysteine proteinases possess a catalytic dyad of cysteine (Cys) and His amino acid residues. Aspartic proteinases are constituted by a catalytic dyad composed of two aspartic acid residues. Metalloprotein proteinases contain a divalent cation, which generally is Zn^{2+}, at the catalytic site. No viral-encoded metalloproeinase has been detected.

TABLE 1

Plant virus groups[1] and genera encoding different types of demonstrated or putative proteinases
(Based on Dougherty and Semler, 1993; and Garcia *et al.*, 1999)

Picorna-like Plant Viruses
   Potyviridae
      Genus *Bymovirus*- - - - - - - - - - -Cysteine, Serine-like
      Genus *Potyvirus** - - - - - - - - - - - -Serine, Cysteine, Serine-like
      Genus *Rhymovirus* - - - - - - - - - - -Serine, Cysteine, Serine-like
   Comoviridae
      Genus *Comovirus*- - - - - - - - - - - -Serine like
      Genus *Nepovirus*- - - - - - - - - - - - Serine like
   Sequiviridae
      Genus *Sequivirus* - - - - - - - - - - - Serine like
      Genus *Waikavirus* - - - - - - - - - - -Serine like

Alpha-like Plant Viruses
      Genus *Benyvirus* - - - - - - - - - - - -Cysteine
      Genus *Capillovirus* - - - - - - - - - Cysteine
      Genus *Carlavirus* - - - - - - - - - - -Cysteine
      Genus *Closterovirus* - - - - - - - - - Cysteine
      Genus *Marafivirus* - - - - - - - - - - Cysteine
      Genus *Trichovirus*- - - - - - - - - - - Cysteine
      Genus *Tymovirus* - - - - - - - - - - - Cysteine

Sobemo-like Viruses
      Genus *Luteovirus* (subgroup II) - -Serine
      Genus *Sobemovirus* - - - - - - - - - -Serine

Miscellaneous RNA Plant Viruses
      Furoviridae- - - - - - - - - - - - - - - -Cysteine?

DNA Plant Viruses
      Badnaviridae - - - - - - - - - - - - - - -Aspartic?
      Caulimoviridae- - - - - - - - - - - - - -Aspartic

    1. Only those virus groups are listed which encode or probably encode (on the basis of sequence analysis) a proteinase
    ?. Class of proteinase shown needs confirmation
    * Possess more than one proteolytic activity

TABLE 2

Polyprotein precursors, cleavage products, and proteinases of some positive-sense RNA
plant viruses (Based on Maia *et al.*, 1996b, with additions and modifications)

Virus	RNA[1]	Precursor[2]	Cleavage products[3]	Proteinases[4]
Picorna-like plant viruses				
CPMV-como	B	200kDa	32kDa/58kDa/VPg/24kDa/87kDa	24kDa and 32kDa[a]
	M	105kDa (95kDa)	58kDa(48kDa)/CP37/CP23	
GFLV-nepo	1	253kDa	~133kDa/VPg/24kDa/RdRp	24kDa[b]
	2	122kDa	28kDa/38kDa/56kDa	
TEV-poty	1	351kDa	P1/HC-Pro/P3/CI/ 6kDa/NIa/NIb/CP	P1[a], HC-Pro[b] and NIa[b]
Alpha-like plant viruses				
BISV-carla[5]	1	223kDa	166kDa/57kDa	166kDa[c]
BYV-clostero	1	295kDa (384kDa)	66kDa/26kDa	26kDa[c]
TYMV-tymo	1	206kDa	140kDa/66kDa	140kDa[c]

1 - Genomic RNAs whose translation proteins undergo proteolytic cleavage
2 - Size of polyprotein
3 - Final protein products
4 - Formed from in-frame initiation at a second AUG

Underlined products - Proteinases
A - accessory protein
B - chymotrypsin-like proteinase
C - papain-like proteinase

Each proteinase of serine, cysteine, and aspartic groups is constituted by two
globular elements having C-terminal and N-terminal halves and occasionally showing
identical secondary and tertiary structures. Each half contributes, across a crevice, the
amino acid residues that cause catalysis. This shared general enzyme structure is,
however, achieved by each type of proteinase class through a unique three-dimensional
manner – as shown by detailed crystallographic studies. The molecular structure of
each proteinase class is conserved even if the proteinases of a particular class are
obtained from different and unrelated organisms. In each class of proteinase, spatial

orientation of amino acids involved in catalysis is precise and strictly conserved, even in those proteinases (of a particular class) having little homology in their primary amino acid sequence.

Viral proteinases are also referred to on the basis of the terms employed in cellular proteinases. Based on the common properties of cellular and viral proteinases, Gorbalenya *et al.* (1991), Dougherty and Semler (1993), and Koonin and Dolja (1993) place proteinases of positive-strand RNA plant viruses into two main classes - the chymotrypsin-related cysteine and serine proteinases and papain-like cysteine proteinases on the basis of (limited) similarity in sequence and spatial folding to the archetype cell enzymes.

1. Chymotrypsin-like Cysteine and Serine Proteinases

These proteinases constitute a large superfamily, in fact the largest class, of virus-encoded proteinases related to chymotrypsin-like cellular serine proteinases and possess chymotrypsin-like structure with catalytic site being composed of a triad of His, Asp, and Ser. These three amino acids (His57, Asp102, and Ser195 as per chymotrypsin numbering) of the catalytic triad function during cleavage of the peptide bond in cellular proteinases. The characteristic feature of these proteinases is the presence of a reactive serine at active site; serine behaves as a nucleophile during cleavage of the peptide bond. Proteinase-substrate interactions are extremely specific in various serine proteinases so that such interactions probably cause faithful formation of various viral proteins in an ordered temporal sequence during viral infection. Koonin and Dolja (1993) state that sequence conservation in chymotrypsin-like proteinases is mostly restricted to four conserved motifs, out of which three center at the catalytic residues while the fourth distal motif is implicated in substrate binding.

Cysteine substitutes the main catalytic Ser in some of these viral proteinases; cysteine proteinases comprise a unique group since no such chymotrypsin-like proteinases (with Cys at the active site) are found in cellular enzymes. Some cysteine protein-ases are the most important factors in viral gene expression and mediate the processing of majority of viral proteins as in picorna-like plant viruses and potyviruses. Thus, active site of chymotrypsin (found in cellular systems) contains serine while a serine or cysteine is located at active site of chymotrypsin-like proteinases found in viruses. Chymotrypsin-like cysteine proteinases containing Cys in active triad (of His, Asp, and Cys) perform a central role in polyprotein processing in bymo-, como-, nepo-, and potyviruses. The polyproteins of potyviruses are cleaved both *in cis* and *in trans* by chymotrypsin-like cysteine proteinase domain present in NIa protein. The polyproteins encoded by bipartite RNA genomic como- and nepoviruses undergo proteolytic processing exclusively by chymotrypsin-like cysteine proteinases, akin to potyvirus NIa proteinase.

The 24-kDa protein of GFLV, encoded by RNA1, is the chymotrypsin-like cysteine proteinase. Other plant viruses predicted, on the basis of sequence data analysis, to have chymotrypsin-like proteinase containing cysteine at the active site are benyviruses (BNYVV), capilloviruses (ASGV), carlaviruses (PVM), foveaviruses (*Apple stem pitting foveavirus*), marafiviruses (OBDV), sequiviruses (PYFV), trichoviruses (ACLSV), and waikaviruses (RTSV) (Rozanov *et al.*, 1995; Maia *et al.*, 1996a; Morozov and Solovyev, 1999) Chymotrypsin-like proteinases containing serine at active site are suggested to occur in luteo- and sobemoviruses (Maia *et al.*, 1996b).

2. Papain-like Cysteine Proteinases

Papain-like cysteine proteinases (PCPs) resemble the classic cellular papain-like thiol proteinases but the actual sequence similarity is rather low. The nature and orientation of the catalytic dyad of Cys and His residues in these viral proteinases also resemble the cellular papain-like thiol proteinases. The conservation among viral papain-like proteinases exists nearly exclusively around the catalytic cysteine residue; no other conserved motif exists so that PCP domains show only marginal sequence conservation and may occupy different positions in polyproteins of even closely related viruses (Koonin and Dolja, 1993). These proteinases are about as common among positive-strand RNA viruses as the chymotrypsin-like proteinases and are dealt with in detail later.

II. SERINE AND SERINE-LIKE PROTEINASES

The serine proteinase-substrate interactions are extremely specific for viral proteins. Such interactions probably cause faithful formation of various viral proteins in an ordered temporal sequence during viral infection. The catalytic triad (His, Asp, and Ser amino acids, with Ser behaving as nucleophile during cleavage of the peptide bond by serine and serine-like proteinases) performs the cleaving function and is arranged in a conserved spatial structure – being fixed in a specific place by the three-dimensional structure of alpha-carbon backbone of proteinase. The serine residue is exceptionally reactive and donates an electron to carboxyl carbon of the peptide bond to be hydrolyzed. This forms an acyl serine and the active site His residue donates a proton to the departing amino group, followed by hydrolysis of the acyl enzyme, release of carboxylic acid product, and regeneration of the active site. All serine-like proteinases seem to be structurally and functionally similar to serine proteinases but instead possess a nucleophilic Cys amino acid at the active site. They are chymotrypsin-like proteinases.

Picornavirus 3C or 3C-related proteinases also bear a substrate-binding pocket that determines the cleavage site specificity. In particular, a His residue located in C-terminal region of proteinase is directly involved in recognition of a glutamine (or glutamate) residue at -1 position of cleavage sites of poty- and comovirus polyproteins. Proteinases of nepovirus subgroups A and B do not contain a His residue in their substrate-binding pockets and recognize cleavage sites that differ from the (Gln,Glu)/(Gly.Ser,Met) consensus of the polyproteins.

A. Family *Potyviridae*

Potyviruses extensively use proteinases during their genome expression and proteolytic processing of potyviral polyproteins has been studied in great detail by employing three translation systems: *in vitro* systems in which cell-free translation systems are used, and *in vivo* heterologous systems (*Escherichia coli*) or *in vivo* homologous systems (infected plants and transgenic plants) (Riechmann *et al.*, 1992). All potyviruses possess identical genome organisation with genomic RNA of each potyvirus containing a single ORF which is translated into a single polyprotein that is processed to form intermediate precursors and functional and mature structural and nonstructural final individual proteins by three virus-encoded proteinases (Riechmann *et al.*, 1992; Lindbo and Dougherty, 1994; Shukla *et al.*, 1994). For example, TEV genome encodes a single polyprotein precursor of 351-kDa, which is processed by three viral proteinases to form intermediate precursors and final individual protein products (Revers *et al.*, 1999). The three proteinases are nuclear inclusion a (NIa) proteinase, helper component proteinase (HC-Pro), and the 35-kDa N-terminal P1 proteinase.

The three potyviral proteinases include two serine and serine-like proteinases (P1 proteinase and NIa proteinase, which are chymotrypsin-like serine and cysteine proteinases, respectively) and one papain-like cysteine proteinase (HC-Pro). The P1 and HC-Pro proteinases are located at N-terminus of the polyprotein and auto-catalytically (that is, *in cis*) cleave themselves at their respective C-termini while NIa proteinase processes all other intra- and intermolecular polyprotein sites. Thus, the N-terminal 87-kDa TEV polyprotein undergoes two autoproteolytic cleavages caused by P1 and HC-Pro. The two serine proteinases are described below while the third (cysteine proteinase) is discussed later. All the three proteinases are multifunctional proteins and perform other important functions in viral life cycle. Primary amino acid sequence of Gly-Xaa-Ser-Gly is typical of serine proteinases, is well conserved, and is found in all potyviruses examined so far.

1. P1 Proteinase

The P1 proteinase of TEV is a chymotrypsin class of serine proteinase; is produced from N-terminal end of the complete polyprotein through a single monomolecular autoproteolytic (in *cis*) cleavage of a Tyr-Ser (Tyrosine304 and Serine305) dipeptide between itself (C-terminus of P1) and HC-Pro (N-terminus of HC-Pro); has two functional sequences - the N-terminal and the C-terminal; the N-terminal domain has a variable amino acid sequence in different potyviruses and so is not conserved; in contrast, the C-terminal segment is highly conserved and contains the P1 domain involved in proteolysis; proteolytic activity occurs within C-terminal 148 amino acid residues and contains the proposed functional catalytic triad (His214, Asp223, and Ser256) which is the conserved motif and is the actual center of enzyme; is unable to act in *trans*; is found in single component potyviruses; and is possibly absent in bipartite bymovirus potyviruses. The TVMV P1 proteinase causes a cleavage between phenylalanine and serine amino acid residues. Proteolytic domain of *Wheat streak mosaic rymovirus* (WSMV) is located in C-terminal region and mature P1 protein would be 48 amino acid residues longer that P1 of TEV (Choi *et al.*, 2002). In TEV, amino acids between 157-188 and 304 are essential for proteolytic activity. Proteolytic cleavage of P1 from HC-Pro is essential for TEV viability in plants (Verchot and Carrington, 1995).

Although potyviral P1 proteins are highly variable, yet the characteristic signature of serine proteinases (Gly-Xaa-Ser-Gly) is well conserved in all of them. In TEV P1, this signature sequence of serine proteinases always occurs at amino acid residue positions of 254 to 257. In addition, conserved His214 and conserved Asp223 residues are also necessary for proteolytic activity and are located upstream of the active site. The characteristic signature of serine proteinase is also present in *Brome streak mosaic rymovirus* but is absent in the bipartite bymovirus.

Choi *et al.* (2002) found that cleavage site of P1 proteinase of WSMV is located downstream of amino acid residue 348 and upstream of amino acid residue 353, with the peptide bond between amino acid residues Tyr352 and Gly353 (Tyrosine352 and Glycine353) being the most probable site of hydrolysis; contains the conserved triad His257, Asp267, Ser305; contains Phe-Ile-Val-unknown amino acid-Gly325-329 residues upstream of the cleavage site that are typical of serine proteinases and are required for P1 proteolysis of TEV; and WSMV P1 proteinase cleavage site is processed *in vivo*. The P1 protein also has an accessory role in potyvirus genome amplification that can be complemented in *trans* in cells expressing transgenic P1 protein and binds single-strand RNA nonselectively *in vitro*. These non-proteolytic functional domains are not yet mapped precisely.

2. NIa Proteinase

It, along with RNA replicase NIb, forms the crystalline inclusion bodies inside the nucleus of cells infected with some potyviruses. The TEV NIa proteinase is initially formed as a high molecular weight precursor; the 49-kDa NIa polyprotein being the most common while a 55-kDa precursor is also formed. The 27-kDa NIa enzyme is the C-terminal part of the 49-kDa NIa polyprotein and is considered to be similar to the serine proteinases in its structure. Thus, TEV cysteine proteinase domain is located in the 27-kDa NIa protein. The NIa proteinase is a key enzyme for potyvirus genome expression, normally occurs during late stages of infection, has very high substrate specificity, shares structural motifs with cellular serine chymotrypsin proteinases and so is a chymotrypsin-like proteinase but in which cysteine has replaced serine in their active center and bears structural resemblance to picornavirus 3C proteinase as well as to comovirus 24-kDa proteinase, and cleaves the polyprotein both in *cis* and in *trans* at six, or possible seven, cleavage sites present in C-terminal 264-kDa polyprotein. Since most of the TEV-encoded proteins accumulate in the cytoplasm, presumably the proteolytic functions of NIa are carried out prior to its translocation to nucleus. Interestingly, one of the functions of NIa multifunctional protein is to act as VPg (Murphy *et al.*, 1990) and this linkage of NIa to viral RNA during synthesis most likely occurs in cytoplasm.

Triad of His46, Asp81, and Cys151 constitutes the active site of TEV 27-kDa NIa and splits polyproteins at specific sites, which are characterized by highly conserved heptapeptide sequences. The suggested catalytic triad is conserved in various genera of

Potyviridae; however, amino acid sequences around the catalytic triad show much divergence. The N-domain of NIa constitutes VPg while C-domain causes all the inter- and intramolecular proteolytic activities leading to most of the proteolytic processing of polyprotein so that the C-terminus directs both *trans* and autoproteolytic activities. Controls for regulation of cleavage during genome expression are proposed to occur at two levels: at the level of primary amino acid sequence of cleavage site and at the level of precursor forms of NIa proteinase that directs subcellular localization and has a possible regulatory role.

Garcia *et al.* (1999) discuss in detail the cleavage sites of potyviruses. The NIa proteinase cleavage sites of polyproteins of TEV and PPV are marked by conserved heptapeptide sequences Gln/Glu-His–Ile/Val/Leu-Tyr-Xaa-Glu/Gln-Ser/Gly in which cleavage site is located between Gln and Ser/Gly residues. Cleavage sites of other viruses of picornaviral superfamily are also generally between Gln and Ser/Gly residues and this is a conserved feature of the most cleavage sites acted upon by NIa potyviral and 24-kDa comovirus proteinases.

In the above heptapeptide sequences, position 1 always bears glutamine (Gln) or rarely glutamic acid (Glu); and position 4 is normally occupied by hydrophobic valine (Val) residue but rarely by some other hydrophobic residue. Other places of the heptapeptide are less conserved; position 2 mostly carries His or an aromatic residue, and position 6 commonly bears acidic (particularly Glu) or amide residue (Gln), and Xaa is an unknown amino acid. Presence of particular amino acids at positions 6, 4, 3, 1, and 1` is essential for proper functioning of the cleavage site while amino acids at positions 5 and 2 affect the reaction profile of the cleavage. In TEV, positions 6 and 3 are more conserved than positions 4 and 2 but reverse occurs in other potyviruses – indicating that importance of a particular amino acid residue of the heptapeptide may vary in different potyviruses. The cleavage sequences of other potyviruses are similar but distinct. Polyproteins encoded by BrSMV RNA and BYMV and barley mild mosaic bymovirus RNA1 (all of family *Potyviridae*) contain sequences similar to those of the potyvirus NIa cleavage sites at expected positions. Despite the considerable similarity of both NIa proteinases and their recognition sequences among various potyviruses, NIa cleavage site of each potyvirus is seemingly well recognised exclusively and efficiently only by its own specific proteinases. This substrate specificity is apparently determined by several protein domains, including the His residue of the substrate-binding pocket.

To sum up, the NIa cysteine chymotrypsin-like proteinase induces the third cleavage, which occurs within the C-terminus of the remainder two-thirds of the polyprotein and causes all other cleavages in the polyprotein and therefore functions in *cis* and in *trans*.

B. Family *Comoviridae*

Genomes of both *Comovirus* and *Nepovirus* genera consist of two positive-strand RNA molecules that encode large polyproteins that are subsequently split proteolytically by virus-encoded proteinases. The comovirus RNA B and nepovirus RNA1 translate all the proteins needed for RNA replication including the viral proteinase while comovirus RNA M and nepovirus RNA2 encode capsid protein and virus movement protein. But, unlike nepovirus RNA2, translation of comovirus RNA M occurs from two alternative in-frame AUG codons producing two co-terminal polyproteins. Serine-like proteinases are encoded by CPMV and GFLV.

1. Comovirus Proteinase

CPMV B RNA encodes a polyprotein that contains proteins participating in viral RNA replication. An internal gene of B RNA encodes a viral proteinase - the 24-kDa chymotrypsin-like proteinase that cleaves the polyprotein encoded by B RNA as well as M RNA. The proteinase gene is located as a part of the VPg-proteinase-replicase ORF segment as in all the other members of the picornaviral group. The proposed catalyst triad is made up of His987, Glu1023, and Cys1113. Cys1131 residue belongs to the substrate-binding pocket and interacts with Gln at cleavage site P1 position. The 24-kDa CPMV proteinase recognises the dipeptide Gln-Gly, Gln-Ser, and Gln-Met cleavage sites. Identical dipeptides occur at the putative cleavage sites of other sequenced comoviruses like *Cowpea severe mosaic virus* and *Red clover mottle virus*.

Upstream sequences are required for efficient cleavage of 24-kDa proteinase at its C-terminal from the B RNA-encoded polyprotein. The cleavage is initiated in *cis* between a Gln-Gly dipeptide, forms the 32- and 170-kDa protein products, and is apparently cotranslational with the formation of the 32-kDa protein. This 32-kDa protein regulates activity of viral proteinase, is located at N-terminus of B RNA-encoded polyprotein, remains associated with 170-kDa polyprotein (that contains the 24-kDa proteinase) through hydrophobic bonds, regulates 170-kDa polyprotein processing which is slow in its absence but fast and complete in its presence, is released from the N-terminal of the polyprotein by an intramolecular cleavage, possibly serves as a cofactor of the cysteine proteinase, and influences the turnover of the M RNA- and B RNA-encoded polyproteins. Besides the 24-kDa proteinase that mediates cleavage of M RNA-encoded polyprotein at Gln-Met dipeptide located between the two overlapping movement proteins and the polyprotein precursor of the structural (capsid) proteins, the 32-kDa protein is also essential for cleavage of this polyprotein leading to formation of 58-kDa and 48-kDa movement proteins (55- and 45-kDa movement proteins according to some), and of 60-kDa (66-kDa) capsid protein precursor. The 60-kDa precursor is then cleaved by 24-kDa proteinase but in this case the 32-kDa protein is not needed.

Involvement of 32-kDa-protein in proteolysis by 24-kDa proteinase (encoded by the same genomic RNA B component) as well as proteolysis of M RNA-encoded polyprotein is a unique phenomenon during comovirus polyprotein processing. This protein is apparently essential for expression of CPMV proteins and is the main regulator of CPMV life cycle during infection. Thus, proteolytic processing of comovirus polyproteins requires an additional scaffolding or chaperone-like protein besides the proteinase enzyme. It appears from above that a two-component proteinase system functions in CPMV: the 24-kDa protein is the proper proteinase inducing various cleavages while 32-kDa protein is a chaperone-like protein which ensures proper association or alignment of substrate and proteinase. Such two-component proteinase system is not operative in nepoviruses.

The 24-kDa proteinases of various comoviruses exhibit large sequence conservation but still possess great specificity so that such proteinase of one comovirus fails to process cleavage sites of another comovirus. This specificity is controlled by interaction between tertiary structures of substrate and the substrate-binding region of the proteinase. The amino acid sequence of the junction site partly determines this cleavage specificity.

2. Nepovirus Proteinase

The large polyprotein, encoded by each nepoviral RNA, is proteolytically processed by a viral serine-like proteinase (Pro), encoded by RNA1, to release mature proteins. The nepovirus proteinase occupies the same genomic place and the same size (24-kDa) as the comovirus proteinase and, like comovirus proteinase, activity of nepovirus 24-kDa proteinase is also modulated by sequences surrounding it. However, unlike comoviruses, no virus-encoded cofactor is essential for *in vitro trans* processing of nepovirus RNA2-encoded polyprotein

The predicted catalytic triad of GFLV proteinase is constituted by His1284, Glu1328, and Cys1420 while the catalytic triad of *Tomato ringspot nepovirus* (TomRSV) proteinase consists of His, glutamate (or aspartate), and Cys. Thus, the 24-kDa TomRSV proteinase is different from those of the GFLV and other related nepoviruses but, instead, shows more similarity to CPMV 24-kDa proteinase as it has a His in the predicted substrate-binding region.

The 3C TomRSV proteinase has similarities to chymotrypsin-like serine proteinases. The consensus amino acid sequence for TomRSV cleavage sites is (Cys or Val) Gln/Gly or Ser (Wang *et al.*, 1999; Carrier *et al.*, 1999). Some cleavages by nepovirus proteinase take place at Arg-Ala, Arg-Gly, and Arg-Ala dipeptides even though cleavages at Cys-Ala, Cys-Ser, and Gly-Glu dipeptides are also reported.

a. Proteolytic Processing

The TomRSV is the only subgroup C nepovirus in which proteolytic processing of RNA1- and RNA2-encoded polyproteins (P1 and P2 polyproteins, respectively) are worked out. The specificity of cleavage site in proteinase of this virus differs from that of the proteinases of other two subgroup (A and B) nepoviruses. Cleavage sites have been identified at N-terminus of each of TomRSV polyprotein; these cleavage sites are not present in polyproteins of subgroup A and B nepoviruses (Wang and Sanfaçon, 2000; Carrier *et al.*, 2001).

i. P1 Polyprotein of *Tomato ringspot nepovirus*

The polyprotein (P1) encoded by nepovirus RNA1 contains proteins that participate in viral RNA replication. Five cleavage sites occur on RNA1-encoded P1 polyprotein to demarcate the putative NTP-binding protein (NTβ), VPg, serine-like proteinase, RNA-dependent RNA polymerase (RdRp) and two additional proteins (x1 and x2) located at N-terminus of polyprotein (Wang *et al.*, 1999; Wang and Sanfaçon, 2000). Viral proteinase, as is apparent from above, is encoded by an internal segment of the larger RNA (nepovirus RNA 1). The proteinase gene is located as a part of the VPg-proteinase-replicase ORF organisation. As in comoviruses, it coincides with that of other members of the picornaviral group.

Wang *et al.* (1999) show that TomRSV 3C-like proteinase processes P1 polyprotein precursor *in vitro* and in *E. coli* and that TomRSV proteinase could cleave the precursor and release the predicted mature proteins or intermediate precursors. Although processing was detected at all three predicted cleavages sites (NTβ-VPg, VPg-Pro, Pro-RdRp), processing at VPg-Pro cleavage site was inefficient. Wang *et al.* (1999) also found three new dipeptide cleavage sites (Gln/Ser between NTβ-VPg, Gln/Gly between VPg-Pro, and Gln/Ser between Pro-RdRp). Glutamine at −1 position of TomRSV proteinase plays a critical role in cleavage site specificity of proteinase. Specificity of cleavage by poty- and comovirus proteinase is for peptide bonds between glutamine (or glutamic acid) and residues with small side chains. Thus, TomRSV cleavage sites are more related to cleavage sites of poty- and comovirus proteinases than to cleavage sites of subgroups A and B nepovirus proteinases (Hans and Sanfaçon, 1995). Besides the conserved dipeptide, primary sequence and secondary structure around cleavage site are important for determining efficiency of cleavage by proteinase.

Cleavage sites on TomRSV RNA1-encoded polyprotein [including NTP-binding (NTB) protein-VPg cleavage site] are not processed *in trans* by proteinase *in vitro* (Carrier *et al.*, 1999; Wang and Sanfaçon, 2000). Thus, NTB and NTB-VPg proteins found in infected plant extracts are most probably produced by alternative *cis*-processing pathway of P1 polyprotein.

ii. P2 Polyprotein of *Tomato ringspot nepovirus*

Nepovirus RNA2 encodes a precursor polyprotein that is constituted by structural protein(s) and movement protein. Two cleavage sites occur in RNA2-encoded polyprotein (P2) allowing definition of capsid protein (CP) and movement protein (MP) domains. Cleavage site between TomRSV MP and CP is Gln/Gly (Hans and Sanfaçon, 1995). Carrier *et al.* (2001) identified a third cleavage site in N-terminal region of P2. This cleavage occurs at dipeptide Gln^{301}/Gly resulting in predicted release of two proteins from N-terminal region of P2 – a 34-kDa protein that is located at N-terminus of P2 by assumed *trans* initiation at first AUG codon and a 71-kDa protein that is situated immediately upstream of MP domain. TomRSV proteinase recognises this third cleavage site on P2 *in vitro*.

This is unlike other nepoviruses. Only one protein domain exists in equivalent area of P2 polyprotein of other characterized nepoviruses. Proteolytic processing of P2 polyprotein of GFLV (of nepovirus subgroup A) and of *Tomato black ring virus* and *Grapevine chrome mosaic virus* (both of nepovirus subgroup B) occurs at only two cleavage sites. Presence of two protein domains in N-terminal region of TomRSV P2 may be a unique feature of TomRSV or a feature common to other nepoviruses of subgroup C (Carrier *et al.*, 2001).

The TomRSV P1 polyprotein cleaving at five sites and TomRSV P2 polyprotein cleaving at three cleavage sites is in contrast to the four and two cleavage sites in P1 and P2 polyproteins, respectively, of nepoviruses of subgroups A and B.

b. Genome-Linked Protein (VPg)

Presence of VPg domain on partially processed precursors of TomRSV proteinase affects the proteinase activity by increasing *cis* cleavage at Pro-Pol site and by decreasing the *trans* proteinase activity on cleavage sites of RNA2-encoded polyprotein (Chisholm *et al.*, 2001). The slow release of mature TomRSV proteinase from VPg-Pro precursor may be the mechanism that favours *cis*-cleavage of P1 polyprotein at Pro-Pol site during early stages of infection to permit release of proteins taking part in viral replication. In contrast, *trans*-cleavage of P2 polyprotein is more efficient during later stages of infection that permits the release of MP and CP and later of other proteins.

The above differential proteolytic activities of precursor and mature forms of proteinase also exist in other nepoviruses. Thus, GFLV VPg-Pro is less active than Pro on cleavage sites of RNA2-encoded polyprotein and more active than Pro on cleavage sites of RNA1-encoded polyprotein. In contrast with TomRSV proteinase, presence of VPg on precursor containing GFLV proteinase did not influence *cis*-processing of Pro-Pol cleavage site but instead, increased *trans*-processing at cleavage site upstream of NTB domain. The VPg domain also modulates activity of proteinase in TBRV (Hemmer *et al.*, 1995). Presence of VPg domain on precursors containing TBRV proteinase increased *trans*-cleavage of cleavage sites of RNA2-encoded polyprotein and increased *cis*-cleavage at Pro-Pol site.

Thus, nepoviruses have developed common mechanisms for regulating processing of RNA1- and RNA2-encoded polyproteins through accumulation of proteinase precursors containing VPg domain. But these results also indicate that effect of VPg domain on the proteinase activity varies from one nepovirus to another. In potyviruses, although both VPg-Pro (NIa) and Pro (NIa-Pro) accumulate in plants, no differential activity occurs between them. Thus, modulation of proteinase activity by VPg domain could be a characteristic feature of nepoviruses (Chisholm *et al.*, 2001).

C. Family *Sequiviridae*

This family contains two monopartite genera: *Sequivirus* (contains PYFV) and *Waikavirus* (contains *Maize chlorotic dwarf waikavirus* – MCDV and RTSV). Their proteinases have been identified by sequence alignments. The predicted catalytic triads of PYFV and RTSV are constituted by His, Glu, and Cys while proposed active site residue of MCDV is Asp. The putative substrate-binding regions also differ. The His of proteinases, which cleave after a Gln residue, exist in RTSV and MCDV sequences but Leu (as in most of the nepoviruses) occurs in PFYV. It is suggested that CPs of RTSV and MCDV possibly originate by proteolytic cleavages at Gln-X dipeptide while no consensus sequence exists in cleavages sites of capsid proteins of PYFV. The RTSV possibly employs a chymotrypsin-like proteinase for processing its polyprotein that contains the conserved structural and nonstructural proteins. This virus also employs a subgenomic RNA for expressing the unique 3'-proximal ORF(s).

D. Putative Serine Proteolytic Activities

Computer generated models suggest that RNA genomes of luteoviruses and sobemoviruses encode chymotrypsin-like serine proteinases (Maia *et al.*, 1996b). These plant viruses employ proteolytic activity for expressing their respective genomes but no definite experimental and biochemical studies have yet been conducted. SBMV, BWYV, and PLRV possibly encode a putative serine proteinase from an internal segment of the polyprotein.

E. Conclusions

- Much variability exists in specificity of cleavage sites.
- The proteinase of comovirus group, proteinase of nepovirus group, NIa proteinase of potyviruses, and 3C proteinase of picornaviruses are all homologous to each other and their basic structure and catalytic activity are conserved. Active-site Ser, existing in all cellular proteinases, is substituted by a Cys residue in proteinases of this group.
- In proteinases of como- and potyviruses, an identical position is occupied by the conserved His and is characteristically located 12 to 18 amino acids downstream of the active-site Cys. This His is suggested to be the most important residue for substrate recognition and forms site S1 of substrate-binding pocket.
- Potyvirus and comovirus proteinases mostly recognize a substrate containing Gln (or less frequently Glu) in position 1.
- Viral proteinases of como- and nepoviruses, despite similarities, are distinct in their characteristic substrate recognition and cleavage regulation properties.
- Nepovirus proteinases may be distinguished from other homologous proteinases by the presence of Leu (leucine) at amino acid position 197

III. PAPAIN-LIKE CYSTEINE PROTEINASES

Proteinases containing cysteine at the active center are of two types: chymotrypsin-like cysteine proteinases that contain an active triad of amino acids (Cys, His, and Asp) (already dealt with) and papain-like cysteine proteinases that contain an active dyad of Cys and His that interact with one another

The PCPs resemble the classic cellular papain-like thiol proteinases but the actual sequence similarity is rather low. The nature and orientation of the catalytic dyad of Cys and His in these viral proteinases also resemble the cellular papain-like thiol proteinases. His is located in C-terminal position to Cys residue while the opposite applies to the corresponding residue in chymotrypsin-like cysteine proteinases. The conservation among viral papain-like proteinases exists nearly exclusively around the catalytic Cys residue; no other conserved motif exists so that PCP domains show only marginal sequence conservation and may occupy different positions in polyproteins of even closely related viruses (Koonin and Dolja, 1993). These proteinases are about as common among positive-strand RNA viruses as the chymotrypsin-like proteinases but are far more variable in their sequences and in genomic location than chymotrypsin-like proteinases of the same virus group. The cleavage site of PCP has the consensus sequence of UG/X (when U is an hydrophobic amino acid residue and X is Gly, Ala, or Val). According to Dougherty and Semler (1993), the sulfhydryl group of Cys residue during proteolysis behaves as a nucleophile and initiates attack on the carbonyl carbon of peptide bond that has to be hydrolyzed. The carbonyl carbon of substrate and sulfhydryl group of active-site Cys together form an acyl enzyme. The active-site His protonates the departing amide. Hydrolysis of carbonyl carbon then occurs from thiol group of proteinase and regeneration of active-site residues takes place."

The PCP was first recognised in the HC-Pro of a potyvirus (Oh and Carrington, 1989) and later predicted/confirmed experimentally in other alpha-like plant virus-encoded proteins of benyviruses (BNYVV), capilloviruses (ASGV), carlaviruses (BlSV), closetroviruses (L-Pro of BYV and CTV), marafiviruses (OBDV), potyviruses (TEV), trichoviruses (ACLSV), and tymoviruses (TYMV) (Agranovsky *et al.*, 1994; Lawrence *et al.*, 1995; Karasev *et al.*, 1995; Rozanov *et al.*, 1995; Garcia *et al.*, 1999; Agranovsky and Morozov, 1999; Karasev, 2000; Peng *et al.*, 2001, 2003). Because of the similarity of PCP domains of above plant viruses to that of the TYMV, they are also called tymo-like cysteine proteinases. PCP domains show only marginal sequence conservation and may occupy different positions in polyproteins of even closely related viruses (Koonin and Dolja, 1993). Experimental evidence for the functioning of PCP is available only for TYMV and BlSV. Papain-like cysteine proteinases of closteroviruses (BYV and CTV) bear striking similarity to potyviral HC-Pro while that of different plant alpha-like viruses are distantly related to HC-Pro. The catalytic dyad of TYMV proteinase is constituted by Cys783 and His869 (Rozanov *et al.*, 1995).

The L-Pro of BYV is expressed via translation of genomic RNA from the very onset of infection. The PCP domain is located between helicase and RNA-dependent RNA polymerase (RdRp or POL) domains in the polyprotein of BNYVV, but is found between methyltransferase and helicase amino acid sequences in polyproteins of carlaviruses and tymoviruses. Hence, in all the reported cases, the PCPs of plant viruses possess only the *cis*-activity and cleave at a single site downstream of the proteinase active center. Regardless of its position, the PCP of beny-, carla-, and tymoviruses mediates the splitting of the methyltransferase-helicase-containing and polymerase-containing fragments. Thus, although the genome structures of the plant viruses, possessing papain-like cysteine proteinases, are very different from each other but their large replication proteins invariably exhibit the same modular organisation: methyltransferase-like protein – proteinase – helicase-like protein – polymerase-like protein.

The PCPs are placed into two main classes – the 'main' proteinases and the 'accessory or leader' proteinases (Gorbalenya *et al.*, 1991; Peng *et al.*, 2001). The 'main' proteinases are required for processing of nonstructural polyproteins during viral RNA replication by conducting one or generally more than one cleavages, retain proteolytic activity after their release from a precursor, cleave virus and cell proteins in *trans*, and constitute parts of the arrays of domains (which include viral polymerase and helicase) that mediate viral RNA replication and expression (Dougherty and Semler, 1993; Strauss and Strauss, 1994; Kadaré *et al.*, 1995). Leader proteinases comprise the N-terminal domain of a polyprotein and typically perform only a single autocatalytic C-terminal cleavage in *cis* to release themselves from the rest of the polyproteins (Gorbalenya *et al.*, 1991; Dougherty and Semler, 1993; Agranovsky *et al.*, 1994; Strauss and Strauss, 1994). Hence, these domains are located outside the arrays of domains directly involved in genome replication and expression and occur at the N-terminus (in bymoviruses) or near N-terminus (in helper component proteinase of potyviruses), justifyng their name 'leader'.

PCPs of the positive-strand RNA viruses are multifunctional proteins. They are involved not only in polyprotein processing but also in genome amplification, virus pathogenicity, spread in the infected organism, and suppression of host defences (Koonin and Dolja, 1993; Dougherty and Semler, 1993; Strauss and Strauss, 1994; Suzuki *et al.*, 1999).

Both the potyviral HC-Pro and closteroviral L-Pro have the same two-domain structure: a non-proteolytic N-terminal domain and the C-terminal papain-like domain. BYV L-Pro and CTV L1 belong to the same functional class while CTV L2 belongs to another class (Peng *et al.*, 2001). The 66-kDa leader proteinase (L-Pro) of BYV is encoded in a 5`-proximal part of ORF1a, and has a non-conserved N-terminal domain (Peng *et al.*, 2003). The C-terminal domain of L-Pro is conserved, is essential, and sufficient for its autocatalytic release from viral polyprotein (Peremyslov *et al.*, 1998; Peng and Dolja, 2000; Peng *et al.*, 2003).

A. 'Main' Papain-like Cysteine Proteinases

'Main' PCPs were predicted, and confirmed experimentally in some cases, in several plant viruses of alpha-like virus supergroup (Garcia *et al.*, 1999; Agranovsky and Morozov, 1999). Because of their similarity to TYMV proteinases, they are also called tymo-like cysteine proteinases.

The suggested TYMV 'main' PCP conducts intramolecular cleavage of the 206-kDa polyprotein *in vitro* into an N-terminal 140-kDa protein (containing methyltransferase, proteinase, and NTPase/helicase domains) and a C-terminal 66-kDa protein having an RdRp-like amino acid sequence. This cleavage is functional *in vivo* (Prod'homme *et al.*, 2001) and both cleavage products appear to be essential for TYMV RNA replication. The methyltransferase and proteinase enzymatic modules are separated by an ~200–amino acid-long proline-rich sequence with a predicted secondary structure that may constitute a hinge between these two domains. Proteinase is located in central region of the 206-kDa polyprotein, possibly in amino acid sequences from 555 to 1051/is mapped just upstream of helicase domain and is delimited to amino acid residues 731 to 885; cleavage occurs between Ala1259 and Thr1260 according to one report while another report regards cleavage to occur between Asn and Gly at a cleavage site between amino acid residues 1253 to 1261; and the predicted catalytic dyad of proteinase is formed by Cys783 and His869. The TYMV proteinase (unlike closterovirus leader and the potyvirus HC-Pro proteinase) does not cleave polyprotein at the end of proteinase domain but further downstream between helicase and polymerase domains.

B. Leader Proteinases / Papain-like Leader Proteinases

Papain-like leader proteinases have been characterized in members of *Potyviridae* (HC-Pro) and *Closteroviridae* [closteroviral leader proteinase (L-Pro)] (Revers *et al.*, 1999; Karasev, 2000); members of these families encode one or two papain-like leader proteinases. As already mentioned, potyviral HC-Pro and closteroviral L-Pro possess a two-domain structure – a conserved C-terminal papain-like domain and a large and variable N-terminal nonproteolytic domain. The N-terminal domain in potyviruses and closteroviruses have specialized functions in efficient virus RNA amplification, suppression of RNA silencing, virus invasion and cell-to-cell spread in the plants, and aphid transmission (Kasschau and Carrington, 1995, 1998; Kasschau *et al.*, 1997; Peng *et al.*, 2001). In polyprotein of BNYVV, PCP is located between helicase and polymerase domains while in those of tymoviruses and carlaviruses it is found between the MT and HEL domains. In all the reported cases, the PCPs of plant viruses possess only the *cis* activity and cleave at a single site downstream of the proteinase active center.

The reported/predicted papain-like leader proteinases are: L-Pro (HC-Pro) of potyviruses (PPV, TEV, and TVMV), L-Pro of bymovirus (BaYMV) (closely related to potyviruses), L-Pro of BYV, L1 and L2 of CTV, and P-Pro of LIYV.

1. Potyviruses

The HC-Pro of potyviruses was the first viral proteinase shown to be cysteine-type papain-like proteinase similar to a cellular cysteine proteinase. It exerts a single *in cis* cleavage to detach its own C-terminus from the rest of the polyprotein, catalytic dyad is constituted by the conserved Cys and His amino acids that seem to form the active site, is well conserved in and is encoded by all potyviruses examined. The HC-Pro domain is also present at homologous polyprotein regions in BrSMV and *Sweet potato mild mottle ipomovirus* (SPMMV) (both of family *Potyviridae*). The N-terminus of 28-kDa protein encoded by RNA2 of bymoviruses has much similarity with HC-Pro domain.

Potyvirus HC-Pro is a nonstructural protein, is adjacent to C-terminus of P1 proteinase in TEV, is the second protein from N-terminal of polyprotein, and is divisible into three functional regions: N-terminal (amino acids 1 to 88), central (amino acids 89 to 321), and C-terminal (amino acids 322 to 456) region, seems to be expressed initially as a segment of genome-encoded N-terminal 87-kDa polyprotein, and causes a single monomolecular cleavage. Size of biologically active HC-Pro is between 100- and 150-kDa for PVY and TVMV (Thornbury *et al.*, 1985). These biologically active forms can be resolved as 58-kDa (PVY) and 53-kDa (TVMV) proteins by denaturing gel electrophoresis, suggesting that HC-Pro is a homodimer. The TEV potyviral HC-Pro is commonly regarded to be a ~50 to 55 -kDa protein. A major domain responsible for homodimerization is located in N-terminal region of HC-Pro in the first 72 amino acids in *Lettuce mosaic virus* (LMV) (Urcuqui-Inchima *et al.*, 1999a).

HC-Pro of TEV is PCP and the catalytic dyad is composed of Cys649 and His722 (Oh and Carrington, 1989). It is the C-terminal half of HC-Pro that contains a papain-like proteinase that cleaves HC-Pro from P3 protein (Carrington and Herndon, 1992). PCP cleaves at a Gly-Gly dipeptide to define its own C-terminus and N-terminus of P3 protein (Carrington *et al.*, 1993). The TEV HC-Pro cleaves itself and upstream sequences from the growing polyprotein and this is the only cleavage event that it activates. The cleavage is rapid, monomolecular, occurs cotranslationally. Genus *Bymovirus* is bi-component and could show proteolytic activity, which is seemingly associated with the N-terminal 24-kDa product of RNA2 in BYMV. Catalytic dyad in TEV is constituted by Cys and His. Like P1 proteinase, the domain of HC-Pro involved in proteolysis is also located in its C-terminal half.

Natural sequences occurring at HC-Pro cleavage sites are conserved in various potyviruses. A requirement for Tyr at position 4, Leu at position 2, and Gly at position 1 is nearly absolute in potyviruses and cleavage site recognition is facilitated by an extensive cleavage site sequence. Whether the HC-Pro requires the entire protein or only the N-terminal part for performing the above functions is still unresolved. TEV HC-Pro cleavage *in vitro* occurs at a glycine (Gly) dipeptide Gly-Gly (Gly↓Gly) (amino acid residues 673-674). There is good consensus at the possible cleavage sites of HC-Pro of different potyviruses [an ipomovirus (SPMMV) and a rymovirus (BrSMV)].

HC-Pro – HC-Pro self-interactions are reported in TEV by using biochemical methods and are reported in *Potato virus A* (PVA), PSbMV and LMV (all potyviruses) by using yeast two-hybrid system. The N-terminal half (amino acids 1 to 228) of HC-Pro of LMV is capable of self-interaction but only 72 amino acids (Urcuqui-Inchima *et al.*, 1999a) or 83 N-terminal amino acids (Urcuqui-Inchima *et al.*, 1999b) are sufficient for this purpose. The highly conserved Cys25 and Cys53 residues within Cys-rich domain, as well as His23, are involved in LMV HC-Pro self-interaction. In PVA, domains located between amino acids 112 and 135 near N-terminus and bet-ween amino acids 329 and 457 at C-terminus are needed for HC-Pro-HC-Pro interaction (Guo *et al.*, 1999). It is, therefore, possible that in different potyviruses different regions of HC-Pro are required for self-interaction. However, the precise biological role of the N-terminal of HC-Pro, with respect to its capacity to self-interact, is still to be established. (Urcuqui-Inchima *et al.*, 1999b).

Apart from its role in HC-Pro interaction, the N-terminus of HC-Pro has been assigned other functions as well. The Lys50 in N-terminus of HC-Pro is essential for transmission and infectivity of TVMV but does not seem to play a role in self-interaction while Cys25 and Cys53 residues (within the Cys-rich domain) and His23 are essential for TVMV viability. In contrast, the entire N-terminal region is dispensable for viability of TEV (Dolja *et al.*, 1993; Verchot and Carrington 1995) and LMV. It suggests that ability of HC-Pro to dimerize, important for aphid transmission, is less important for other functions of this protein. This possibly answers the question whether HC-Pro is biologically active as a monomer or as a dimer because early purification experiments indicated that the soluble HC-Pro is probably present as a dimer in infected plants (Thornbury *et al.*, 1985).

A major domain is located in N-terminal region of LMV HC-Pro in the first 72 amino acids (Urcuqui-Inchima *et al.*, 1999a); this domain is crucial for viral transmission by aphids (Atreya *et al.*, 1992). The Lys-Ile-Thr-Cys (KITC) motif at amino acid positions 52-55 in LMV and 50-53 in PVY (Urcuqui-Inchima *et al.*, 1999a) is the determinant of this activity (Atreya *et al.*, 1992; Atreya and Pirone, 1993; Dolja *et al.*, 1993, 1997).

2. Closteroviruses

Family *Closteroviridae* contains two genera: *Closterovirus* and *Crinivirus* with BYV and LIYV, respectively, as the type members. Closterovirus leader proteinase is about 60/66-kDa and Cys509 and His569 possibly constitute the catalytic dyad of BYV. The 5`-terminal ORF1 encodes a large polyprotein whose N-terminal part encompasses the leader proteinase (Agranovsky *et al.*, 1994), which is traditionally abbreviated as L-Pro in BYV (Peremyslov *et al.*, 1998).

LIYV (*Crinivirus*) has two RNAs as genome; RNA1 encodes the leader proteinase (abbreviated as P-Pro – Klaassen *et al.*, 1995) and replicase while RNA2 specifies most of the viral subgenomic RNAs. One or two related PCP domains, located at the N-terminus of ORF1a product, cause release of leader protein from rest of the polyprotein containing the methyltransferase and helicase domains.

The leader proteinase (L-Pro) of BYV is a 66-kDa multifunctional protein, is associated with closterovirus-induced vesicle aggregates (Zinovkin *et al.*, 2003), is a papain-like cysteine proteinase (Agranovsky *et al.*, 1994; Peremyslov *et al.*, 1998), is encoded by a 5`-proximal part of BYV ORF1a which also encodes methyltransferase and RNA helicase, self-processes itself leading to autocatalytic processing at Gly^{588}/Gly^{589} bond in 1a polyprotein (Zinovkin *et al.*, 2003), the predicted catalytic dyad is formed by Cys783 and His869 or Cys509 and His569 but is constituted by Cys509 and His569 according to Agranovsky *et al.* (1994), possesses a non-conserved and nonproteolytic N-terminal domain that functions in RNA amplification and is required for efficient accumulation of BYV RNAs (Peng and Dolja, 2000), and contains a conserved papain-like C-terminal domain that is required for autoproteolysis as well as for RNA amplification (RNA replication-related function) and these two functions are genetically separable and independent of each other (Peng *et al.*, 2003). The variable N-terminal domain of L-Pro plays a prominent role in RNA replication and amplification (Peng and Dolja, 2000). A short RNA element, located within the 5`-terminal region of L-Pro ORF, is indispensable for genome replication. The 1a BYV protein, apart from cleavage by L-Pro that is expected to produce 66-kDa and 229-kDa fragments, undergoes additional processing by a virus-encoded or cellular proteinase (Erokhina *et al.*, 2000). Release of BYV L-Pro is essential for genome replication.

CTV and some other closteroviruses encode a tandem of leader proteinases (L1 and L2) that are mechanistically distinct and act synergistically (Karasev *et al.*, 1995).

LIYV also encodes a leader proteinase. Leader proteinases show distinct patterns of subcellular localization: most of the CTV L1 and L2 proteinases are uniformly distributed in cytoplasm and nucleus; leader proteinase (P-Pro) of LIYV is almost exclusively localized to nuclei; while leader proteinase of BYV is predominantly in cytoplasmic inclusion bodies. This shows that these proteinases have special in-borne signals for their interaction with particular cellular components. The homologous papain-like leader domains of L-Pro, L1, and L2 (of closteroviruses) and HC-Pro (of potyviruses) are functionally specialized (Peng *et al.*, 2001).

Apart from having the common characteristics of being involved in autocatalytic processing and possessing a proteolytic activity causing polyprotein processing, the leader proteinases/domains of closteroviruses and potyviruses are implicated in efficient genome amplification, virus invasion, and virus movement from cell-to-cell (Kasschau and Carrington, 1995; Kasschau *et al.*, 1997; Peng and Dolja, 2000; Peremyslov *et al.*, 1998).

IV. ASPARTIC PROTEINASES

Cellular and viral aspartic proteinases, although related, have differences also: viral enzymes are composed of only 99 to 125 amino acid residues while cellular enzymes are composed of about 325 amino acid residues. Viral aspartic proteinases are found only in plant pararetroviruses in which they perform a highly sequence-specific processing function having an important role in viral life cycle. The active form of proteinase enzymes is a dimer formed by two identical subunits. Their catalytic mechanism is not yet fully understood although an acid-base catalysis has been suggested and no covalent enzyme-substrate intermediates are formed.

Plant pararetroviruses belong to family *Caulimoviridae* and contain two genera: *Caulimovirus* and *Badnavirus*. *Caulimovirus* (CaMV) genome encodes capsid protein, proteinase, reverse transcriptase (RT), and RNase H in the same order as animal retroviruses. Two RNAs (19S and 35S) are responsible for CaMV gene expression and are synthesized in infected plants. The 19S RNA is a monocistronic mRNA for ORF VI while the 35S RNA is the mRNA for all the remaining seven ORFs. A multifunctional protein is encoded by the 19S RNA and one of the functions of this protein is the regulation of translation. It does so by binding to the long (~600 nucleotides) 5` untranslated leader of the 35S mRNA. Ribosomes bind to this region, translate 35S mRNA in a unique manner, called the relay-race model, by which the individual translation products of each of the seven ORFs is expressed. ORF IV is expressed as a polyprotein, which is the viral capsid protein precursor.

The ORF V (the *pol* gene) is expressed as a non-structural 80-kDa polyprotein that contains the proteinase-RT-RNase H domains and is processed by proteolysis to yield an N-terminal 22-kDa protein that contains the proteinase domain and a C-terminal 58-kDa protein that contains RT and RNase H domains. The 22-kDa protein has a conserved amino acid sequence of aspartic proteinase, represents half of the of the active proteinase molecule so that dimerization of this protein is essential for its activity so that CaMV proteinase is active as a dimer and its proteolytic activity has been demonstrated. The C-terminal 58-kDa protein has conserved amino acid motifs suggestive of RT and RNase II functions and these activities have been detected in extracts of infected cells. The ORFs IV and V can be regarded as being equivalent to the retroviral *gag* and *pol* genes.

Very little is known about the proteolytic activity during replication of badnaviruses, genus *Badnavirus*. *Commelina yellow mottle badnavirus* possesses a large ORF that can potentially code for a polyprotein of ~200-kDa and also possesses a nucleotide sequence homologous with gene sequences of retroviral proteinases. An aspartic acid-type proteinase is suggested to be concerned with gene expression of this badnavirus but no direct evidence is available about this. Some experimental evidence shows the presence of aspartyl proteinase in *Rice tungro bacilliform badnavirus* (RTBV). The movement protein, capsid protein, proteinase motif, RT motif, and RNase H domain are present as part of an anticipated single polyprotein that also contains additional sequences of unknown function. Aspartyl proteinase of this badnavirus cleaves the polyprotein at two places: upstream of RT and downstream of RNase H domain.

V. FUNCTIONS OF PROTEINASES

The basic and primary function of proteinases is gene expression through post-translational proteolytic cleavage of a high molecular weight polyprotein, synthesized by translation of genomic RNA, into functional proteins. This is called the proteinase function. Thus, the similar ~140-residue PCP domains (which are embedded in the replicative cores) of BYV L-PCP and potyvirus HC-Pro control multiple *in cis* and *in trans* cleavages that produce mature proteins possessing HEL, MT, and RdRp sequences. In HC-Pro, several of these important functions have been mapped to specific domains (Maia *et al.*, 1996; Kasschau *et al.*, 1997). The eleven functional proteins of *Clover yellow vein potyvirus* are generated by viral polyprotein processing by three virus-encoded proteinases (P1, HC-Pro, and NIa) (Riechmann *et al.*, 1992).

However, all proteinases (serine and serine-like proteinases, papain-like cysteine proteinases, BYV L-PCP, potyviral HC-Pro, L-Pro, and others) are generally multifunctional proteins. The reported functions of proteinases of various plant viruses are as below.

Proteinases possess autoproteolytic activity by which they release themselves from polyproteins. HC-Pro autoproteolytically releases its own C-terminus from polyprotein; this activity requires C-terminal part of HC-Pro. The proteolytic activity in TEV-infected plants is initially associated with a 49-kDa polyprotein precursor, which can be internally processed at a suboptimal cleavage site to generate VPg and a functional 27-kDa proteinase domain. A self-cleavage site exists within the C-terminus of 27-kDa TEV proteinase (Parks *et al.*, 1995).

Gene expression/genome amplification is a crucial event in the life cycle of plant viruses. Cleaving of a polyprotein by a proteinase is not a simple enzyme-substrate reaction. Post-translational and differential proteolytic processing is one of the mechanisms by which gene/genome expression is regulated by some viruses. Como- and potyviruses cannot regulate gene expression at transcriptional stage (that is, subgenomic RNAs are not transcribed) and practically all gene products are produced at the same time by forming a polyprotein. However, time, rate, and efficiency of cleaving of different protein products from the polyprotein could have considerable variation. Proteolytic processing of polyproteins determines the time of appearance of each gene product. Some gene products or intermediate products or precursors may appear early or accumulate at a much faster rate and in greater quantity while other gene products appear late or accumulate at a slower rate and in much less quantity during replication cycle of these viruses. Differential processing of the polyprotein may account for all the above variations in product formation and accumulation. However, some cellular processes or factors may also be involved and contribute to differential production of a particular protein. Post-translational modifications of polyproteins can serve several other functions: the tethered proteins can move together to the appropriate assembly site, and activation of enzymes is delayed till the unfolding and assembly start. Together, all the above events regulate activity and concentration of important viral proteins and cause optimization of virus production by controlling virus replication and maturation.

Genetic evidence implicates the highly conserved motif Iie-Gly-Asn (IGN) in the central region of HC-Pro during potyviral genome amplification. (Cronin *et al.*, 1995; Kasschau *et al.*, 1997). The RNA-binding activity of domain B of HC-Pro is postulated to be involved in genome amplification (Urcuqui-Inchima *et al.*, 2000). Like HC-Pro, BYV L-PCP also affects virus RNA amplification (Peremyslov *et al.*, 1998; Peng and Dolja, 2000; Peng *et al.*, 2003). TEV P1 acts in *trans* as an accessory factor for genome amplification (Verchot and Carrington, 1995).

HC-Pro is required for efficient acquisition and transmission of potyviruses from plant to plant by aphids because HC-Pro has aphid stylet-binding properties (Blanc *et al.*, 1997, 1998; Peng *et al.*, 1998; Wang *et al.*, 1998). The N-terminus of HC-Pro, required for aphid transmission (Pirone and Blanc, 1996) of all potyviruses, possesses a region containing highly conserved Lys-Ile-Thr-Cys (KITC) sequence and Lys of this sequence is involved in interaction between HC-Pro and aphid mouthparts (Blanc *et al.*, 1998). The K^{50} in HC-Pro is an essential residue for transmission and infectivity (Atreya and Pirone, 1993). The zone, comprised of the first 72 amino acids in N-terminal region of HC-Pro in LMV, is crucial for aphid transmission (Atreya *et al.*, 1992). The KITC motif at amino acid positions 52-55 in LMV and 50-53 in PVY is the determinant of this activity (Atreya and Pirone, 1993; Dolja *et al.*, 1993, 1997). The ability of HC-Pro to dimerize is important for aphid transmission.

In all potyviruses, the KITC region (which in PVY encompasses amino acids 23 to 56) is rich in Cys residues and contains a highly conserved His residue at position 23. The N-terminus of potyvirus HC-Pro contains Cys-rich domain. Urcuqui-Inchima *et al.* (1999b) report that the first ~60 amino acids of N-terminal part of potyvirus HC-Pro include conserved residues comprising a Cys-rich region; that the domain in PVY that is sufficient for self-interaction was mapped to the 83 N-terminal amino acids of HC-Pro. (Urcuqui-Inchima *et al.*, 1999b); and that, in agreement with the suggestions of

Atreya and Pirone (1993), N-terminus may have important structural features for the proper folding of an active HC-Pro.

Besides interacting with aphids, HC-Pro also possesses a sequence that interacts with virus particles and thereby ensures their aphid transmission. The Pro-Thr-Lys (PTK) motif, situated in domain B of HC-Pro, is involved in interaction of HC-Pro with virus particle supposedly through protein-protein interaction between CP and HC-Pro (Peng *et al.*, 1998). TEV HC-Pro also performs a helper-component function, which is essential for virus transmission by aphids by supposedly mediating virus association with stylet of the vector.

Urcuqui-Inchima *et al.* (2000) show that N- and C-terminal regions of HC-Pro are dispensable for RNA binding, that two independent RNA-binding domains (designated A and B domains) are located in central part of HC-Pro, both domains are capable of RNA binding independently of each other and could be functionally different. Viral proteins containing two different RNA-binding domains have been reported [in TMV-MP (Citovsky *et al.*, 1992) and in TSWV-nucleocapsid]. Domain B also appears to contain a ribonucleoprotein (RNP) motif typical of a large family of RNA-binding proteins involved in several cellular processes.

The RNP motif [also called RRM (RNA recognition motif)] is the hallmark of a large family of RNA-binding proteins that are involved in several cellular processes but is mostly involved in RNA processing, transport process, gene expression, and development. RNP domain is composed of about 95 amino acids. It characteristically possess a $\beta1$-$\alpha1$-$\beta2$-$\beta3$-$\alpha2$-$\beta4$ secondary structure forming a four stranded anti-parallel β-sheet connected by two α-helices. Two short motifs, RNP-2 and RNP-1, with conserved amino acids are located on $\beta1$ and $\beta3$ strands, respectively.

Urcuqui-Inchima *et al.* (2000) suggest that HC-Pro domain B appears to contain a ribonucleoprotein (RNP) motif, that HC-Pro might belong to the large family of RNA-binding proteins possessing RNP domains, and that the binding activity of domain B is associated with one aromatic amino acid at its position 288. Two sequence stretches present in equivalent positions in domain B of HC-Pro share similarities with characterized RNP-2 and RNP-1 motifs. Indeed, in first motif (amino acids 246-251), 5 out of 6 amino acids are identical to consensus sequences found in this family of proteins. The second motif (amino acids 282-289) is more distantly related to RNP-1 and possess four out of eight conserved residues, one of which is involved in RNA binding. The secondary structure of domain B is composed of four β strands and one α-helix in the same arrangement as classical RNP folding, with the putative RNP2 and RNP1 located on $\beta1$and $\beta3$, respectively, as expected. Several amino acids of domain B are highly conserved among potyvirus HC-Pro (Cronin *et al.*, 1995). Domain B of HC-Pro is also possibly involved in genome amplification (Urcuqui-Inchima *et al.*, 2000) and in interaction of HC-Pro with virus particle during transmission by vector (Peng *et al.*, 1998).

Amino acids located upstream or within domain A of HC-Pro have been implicated in increased pathogenicity of PVX in co-infection with potyviruses (Pruss *et al.*, 1997; Shi *et al.*, 1997).

HC-Pro is also necessary for long-distance and systemic movement of virus within infected plants. A central Cys-Cys/Ser-Cys (CC/SC) motif is necessary for systemic spread in infected plants and for gene amplification (Cronin *et al.*, 1995; Kasschau *et al.*, 1997). Consistent with this is the fact that HC-Pro has plasmodesmata gating properties (Rojas *et al.*, 1997). Like HC-Pro, BYV L-PCP also affects virus cell-to-cell movement (Peng and Dolja, 2000; Peng *et al.*, 2003).

HC-Pro performs still other functions. It is also involved in symptom expression and appears as a 'general' pathogenicity factor; in fact, both P1 and HC-Pro can act as pathogenicity enhancers of several heterologous plant viruses, probably as *trans*-activators of viral replication (Verchot and Carrington, 1995; Pruss *et al.*, 1997; Shi *et al.*, 1997). HC-Pro also has nucleic acid-binding properties (Maia and Bernardi, 1996); it binds non-specifically to single-stranded nucleic acids but has preference for RNA. Proteinases are involved in maturation of structural proteins during assembly of virus particles. HC-Pro can suppress gene silencing in plants (Shi *et al.*, 1997; Brigneti *et al.*, 1998; Kassachau and Carrington, 1998). P1/HC-Pro coding region of a potyvirus is a general suppressor of post-transcriptional gene silencing (Anandalakshmi *et al.*, 1998; Brigneti *et al.*, 1998; Marathe *et al.*, 2000). This property of HC-Pro could explain its role in prolonging potyvirus genome amplification and heterologous virus accumulation. But the precise biochemical role of HC-Pro in replication remains unknown.

Thus, proteinases (leader or otherwise) are implicated in genome replication and genome amplification, synthesis of subgenomic RNAs, and in various aspects of virus-host interactions like virus pathogenicity, long-distance virus transport and spread in plants, aphid transmission, and suppression of RNA silencing in host defence/ suppression of host defences. BYV L-Pro is suggested to be required for the virus to establish itself in inoculated plant tissue. Some of these functions are performed by the N-terminal domain of papain-like proteinases and the others by the C-terminal domain. The N-terminal domain of L-Pro is involved in BYV RNA amplification and accumulation and a short RNA element located within the 5`-terminal region of L-Pro is essential for genome replication (Peng and Dolja, 2000; Peng *et al.*, 2003). The C-terminal domain is also required for efficient RNA amplification (Peng *et al.*, 2001, 2002) and the two roles are genetically separable.

Peng *et al.* (2001) found that L-Pro, L1, L2, and HC-Pro proteinases are mechanistically distinct and that their homologous papain-like domains are functionally specialized. Functional profiles of L-Pro overlap that of potyviral HC-Pro (Dolja *et al.*, 1997; Peremyslov *et al.*, 1998). It can be suggested that the genome amplification and viral movement in plants may be two different faces of the capability of HC-Pro to suppress host defences by effecting host pathogenicity (Pruss *et al.*, 1997) and the capacity of HC-Pro to suppress gene silencing (Shi *et al.*, 1997; Brigneti *et al.*, 1998; Kasschau and Carrington, 1998).

Aspartic proteinases, encoded by viral genomes of all caulimoviruses, possess the essential sequence of Asp-Thr(Ser)-Gly in their active site and possess distinct substrate specificities. These viral proteinases become enzymatically active only after dimerization of two identical molecules. CaMV proteinase processes the polyprotein from which it is derived. In addition, it processes a 55-kDa capsid protein precursor, which is encoded by ORF IV; the 55-kDa protein precursor is cleaved in cell-free extracts into 11- and 44-kDa proteins. The CP is obtained by modification of the 44-kDa protein. However, CP exists in multiple forms because of post–translational modifications such as proteolytic processing, glycosylation, and phosphorylation. It is difficult to determine the exact site(s) of cleavage because of these modifications.

VI. REFERENCES

Agranovsky, A. A., and Morozov, S. 1999. Gene expression in positive-strand RNA viruses. In: Mandahar, C. L. (Ed.). Molecular Biology of Plant Viruses. Kluwer Academic Publ. Boston / Dordrecht / London. p. 99-119.

Agranovsky, A. A., Koonin, E. V., Boyko, V. P., Maiss, E., Frötschl, R., Lunina, N. A., and Atabekov, J. G. 1994. *Beet yellows closterovirus*: Complete genome structure and identification of a leader papain-like thiol proteinase. *Virology* 198: 311-324.

Anandalakshmi, R., Pruss, G. J., Ge, X., Merethe, R., Mallory, A. C., Smith, T. H., and Vance, V. B. 1998. A viral suppressor of gene silencing in plants. *Proc. Natl. Acad. Sci. USA* 95: 13079-13084.

Atreya, C. D., and Pirone, T. P. 1993. Mutational analysis of the helper component-proteinase gene of a potyvirus: Effects of amino acid substitutions, deletions, and gene replacement on virulence and aphid transmissibility. *Proc. Nat. Acad. Sci. USA* 90: 11919- 11923.

Atreya, C. D., Atreya, P. L., Thornbury, D. W., and Pirone, T. P. 1992. Site-directed mutations in the potyvirus HC-Pro gene affect helper component activity, virus accumulation, and symptom expression in infected tobacco plants. *Virology* 191: 106-111.

Blanc, S., Lopez-Moya, J. J., Wang, R., Garcia-Lampasona, S., Thornbury, D. W., and Pirone, T. P. 1997. A specific interaction between coat protein and helper component correlates with aphid transmission of a potyvirus. *Virology* 231: 141-147.

Blanc, S., Ammar, E. D., Garcia-Lampasona, S., Dolja, V. V., Llave, C., Baker, J., and Pirone, T. P. 1998. Mutations in the potyvirus helper component protein: Effects on interactions with virions and aphid stylets. *J. Gen. Virol.* 79: 3119-3122.

Brigneti, G., Voinnet, O., Li, W.-X., Ji, L.-H., Ding, S.-W., and Baulcombe, D. C. 1998. Viral pathogenicity determinants are suppressors of transgene silencing in *Nicotiana benthamiana*. *EMBO J.* 17: 6739-6746.

Carrier, K., Hans, F., and Sanfaçon, H. 1999. Mutagenesis of amino acids at two tomato ringspot nepovirus cleavage sites: Effect on proteolytic processing in *cis* and in *trans* by the 3C-like proteinase. *Virology* 258: 161-175.

Carrier, K., Xiong, Y., and Sanfaçon, H. 2001. Genome organisation of tomato ringspot nepovirus RNA2: Processing at a third cleavage site in the N-terminal region of the polyprotein *in vitro*. *J. Gen. Virol.* 82: 1785-1790.

Carrington, J. C., and Herndon, K. L. 1992. Characterization of the potyviral HC-Pro autoproteolytic cleavage site. *Virology* 187: 308-315.

Carrington, J. C., Haldeman, R., Dolja, V. V., and Restrepo-Hartwig, M. A. 1993. Internal cleavage and *trans*-proteolytic activities of the VPg-proteinase (NIa) of *Tobacco etch potyvirus in vivo. J. Virol.* 67: 6995-7000.

Carrington, J. C., Kasschau, K. D., and Johansen, L. K. 2001. Activation and suppression of RNA silencing by plant viruses. *Virology* 281: 1-5.

Chisholm, J., Wieczorek, A., and Sanfaçon, H. 2001. Expression and partial purification of recombinant tomato ringspot nepovirus 3C-like proteinase: Comparison of the activity of the mature proteinase and the VPg-proteinase precursor. *Virus Res.* 79: 153-164.

Choi, I.-R., Horken, K. M., Stenger, D. C., and French, R. 2002. Mapping of the P1 proteinase cleavage site in the polyprotein of *Wheat streak mosaic virus* (Genus *Tritimovirus*). *J. Gen. Virol.* 83: 443-461.

Citovsky, V., Wong, M. L., Shaw, A. L., Prasad, B. V. V., and Zambryski, P. 1992. Visualization and characterization of tobacco mosaic virus protein binding to single-stranded nucleic acids. *Plant Cell* 4: 397-411.

Cronin, S., Verchot, J., Haldeman-Cahill, R., Schaad, M. C., and Carrington, J. C. 1995. Long-distance movement factor: A transport function of the potyvirus helper component proteinase. *Plant Cell* 7: 549-559.

Dolja, V. V., Herndon, K. L., Pirone, T. P., and Carrington, J. C. 1993. Spontaneous mutagenesis of a plant potyvirus genome after insertion of a foreign gene. *J. Virol.* 67: 5968-5975.

Dolja, V. V., Hong, J., Keller, K. E., Martin, R. R., and Peremyslov, V. V. 1997. Suppression of potyvirus infection by coexpressed closterovirus protein. *Virology* 234: 243-252.

Dougherty, W. G., and Parks, T. D. 1989. Molecular genetic and biochemical evidence for the involvement of the heptapeptide cleavage sequence in determining the reaction profile at two tobacco etch virus cleavage sites in cell-free assays. *Virology* 172: 145-155.

Dougherty, W. G., and Semler, B. L. 1993. Expression of virus-encoded proteinases: Functional and structural similarities with cellular enzymes. *Microbiol. Rev.* 57: 781-822.

Erokhina, T. N., Zinovkin, R. A., Vitushkina, M. V., Jelkmann, W., and Agranovsky, A. A. 2000. Detection of beet yellows closterovirus methyltransferase-like and helicase-like proteins *in vivo* using monoclonal antibodies. *J. Gen. Virol.* 81: 597-603.

Garcia, J. A., Fernāndez-Fernāndez, M. R., and López-Moya, J. J. 1999. Proteinases involved in plant virus genome expression. In: Proteinases of Infectious Agents. Academic Press, New York. p. 233-263.

Gorbalenya, A. E., Koonin, E. V., and Lai, M. M. C. 1991. Putative papain-related thiol proteinases of positive-strand RNA viruses. *FEBS Lett.* 288: 201-205.

Guo, D., Merits, A., and Saarma, M. 1999. Self-association and mapping of interaction domains of helper component-proteinase of *Potato A potyvirus. J. Gen. Virol.* 80: 1127-1131.

Hans, F., and Sanfaçon, H. 1995. Tomato ringspot nepovirus proteinase: Characterization and cleavage site specificity. *J. Gen. Virol.* 76: 917-927.

Hemmer, O., Greif, C., Dufourcq, P., Reinbolt, J., and Fritsch, C. 1995. Functional characterization of the proteolytic activity of the tomato black ring nepovirus RNA1-encoded polyprotein. *Virology* 206: 362-371.

Hohn, T., and Fütterer, J. 1997. The proteins and functions of plant pararetroviruses: Knowns and unknowns. *Crit. Rev. Plant Sci.* 16: 133-161.

Kadaré, G., Rozanov, M., and Haenni, A.-L. 1995. Expression of the turnip yellow mosaic virus proteinase in *Escherichia coli* and determination of the cleavage site within the 206 kDa protein. *J. Gen. Virol.* 76: 2853-2857.

Karasev, A. V. 2000. Genetic diversity and evolution of closteroviruses. *Annu. Rev. Phytopathol.* 38: 293-324.

Karasev, A. V., Boykov, V. P., Gowda, S., Nikolaeva, O. V., Hilf, M. E., Koonin, E. V., *et al.*, 1995. Complete sequence of the citrus tristeza virus RNA genome. *Virology* 208: 511-520.

Kasschau, K. D., and Carrington, J. C. 1995. Requirement for HC-Pro processing during genome amplification of *Tobacco etch potyvirus. Virology* 209: 268-273.

Kasschau, K. D., and Carrington, J. C. 1998. A counter defensive strategy of plant viruses: Suppression of post-transcriptional gene silencing. *Cell* 95: 461-470.

Kasschau, K. D., Cronin, S., and Carrington, J. C. 1997. Genome amplification and long-distance movement functions associated with the central domain of tobacco etch potyvirus helper component-proteinase. *Virology* 228: 251-262.

Klein, P. G., Klein, R. R., Rodriguez-Cerezo, E., Hung, A. G., and Shaw, J. G. 1994. Mutational analysis of the tobacco vein mottling virus genome. *Virology* 204: 759-769.

Koonin, E. V., and Dolja, V. V. 1993. Evolution and taxonomy of positive-strand RNA viruses: Implications of comparative analysis of amino acid sequences. *Crit. Rev. Biochem. Mol. Biol.* 28: 375-430.

Lawrence, D. M., Rozanov, M. N., and Hillman, B. I. 1995. Autocatalytic processing of the 223-kDa protein of *Blueberry scorch carlavirus* by a papain-like proteinase. *Virology* 207: 127-135.

Lindbo, J. A., and Dougherty, W. G. 1994. Potyviruses. In: Webster, R. G., and Granoff, A. (Eds.).. Encyclopedia of Virology. Academic Press, San Diego, USA. p. 1148-1153.

Maia, I. G., and Bernardi, F. 1996. Nucleic acid binding properties of a bacterially expressed potato virus Y helper component-proteinase. *J. Gen. Virol.* 77: 869-877.

Maia, I. G., Haenni, A.-L., and Bernardi, F. 1996a. Potyviral HC-Pro: A multifunctional protein. *J. Gen. Virol.* 77: 1335-1341.

Maia, I. G., Seron, K., Haenni, A.-L., and Bernardi, F. 1996b. Gene expression from viral RNA genomes. *Plant Mol. Biol.* 32: 367-391.

Marathe, R., Anandalakshmi, R., Smith, T. H., Pruss, G. J., and Vance, V. B. 2000. RNA viruses as inducers, suppressors and targets of post-transcriptional gene silencing. *Plant Mol. Biol.* 43: 295-306.

Mayo, M. A., and Robinson, D. J. 1996. Nepoviruses: Molecular biology and replication. In: Harrison, B. D., and Murant, A. F. (Eds.). The Plant Viruses. Vol. 5. Polyhedral Virions and Bipartite RNA Genomes. Plenum Press, New York. p. 139-186.

Morozov, S., and Solovyev, A. 1999. Genome organization in RNA viruses. In: Mandahar, C. L. (Ed,). Molecular Biology of Plant Viruses. Kluwer Academic Publ., Boston / Dordrecht / London. p. 47-98.

Murphy, J.-F., Rhoads, R. E., Hunt, A. G., and Shaw, J. G. 1990. The VPg of tobacco etch virus RNA is the 49-kDa proteinase or the N-terminal 24-kDa part of the proteinase. *Virology* 178: 285-288.

Oh, C.-S., and Carrington, J. C. 1989. Identification of essential residues in potyvirus proteinase HC-Pro by site-directed mutagenesis. *Virology* 173: 692-699.

Parks, T. D., Howard, E. D., Wolpert, T. J., Arp, D. J., and Dougherty, W. G. 1995. Expression and purification of a recombinant tobacco etch virus NIa proteinase: Biochemical analysis of the full-length and a naturally occurring truncated proteinase form. *Virology* 210: 194-210.

Peng, C. W., and Dolja, V. V. 2000. Leader proteinase of the *Beet yellows closterovirus*: Mutation analysis of the function in genome amplification. *J. Virol.* 74: 9766-9770.

Peng, C. W., Peremyslov, V. V., Mushegian, A. R., Dawson, W. O., and Dolja, V. V. 2001. Functional specialization and evolution of leader proteinase in the family *Closteroviridae. J. Virol.* 75: 12153-12160.

Peng, C. W., Peremyslov, V. V., Snijder, E. J., and Dolja, V. V. 2002. A replication-component chimera of the plant and animal viruses. *Virology* 294: 75-84.

Peng, C. W., Napuli, A. J., and Dolja, V. V. 2003. Leader proteinase of *Beet yellows virus* functions in long-distance transport. *J. Virol.* 77: 2843-2849.

Peng, Y. H., Kadoury, D., Gal-on, A., Huet, H., Wang, Y., and Raccah, B. 1998. Mutations in the HC-Pro gene of *Zucchini yellow mosaic potyvirus*: Effects on aphid transmission and binding to purified virions. *J. Gen. Virol.* 79: 897-904.

Peremyslov, V. V., Hagiwara, Y., and Dolja, V. V. 1998. Genes required for replication of the 15.5-kilobase RNA genome of a plant closterovirus. *J. Virol.* 72: 5870-5876.

Pirone, T. P., and Blanc, S. 1996. Helper-dependent vector transmission of plant viruses. *Annu. Rev. Phytopathol.* 34: 227-247.

Prod'homme, D., Le Panse, S., Drugeon, G., and Jupin, I. 2001. Detection and subcellular localization of the turnip yellow mosaic virus 66K replication protein in infected cells. *Virology* 281: 88-101.

Pruss, G., Ge, X., Shi, X. M., Carrington, J. C., and Vance, V. B. 1997. Plant virus synergism: The potyviral genome encodes a broad-range pathogenicity enhancer that trans-activates replication of heterologous viruses. *Plant Cell* 9: 859-868.

Revers, F., Le Gall, O., Candresse, T., and Maule, A. J. 1999. New advances in understanding the molecular biology of plant/potyvirus interactions. *Mol. Plant-Microbe Interact.* 12: 367-376.

Riechmann, J. L., Lain, S., and Garcia, J. A. 1992. Highlights and prospects of potyvirus molecular biology. *J. Gen. Virol.* 73: 1-16.

Rojas, M. R., Zerbini, F. M., Allison, R. F., Gilbertson, R. L., and Lucas, W. J. 1997. Capsid protein and helper component-proteinase function as potyvirus cell-to-cell movement proteins. *Virology* 237: 283-296.

Rozanov, M. N., Drugeon, G., and Haenni, A.-L. 1995. Papain-like proteinase of *Turnip yellow mosaic virus*: A prototype of a new viral proteinase group. *Arch. Virol.* 140: 273-288.

Shen, P., Kaniewska, M., Smith, C., and Beachy, R. N. 1993. Nucleotide sequence and genomic organisation of *Rice tungro spherical virus. Virology* 193: 621-630.

Shi, X.-M., Miller, H., Verchot, J., Carrington, J. C., and Vance, V. B. 1997. Mutations in the region encoding the control domain of helper component-proteinase (HC-Pro) eliminates *Potato virus X* / potyviral synergism. *Virology* 231: 35-42.

Shukla, D. D., Ward, C. W., and Brunt, A. A. 1994. Genome structure, variation and function. In: Shukla, D. D., Ward, C. W., and Brunt, A. A. (Eds.). The *Potyviridae.* CAB International, Cambridge. p. 74-112.

Strauss, J. H. (Ed.). 1990. Viral proteinases. *Semin. Virol.* 1, No. 1.

Strauss, J. H., and Strauss, E. G. 1994. The alphaviruses; Gene expression, replication and evolution. *Microbiol. Rev.* 58: 491-562.

Suzuki, N., Che, B., and Nuss, D. L. 1999. Mapping of a hypovirus p29 protease symptom determinant domain with sequence similarity to potyvirus HC-Pro protease. *J. Virol.* 73: 9478-9484.

Thornbury, D. W., Hellmann, G. M., Rhoads, R. E., and Pirone, T. P. 1985. Purification and characterization of potyvirus helper component. *Virology* 144: 260-267.

Tijms, M. A., van Dinten, L. C., Gorbalenya, A. E., and Snijder, E. J. 2001. A zinc finger-containing papain-like proteinase couples subgenomic mRNA synthesis to genome translation in a positive-stranded RNA virus. *Proc. Nat. Acad. USA* 98: 1889-1894.

Urcuqui-Inchima, S., Maia, I. G., Drugeon, G., Haenni, A.-L., and Bernardi, F. 1999a. Effect of mutations within the Cys-rich region of potyvirus helper component-proteinase on self-interaction. *J. Gen. Virol.* 80: 2809-2812.

Urcuqui-Inchima, S., Walter, J., Drugeon, G., German-Ratana, S., Haenni, A.-L., Candresse, T., Bernardi, F., and Le Gall, O. 1999b. Potyvirus HC-Pro self-interaction in the yeast two-hybrid system and delineation of the interaction domain involved. *Virology* 258: 95-99.

Urcuqui-Inchima, S., Maia, I. G., Arruda, P., Haenni, A.-L., and Bernardi, F. 2000. Deletion mapping of the potyvirus helper-component-proteinase reveals two regions involved in RNA binding. *Virology* 268: 104-111.

Verchot, J., and Carrington, J. C. 1995. Evidence that the potyvirus P1 proteinase functions in *trans* as an accessory factor for genome amplification. *J. Virol.* 69: 3668-3674.

Wang, A., and Sanfaçon, H. 2000. Proteolytic processing at a novel cleavage site in the N-terminal region of the tomato ringspot nepovirus RNA1-encoded polyprotein *in vitro. J. Gen. Virol.* 81: 2771-1781.

Wang, A., Carrier, K., Chisholm, J., Wieczorek, A., Huguenot, C., and Sanfaçon, H. 1999. Proteolytic processing of tomato ringspot nepovirus 3C-like proteinase precursors: Definition of the domains for the VPg, proteinase and putative RNA-dependent RNA polymerase. *J. Gen. Virol.* 80: 799-809.

Wang, R. Y., Powell, G., Hardie, J., and Pirone, T. P. 1998. Role of the helper component in vector-specific transmission of potyviruses. *J. Gen. Virol.* 79: 1519-1524.

Zinovkin, R. A., Erokhina, T. N., Lesemann, D. E., Jelkmann, W., and Agranovsky, A. A. 2003. Processing and subcellular localization of the leader papain-like proteinase of *Beet yellows closterovirus. J. Gen. Virol.* 84: 2265-2270.

8 SUBGENOMIC RNAs

I. INTRODUCTION

Some reviews on subgenomic RNAs are: Miller *et al.* (1995), Maia *et al.* (1996), Buck (1996), and Miller and Koev (2000). Ayllón *et al.* (2003) also provide detailed information.

The genome of about all eukaryotic RNA viruses is polygenic; however, functionally it is monocistronic since only the 5`-terminal gene can be properly expressed. The translation of all downstream genes, located internally as well as on 3`-end of a genomic RNA, is achieved by one or more strategies like frameshift, formation of subgenomic RNA (sgRNA), internal translation initiation, leaky scanning, readthrough of stop codon, and reinitiation (Gale *et al.*, 2000). Of these, the formation of sgRNAs is the most common strategy of gene expression adopted by plant viruses. Majority of mRNAs are functionally monocistronic so that, as a general rule, each sgRNA translates only one viral gene borne at its 5`-end. However, some sgRNAs, particularly of hordei-viruses (Zhou and Jackson, 1996), luteoviruses (Tacke *et al.*, 1990; Dinesh-Kumar *et al.*, 1992), and tombusviruses (Rochon and Johnston, 1991) express two genes that are nested in different reading frames. Thus, certain sgRNAs direct translation of two and even three different products encoded by adjacent or overlapping open reading frames (Johnston and Rochon, 1996). Both codon context and leader length contribute to efficient expression of two overlapping ORFs of CNV bifunctional sgRNA (Johnston and Rochon, 1996)

The sgRNAs are the less-than-full-length viral RNAs obtained from genomic RNA template (Miller and Koev, 2000). The first report about the existence of an sgRNA was in BMV (Shih *et al.*, 1972). Four years later, Hunter *et al.* (1976) suggested that TMV sgRNA was the mRNA for CP. By now, production of sgRNAs has been reported in numerous plant viruses belonging to families *Bromoviridae, Closteroviridae,* and *Tombusviridae* and to genera *Carlavirus, Furovirus, Hordeivirus, Luteovirus, Polerovirus, Potexvirus, Sobemovirus, Tobamovirus, Tobravirus,* and *Tymovirus* (Table 1). Thus, the viruses of three supergroups (the alpha-, carmo- and sobemoviruses) of positive-strand RNA plant viruses produce sgRNAs despite having considerably different genome organization and translation strategies. The viruses with monopartite as well as divided genomcs utilize sgRNAs. However, positive-strand RNA viruses of picorna-like supergroup (families *Comoviridae* and *Potyviridae* and the genus *Sequivirus* of family *Sequiviridae*) do not produce sgRNAs (Zaccomer *et al.*, 1995). The ambisense RNA plant viruses (of family *Bunyaviridae* and genus *Tenuivirus*) also synthesize sgRNAs. Thus, evolutionarily different RNA viruses employ sgRNA as a major strategy for gene expression (Koonin and Dolja, 1993).

The sgRNAs can be 3`-coterminal and/or 5`-coterminal. The CTV produces both these types of sgRNAs. The 3`-coterminal sgRNAs are considered as the normal sgRNAs and are most common in plant viruses. The 5`-coterminal sgRNAs are not common, are formed by a few plant viruses, and the little information available about them is given below separately. All other information in this chapter pertains to the 3`-coterminal sgRNAs so that the words 'subgenomic RNAs' mean 3`-coterminal sgRNAs.

TABLE 1
Positive-sense RNA plant viruses that produce subgenomic RNAs
(Based on Table 1 of Maia *et al.*, 1996)

Virus	RNA Genome Segments	5`-sequence of g and sgRNA[1]		Subgenomic RNA designation[2]	Location of 5`-end of subgenomic RNA[3]	5`-end[4]	Gene encoded by subgenomic RNA[5]	Virion[6]
		g and sg RNA	Sequence					
Alfamovirus								
AMV	3	G3	GUUUUA			cap		
		3sg1	GUUUUU	RNA4	IR	+	CP	+
Bromovirus								
BMV	3	G3	GUAAAA			cap		
		3sg1	GUAUUA	RNA4	IR	+	CP	+
Carlavirus								
PVM	1	G	NNUAAA					
		Sg1	Post		IR		TGB	+/-
		Sg2	GAAAU		CS		CP, 11kDa	+/-
Carmovirus								
CarMV	1	G	GGGUAA					
		Sg1	GGUAAC		CS		7kDa	+
		Sg2	GUGAAG		CS		CP	+/-
Cucumovirus								
CMV-Q	3	G2	GUUUAU			cap		
		2sg1	GUUUUG	RNA4A	CS	+	2b	+
		Sg3	GUAAUC			cap		
		3sg1	GUUUAG	RNA4	IR	+	CP	+
Furovirus								
SBWMV	2	G1	GUAUUU			cap		
		1sg	post		IR		37kDa	-

BNYVV	4	G2	GUAUUU				cap	
		2sg	post			IR		19kDa	-
		G2	AAAUUC				cap	
		2sg1	nd			CS		TGB:42kDa	+/-
		2sg2	nd			CS		TGB:13kDa	+/-
		2sg3	AAUGUC	2subc		CS	+	14kDa
		G3	AAAAUU				cap	
		3sg	AAUCG	3sub		IR		4.5kDa	-
Hordeivirus								
BSMV	3	G2	GUAAAA	*B*			cap	
		2sg1	post			IR		TGB	-
		2sg2	post			CS			-
		G3	GUAUAG	R			cap	
		3sg	GUUUAA	rsg		IR	+	17kDa	+/-
Ilarvirus								
TSV	3	G3	GUAUUC				cap	
		3sg	nd	RNA4		IR		CP	+/-
Idaeovirus								
RBDV	2	G2	AUAUAU					
		2sg	UAUUUC	RNA3		IR		CP
Luteovirus								
BYDV-PAV	1	G	AGUGAA					
		Sg1	GUGAAG					CP/Rt.17kDa$_N$	-
		Sg2	AGUGAA					6kDa	-
		Sg3	GACGAC					non coding	+/-
BYDV-PAV	1	G	XGUGAA				VPg	
		Sg1	GAUAG	sgRNA1				CP/Rt.17kDa$_N$
		Sg2	GGCAG	sgRNA2				6.7kDa
Machlomovirus								
MCMV	1	G	AGGUAA				cap	
		Sg	AUCAGA			CS			+/-
Necrovirus								
TNV	1	G	AGUAUU				ppA	
		Sg	nd			CS		7.9kDa
		Sg2	nd			CS		CP

Nepovirus								
CLRV	2	G1	GAAAA			VPg		
		1sg1	AAAAGG	RNA 1A	IR		non-coding	
Polerovirus								
PLRV	1	G	CAAAAG			VPg		
		Sg	ACAAAA	subgenomic RNA1	CS		CP/Rt.17kDa$_N$	+
Potexvirus								
PVX	1	G	GAAAAC			cap		
		Sg1	GAAUA		IR		TGB:25kDa	
		Sg2	GAAAU		CS		TGB:12kDa+8kDa	-
		Sg3	GAAAG		CS		CP	-
Sobemovirus								
SBMV-B	1	G	CACAAA			VPg		
		Sg	nd		IR	+	CP	+/-
Tobamovirus								
TMV	1	G	GUAUUU			cap		
		Sg1	CUCCAG		CS	-	30kDa	-
		Sg2	GUUUUA		CS	+	CP	-
Tobravirus								
TRV-PSG	2	G1	AUAAAA			cap		
		1sg1	AUAUUA	RNA3	CS		29kDa	
		1sg2	AUAAAG	RNA5	CS		16kDa	-
		G2	AUAAAA			cap		-
		2sg1	AUAAAU	RNA4	IR	+	CP(PSG)	
Tombusvirus								
CNV	1	G	NNAAAU					
		Sg1	ACCAA		IR		CP	+
		Sg2	GAAUCU		IR		22kDa, 19kDa^N	+
CyRSV	1	G	NGAAAUC					
		Sg1	GACCAA		IR		CP	+
		Sg2	GAACCU		IR		22kDa, 19kDa^N	+

Tymovirus TYMV	1	G	GUAAUC			cap	
		Sg	AAUAGC		CS	+	CP	+
Miscellaneous PEMV	2	G1	NGUGAA				
		1sg			IR	VPg	CP/Rt.	+/-
RTSV	1	G	UGAAAA				
		Sg1	GAUGAC		CS		Mvt or trans-	
		Sg2	GCUGGG		IR		mission	

1. Genomic RNAs (gRNA) are numbered according to decreasing size; e.g. g3 means genomic RNA3; subgenomic RNA of a particular   genomic RNA and subgenomic RNA numbered according to decreasing size e.g., 3sg1 means subgenomic RNA1 of genomic segment 3; 5`-sequence of gRNA and subgenomic RNA are given; underlined sequences means that the sequence has been postulated on the basis of homologies; 'post' indicates postulated subgenomic RNA 'nd' indicates that subgenomic RNA has been detected but not sequenced.
2. Name by which subgenomic RNA is known

3. IR indicates that 5`-end of subgenomic RNA is located in intergenomic region or when its precise position is not known. CS indicates that 5`-and of subgenomic RNA is located in coding sequences.
4. Structure located at 5`-end of RNA; Presence (+) or absence (-) of that particular structure on 5` -end of the relevant subgenomic RNA
5. Name of subgenomic RNA-encoded protein; TGB indicates triple gene block; CP indicates capsid protein; N is nested gene; CP/RT indicates CP/readthrough; MVT indicates movement protein;
6. Virion is extent of encapsidation denoted by +, - or +/-.

A. 5` -Coterminal Subgenomic RNAs

The CTV produces two most abundant populations of small 5` -coterminal sgRNAs of ~800 nucleotides (Mawassi *et al.*, 1995; Che *et al.*, 2001), called low molecular weight tristeza (LMT) RNAs (Mawassi *et al.*, 1995). These sgRNAs are populations with heterogeneous 3` -termini at nucleotides 842, 854 (LMT1), and nucleotides 744 to 746 (LMT2) for CTV-VT strain (Che *et al.*, 2001). All examined wild-type CTV strains produced both populations of LMT RNAs. These sgRNAs have no corresponding complementary negative-strand sgRNAs indicating that LMT RNAs are produced by premature termination near 3` -controller clements during genomic RNA synthesis (Che *et al.*, 2001; Gowda *et al.*, 2001). Thus, a *cis*-acting element [5` -termination region (5` -TR)] in area corresponding to 3` -termini of LMT1 RNAs controls production of LMT1 sgRNAs (Gowda *et al.*, 2003). Similar 5`-terminal positive-strand sgRNAs are produced by some other members of *Closteroviridae*; a similarly sized RNA in BYV (He *et al.*, 1997) and a larger RNA in *Lettuce infectious yellows virus* (Rubio *et al.*, 2000). Out of the four sgRNAs produced by *Citrus leaf blotch virus*

(CLBV), two are 5` -coterminal and two are 3` -coterminal (Vives *et al.*, 2002). The two 5` -coterminal sgRNAs are constituted by 6795 and 5798 nucleotides, respectively, are collinear with genomic RNA, and bear most of the ORF1 and most of movement protein gene (5` -MP sgRNA), respectively.

B. 3`-Coterminal Subgenomic RNAs

Formation of these sgRNAs enables each internal and 3`-proximal cistron to occupy a 5`-proximal position on an sgRNA – meaning that 5`-end of an sgRNA is located in an intercistronic area or in the coding motif of an upstream gene. This implies several things: that the sgRNAs are copies of 3`-part of genomic RNAs and so are identical in nucleotide sequence to the 3`-end of genomic RNA, that they act as mRNAs of the 3`-cistrons of polycistronic RNA of viruses, and have the same 3`-end as the genomic RNA.

Several genes are often present at the 3`-end of genomic RNA so that more than one 3`-collinear sgRNAs are produced with each gene to be expressed being located at 5`-end of a sgRNA. The number of sgRNAs varies in different virus groups from one to ten (Buck, 1996; Agranovsky and Morozov, 1999): one in bromoviruses and dianthoviruses and this is generally the capsid protein sgRNA; two in BaMV, CNV, *Hibiscus chlorotic ringspot carmovirus*, TBSV, and TCV; three in BYDV, and certain strains of TRV; four [out of which two are reported and two are anticipated (Hemmer *et al.*, 2003)] in PCV; and 6 to 10 in closteroviruses (Hilf *et al.*, 1995), which possess the largest genome amongst the alphavirus-like group. The viruses of carmo-, potex-, tobamo-, and tombusvirus groups produce two or three sgRNAs for expression of their capsid protein and movement protein genes while closteroviruses produce several sgRNAs. Different strains of TRV produce one to five sgRNAs.

The ~15 kb genome of BYV specifies seven sgRNAs while ~20 kb genome of CTV specifies 9 to 10 3`-coterminal sgRNAs. The CTV expresses 10 3`-genes through a nested set of 9 or 10 3`-co-terminal sgRNAs (Satyanarayana *et al.*, 2002). The two major sgRNAs found in BaMV-infected cells are 2.0 and 1.0 kb long (Cheng *et al.*, 2002). Two sgRNAs [a *Cherry leaf roll virus* subgenomic RNA (Brooks and Bruening, 1995) and a BYDV (Australian isolate) sgRNA (Kelly *et al.*, 1994)] do not encode any gene although they are expressed at high levels.

The sgRNAs express proteins that are required during intermediate or late stages of infection. This is particularly true of structural (capsid protein) and movement protein(s). The coat protein sgRNA is the most common type of sgRNA found and is derived from the 3`-end of genomic RNA. It occurs in all multicomponent viruses and in many monocomponent viruses like TMV, tymoviruses, sobemoviruses, and several other plant viruses.

Usually, the same first nucleotide occurs in both genomic and sgRNAs of each virus (Maia *et al.*, 1996). Thus, the sgRNA may have the same 5`-end structure (cap-like structure, VPg, or ppX) as the genomic RNA. This sequence similarity at 5`-ends of genomic and 5`-leaders of sgRNAs is characteristic of numerous viruses: BYV, CLBV, *Clover yellow mosaic virus*, MCMV, *Oat chlorotic stunt virus*, *Panicum mosaic virus*, RCNMV, *Sweet potato chlorotic stunt virus*, TCV, and many others.

This sequence similarity possibly helps viral replicase complex to recognise and interact with specific minus-strand signals for initiating synthesis of plus-strand genomic RNA and sgRNAs.

Similarly, the 3`-end of the sgRNA also bears the same structure [poly(A) tail, tRNA-like structure, or OH] as of the corresponding genomic RNA. Only one exception to this generality is known. In BSMV, the 3`-end of sgRNA produced from RNA γ has poly(A) tail while that of the corresponding genomic RNA has tRNA-like structure. This poly(A) tract on sgRNA corresponds to a poly(A) tract on genomic RNA located about 200 nucleotides upstream of the 3`-end.

The sgRNAs are frequently highly efficient mRNAs, are more abundant than genomic RNAs so that they may interfere with viral replication, are often synthesized far in excess of their need for production of the desired quantity of a particular protein, and are often synthesized late in infection and hence encode late viral genes whose products are needed during pathogenesis and particle formation. The sgRNAs located near the 3`-end of the genomic RNA are produced in higher levels as in BMV, CNV, and TMV but not in AMV. The BYDV sgRNA2 occurs 20- to 40-fold molar excess over genomic RNA.

The sgRNAs are mostly encapsidated and exist as small virus particles. Encapsidation of sgRNAs of AMV, BMV and RBDV is as efficient as that of the genomic RNA. The sgRNA2a of PCV is encapsidated. However, sgRNA of PVX is not encapsidated since its encapsidation site is localized at 5`-end of genomic RNA. Of the two sgRNAs formed from the same genomic RNA in TMV and TRV, only the larger sgRNA is encapsidated. Thus, TMV CP sgRNA is capped and is highly expressed while the 30-kDa movement protein sgRNA is possibly uncapped and is less abundant. This happens because encapsidation site in certain TMV strains is localized in sgRNA1 and so is absent in sgRNA2. The same condition has been postulated for TRV. The PCV sgRNA encoding protein p15 (a suppressor of post-transcriptional gene silencing) is not encapsidated (Hemmer *et al.*, 2003). The results of Qu and Morris (2000) suggest but do not prove that sgRNAs of TCV are not capped *in vivo* since domains of capping enzymes (including methyltransferase and guanylyltransferase that are present in BMV) are absent in TCV.

The sgRNA1, derived from genomic RNA1 of bipartite PCV, contains about 750 nucleotides (Hemmer *et al.*, 2003). Peremyslov and Dolja (2002) give the number of nucleotides constituting different sgRNAs of BYV and the proteins encoded by them: sgRNA1 is of 6066 nucleotides and encodes p6; sgRNA2 is of 6001 nucleotides and encodes heat shock protein 70 homologue (HSP70h); subgenomic RNA3 of ~4200 nucleotides and encodes p64; sgRNA4 is of 2654 nucleotides and encodes minor CP (CPm); sgRNA5 is of 1879 nucleotides and encodes CP; sgRNA 6 is of ~1250 nucleotides and encodes p20; and sgRNA 7 is of ~750 nucleotides and encodes p21. The sgRNAs are numbered above according to their lengths; and since the exact 5` -termini of sgRNAs 3, 6, and 7 are not known, their lengths are determined arbitrarily.

The 5`-terminal nucleotide of BYV sgRNAs 2, 4, and 5 is A, while this nucleotide of sgRNA1 is G. The sgRNA1 is the least abundant of all sgRNAs and its encoded protein (p6) itself, rather than the corresponding RNA region, is critical for BYV cell-to-cell movement. The untranslated regions of sgRNAs 1 and 2 are 43 and 142 nucleotides, respectively. Each of these RNAs possesses an identical heptanucleotide (GUG<u>AUG</u>G) that includes the start codon and provides an optimal translation context. An identical octanucleotide (GACUGUGU) is located at a distance of 11 to 12 nucleotides from the start site of each sgRNA.

Peremyslov and Dolja (2002) show that hydrophobic protein p6 is produced in BYV-infected plants; that p6 and HSP70h protein are encoded by separate monocistronic sgRNAs; and that 5`-termini of p6 and HSP70h sgRNAs are localized to BYV nucleotides G9402 and A9467, respectively. Translational control over HSP70h sgRNA is virus-specific.

II. MECHANISMS OF SUBGENOMIC RNA SYNTHESIS

Two general mechanisms have been proposed for sgRNA synthesis on the basis of extensive *in vitro* and *in vivo* studies: the transcription or promotion mechanism and premature termination mechanism (Adkins *et al.*, 1998; Siegel *et al.*, 1998; Miller and Koev, 2000). Both these well-established mechanisms possess some common characteristics: the viral replicase has to precisely interact with the negative-strand template, either internally at the full-length minus strand for promotion or at the 3`-end of the sgRNA minus strand for transcription of sgRNA plus-strands; the template nucleotide used to initiate sgRNA synthesis is considered as the +1 nucleotide; generally, alpha-like viruses are regarded to follow the +1 pyrimidine and +2 adenylate (in reference to the negative strand of genome) as initiation nucleotides for synthesis of genomic positive strand and sgRNA (Adkins *et al.*, 1998); and mostly both the genomic RNA and sgRNA of each virus possess the same first nucleotide (Maia *et al.*, 1996; Adkins *et al.*, 1998). In addition, to the synthesis of genome-length RNAs, minus-strand BMV RNA3 directs the transcription of a sgRNA4 (Miller *et al.*, 1985). Formation of these two RNAs from the same BMV RNA3 template mutually interferes with each other (Grdzelishvili *et al.*, 2005). Initiation of sgRNA at subgenomic promoter may at least temporarily block genomic RNA synthesis from the same negative-strand RNA templates. Vives *et al.* (2002) suggest that, like alphaviruses, internal genes of CLBV are expressed through 3`-coterminal sgRNAs from the minus genomic RNA strand while the two 5`-coterminal sgRNAs could be produced by early termination of genomic RNA during plus-strand RNA synthesis.

Mutagenesis/modification of +1 nucleotide prevents or greatly reduces sgRNA synthesis *in vivo* and *in vitro* in the promotion mechanism as in AMV, BMV, PVX, and TCV (Adkins *et al.*, 1997; Adkins and Kao, 1998; Koev and Miller, 2000) but inhibits accumulation/formation of sgRNA plus strand without affecting minus-strand synthesis as in TBSV(Choi *et al.*, 2001).

Post-transcriptional mechanism is the third mechanism of sgRNA formation but has no support among plant viruses.

A. Transcription or Promotion Mechanism

The sgRNA is formed through partial transcription by viral RdRp on an internal (intergenic) initiation site (called subgenomic promoter / internal promoter) located on a complementary negative-strand RNA template of full-length genomic RNA; that is, initiation of sgRNA synthesis takes place from a fully formed minus-strand RNA (Miller *et al.*, 1985; Marsh *et al.*, 1988; Gargouri *et al.*, 1989; Buck, 1996; Wang and Simon, 1997; Adkins and Kao, 1998; Siegel *et al.*, 1997, 1998; Miller and Koev, 2000).

The above model has also been expounded in several other ways: specific promoter elements on genomic negative strand permit precise replicase complex recognition for initiation of sgRNA synthesis at internal sites, followed by continued synthesis to the terminus of the template/the viral replicase complex precisely interacts with negative-stranded template internally at the full-length minus strand and has to interact with numerous internal *cis*-acting elements on genomic RNA minus strand to initiate sgRNA synthesis/the sgRNAs are formed by transcription from full-sized minus-sense RNA strand and represent the 5`-terminally truncated 3`-coterminal copies of genomic RNA/intergenic initiation of a sgRNA occurs on a negative-strand RNA template/sgRNA is formed by partial transcription of selected segment(s) of minus strand of genomic RNA/sgRNAs are synthesized by viral replicase from internal promoters (hence also called 'promotion mechanism') located on complementary minus-strand/the replicase initiates sgRNA internally on complementary negative strand of genomic RNA/replicase initiates positive-strand sgRNA transcription internally on a negative strand copy of genomic RNA/synthesis of sgRNA occurs by internal initiation on negative-strands/the *de novo* internal initiation of sgRNA formation occurs at a subgenomic promoter/the sgRNA is formed by transcription from full-sized minus-strand RNA and represents the 5`-terminally truncated 3`-coterminal copies of genomic RNA/sgRNA is formed by initiation at internal promoter sequences on genomic negative-stranded RNA or by internal initiation of transcription on full-length negative-strand RNA.

This most widely recognised model is supported by experimental evidence and has been unequivocally demonstrated in great majority of plant viruses like viruses of the alpha-like, carmo- and sobemo-like supergroups and has been demonstrated un-equivocally in many plant viruses including AMV, BNYVV, BMV, CTV (Gowda *et al.*, 2003), CMV, PVX, TBSV, TCV, and TYMV. Relatively large amounts of complementary CTV sgRNAs are produced (Ayllón *et al.*, 2003).

The RdRp domain of CTV falls within the alphavirus supergroup so that it is expected that viral replicase complex follows a promotion mechanism (internal initiation) for sgRNA synthesis. Moreover, alpha-like viruses that produce sgRNAs by promotion are expected to follow the +1 pyrimidine and +2 adenylate rule (with reference to minus strand) as initiation nucleotides for positive-strand RNAs (Adkins *et al.*, 1998). Internal initiation in PVX and TBSV depends on interaction between *cis* regions and sgRNA promoter region in the genome even though the *cis* region and the sgRNA promoter region are distantly located from each other.

van der Vossen *et al.* (1995) proposed a model of AMV sgRNA replication. The RdRp is initially bound to two internal sites, the subgenomic promoter and a cryptic promoter sequence near the 3`-end of the template, on negative-strand AMV RNA3. The binding possibly induces a conformational change in RNA and is followed by initiation of sgRNA synthesis at 3`-end and/or at subgenomic promoter, depending on the presence of some other factors. This mechanism involves looping out of the intervening RNA sequence, which facilitates interaction between the different factors. *trans*-Acting factors, that are located at distinct sites on RNA, may interact by protein-protein bonds, to stimulate gene expression (Maia *et al.*, 1996). The AMV CP plays a definite role during sgRNA synthesis. This was indicated by the fact that mutations in 3`-end of AMV RNA3 that block CP binding led to loss of sgRNA4 accumulation.

Transcription promoters have at least two functions: to recruit viral polymerase and to promote complementary RNA synthesis from the initiating nucleotide.

B. Termination Mechanism

The sgRNAs are formed by premature termination during negative strand synthesis from the full-length genomic RNA template to give a subgenomic-length negative strand followed by production of plus-sense sgRNA from this truncated negative RNA (subgenomic template) by transcription (Sit *et al.*, 1998; White, 2002). Another way to put is: the (positive-strand) sgRNA is transcribed from prematurely terminated subgenomic mRNA-length complementary negative strands rather than the full-length minus strands of genome. Apparently, the negative strands are produced first. In this strategy, replicase complex precisely interacts with negative-stranded template at 3` -end of the sgRNA minus strand for transcription of sgRNA plus strands. The viral replicase complex has to interact with numerous promoters on 3`-end of minus-strand sgRNA to initiate sgRNA synthesis and the process is controlled by the same replicase complex or by different replicase complexes.

This model is suggested to be functional in RCNMV. The *trans* activation between the two RNA segments of the bipartite viral genome of this virus is responsible for this termination. This model possibly also operates in TBSV (Choi *et al.*, 2001, 2002); TBSV sgRNA2 needs *cis* base pairing between sequence elements that are more than a kilobase away from each other (Zhang *et al.*, 1999). Thus, it is believed that TBSV sgRNAs are transcribed from prematurely terminated subgenomic mRNA-length complementary negative strands rather than the full-length minus strands of genome (Choi *et al.*, 2001, 2002; Choi and White, 2002). Ray and White (2003) found that a stem-loop (designated SL1-III) (that is, an hairpin) in region III of negative strand RNA of TBSV could facilitate sgRNA transcription and may be present in the truncated minus strand templates where it could stimulate sgRNA transcription.

Gowda *et al.* (2001) characterized some of the elements for production of the ten 3`-sgRNAs of CTV and found that the 3`-ORF controller elements produce 3` -terminal plus-strand CTV sgRNAs by promotion and negative strand sgRNA independently by termination. Sivakumaran *et al.* (2004) proposed that in BMV sgRNA initiation, the replicase binding could be coupled to premature termination and that there is possibly an induced fit between the core subgenomic promoter and the replicase.

C. Mechanism in *Citrus tristeza closterovirus*

Ayllón *et al.* (2003) were unable to determine whether the 3`-coterminal 10 sgRNAs of CTV were produced through either of the above two mechanisms (that is, production of sgRNAs by internal promotion from full-length template minus strand or production of negative-strand sgRNAs by termination and of sgRNA plus strands by transcription – that is, by transcription from minus-strand sgRNAs). They suggested that CTV seems to employ an alternative or modified mechanism for formation of its sgRNAs.

All CTV sgRNAs mapped so far [of CP, p13, p18 (all mapped by Ayllón *et al.* (2003) and of p20 and p23 (mapped by Karasev *et al.*, 1997)] have an adenylate as 5`-terminus. The 5`-termini of p13, p18, p20, and p23 sgRNAs of other CTV strains also map to a +1 adenylate (Karasev *et al.*, 1997; Yang *et al.*, 1997). This indicates that CTV replication complex appears to initiate sgRNA synthesis with a purine (pyrimidine in the negative strand), preferably with an adenylate, as the initiating nucleotide of sgRNA synthesis. This is supported by the observation that, during *in vitro* studies, the RdRps of BMV, CCMV, and CMV showed flexibility in selecting +1 sites for sgRNA synthesis since they could use alternative positions and/or nucleotides when the +1 site was modified (Siegel *et al.*, 1997; Adkins and Kao, 1998; Adkins *et al.*, 1997, 1998; Stawicki and Kao, 1999; Chen *et al.*, 2000; Sivakumaran *et al.*, 2002). In all these cases, RdRps always initiated sgRNA synthesis with a guanylate or adenylate as authentic or alternative 5`-termini, respectively. Some closteroviruses [*Sweet potato chlorotic stunt crinivirus* (Kreuze *et al.*, 2002) and BYV (Peremyslov and Dolja, 2002)] can also initiate genomic and sgRNA synthesis with guanylate. This shows that, besides adenylate, CTV replicase complex could also initiate sgRNA synthesis with other nucleotides. Moreover, Ayllón *et al.* (2003) suggested that CTV replicase complex could initiate sgRNA synthesis within a range of several nucleotides around the original +1 site (at positions ranging from -5 to +6 relative to the native +1 position) and that the context of the initiation site is able to modulate efficiency of sgRNA synthesis.

D. Post-transcriptional Mechanism

The processing of full-length positive-strand RNA by specific cleavage and splicing of genomic RNA is still another method for production of sgRNAs. This mechanism can most likely be discarded since there is hardly any information in its favour and is not supported by significant experimental data.

III. REPLICATION OF SUBGENOMIC RNA

The sgRNAs cannot undergo autonomous replication since subgenomic RNAs do not contain, at their 3`-end, the elements required for production of complementary sgRNA chains. Various explanations have been put forward to account for this lack of autonomous replication of sgRNAs: the sequence contained within the sgRNA is insufficient for replication of sgRNA; the sgRNA may not be available for replication, and the sgRNA may be produced late in infection or at a time when negative-strand synthesis has ceased. Experimental evidence indicates that the first explanation is the most likely reason for the absence of sgRNA replication.

During replication of genomic RNA of positive-strand viruses, the 3`-end of negative-strand of genomic RNA in the double-stranded replicative form (RF) RNA possesses an extra and unpaired G residue in contrast to the positive-strand of the RF RNA. This extra nucleotide is essential for the synthesis of the new positive-strand genomic RNA. However, the 3`-end of negative-strand of sgRNA in the double-stranded RF RNA of sgRNA lacks this extra nucleotide so that both strands of the double-stranded RF RNA are perfectly matching at this terminus. This abolishes the production of positive-strand progeny sgRNA via replication from a template sgRNA (Maia *et al.*, 1996). Thus, this hypothesis implies that sgRNA does not appear to undergo replication. Accordingly, the double-stranded RNAs (representing the putative RF RNAs of the corresponding sgRNAs of some plant viruses like BSMV, BYDV, PVX, RCNMV, and TNV) must be artifacts.

IV. SUBGENOMIC RNA PROMOTERS

Miller and Koev (2000) refer to all nucleotide sequences that give rise to sgRNAs as sgRNA promoters or subgenomic promoters (sg-promoters) or transcriptional control elements. The sg-promoters are the *cis*-acting elements that are required for transcription of viral RNA leading to formation of sgRNAs. The 5` -termini of the sgRNAs of bromo-, carmo-, clostero-, diantho-, cucumo-, hordei-, luteo-, tymo-, tobra-, and tombusviruses have been mapped so that the *cis*-acting sequences directing sgRNA synthesis could thus be identified. They have been mapped and characterized in AMV, BMV, BNYVV, BYDV, CMV, CNV, RCNMV, and TCV. These promoter sequences, when inserted into a new site in a viral RNA, lead to production of new sgRNA species. The nucleotide sequences at 5` -termini of CP sgRNAs of bromo-, clostero-, cucumo-, and tobamoviruses, which possess closely related replicases, also show some similarity. Agranovsky *et al.* (1994), therefore, suggested parallel evolution of sg-promoter-binding domains in virus replicases and the recognition signals in the genomic RNAs.

Nucleotide sequences that correspond to sgRNA promoter regions possess common characteristics in all viral groups: First, the first nucleotide of genomic RNAs and sgRNAs are generally the same. This conservation is important since its mutation completely stopped sgRNA formation *in vivo* in AMV. Second, mostly, the homologies between genomic RNAs and sgRNAs extend further down stream up to 20 nucleotides from 5`-end. These homologous nucleotide sequences are mostly AU- or U-rich and are essential for correct sgRNA initiation in AMV and BMV. Third, these sequence homologies occur near, but mostly upstream, of the 5`-sgRNA start site as in potex-, tobra-, and tymoviruses and possibly correspond to sg-promoters.

The sg-promoter motif, in most of the papers, is regarded to be present on the positive RNA strand though in fact the sequence recognized by RdRp is located on the negative RNA strand. The numbering of sgRNA promoter sequence starts at the +1 nucleotide of sgRNA; the transcription start site is taken as the +1 nucleotide. There are two differing opinions about the existence of sequence homologies between sg-promoters of different viruses: no sequence homologies exist between their sg-promoters; and sequence homologies do occur (mostly upstream) near the 5`-sgRNA start site in a virus group as in potex-, tobra-, and tobamoviruses. The first nucleotide of sgRNAs and genomic RNAs is generally the same. This conservation of the first nucleotide is of considerable importance during *in vivo* production of sgRNA in AMV.

Koev and Miller (2000) postulate that an RNA virus could have divergent promoters for differential temporal and quantitative regulation of sgRNA formation. Moreover, overlapping of sg-promoters, important ORFs and *cis*-acting elements in a small virus may cause formation of dissimilar promoters. Thus, sgRNA1 promoter sequence has to accommodate an essential part of the replicase coding region and a part of 5`-UTR of sgRNA1 that controls translation (Allen *et al.*, 1999). The region that controls sgRNA2 promoter overlaps 3`-translation enhancer and part of ORF6 as well as sgRNA2 5`-UTR. Koev and Miller (2000) have thus proposed that different sgRNA promoters arose independently within RNA sequences that have other functions and that this versatility of promoter sequences facilitates small genome size by permitting overlapping functions on an RNA sequence.

High sequence similarities, existing between 5`-termini of genomic and sgRNAs in several luteoviruses (Miller *et al.*, 1995), suggest that these sequences may be part of genomic and sg-promoters. The putative sg-promoter is nearly identical to 5`-terminus of genomic RNA1 of RCNMV (Zavriev *et al.*, 1996).

The sg-promoters range in size from 24 nucleotides in TCV (Wang and Simon, 1997) to more than 100 nucleotides in AMV, BYDV, BNYVV, BMV, CMV, CNV, and other plant viruses. A part of each promoter is present immediately upstream of the 5`-end of the sgRNA. But, in some cases, essential components of some sgRNA promoters are present downstream of the 5`-end of the sgRNA (as in BNYVV and two of the three sgRNA promoters of BYDV) or very distantly upstream or, in exceptional cases, on a separate RNA entity.

There can be still more diversity and complexity in sgRNA promoters of plant viruses. The sgRNA promoters from related viruses or even within a virus often possess few or no common characters as in TMV and CNV. In other cases, viruses may possess short homologous oligonucleotides at the 5`-ends of their sgRNAs but with different adjacent sequences as in BNYVV and TNV.

Distal *cis*-elements regulate sgRNA synthesis. Some of these elements influence/modulate sgRNA transcription/synthesis by long-distance base-pairing to the regions near the transcription start site; base pairing between an area at 5`-end of PVX genomic RNA and regions upstream of the transcription start site of sgRNAs was essential for sgRNA transcription (that is, for formation of sgRNAs) of PVX (Kim and Hemenway, 1999). In TBSV, base pairing between a region immediately upstream of the sgRNA2 start site and a region located about 1000 nucleotides was required for sgRNA synthesis (Zhang *et al.*, 1999). In RCNMV, base pairing occurs between two different RNAs (RNA1 and RNA2) to form the sgRNA promoter (Sit *et al.*, 1998). This sgRNA promoter is split between two RNA molecules so that RCNMV is the only plant virus in which *trans*-activation of sgRNA transcription takes place.

The sg-promoters may vary greatly among and within viruses. The three sgRNA promoters of BYDV possess different primary and secondary structures and locations with regard to the start site (Koev and Miller, 2000a). Presence of the hexanucleotide GUGAAG [in positive-sense] at 5`-termini of sgRNAs 1 and 2 are the only common feature among them. The sgRNA1 promoter is placed generally upstream of the initiation site and it contains an essential stem-loop and one which obstructs sgRNA synthesis while sgRNA2 and sgRNA3 core promoters occur downstream of sgRNA start sites.

What causes the presence of divergent promoters within the same RNA virus? One reason could be the limited size of viral genomes which necessitates overlapping of genes and *cis*-acting motifs. The limited size could have resulted in the following situations: it could compel the sg-promoters to share locations with other essential motifs or ORFs on the genome and/or sg-promoters could have arisen within an already existing functional sequence like an ORF. This indicates that the sg-promoters may have evolved independently.

Gowda *et al.* (2003) suggest that results of various workers indicate that different sgRNA controller (promoter) elements had different origins in modular evolution of closteroviruses.

On the basis of *in vitro* studies, two different and contrasting mechanisms have been suggested for recognition of BMV subgenomic core promoter by replicase. The first model anticipates that the replicase recognises at least four key nucleotides in the core promoter, followed by an induced fit, wherein some of the nucleotides base pair to the initiation site of RNA synthesis (Adkins and Kao. 1998). This is the mechanism of sequence-specific recognition of BMV subgenomic core promoter by replicase. The second model postulates that a short RNA hairpin in the core promoter serves as the recognition area for the replicase and that at least some of the key nucleotides help form a stable hairpin (Haasnoot *et al.*, 2000, 2002). Jaspars (1998) identified a short RNA hairpin in subgenomic core promoter of plant RNA viruses. Haasnoot *et al.*, (2000) characterized the sequence forming this hairpin in BMV and proposed that it has a trinucleotide loop and is required to direct BMV sgRNA synthesis *in vitro* and proposed that recognition of sg-promoter in a manner identical to the recognition of the genomic minus-strand core promoter.

Sivakumaran *et al.* (2004) found that the hairpin structure is not essential for specific recognition by BMV replicase *in vitro*; proposed that the hairpin structure may be stabilized by its interaction with the replicase for which the key nucleotides are essential but not the hairpin; and suggested that BMV sgRNA transcription is a multistep process consisting of three steps - binding of replicase to the sgRNA core promoter sequence of minus-strand RNA3, subsequent formation of a hairpin structure in the presence of the replicase (that is, formation of the hairpin as a result of this binding), and initiation of sgRNA synthesis.

The RNA recognition elements in BMV subgenomic core promoter have been characterized by using minimal length promoter templates called proscripts and enriched viral replicase extracted from infected plants (Adkins *et al.*, 1997; Siegel *et al.*, 1997, 1998; Haasnoot *et al.*, 2000, 2002). It has been determined that RNA synthesis *in vitro* from proscripts requires at least 4 nucleotides at positions -17G, -14A,

–13C, and –11G relative to initiation cytidylate (+1C) (Adkins *et al.*, 1997; Siegel *et al.*, 1997, 1998). Sivakumaran *et al.* (2004) concluded that key nucleotides in BMV subgenomic core promoter direct replicase recognition, that the sgRNA core promoter can induce the BMV replicase in interactions needed for sgRNA transcription *in vitro*, and that the formation of a stem-loop is required at a step after replicase binding.

The BMV sg-promoter consists of an A/U-rich sequence, a poly(U) tract, and a 20-nucleotide core sg-promoter (Adkins *et al.*, 1997, 1998; French and Ahlquist, 1987; Marsh *et al.*, 1988; Miller *et al.*, 1985). The core promoter positions the replicase for accurate sgRNA initiation and resembles simple DNA promoters.

A. Structure and Recognition by Viral Polymerase

The sg-promoters of BMV RNA4 and AMV RNA4 (Maia *et al.*, 1996; Siegel *et al.*, 1999) have been studied in detail and found to possess a common organization. The sg-promoters of these two plant viruses and, possibly of other plant viruses as well, have two regions: a 'core promoter' and the flanking one or more 'enhancer region(s)' that overlap the transcriptional start site in most of the cases. The smallest region, that is competent to promote RNA synthesis at a certain basal level, is called the core promoter, which, in general, has a low accuracy of function. Enhancer regions provide accuracy of replication initiation, give yield of sgRNA that is comparable to the yield obtained *in vivo*, can be several in number, and are placed on the genome at appropriate distances from each other. The core promoter and the enhancer regions seem to have varying boundaries in different viral genomes depending upon their examination *in vitro* and/or *in vivo*.

Nucleotide constituents of the functional sgRNA promoters has been worked out and are generally composed of up to about 150 nucleotides. They vary *in vivo* in length from 24 nucleotides in *Sindbis virus* (an animal virus) to more than 100 nucleotides in BNYVV. Minimal BYDV sgRNA2 promoter is about 143 nucleotides long while the minimal sgRNA3 core promoter is 44-nucleotide long (Koev *et al.*, 1999). The putative sg-promoter of RCNMV is 14 nucleotides long.

Presence of stable hairpin structures upstream of the transcription initiation site of sgRNA formation has been detected or proposed for sgRNA promoters of BYDV, *Bromoviridae*, RCNMV, TCV, and TYMV (Schirawski *et al.*, 2000). TCV possesses two sgRNAs and their sg-promoters possess identical and extensive stable hairpin structures located just upstream of the transcription start site, in addition to the conserved GGG at the initiation site (Wang and Simon, 1997). In contrast, no secondary structure homology exists between the three promoters of the three sgRNAs of BYDV. This is rather exceptional.

Viral polymerases recognize sgRNA promoters. Consequently, there must be certain characters in sg-promoters that are responsible for this specificity, particularly in cases where more than one sgRNAs exist. In such cases, short stretches of homologous nucleotides sequences do occur near their transcription initiation sites (Maia *et al.*, 1996) but, by themselves, are not sufficient for this purpose. Instead, they act as parts of larger sg-promoters. Only little information is available in this connection. Adkins and Kao (1998) speculated, on the basis of flexibility in promoter recognition, that RNA induces the replicase to recognise different nucleotide contacts in promoter. This possibly explains the fact as to why a replicase is able to recognise different promoters in the same viral RNA and also the completely different tRNA-like structure located at 3`-end of BMV positive-strand RNA.

The replicase can recognise at least three different types of competing sequences during RNA replication. They are origins of genomic RNA negative strand synthesis, origins of genomic RNA positive strand synthesis, and origins of sgRNA synthesis (that is, sgRNA promoter). Obviously, RNA initiation at these three structurally different initiation sites has to be very precisely regulated. Miller and Koev (2000) indicate that sg-promoters in general do not resemble the genomic origins so that the replicase either has a multifunctional binding site, which has separate RNA binding sites for each promoter or separate RNA-binding proteins occur for sgRNA promoters. The viral RdRp is essential for sgRNA transcription; however, other proteins in the replicase complex (holoenzyme) or separately seem to modulate sgRNA synthesis.

B. Subgenomic RNA Promoters of Some Plant Viruses

The core promoter of sgRNA of AMV maps to positions -8 to -55 *in vitro* with an upstream enhancer region from -55 to -136 nucleotide. For full activity *in vivo*, additional sequences between positions +12 and -136 are essential. The conserved G at -12 (equivalent to G -11 in BMV) is not required in AMV sg-promoter (Haasnoot *et al.*, 2000). Moreover, a hairpin loop that contains many of the consensus bases is essential in both AMV and BMV sgRNA promoters. The sg-promoter for AMV RNA synthesis requires ~40 nucleotides and forms a single triloop hairpin E (hpE). The minus-strand AMV RNA3 contains a sequence of 37 nucleotides that acts as a core promoter for sgRNA synthesis by purified AMV RdRp *in vitro* (Haasnoot *et al.*, 2000). This sequence bears a hairpin structure, the sg-promoter, which is located 5 nucleotides 3`-from start site of RNA4 synthesis. Structurally, hpE from negative strand promoter and sg-promoter are similar, both hairpins are required for promoter activity. Bol (2003) proposed that hairpin hpE and sg-promoter are the primary elements recognized by RdRp in negative-strand and sgRNA promoter, respectively.

The sg-promoter of BMV RNA4 is the best-characterized sg-promoter. It is transcribed from genomic RNA3, is constituted by about 70 nucleotides organised as four functional domains/sgRNA promoter elements: the core region (-20 to +1) and three enhancer regions (the upstream AU-rich region, the polyuridylate tract, and the downstream AU-rich region). One of the enhancer regions is placed at 16th nucleotide downstream of the sgRNA start site (+1 to +16) and ensures accurate initiation of the sgRNA. The poly(A) stretch and three repeats of UUA are the two remaining enhancer regions. The poly(A) stretch is present in all bromoviruses (-20 to -37) and is essential for sgRNA synthesis. The three UUA repeats exist between positions -38 to -48. All the four functional domains are essential for sgRNA formation. The upstream AU-rich region is implicated in sgRNA synthesis at high levels both *in vitro* and *in vivo* while the downstream AU-rich region is involved mainly in determining or permitting correct initiation of sgRNA synthesis. The polyuridylate tract is generally considered as a spacer that permits RdRp a better approach to the actual sgRNA promoter elements. The polyuridylate tract binds to some component(s) necessary for RdRp activity and acts as a sequence-specific and position-dependent up-regulator for correct initiation of BMV sgRNA synthesis and reduces the previously reported size of the functional core promoter by 12 nucleotides. The CCMV sgRNA promoter also needs the same four bases as BMV and additional bases at -20, -16, -15, and -10 (Adkins and Kao, 1998).

Initiation sequences for BMV sgRNA synthesis by RdRp is AUAC, in which initiation starts at C residue. The A residue at position 2 (the second A of AUAC) is important for sgRNA synthesis (Adkins *et al.*, 1998). Deiman *et al.* (2000) suggest that U residue of AUAC is not specifically involved in RdRp binding. A 21-nucleotide BMV promoter element was sufficient for the basal level of sgRNA synthesis by viral RdRp. The only bases required are at positions -17, -14, -13, and -11 relative to the start site at +1. These form part of a consensus sequence that is shared by sg-promoters of *Bromoviridae* and alphaviruses: 3`-AGANNCNG(N)₅A(N)₃₋₅XA-5`, where X (which is generally a U or a C) is the initiation site (Siegel *et al.*, 1997).

BYDV has three sgRNAs (1, 2, and 3) but none of these is essential for RNA replication. Each sgRNA has its own sg-promoter. These three sgRNA promoters are different and unrelated to each other, and have very little structural or sequence similarity (Koev and Miller, 2000). This suggests a very complex system for recognition of the sg-promoter as well as regulation of sgRNA production. The BYDV sgRNA1 promoter folds into two stem-loops. Its secondary structural elements are necessary for its activity (Koev *et al.*, 1999). This promoter also contains a helical structure that is indispensable for the synthesis of genomic RNA and this region can form identical secondary structures in both positive- and negative-strands (Koev *et al.*, 1999). Moreover, the 3`-translation enhancer region overlaps with sgRNA2 promoter (Koev and Miller, 2000).

Ayllón *et al.* (2003) characterized two different CTV sgRNA controller elements to a region approximately 50 nucleotides upstream of the controlled ORF. CTV CP core controller element mapped to positions -47 to -5 and did not contain +1 nucleotide corresponding to 5`-terminus of sgRNA (Gowda *et al.*, 2001). The levels of CTV sgRNAs are based on promoter strength and position within the genome (Satyanarayana *et al.*, 1999; Gowda *et al.*, 2001, 2003). The sequence context of initiation site of CTV sgRNAs, which is outside the core promoter, appears to be an additional regulatory mechanism to control sgRNA synthesis. Mutation of the +1 A to C or U of the highly abundant P20 sgRNA led to decreased sgRNA production, whereas mutation of the +1 A to G resulted in increased levels of sgRNA accumulation. Thus, even the most abundant sgRNA appears to be capable of increasing synthesis even further.

The TCV has two sgRNAs; the 1.7 kb sgRNA that encodes movement proteins and the 1.45 kb sgRNA that encodes CP and, therefore, it has two sgRNA promoters. These sg-promoters have 40% overall sequence similarity including the short conserved sequences at the transcription start sites and it was predicted that similar hairpins form immediately upstream of sgRNA 5`-end (Wang *et al.*, 1999). The *in vivo* 96-nucleotide sgRNA promoter (from -90 to +6 nucleotide position) of 1.45 kb sgRNA of TCV is located between positions 2517 and 2612, and form a putative large hairpin that is consistent with secondary structural elements contributing to the activity of the promoter. This 96-nucleotide sgRNA promoter of 1.45 kb subgenomic RNA has a minimal 30-nucleotide core sequence as essential for promoter activity. The core consists of a 21-nucleotide hairpin and a 9-nucleotide flanking sequence. The 9-nucleotide sequence is important for sgRNA promoter function. The transcription start site is located within this 9-nucleotide region at position 2607. Wang *et al.* (1999) suggested that the sequence CCCAUUA contains the transcription start site and is essential for efficient transcription of 1.45 kb sgRNA by TCV RdRp *in vivo*. The CCCAUUA sequence is identical to the consensus sequence (C)CC A/U A/U A/U at 3`-terminus of negative-strand TCV genomic and satRNAs and other carmovirus genomic RNAs. The TCV 94-nucleotide *in vivo* sgRNA promoter (from -90 to +4 nucleotide position) is the 1.7 kb sgRNA promoter.

The RCNMV is a bipartite virus in which RNA1 encodes replicase and CP while RNA2 encodes movement protein. Synthesis of its CP sgRNA requires base pairing between sg-promoter in viral RNA1 negative-strand and a 34-nucleotide region in RNA2 (Sit *et al.*, 1998). This is a unique method of RNA-mediated *trans*-activation of sgRNA transcription. The sgRNA promoter of BNYVV is located at positions from -16 to between +100 and 208 *in vivo. In vivo* studies placed sgRNA promoter of CMV between positions -70 and +20, which include an ICR2-like motif upstream of transcription site. This is unlike a similar ICR2 motif in BMV that is outside the sgRNA promoter.

C. Location

Great majority of the sgRNA promoters in RNA viruses [AMV, BMV, BYDV subgenomic RNA1 (Koev *et al.*, 1999), CMV, CNV (Johnston and Rochon, 1995), and RCNMV (Zavreiv *et al.*, 1996)] are located upstream of the transcription start site. In contrast, promoters for BYDV sgRNAs 2 and 3 are located downstream of the start site. Similarly, the sgRNA promoter of BNYVV RNA3 is located mainly downstream of the transcription start site. Location of transcriptional control element (i.e., sg-promoter) on genome is not always restricted to the areas that are collinear with the sgRNA 5`-ends. This lack of collinearity indicates that some long-distance interactions are also involved in transcriptional regulation as proposed for PVX (Miller *et al.*, 1998; Kim and Hemenway, 1999) and TBSV (Zhang *et al.*, 1999). Transcriptional control in bipartite RCNMV is unusual: base-pairing between the regulatory factors of RNAs 1 and 2 controls sgRNA transcription from RNA 1 template (Sit *et al.*, 1998).

The sg-promoter of BMV RNA3 is situated in 250-nucleotide-long intergenic region (IR) between movement protein and CP genes. A core sg-promoter extends about 20 bases upstream of sgRNA initiation site (Marsh *et al.*, 1988). This core promoter directs synthesis of sgRNA in low levels but, for full sg-promoter activity, upstream sequences including a poly(A) tract and a AU-rich sequence are needed. The sg-promoter closest to 3`-end of RNA is most active. This suggests that activity of sg-promoter is position-dependent, which is also true of the CMV sgRNA promoter. Unlike BMV and CMV, higher levels of AMV sgRNA synthesis take place when sg-promoter was nearer to 5`-end of genomic RNA.

The sg-promoter in AMV RNA3 also contains a core promoter, comparable to BMV core promoter, and two enhancer areas that are located in either side of transcriptional start site. This sg-promoter is located in IR region which is only 49 nucleotides-long and is thus too small to accommodate the sg-promoter which overlaps into the coding region of 3a movement protein whose C-terminus is essential for sg-promoter activity (van der Vossen *et al.*, 1995). The *in vivo* studies indicated that core promoter is located between -26 and +1; this resembles the core region of BMV. The two enhancer regions, one being upstream (-136 to -94) and the second being downstream (+1 to +12), are essential for maximum yields of sgRNA *in vivo*.

The sg-promoter of CMV is situated between -70 to +20 region, carries ICR-like sequence, and is enough by itself for production of wild-type levels of subgenomic RNA in vivo. The sg-promoter of subgenomic RNA2 of CNV is situated between –20 and +6 nucleotides. This core region of 26 nucleotides in sgRNA promoter spans the transcription start site. It is enough by itself for production of wild-type levels of the sgRNA2. No enhancer regions are possibly involved in synthesis of this sgRNA. The whole of sgRNA promoter of TYMV is situated within a zone of 494 nucleotides that contain the tymobox as the central element (Schirawski *et al.*, 2000).

The three sgRNA promoters of BYDV have been investigated in detail. Koev *et al.* (1999) demonstrated that sgRNA 1 promoter maps to a 98 nucleotide region with majority of its sequences being located upstream of the transcription start site. BYDV sgRNA2 promoter is placed immediately downstream of the putative sgRNA2 start site. Majority of the sgRNA3 core promoter sequences is located downstream of sgRNA3 start site. Thus, the minimal core domains of sgRNAs 2 and 3 promoters are both located downstream of transcription initiation site in contrast to sgRNA 1 promoter. Some additional sequences upstream of the start site increased sgRNA promoter activity in both sgRNAs 2 and 3.

Area surrounding the initiation site for sgRNA synthesis in TYMV genomes and 14 other tymoviruses contains two highly conserved sequence blocks within a zone of about 40 nucleotides. This area is called the tymobox region (Ding *et al.*, 1990). The 5`-terminal block of this region is the tymobox itself. It is 16-nucleotide-long, is identical in nucleotide sequences in 11 of the 15 tymoviruses, and after 7 or 8 nucleotides downstream is followed by 4-nucleotide initiation box. Initiation of sgRNA synthesis, in genome of three tymoviruses for which this information has been collected, starts in the initiation box at the triplet AAU and is preceded by a C residue. Thirteen of the 15 tymovirus RNA genomes contain this common CAAU sequence. Ding *et al.* (1990) proposed that the two conserved sequences (the tymobox and the initiation box) possibly function as a sgRNA promoter. Later, Schirawski *et al.* (2000) established that the highly conserved tymobox and initiation box together are part of the sg-promoter and are essential for promoter function, that the promoter could operate ectopically and in *trans*, and that this sg-promoter is situated on a 494-nucleotide RNA fragment. In other words, this RNA fragment contains the sg-promoter which is made up of the tymobox and the initiation box, both of which are highly conserved sequences. The tymobox region of TYMV RNA performs two functions: it encodes RdRp and regulates sgRNA production.

The CTV sgRNAs of CP and p13 are found in intermediate abundance, the sgRNA of p18 is found in low abundance, while sgRNAs of p20 and p23 are found in high abundance. All these five sgRNAs of CTV initiate with an adenylate. On this basis, Ayllón *et al.* (2003) concluded that the context of the initiation site can modulate efficiency of sgRNA synthesis so that the context of initiation site appears to be a regulatory mechanism for levels of sgRNA production.

D. Role

The exact role of the different sg-promoter domains is not yet known but it can be safely assumed that regulation of sgRNA synthesis by specific promoter elements is a major step in control of viral gene expression. The sg-promoter may be the main RdRp recognition site (Olsthoorn *et al.*, 2004). When a sg-promoter was inserted into a new location on the relevant RNA, a new sgRNA was produced in AMV, BMV, CMV, and TMV. This clearly demonstrated two things: that the sg-promoter is a functional entity, and that the formation of sgRNA depends on the presence of the relevant sg-promoter on genomic RNA. Wierzchoslawski *et al.* (2003) found an increased homologous RNA recombination activity within the sgRNA promoter region in BMV RNA3. Wierzchoslawski *et al.* (2004) investigated this in detail and proposed a dual mechanism whereby recombination is primed at the poly(A) tract by the predetached nascent plus strand while transcription is initiated *de novo* at the sgRNA core promoter; demonstrated that transcription and recombination utilized different sgRNA promoter sequences (and thus mechanisms), although partial overlapping of the active sequences involved in the two processes is possible; finely mapped the crossovers occurring within the sgRNA promoter; and also verified the participation of sgRNA promoter regions in transcription.

V. SUBGENOMIC RNAs OF SOME PLANT VIRUSES

The CP of five genera (*Alfamovirus*, *Bromovirus*, *Cucumovirus*, *Ilarvirus*, and *Oleavirus* – all of which have a tripartite RNA genome) of family *Bromoviridae* is translated from a sgRNA, called RNA4. This sgRNA in all the five genera is coterminal with the 3`-800 to 1,000 nucleotides of RNA3. The single sgRNA of BMV is derived from 3`-region of BMV RNA3, is called RNA4, is 876 nucleotides in length, is encapsidated together with RNA3, and encodes CP. An encapsidated CP sgRNA of *Narcissus mosaic potexvirus* (NMV) is present in infected *Gomphrena globosa*. This sgRNA is about 840 nucleotides long and is encapsidated in a short virus particle approximately 100 nm long. About 50% of the NMV particles obtained from local lesions two weeks after inoculation (the early stages of infection) were short particles compared to about 10% of full-length 550 nm-long NMV particles. However, the situation was reverse in a systemically infected leaf two years after inoculation in which full-length particles predominated (Short and Davies, 1983).

SBMV contains heterogeneous sgRNAs. When the total virion RNA is analysed by sucrose gradient centrifugation, the sgRNAs appear as minor ill-defined entities either preceding or contiguous with the 25S genomic peak. The sgRNAs disaggregate upon heat-denaturation and can then be separated effectively from genomic RNA by velocity centrifugation or gel electrophoresis. The mean contour length of sgRNA is 0.31 ± 0.17 µm (calculated molecular weight 1.31×10^6). The most prominent sgRNA has a molecular weight of 0.38×10^6.

TMV produces three functional sgRNAs, all of which are associated with polyribosomes (Palukaitis *et al.*, 1983; Carr, 2004). One sgRNA, commonly known as LMC (low molecular mass component) or CP sgRNA, encodes the 3`-proximal 17.5-kDa CP, has 0.23×10^6 molecular weight, is about 700 nucleotides long, is not encapsidated in common TMV and occurs free because of the absence of an assembly site but is encapsidated in other TMV strains like cowpea strain (C_c), the watermelon strain of *Cucumber green mottle mosaic virus* (CGMMV), and *Cucumber virus 4* in which such short particles are observed in preparations of these TMV-related viruses. The CP sgRNA of cowpea strain is 711 nucleotides long, is capped, and is encapsidated in short rodlets, which exist in preparations of this strain from infected plants. The second functional sgRNA is called I_2 (intermediate-class RNA2), has 0.5×10^6 molecular weight, is uncapped at its 5`-end, encodes the 30-kDa movement protein in *in vitro* systems, and also expresses CP in a crucifer-infecting TMV strain (cr-TMV) via an internal ribosome entry site (IRES) upstream of the CP gene (Ivanov *et al.*, 1997) but not in some other TMV strains. The third functional sgRNA is called I_1 sgRNA, has about 1.0×10^6 molecular weight, encodes a major protein in rabbit reticulocyte lysate system, this major protein is of 54-kDa molecular mass with a predicted amino acid sequence that is identical to that of the readthrough portion of the 183-kDa replicase protein, and represents the 3`-half of genomic RNA. The 5`-end of the I_1 sgRNA is at nucleotide 3405 of genomic RNA with the first 5`-proximal AUG codon at nucleotides 3495 to 3497 so that a 90-nucleotide noncoding region precedes the I_1 RNA coding sequence. Sulzinsky *et al.*, (1985) therefore proposed that, after the 90 base sequence of untranslated region, the 54-kDa protein is initiated by the AUG at residues 3495 to 3497 and is terminated at residue 4915. There are, however, some serious doubts about the expression of I_1 sgRNA since the 54-kDa protein has never been detected *in vivo* in TMV-infected plants, in transgenic plants constitutively expressing 54-kDa protein, in TMV replicase preparations, and in TMV-infected tobacco protoplasts (Carr, 2004). Reports also exist suggesting that six other 3`-coterminus sgRNAs are found in TMV-infected tobacco leaves; some of these sgRNAs are encapsidated and thus occur in TMV preparations.

TRV produces one to five sgRNAs (RNAs 3, 4, 5, 6, and 7) in different strains. These sgRNAs are generally encapsidated. RNA3 should be encapsidated in a rod of about 43 nm length and rods of this length occur in several TRV strains. At least five sgRNAs (RNAs 3-7) exist in TRV strain CAM (TRV-CAM) - infected tissues. The sgRNA3 of CAM (and also of strains Lisse and PRN) is about 1.6 kb, has a molecular weight of about 0.5×10^6, does not encode CP but encodes the 30-kDa protein, is made up of up to 1750 nucleotides, seems to be capped, is generated from RNA2, and is normally encapsidated except in CAM. Out of the five sgRNAs found in CAM-infected tissues, only RNA5 is generated from RNA1 while the remaining four RNAs (3, 4, 6, and 7) originate from 3`-terminus of RNA2. These four sgRNAs seem to possess some overlapping sequences since the sum of their molecular weights is greater than the molecular weight of RNA2. Template activity of only RNA3 is known. The TRV SYM strain contains two sgRNAs, RNAs 3 and 4. The sgRNA3, unlike the examples mentioned above, is the CP mRNA, has 0.6×10^6 molecular weight, contains 1750 nucleotides, is derived from RNA2, and is encapsidated in 7 nm VS particles. The SYM RNA4 is derived from RNA1, is encapsidated in 48 VS particles, has 0.54×10^6 molecular weight, contains 1550 nucleotides, and is translated into a 29-kDa protein, which is not related to CP.

Three sgRNAs (RNAs 3, 4, and 5) are present in TRV strain PSG-infected tissues. The sgRNA3 is 1.6 kb, is the putative mRNA for 29-kDa protein, is derived from RNA1, and is coterminus with its 3`-end which would place its 5`-end just upstream of the 29-kDa protein on RNA1. The RNA4 is 1431 nucleotides long, is derived from RNA2 but lacks its 5`-terminal 474 nucleotides, and is the putative CP mRNA. The sgRNA5 is the putative mRNA of 16-kDa protein, is estimated to be 700 nucleotides long which would place its 5`-end just upstream of the 16-kDa protein in RNA1, and is most probably 3`-coterminal with the RNA1 but could also have been derived from RNA2. The sequence of sgRNA4 begins at nucleotide 475 in RNA2, immediately downstream of the 8th AUG codon. Three sgRNAs (RNAs 3-5) have also been extracted from virus preparations of TRV strains TCM, TAK, ORY, and CAM. The sgRNA4 is 400 to 500 nucleotides shorter than RNA2 in all these strains and also in PSG strain. This suggests that CP cistron on RNA2 is located at a fixed position with reference to its 5`-end even though the length of RNA2 is highly variable.

TCV has five ORFs out of which three genes are translated from two sgRNAs (White *et al.*, 1995). Two small nested ORFs in the middle of genome encode two proteins (p8 and p9) that are the viral movement proteins. Both these are translated from a 1.7 kb sgRNA by leaky scanning. The TCV CP (p38) is encoded by the most 3`-proximal ORF and is translated from a 1.45 kb sgRNA. TYMV forms at least nine sgRNAs; one is the CP sgRNA and the eight are of other proteins. Five of the eight sgRNAs have been associated with particular polypeptides synthesized in *in vitro* systems. The two 3`-coterminal sgRNAs of *Hibiscus chlorotic ringspot carmovirus* are of 1.7 and 1.5 kb, respectively (Huang *et al.*, 2000). TBSV CP is translated from gene *p41* on sgRNA1 and p22 and p19 are expressed from genes *p22* and *p19* on sgRNA2.

VI. EXPRESSION OF SUBGENOMIC RNAs

Expression of genes carried by sgRNAs occurs by the same translation strategies as those of the genomic RNAs. Normally, sgRNAs express only a single gene but sgRNAs of several viruses expresses multiple genes. CNV has bifunctional sgRNAs, which express nested genes by leaky scanning. BYDV sgRNA carries three genes (CP gene, CP-readthrough cistron, and nested 17-kDa gene) that are expressed through readthrough and leaky scanning (Dinesh-Kumar *et al.*, 1992).

VII. FUNCTIONS OF SUBGENOMIC RNAs

A. Temporal Regulation of Gene Expression

In viruses that form several sgRNAs, levels, kinetics, amounts, and time course of accumulation of different sgRNAs are regulated at transcription level by the viruses themselves (Kim and Hemenway, 1997; Navas-Castello *et al.*, 1997; Wang and Simon, 1997; Hagiwara *et al.*, 1999). Formation of sgRNAs results in temporal regulation of gene expression. Synthesis of different viral mRNAs (and proteins) often follows distinct time patterns since expression of sgRNAs must be correlated with the time and extent of utilization of a particular protein in viral life cycle. For example, synthesis of different CTV sgRNAs at different times and in different amounts regulates formation of different and individual gene products (Navas-Castillo *et al.*, 1997). This regulation of gene expression is made possible by quantitative regulation of activities of their respective sgRNA promoters. The core promoter and additional sequences regulating transcription levels may be located in adjacent areas of template (Kim and Hemenway, 1997; Siegel *et al.*, 1997; Wang and Simon, 1997; Koev *et al.*, 1999) or on a distinct RNA strand (Sit *et al.*, 1998).

TMV regulates the time of production and accumulation of two of its sgRNAs. Expression of proteins involved in viral replication and movement is transient at early stages of infection while CP expression occurs during late phases of infection (Dawson, 1992). Movement protein is synthesized from sgRNA1 in low amount and is time-dependent while CP synthesis from sgRNA2 occurs later but is then produced at a very high and constant rate. This is attributed to positions of genes relative to 3`-end of genomic RNA. The sg-promoters are possibly the most important factors for determining temporal regulation of TMV gene expression.

Temporal regulation of gene expression also occurs in BYV (Hagiwara *et al.*, 1999): some genes [HSP70 homologue (HSP70h), minor CP (CPm), CP, and 21-kDa protein (p21)] are expressed early while others [64-kDa protein (p64), and 20-kDa protein (p20)] are expressed late in viral life cycle. Earliest activation was of HSP70 and CP promoters; p20 had delayed activation; sgRNAs of p21 and CPm accumulated early and at a relatively fast rate while sgRNA of p64 accumulated at a much slower rate. It is significant that the order of gene activation bears no relationship to the order of organization of ORFs on genomic RNA of BYV.

The viral proteins encoded by BSMV genome at different times are: the polymerase and the triple gene block proteins expressed transiently at early phases of infection, the methyltransferase-helicase is expressed early but later declines gradually, and CP and regulatory cysteine-rich protein expressed stably throughout the course of infection (Donald *et al.*, 1993). Similarly, the production of the two co-linear sgRNAs of BYDV, CLRV, and CNV is also regulated by different mechanisms. In contrast, the two sgRNAs of TCV do not show any temporal regulation (Wang and Simon, 1997).

B. Role in Translation

Formation of sgRNAs is an important method of regulating gene expression and the regulation is by preferential sgRNA initiation. Genomic RNA of many plant viruses contain at least two different initiation sites and preferential initiation from one or the other site can play an important role in gene expression. Preferential formation of CP sgRNA followed by its preferential translation due to preferential initiation is one of the most common examples. The sgRNAs and their translation products have been detected both *in vivo* and *in vitro* systems so that role of these RNAs during replication of plant viruses *in vivo* is not in doubt.

C. Role in Infectivity

Virus particles of three-component viruses contain at least three large major RNA molecules (RNAs 1, 2, and 3) but often a small minor RNA (RNA4) of about 0.3×10^6 molecular weight also occurs. All the three major RNA species are separately encapsidated while RNA4 is either encapsidated separately or together with RNA3. The RNA3 has two cistrons. However, CP cistron on RNA3 is closed and is expressed in an *in vitro* translation system through RNA4. Consequently, RNA4 is derived from 3`-end of RNA3 and is the CP subgenomic mRNA.

All the four RNAs or only the major three RNAs are needed for infectivity. Thus, RNA4 may or may not influence infectivity. It does not do so in bromoviruses and cucumoviruses so that their unfractionated RNA preparations, even when freed of this, remain infective. Infectivity of AMV and ilarviruses, on the other hand, is completely lost in its absence that is restored on its addition or on the addition of its translation product, the CP. This difference has divided these multicomponent viruses into two categories: protein-independent genome/viruses in which different RNAs are present in similar isometric capsids to produce identical nucleoprotein components which sediment together and have been called isocapsid viruses; protein-dependent genome/viruses in which different RNAs are present in bacilliform or isometric capsids of different sizes so that they sediment as clearly defined centrifugal multicomponents and have been called heterocapsid viruses. Bromoviruses and cucumoviruses are the isocapsid viruses while ilarviruses and AMV are the heterocapsid viruses. A remarkable feature of heterocapsid viruses (excluding hordeiviruses) is that a mixture of RNAs 1, 2, and 3 of any one of these viruses may or may not be activated by addition of heterologous RNA4 or heterologous CP.

VIII. REFERENCES

Adkins, S., and Kao, C. C. 1998. Subgenomic RNA promoters dictate the mode of recognition by bromoviral RNA-dependent RNA polymerase. *Virology* 252: 1-8.

Adkins, S., Siegel, R. W., Sun, J. H., and Kao, C. C. 1997. Minimal templates directing accurate initiation of subgenomic RNA synthesis *in vitro* by the brome mosaic virus RNA-dependent RNA polymerase. *RNA* 3: 634-647.

Adkins, S., Stawicki, S. S., Faurote, G., Siegel, R. W., and Kao, C. C. 1998. Mechanistic analysis of RNA synthesis by RNA-dependent RNA polymerase from two promoters reveals similarities to DNA-dependent RNA polymerase. *RNA* 4: 455-470.

Agranovsky, A. A., and Morozov, S. 1999. Gene expression in positive-strand RNA viruses. In: Mandahar, C. L. (Ed.). Molecular Biology of Plant Viruses. Kluwer Academic Publ. Boston / Dordrecht / London. p. 99-119.

Agranovsky, A. A., Koenig, R., Maiss, E., Boyko, V. P., Casper, R., and Atabekov, J. G. 1994. Expression of the beet yellows closterovirus capsid protein and p24, a capsid protein homologue, *in vitro* and *in vivo. J. Gen. Virol.* 75: 1431-1439.

Allen, E., Wang, S., and Miller, W. A. 1999. Barley yellow dwarf virus RNA requires a cap-independent translation sequence because it lacks a 5`-cap. *Virology* 253: 139-144.

Ayllón, M. A., Gowda, S., Satyanarayana, T., Karasev, A. V., Adkins, S., Mawassi, M., Guerri, J., Moreno, P., and Dawson, W. O. 2003. Effects of modification of the transcription initiation site context on citrus tristeza virus subgenomic RNA synthesis. *J. Virol.* 77: 9232-9243.

Bol, J. F. 2003. *Alfalfa mosaic virus*: Coat protein-dependent initiation of infection. *Mol. Plant Pathol.* 4: 1-8.

Brooks, M., and Bruening, G. 1995. A subgenomic RNA associated with cherry leafroll virus infections. *Virology* 211: 33-41.

Buck, K. W. 1996. Comparison of the replication of positive-stranded RNA viruses of plants and animals. *Adv. Virus Res.* 47: 159-251.

Carr, J. P. 2004. *Tobacco mosaic virus. Annu. Plant Rev.* 11: 27-67.

Che, X., Piestun, D., Mawassi, M., Satyanarayana, T., Gowda, S., Dawson, W. O., and Bar-Joseph, M. 2001. 5`-coterminal subgenomic RNAs in *Citrus tristeza virus*-infected cells. *Virology* 283: 374-381.

Chen, M.-H., Roossinck, M. J., and Kao, C. C. 2000. Efficient and specific initiation of subgenomic RNA synthesis by cucumber mosaic virus replicase *in vitro* requires an upstream RNA stem-loop. *J. Virol.* 74: 11201-11209.

Cheng, J.-H., Peng, C.-W., Hsu, Y.-H., and Tsai, C.-H. 2002. The synthesis of minus-strand RNA of *Bamboo mosaic potexvirus* initiates from multiple sites within the poly(A) tail. *J. Virol.* 76: 6114-6120.

Choi, I.-R., and White, K. A. 2002. An RNA activator of subgenomic mRNA1 transcription in *Tomato bushy stunt virus. J. Biol. Chem.* 277: 3760-3766.

Choi, I-R., Stenger, D. C., and French, R. 2000. Multiple interactions among proteins encoded by the mite-transmitted *Wheat streak mosaic tritimovirus. Virology* 276: 185-198.

Choi, I.-R., Ostrovsky, M., Zhang, G., and White, K. A. 2001. An RNA activator of subgenomkic mRNA1 transcription in *Tomato bushy stunt virus. J. Biol. Chem.* 277: 3760-3766.

Choi, I.-R., Horken, K. M., Stenger, D. C., and French, R. 2002. Mapping of the P1 proteinase cleavage site in the polyprotein of *Wheat streak mosaic virus* (Genus *Tritimovirus*). *J. Gen. Virol.* 83: 443-461.

Dawson, W. O. 1992. Tobamovirus-plant interactions. *Virology* 186: 359-367.

Deiman, B. A. L. M., Verlaan, P. W. G., and Pleij, C. W. A. 2000. *In vitro* transcription by the turnip yellow mosaic virus RNA polymerase: A comparison with alfalfa mosaic virus and brome mosaic virus replicases. *J. Virol.* 74: 264-271.

Dinesh-Kumar, S. P., Brault, V., and Miller, W. A. 1992. Precise mapping and *in vitro* translation of a trifunctional subgenomic RNA of *Barley yellow dwarf virus. Virology* 187: 711-722.

Ding, S., Howe, J., Keese, P., Mackenzie, A., Meek, D., Osorio-Keese, M., Skotnicki, M., Srifah, P., Torronen, M., and Gibbs, A. 1990. The tymobox, a sequence shared by most tymoviruses, its use in molecular studies of tymoviruses. *Nucl. Acids Res.* 18: 1181-1187.

Donald, R. G. K., Zhou, H., and Jackson, A. O. 1993. Serological analysis of barley stripe mosaic virus-encoded proteins in infected barley. *Virology* 195: 659-668.

French, R., and Ahlquist, P. 1987. Intercistronic as well as terminal sequences are required for efficient amplification of brome mosaic virus RNA3. *J. Virol.* 6: 1457-1465.

French, R., and Ahlquist, P. 1988. Characterization and engineering of sequences controlling *in vitro* synthesis of brome mosaic virus subgenomic RNA. *J. Virol.* 62: 2411-2420.

Gale, M., Jr., Tan, S.-L., and Katze, M. G. 2000. Translational control of viral gene expression in eukaryotes. *Microbiol. Mol. Biol. Rev.* 64: 239-280.

Gargouri, R., Joshi, R. L., BoL, J. F., Astier-Manifacier, S., and Haenni, A.-L. 1989. Mechanism of synthesis of turnip yellow mosaic virus coat protein subgenomic RNA *in vivo. Virology* 171: 386-393.

Gowda, S., Satyanarayana, T., Ayllòn, M. A., Albiach-Marti, M. R., Mawassi, M., Rabindran, S., Garnsey, S. M., and Dawson, W. O. 2001. Characterization of the *cis*-acting elements controlling subgenomic mRNAs of *Citrus tristeza virus*: Production of positive- and negative-stranded 3`-terminal and positive-stranded 5`-termianl RNAs. *Virology* 286: 134-151.

Gowda, S., Ayllòn, M. A., Satyanarayana, T., Bar-Joseph, M., and Dawson, W. O. 2003. Transcription strategy in a closterovirus: A novel 5`-proximal controller element of *Citrus tristeza virus* produces 5`- and 3`-terminal subgenomic RNAs and differs from 3`-open reading frame controller elements. *J. Virol.* 77: 340-352.

Grdzelishvili, V. Z., Garcia-Ruiz, H., Watanabe, T., and Ahlquist, P. 2005. Mutual interference between genomic RNA replication and subgenomic mRNA transcription in *Brome mosaic virus*. *J. Virol.* 79: 1438-1451.

Haasnoot, P. C. J., Brederode, F. Th., Olsthoorn, R. C. L., and Bol, J. F., 2000. A conserved hairpin structure in *Alfamovirus* and *Bromovirus* subgenomic promoters is required for efficient RNA synthesis *in vitro*. *RNA* 6: 708-716.

Haasnoot, P. C. J., Olsthoorn, R. C. L., and Bol, J. F., 2002. The brome mosaic virus subgenomic promoter hairpin is structurally similar to the iron-responsive elements and functionally equivalent to the minus-strand core promoter stem-loop C. *RNA* 8: 110-122.

Hagiwara, Y., Peremyslov, V. V., and Dolja, V. V. 1999. Regulation of closterovirus gene expression examined by insertion of a self-processing reporter and by Northern hybridization. *J. Virol.* 73: 7988-7993.

He, X.-H., Rao, A. L. N., and Creamer, R. 1997. Characterization of *Beet yellows closterovirus*-specific RNAs in infected plants and protoplasts. *Phytopathology* 87: 347-352.

Hemmer, O., Dunoyer, P., Richards, K., and Fritsch, C. 2003. Mapping of viral RNA sequences required for assembly of *Peanut clump virus* particles. *J. Gen. Virol.* 84: 2585-2594.

Hilf, M. E., Karasev, A. V., Pappu, H. R., Gumpf, D. J., Niblett, C. L., and Garnsey, S. M. 1995. Characterization of citrus tristeza virus subgenomic RNAs in infected plants. *Virology* 208: 576-582.

Huang, M., Koh, D. C. Y., Weng, L. J., Chang, M. L., Yap, Y. K., Zhang, L., and Wong, S. M. 2000. Complete nucleotide sequence and genome organisation of *Hibiscus chlorotic ringspot virus*, a new member of genus *Carmovirus*: Evidence for presence and expression of two novel open reading frames. *J. Virol.* 74: 3149-3155.

Hunter, T. R., Hunt, T., Knowland, J., and Zimmern, D. 1976. Messenger RNA for the coat protein of *Tobacco mosaic virus*. *Nature* (*London*) 260: 759-764.

Ivanov, P. A., Karpova, O. V., Skulachev, M. V., Tomashevskaya, O. L., Rodionova, N. P., Dorokhov, Y. L., and Atabekov, J. G. 1997. A tobamovirus genome that contains an internal ribosome entry site functional *in vitro*. *Virology* 232: 32-43.

Jaspars, E. M. J. 1998. A core promoter hairpin is essential for subgenomic RNA synthesis in *Alfalfa mosaic alfamovirus* and is conserved in other *Bromoviridae*. *Virus Genes* 17: 233-242.

Johnston, J. C., and Rochon, D. M. 1995. Deletion analysis of the promoter for the cucumber necrosis virus 0.9-kb subgenomic RNA. *Virology* 214: 100-109.

Johnston, J. C., and Rochon, D. M. 1996. Both codon context and leader length contribute to efficient expression of the two overlapping open reading frames of cucumber necrosis virus bifunctional subgenomic mRNA. *Virology* 221: 232-239.

Karasev, A. V., Hilf, M. E., Garnsey, S. M., and Dawson, W. O. 1997. Transcriptional strategy of closteroviruses: Mapping the 5`-termini of the citrus tristeza subgenomic RNAs. *J. Virol.* 71: 6233-6236.

Kelly, L., Gerlach, W. L., and Waterhouse, P. M. 1994. Characterization of the subgenomic RNAs of an Australian isolate of *Barley yellow dwarf luteovirus*. *Virology* 202: 565-573.

Kim, K.-H., and Hemenway, C. 1997. Mutations that alter a conserved element upstream of the potato virus X triple gene block and coat protein genes affect subgenomic RNA accumulation. *Virology* 232: 187-197.

Kim, K.-H., and Hemenway, C. 1999. Long-distance RNA-RNA interactions and conserved sequence elements affect potato virus X plus-strand RNA accumulation. *RNA* 5: 636-645.

Koev, G., and Miller, W. A. 2000. A positive-strand RNA virus with three different subgenomic RNA promoters. *J. Virol.* 74: 5988-5996.

Koev, G., Mohan, B. R., and Miller, W. A. 1999. Primary and secondary structural elements required for synthesis of barley yellow dwarf virus subgenomic RNA1. *J. Virol.* 73: 2876-2885.

Koonin, V., and Dolja, V. V. 1993. Evolution and taxonomy of positive-strand RNA viruses: Implications of comparative analysis of amino acid sequences. *Crit. Rev. Biochem. Mol. Biol.* 28: 375-430.

Kreuze, J. F., Savenkov, E. I., and Valkonen, J. P. T. 2002. Complete genome sequence and analysis of the subgenomic RNAs of *Sweet potato chlorotic stunt virus* reveal several new features for the genus *Crinivirus*. *J. Virol.* 76: 9260-9270.

Maia, I. G., Haenni, A.-L., and Bernardi, F. 1996. Potyviral HC-Pro: A multifunctional protein. *J. Gen. Virol.* 77: 1335-1341.

Marsh, L. E., Dreher, T. W., and Hall, T. C. 1988. Mutational analysis of the core and modular sequences of BMV RNA3 subgenomic promoter. *Nucl. Acids Res.* 16: 981-995.

Mawassi, M., Gafny, R., Gagliardi, D., and Bar-Joseph, M. 1995. Populations of *Citrus tristeza virus* contain smaller than full-length particles which encapsidate subgenomic RNA molecules. *J. Gen. Virol.* 76: 651-659.

Miller, W. A., and Koev, G. 2000. Synthesis of subgenomic RNAs by positive-strand RNA viruses. *Virology* 273: 1-8.

Miller, W. A., Dreher, T. W., and Hall, T. C. 1985. Synthesis of brome mosaic virus subgenomic RNA *in vitro* by internal initiation on (-)-sense genomic RNA. *Nature (London)* 313: 68-70.

Miller, W. A., Dinesh-Kumar, S. P., and Paul, C. P. 1995. Luteovirus gene expression. Crit. Rev. Plant Sci. 14: 179-211.

Miller, W. A., Plante, C. A., Kim, K. H., Brown, J. W., and Hemenway, C. L. 1998. Stem-loop structure in the 5`-region of potato virus X genome required for plus-strand RNA accumulation. J. Mol. Biol. 284: 591-608.

Navas-Castillo, J., Albiach-Marti, M. R., Gowda, S., Hilf, M. E., Garnsey, S. M., and Dawson, W. O. 1997. Kinetics of accumulation of citrus tristeza virus RNAs. Virology 228: 92-97.

Olsthoon, R. C. L., Haasnoot, P. C. J., and Bol, J. F. 2004. Similarities and differences between the subgenomic and minus-strand promoters of an RNA plant virus. *J. Virol.* 78: 4048-4053.

Peremyslov, V. V. and V. V. Dolja, 2002. Identification of the subgenomic mRNAs that encode 6-kDa movement protein and Hsp 70 homologue of *Beet yellows virus. Virology* 295: 299-306.

Qu, F., and Morris, T. J. 2000. Cap-independent translational enhancement of turnip crinkle virus genomic and subgenomic RNAs. *J. Virol.* 74: 1085-1093.

Palukaitis, P., Gacia-Arenal, F., Sulzinski, M. A., and Zaitlin, M. 1983. Replication of *Tobacco mosaic virus.* VII. Further characterization of single- and double-stranded virus-related RNAs from TMV-infected plants. *Virology* 131: 533-545.

Ray, D., and White, K. A. 2003. An internally located RNA hairpin enhances replication of tomato bushy stunt virus RNAs. *J. Virol.* 77: 245-257.

Rochon, D. M., and Johnston, J. C. 1991. Infectious transcripts from cloned cucumber necrosis virus cDNA: Evidence for a bifunctional subgenomic mRNA. *Virology* 181: 656-665.

Rubio, L., Yeh, H. H., Tian, T., and Falk, B. W. 2000. Heterogeneous population of defective RNAs associated with *Lettuce infectious yellows virus. Virology* 271: 205-212.

Satyanarayana, T., Gowda, S., Boyko, V. P., Albiach-Marti, M. R., Mawassi, M., Navas-Castillo, J., Karasev, A. V., Dolja, V. V., Hilf, M. E., Lewandowski, D. J., Moreno, P., Bar-Joseph, M., Garnsey, S. M., and Dawson, W. O. 1999. An engineered closterovirus RNA replicon and analysis of heterologous terminal sequences for replication. *Proc. Natl. Acad. Sci. USA* 96: 7433-7438.

Satyanarayana, T., Gowda, T., Ayllòn, M. A., Albiach-Marti, M. R., and Dawson, W. O. 2002. Mutational analysis of the replication signals in the 3`-nontranslated region of *Citrus tristeza virus. Virology* 300: 140-152.

Schirawski, J., Voyatzakis, A., Zaccomer, B., Bernardi, F., and Haenni, A.-L. 2000. Identification and functional analysis of the turnip yellow mosaic tymovirus subgenomic promoters. *J. Virol.* 74: 11073-11080.

Shih, D.-S., Lane, L. C., and Kaesberg, P. 1972. Origin of the small component of brome mosaic virus RNA. *J. Mol. Biol.* 64: 353-362.

Short, M. N., and Davies, J. W. 1983. *Narcissus mosaic virus:* A potexvirus with an encapsidated subgenomic messenger RNA for coat protein. *Biosci. Rep.* 3: 837-846.

Siegel, R. W., Adkins, S., and Kao, C. C. 1997. Sequence-specific recognition of a subgenomic RNA promoter by a viral RNA polymerase. *Proc. Natl. Acad. Sci. USA* 94: 11238-11243.

Siegel, R. W., Bellon, L., Beigelman, L., and Kao, C. C. 1998. Moieties in an RNA promoter specifically recognised by a viral RNA-dependent RNA polymerase. *Proc. Natl. Acad. Sci. USA* 95: 11613-11618.

Sit, T. L., Vaewhongs, A. A., and Lommel, S. A. 1998. RNA-mediated transactivation of transcription from a viral RNA. *Science* 281: 829-832.

Sivakumaran, K., Chen, M.-H., Roossinck, M. J., and Kao, C. C. 2002. Core promoter for initiation of cucumber mosaic virus subgenomic RNA4A. *Mol. Plant Pathol.* 3: 43-52.

Sivakumaran, K., Choi, S.-K., Hema, M., and Kao, C. C. 2004. Requirements for brome mosaic virus subgenomic RNA synthesis *in vivo* and replicase-core promoter interactions *in vitro*. *J. Virol.* 78: 6091-6101.

Stawicki, S. S., and Kao, C. C. 1999. Spatial perturbations within an RNA promoter specifically recognized by a viral RNA-dependent RNA polymerase (RdRp) reveal that RdRp can adjust its promoter binding sites. *J. Virol.* 73: 198-204.

Sulzinsky, M. A., Gabard, K. A., Palukaitis, P., and Zaitlin, M. 1985. Replication of *Tobacco mosaic virus*. VIII. Characterization of a third subgenomic TMV RNA. *Virology* 145: 132-140.

Tacke, E., Prüfer, D., Salamini, F., and Rohde, W. 1990. Characterization of a potato leafroll luteovirus subgenomic RNA: Differential expression by internal translation initiation and UAG suppression. *J. Gen. Virol.* 71: 2265-2272.

van der Vossen, E. A. G., Notenboom, T., and Bol, J. F. 1995. Characterization of sequences controlling the synthesis of alfalfa mosaic virus subgenomic RNA *in vivo*. *Virology* 212: 663-672.

Vives, M. C., Galipienso, L., Navarro, L., Moreno, P., and Guerri, J., 2002. Characterization of two kinds of subgenomic RNAs produced by *Citrus leaf blotch virus*. *Virology* 295: 328-336.

Wang, J., and Simon, A. E. 1997. Analysis of the two subgenomic RNA promoters for *Turnip crinkle virus in vivo* and *in vitro*. *Virology* 232: 174-186.

Wang, J., Carpenter, C. D., and Simon, A. E. 1999. Minimal sequence and structural requirements of a subgenomic RNA promoter for *Turnip crinkle virus*. *Virology* 253: 327-336.

White, K. A. 2002. The premature termination model: A possible third mechanism for subgenomic mRNA transcription in (+)-strand RNA viruses. *Virology* 304: 147-154.

White, K. A., Skuzeski, J. M., Li, W., Wei, N., and Morris, T. J. 1995. Immunodetection, expression strategy and complementation of turnip crinkle virus p28 and replication components. *Virology* 211: 525-534.

Wierzchoslawski, R., Dzianott, A., Kunimalayan, S., and Bujarski, J. J. 2003. A transcriptionally active subgenomic promoter supports homologous corss-overs in a plus-strand RNA virus. *J. Virol.* 77: 6769-6776.

Wierzchoslawski, R., Dzianott, A., and Bujarski, J. J. 2004. Dissecting the requirement for subgenomic promoter sequences by RNA recombination of *Brome mosaic virus in vivo*: Evidence for functional separation of transcription and recombination. *J. Virol.* 78: 8552-8564.

Yang, G., Mawassi, M., Gofman, R., Gafny, R., and Bar-Joseph, M. 1997. Involvement of subgenomic mRNA in the generation of a variable population of defective *Citrus tristeza virus* molecules. *J. Virol.* 71: 9800-9812.

Zaccomer, B., Haenni, A.-L., and Macaya, G. 1995. The remarkable variety of plant RNA virus genomes. *J. Gen. Virol.* 76: 231-247.

Zavreiv, S. K., Hickey, C. M., and Lommel, S. A. 1996. Mapping of the red clover necrotic mosaic virus subgenomic RNA. *Virology* 216: 407-410.

Zhang, G., Slowinski, V., and White, K. A. 1999. Subgenomic mRNA regulation by a distal RNA element in a (+)-strand RNA virus. *RNA* 5: 550-561.

Zhou, H., and Jackson, A. O. 1996. Expression of the barley stripe mosaic virus RNA β 'triple gene block'. *Virology* 216: 367-379.

9 GENE EXPRESSION

I. INTRODUCTION

The conventional and aberrant strategies of gene expression (translation) in positive-strand RNA plant viruses have been reviewed in around twenty five reviews starting from 1977 (Fraenkel-Conrat *et al.*, 1977) till date and are referred to at appropriate places.

Virus genomic RNA is essentially a polycistronic messenger that has to function within cells that utilize monocistronic mRNAs. The eukaryotic 80S ribosomes initiate protein biosynthesis from translation initiation site (the shortest region of mRNA containing all the nucleotide sequences essential for specific interaction with ribosomes and initiation of translation – Atabekov and Morozov, 1979) located towards the 5`-terminus with the internal cistrons being normally closed to translation. This anomaly has forced the plant viruses to develop various methods to somehow overcome the limitations imposed by rules of gene expression of the eukaryotic hosts (Bushell and Sarnow, 2002). Viruses do this by evolving different gene expressions strategies: readthrough of stop codons, proteolysis of polyproteins into functional proteins, ribosomal frameshifting, internal ribosome entry sites, and formation of subgenomic RNAs. Nucleotide sequence of RNA genomes of many plant viruses is known. This, and the knowledge of the location of ORFs, translation initiation codons (AUG), and stop or termination codons (UAG) in RNA, enables the virologists to predict the number, molecular weight, and amino acid sequence of the possibly encoded polypeptides.

Protein synthesis is the last step in gene expression and is carried out by 80S cytoplasmic ribosomes that can be regarded as a massive collection of enzymes that catalyze mRNA-directed protein synthesis. Thus, plant viral genomic and/or subgenomic mRNA is associated with 80S cytoplasmic polyribosomes. Coat protein subgenomic RNA in TMV-infected mesophyll protoplasts is associated with small polysomes (monopolysomes to tetrapolysomes) while full-length genomic TMV RNA is associated with larger polyribosomes containing 5 to 7 ribosomes. Functionally, ribosomes are polymerases since, like DNA and RNA polymerases, they also catalyze synthesis of biopolymers of a single chemical class. They interact with a nucleic acid template, during which aminoacyl tRNAs are the substrates consumed by ribosomes while proteins are the products formed.

Ribosomes contain three binding sites for tRNAs – site A, to which aminoacyl tRNAs are delivered in an mRNA-directed fashion; site P, where peptidyl tRNAs

223

reside; and site E through which deacylated tRNAs pass as they are released from ribosomes. Nascent polypeptides are elongated by a cyclic process that starts with a molecule of mRNA bound to ribosomes, a deacylated tRNA in E site, a peptidyl tRNA in P site, and a vacant A site.

A. Translation Initiation Factors

Eukaryotic (translation) initiation factors (eIFs) play crucial roles in translation initiation. Bushell and Sarnow (2002) review these with respect to animal viruses while Ahlquist *et al.* (2003) review these and other host factors with regard to plant viruses. Various eIFs and their functions during translation initiation, mentioned here, are essentially after Hellen and Sarnow (2001). The eIF1 has a mass of 12.6-kDa, only one subunit is needed for translation initiation, and it enables the ribosomes to scan and destabilize the aberrant initiation complexes. The eIF1A has a mass of 16.5-kDa, only one subunit is needed for translation initiation, and promotes binding of Met-tRNA to 40S ribosomal subunit as well as ribosome scanning. The eIF2 directs the ternary complex to 40S ribosomal subunit to form the 43S pre-initiation complex that includes eIF3, phosphorylation of eIF2 dramatically alters the efficiency and rate of mRNA translation and is the first major point of control over translation initiation process, has a mass of 126-kDa, its three subunits are needed for translation initiation, performs GRP-dependent binding of Met-tRNA to 40S subunit, and acts as GTPase. The eIF2B has a mass of 261-kDa; its five subunits are needed for translation initiation, and functions as a guanine nucleotide exchange factor for eIF2. The eIF3 is of ~700-kDa, its eleven subunits are needed for translation initiation, causes ribosomal dissociation, facilitates binding of 43S pre-initiation complex to mRNA via the cap-binding complex (eIF4F), and promotes binding of mRNA and Met-tRNA to 40S ribosomal subunit.

The eIF4F is a complex factor, is essential for recognition of cap structure, links mRNAs to ribosomes, promotes search for translation start site, its components interact with PABP, interaction between eIF4F and PABP increases the affinity of PABP for poly(A) tail and also the affinity of eIF4F for cap structure of mRNAs, this association of PABP with eIF4F leads to circularization of the genomic mRNAs and synergistically stimulate translation (Wilkie *et al.*, 2003), and its cleavage from mRNAs inhibits formation of initiation complex and blocks cell protein synthesis. Indeed, recruitment of mRNAs by eIF4F complex is one of the first steps of translation. Thus, eIF4F complex is the cap-binding protein macromolecular complex and, for translation initiation on most of the capped viral and cellular mRNAs in eukaryotes, the 80S ribosomes are recruited by this complex (along with other factors) for scanning from 5`-end of mRNA to the initiating AUG. The eIF4F complex is comprised of three subunits - (cap-binding subunit) eIF4E, (the helicase subunit) eIF4A, and (the multiadapter subunit) eIF4G. Thus, it is a heterotrimer composed of eIF4E/4A/4G, binds m^{7}G cap, acts as RNA helicase, has a mass of 223-kDa, and its three molecules are required for translation initiation. The eIF4F complex facilitates the docking of 43S complex (comprising eIF3 and 40S ribosomal subunit charged with eIF2-GTP-Met-tRNA) via interaction of eIF3 (of the 43S complex) with eIF4G (of the heterotrimeric eIF4F initiation factor). Thus, eIF4F facilitates binding of mRNA to 43S pre-initiation complex.

The eIF4A is 44-kDa, only one subunit is required for translation initiation, and acts as RNA-dependent ATPase as well as RNA helicase. The eIF4A subunit (of eIF4F) is a DEAD-box RNA helicase/RNA dependent ATPase (ATPase/RNA helicase) that exists both free and as part of eIF4F. The helicase activities of eIF4A (and consequently of eIF4F) are strongly enhanced by eIF4B. The DEAD-box proteins, like eIF4A, form a large family of established and putative RNA helicases that share seven conserved amino acid motifs, occur in bacteria and eukaryotes and in viruses infecting them. The eIF4B is 70-kDa, promotes RNA helicase activity of eIF4A and eIF4F, and only one molecule is needed for translation initiation.

The eIF4E is the 5` cap-binding subunit of translation initiation factor eIF4F, recognizes and binds to both cellular and viral mRNAs m^7G cap by being assembled around the mRNA cap structure, its presence in the cells is essential for assembly of eIF4F, possesses a mass of 26-kDa, and only one molecule is required for translation initiation. The affinity of eIF4E for m^7G cap constitutes a major control point in translation initiation pathway and is subject to variation through phosphorylation of eIF4E. Accordingly, the eIF4E is a pivotal translation initiation factor and, as compared to other translation initiation factors, is present in limiting amounts in cell. The eIF4E also interacts with eIF4A and eIF4G (leading to the formation of the eIF4F macromolecular complex which facilitates binding of mRNA to 43S pre-initiation complex). Affinity of eIF4E for mRNA 5` cap is increased by phosphorylation on serine 209. Phospho-eIF4E is stimulatory for mRNA translation. Viruses have targeted the processes of eIF4E regulation to facilitate translational selectivity for viral mRNAs during infection.

The eIF4G subunit (of eIF4F) has been variously called as the multiadapter subunit, the bridging protein/subunit or a molecular scaffold, is of 154-kDa, and only one molecule is needed for translation initiation (Gallie and Browning, 2001). It is a large polypeptide and acts as a molecular scaffold that binds RNA, eIF3, eIF4A, eIF4E, and PABP and coordinates their activities. Binding of this factor to both eIF4E (bound to cap structure) and PABP [bound to 3` poly(A) tail] makes it behave as an adapter bringing together m^7G cap and poly(A) tail of an mRNA leading to circularization of the mRNA, and thereby promoting ribosome activity (of initiation factor eIF3). A main feature of host shut-off involves disruption of eIF4F function through virus-mediated cleavage of eIF4G from eIF4F complex. The current model for eIF4F function in cap-dependent translation proposes that eIF4G serves as a molecular bridge for assembly of cap-binding complex upon 5`-end of mRNA and facilitates its interaction with 43S pre-initiation complex.

The eIF5 activates GTPase activity of eIF2, only one molecule is required for this function, and is of 49-kDa. The eIF5B functions in ribosomal subunit joining, acts as GTPase, only one molecule is required, and is of 139-kDa.

1. Translation Initiation Factors and Plant Viruses

Viral replication requires a large amount of energy and thereby requires total metabolic control of cellular resources. For this, viruses have evolved certain mechanisms that supersede cellular mRNA translation (Bushell and Sarnow, 2002; Ahlquist *et al.*, 2003) through recruitment and/or modification of functions of translation initiation factors. Some of these modifications are - inactivation of eIF4E translation factor, cleavage of eIF4G, cleavage of PABP that disrupts the closed-loop translation complex, and modification of eIF1 factor. Effects of such modifications range from alteration in efficiency of cap-dependent translation and translation elongation to altering the rate of innate antiviral response of host cell.

Translation initiation factors participate in viral RNA synthesis by association with replicase or *cis*-acting elements connected with RNA replication as in BMV and TMV or by direct binding to viral RNA as in TYMV. This indicates that translation factors are directly associated with viral RNA synthesis, that translation machinery is required for viral RNA synthesis, and that viruses subvert this machinery for replication and transcription of their genomic RNA. Surprisingly, translation factors have no connection with regulation of viral RNA translation since they fail to bind to viral motifs that directly participate in translation process.

Translation initiation factors are also associated with active viral RNA replication complexes. For example, virus-specific RdRp complexes, isolated from BMV- or TMV-infected cells, are associated with subunits of host translation initiation factor eIF3. A 41-kDa subunit of eIF3 purifies along with BMV RdRp and also interacts with 2a polymerase of BMV. Moreover, addition of eIF3 or 41-kDa-subunit stimulated negative-strand RNA synthesis by BMV RdRp (Quadt *et al.*, 1993). Similarly, TMV RNA replication proteins interacted with and co-immunoprecipitated with host GCD10 subunit of eIF3 which was involved in negative-strand RNA synthesis by viral polymerase (Osman and Buck, 1997; Taylor and Carr, 2000). Moreover, the eIF3 host factors may have functions related to aminoacylatable 3`-tRNA-like elements of TMV and BMV genomic RNAs (where synthesis of negative-strand RNA is initiated) or tRNA TΨC stem-loops of BMV replication elements that act as template-recruitment signals (which are the *cis*-acting recognition elements), or to both. The 5`-genome-linked proteins of some potyviruses bind to cellular cap-binding protein eIF4E or eIF(iso)4E, and eIF(iso)4E mutations block replication of a potyvirus (Lellis *et al.*, 2002).

II.   CANONICAL TRANSLATION

Nearly all eukaryotic cellular mRNAs are capped (Furuchi and Shatkin, 2000) and polyadenylated. These two structural elements [5`-terminal cap (capped 5`-UTR) and 3`-terminal poly(A) tail (polyadenylated 3`-UTR)] are crucial to translation. Cap structure (more precisely, eIF4E) and poly(A) tail (PABP) together function synergistically to perform several essential functions: recruit translational machinery, greatly enhance translation initiation, increase efficiency of translation, influence translational regulation, and stabilize replication complex (Gallie, 1991, 1996; Tarun and Sachs, 1995; Preiss and Hentze, 1998, 1999; Le *et al.*, 2000; Wilkie *et al.*, 2003).

The cap structure serves as binding site for translation initiation factor eIF4F; more appropriately, cap structure interacts with subunit eIF4E of the heterotrimeric eIF4F. The eIF4F also assists the binding of 40S ribosomes to mRNAs. The poly(A) tail serves as the binding site for the PABP. These two translation factors (eIF4E and PABP) act synergistically and possibly also physically interact with each other. Both the eIF4E (bound to 5`-cap) and PABP [bound to poly(A) tail], bind to/interact with the scaffolding protein eIF4G, forming a closed loop translation complex by causing circularization of mRNAs (Gale *et al.*, 2000; Prevot *et al.*, 2003). The closed loop is thought to stabilize mRNA and 5`-cap complex, provides for the efficient recruitment and recycling of ribosomal subunits, and also seems to enhance recruitment of the 43S ribosomal initiation complex to the 5` untranslated region (5`-UTR) of the message (Gallie, 1991, 1998; Sachs *et al.*, 1997; Sachs, 2000; Magnus *et al.*, 2003). Thus circularization of RNA is regarded as an essential prerequisite for efficient translation initiation as well as for translation of most mRNAs. Circularization of RNA also occurs in animal viruses, and facilitates multiple rounds of translation (Sachs, 2000). The 5`-cap structure can also be absent so that translation in such situations is cap-independent in contrast to the normal cap-dependent translation (Jackson *et al.*, 1995).

The mRNAs possess certain structural features that enable them to exercise some control over translation process. Such characteristics, reviewed several times and briefly given below after Gale *et al.* (2000), are also operational in viral mRNAs and are - length and structural complexity of 5`-UTR influences scanning process since long 5`-UTR could impede initiation; length and structural complexity of 3`-UTR could affect efficiency of translation and interaction with *trans*-acting factors; the 3`-end mediates closed-loop translational complex through interaction with PABP; secondary structure of 5`-UTR is an important factor since complex 5`-UTR structures could interfere with scanning process; sequence context and position of initiation codon impart ribosome selectivity for first AUG while 'weak' AUG start codon leads to and promotes leaky scanning process; and upstream ORF could impede ribosome scanning to downstream cistrons. Stability and accessibility of cap structure and cap-binding complex perform multiple functions - they promote stability of mRNA as well as interaction with eIF4F, facilitate translation of most cellular mRNAs, and accessibility to initiation factors can influence efficiency of translation. Length of poly(A) tail imparts stability and translational efficiency to mRNAs while its interaction with poly(A)-binding protein mediates association with cap-binding complex on mRNA. *trans*-Acting factors like specific RNA sequences and/or structural features promote interaction with RNA-binding proteins, which may influence translational efficiency.

Translation of mRNAs is a complex process comprising of several steps in which many proteins take part as enzymes or factors and which is subjected to strict regulatory controls. Some basic points in the generally agreed conventional scanning and translation model in eukaryotes are mentioned here.

A. Translation Initiation

Translation initiation (that is, initiation of protein synthesis) in eukaryotic cells (Kozak, 1999) generally begins from the first 5`-proximal translation start codon AUG in the mRNA template. Majority of eukaryotic cell mRNAs are mono-cistronic so that translation initiation at 5`-end is no problem. In most eukaryotic mRNAs, translation initiation commences with recruitment of cap-binding eIF4F protein complex to capped 5`-end of mRNA and with the formation of initiator methyl-tRNA (Met-tRNA). The latter step is facilitated by formation of an eIF2-GTP-Met-tRNA ternary complex. The eIF2 directs the ternary complex to 40S ribosomal subunit to form the 43S pre-initiation complex that includes eIF3. The eIF3 facilitates binding of 43S pre-initiation complex to mRNA through the cap-binding complex (eIF4F). Thus, most mRNAs in eukaryotic cells recruit ribosomes by a mechanism in which a 43S complex (composed of a 40S ribosomal subunit bound to eIF2-GTP/Met-tRNA, eIF1A, and eIF3) is recruited to capped 5`-end (Kozak, 1989; Donahue, 2000; Hershey and Merrick, 2000). Binding of 43S complex to mRNA involves recognition of 5`-capped mRNA by eIF4E (cap-binding subunit of eIF4F) and is greatly enhanced by PABP bound to 3`-poly(A) tail (Sachs and Varani, 2000).

After association with capped 5`-end of mRNA, the 43S pre-initiation complex begins scanning from 5`-end to 3`-direction of mRNA or the site of ribosomal entry (as in cap-independent translation) and continues scanning until the Met-tRNA is reached and interacts with an appropriate translation initiator codon AUG (Kozak, 1986a, 1991). At this point, the anticodon in initiator tRNA (Met-tRNA), which is positioned in ribosomal P-site, engages in base pairing with start codon in mRNA. Once the Met-tRNA associates with initiator AUG codon, GTP is hydrolyzed from ternary complex, bound initiation factors are released, and the large ribosomal 60S subunit joins the pre-initiation complex to form the 80S ribosome that decodes the mRNA into protein(s) – leading to translation initiation and start of protein synthesis till a stop codon where translation terminates (Kozak, 1989; Hershey and Merrick, 2000). In addition, PABP interacts with eIF4G.

The AUG site selection (Jackson, 2000) for translation initiation and the efficiency of translation initiation is greatly dependent upon and influenced by the context of an AUG codon (Kozak, 1986b). The most efficiently used AUG triplet is embedded within the canonical sequence/context ACC<u>AUG</u>G (the initiation codon is underlined) in which a purine and a guanine at -3 and +4 positions, respectively, are the most important positions (Lütcke *et al.*, 1987; Kozak, 1991, 1999; Lukaszewicz *et al.*, 2000).

1. Abberant Translation Initiation

a. Leaky scanning

The scanning ribosome predominantly initiates translation at the 5`-most AUG in the template (Kozak, 1989) and has low re-initiation capacity. Leaky scanning is with reference to translation initiation of polycistronic mRNAs and is greatly facilitated if AUG start codon at 5`-end ORF has less favourable context than AUG of the second ORF. In such cases, the pyrimidine residues are at positions –3 or +4, when adenosine in AUG triplet is regarded at being position +1, with the result that the scanning ribosomal complex could bypass the embedded AUG codon (Kozak, 1989). The 5`-proximal AUG codon is thus a silent initiation codon so that it is  bypassed by a portion of 40S ribosome subunits to initiate translation at a downstream AUG, which is placed in a more favourable context for translation initiation. Therefore, translation, instead of starting at first 5`-proximal AUG initiation codon, starts from the next or some other AUG triplet and is internal initiation of protein synthesis. Thus, leaky scanning mechanism is primarily based on discrimination between two or more initiation sites having unequal contexts, the downstream codon most closely matches the canonical initiator AUG, and the two independently initiated proteins are synthesized from the same mRNA (Kozak, 1986a, 1986b, 1989).

b. Ribosomal Shunting and Internal Translation Initiation

Very long and highly structured 5`-untranslated regions (UTRs), that contain multiple AUG codons, occur in a small proportion of eukaryotic mRNAs (Kozak, 1991). Normally, deletion of such highly structured or AUG-rich 5`-UTRs from mRNAs enhances translation of encoded genes but this is not the case in certain viral and cellular mRNAs that can be translated very efficiently even if they contain long structured 5`-UTRs. Two such translation initiation mechanisms are ribosomal shunting and internal initiation. These mechanisms are operationally distinct from the canonical translation initiation mechanism.

Ribosomal shunting, for mediating translation initiation, is not functional in positive-strand RNA viruses but is operative in a few other plant viruses including CaMV 35S mRNA (Fütterer *et al.*, 1994) and in cellular mRNA that encodes heat shock protein 70. The ribosome subunits bind mRNA in a 5`-cap-dependent manner and begin downstream scanning till they reach a stable RNA structure that can stop the scanning ribosomes or dissociate them from mRNA. This arrest leads to an intra-molecular shunting of ribosomal subunits to a downstream landing site, thereby bypassing the RNA structure. The ribosomes then resume scanning until the next appropriate AUG start codon is encountered.

Certain mRNAs contain internal ribosome entry sites (IRES). These sites are generally in the 5`-UTR and enable 5`-end-independent translation initiation to take place. Many of the regulatory mechanisms, that control recruitment of most mRNAs to translation apparatus, are not applicable to IRES-containing mRNAs. Jang *et al.*, (1988) and Pelletier and Sonenberg (1988) discovered that picornaviral mRNAs undergo translation by a mechanism that enables ribosomes to initiate translation effectively on highly structured areas (the IRES sites) that are located within the 5`-UTR. All picornaviral mRNAs contain IRES elements. Picornaviral 5`-UTRs possess a length varying from 610 to 1500 nucleotides, are highly structured and contain multiple nonconserved AUG triplets located upstream of the initiation codon that should act as a strong barrier to scanning ribosomes. Few, if any definite, similarities in sequence, size, or structure, exist between different IRES elements except those between families of related viruses – implying that no universal mechanism occurs for the internal ribosome entry.

Much is known about the sequence and factors responsible for recruiting ribosomal subunits to IRES of two picornaviruses - *Hepatitis C virus* and *Cricket paralysis virus* (Hellen and Sarnow, 2001). All picornavirus IRESs contain an AUG triplet at their 3`-border, 25 nucleotides downstream from the beginning of pyrimidine-rich tract. In *Encephalomyocarditis virus* (EMCV) and *Foot-and-mouth disease virus*, this AUG codon is the initiation codon and ribosomes bind directly to it without scanning.

Virus IRES elements have different structures so that the 40S ribosomes can be recruited by at least three different mechanisms. First, in EMCV, interaction of eIF4G/eIF4A with IRES is aided by canonical eIFs 2, 3, and 4B and is necessary to recruit 40S subunits to IRES. Second, the hepatitis C virus IRES can bind eIF-free 40S subunits without helicase eIF4A or other eIFs, yet recruits the ternary complex to position the initiator tRNA into ribosomal P site. Third, the cricket paralysis virus-like IRES elements can assemble 80S ribosomes without any eIF, including the tRNA. The above division indicates that structural features in IRES elements dictate the requirement for certain eIFs, like eIF4A, that help in the recruitment of 40S ribosomes. Moreover, in all these three mechanisms, ribosomal recruitment involves non-canonical interactions with canonical components of translation apparatus. This shows that most IRES are likely to be refractory to mechanisms controlling 5`-end-dependent translation via dephosphorylation of eIR4E or sequestration of eIF4E by eIF4E-binding proteins.

2. Scanning

The scanning model is mostly used by plant viruses for translation of viral genes. Since viruses depend upon the host machinery for translation of the genetic information, viral genomes have to abide by the cellular rules. Scanning is done by ribosomes from 5`-end of mRNA to translation initiating codon AUG and needs two essential factors – the canonical translation initiation factor eIF4A and a DEAD (Asp-Glu-Ala-Asp)- box RNA helicase for unwinding secondary structure in 5`-noncoding region (McCarthy, 1998). The DEAD-box proteins like eIF4A constitute a large family of established and putative RNA helicases (Gorbalenya and Koonin, 1993), which have been detected in bacteria, eukaryotes, and their viruses. A second DEAD-box protein (Ded1p encoded by essential *DED1* gene) is also a general translation initiation protein (Noueiry *et al.*, 2000).

B. Elongation Step of Translation

The mRNA is associated with multiple 80S ribosomes (polyribosomes) as amino acid residues are sequentially placed on carboxyl end of the growing peptide chain. Elongation control mechanisms include ribosomal frameshifting.

C. Termination of Translation

Translation termination occurs when translating 80S ribosome encounters an in-frame termination codon within the template mRNA. Termination codon is recognised by a release factor, which mediates hydrolysis of peptide chain from the bound tRNA. This releases the nascent polypeptide from 80S ribosome and leads to an eventual dissociation of ribosomal subunits. Once termination has occurred, the 40S subunit is free to again continue scanning mRNA. In multicistronic transcripts, termination can be followed by reinitiation at downstream cistron, subject to availability of ternary complex. However, reinitiation is usually inefficient and presence of an upstream ORF can confer limitations to translation efficiency of the major downstream ORF. This termination-reinitiation translational control mechanism is very prevalent among viruses and is used to control synthesis of specific viral gene products.

D. Interference by Viral RNAs

Viruses employ canonical translation factors and machinery of host cells to facilitate completion of their translational programming. They can disrupt major translation checkpoints and signals so that the cells often fail to respond to normal translation-modulatory signals but instead respond to regulatory signals superimposed by the infecting viruses; thus, RNA viruses interfere with some of the key steps in translation initiation. Following three situations make the host translation machinery available to viruses for translation of their genomes.

Proteolysis of eIF4G and PABP and sequestration of eIF4E is the first method. Picornaviruses (except EMCV) seem to proteolytically cleave eIF4G (henceforth called eIF4GI and eIFG4GII) and PABP, thereby destroying both the eIF4F complex and end-on-end communication in cellular mRNAs. Viral RNA possibly efficiently competes with cellular mRNAs for the limited amounts of truncated eIF4GI and eIF4GII found in infected cells. Viral IRES are hypothesized to preferentially recruit the COOH-terminal fragments of eIF4GI and eIF4GII, which contain binding sites for eIF4A and eIF3 – leading to inhibition of host cell translation. EMCV undertakes sequestration of eIF4E by eIF4E-binding proteins (4E-BPs) (a group of proteins that sequesters eIF4E component of eIF4F) – again leading to inhibition of host cell translation. Therefore, lowering the abundance of cap binding protein complex [whether by cleavage of eIF4GI and eIF4GII or by sequestration of eIF4E] selectively inhibits translation of capped host cell mRNAs without inhibiting translation of IRES-containing picornaviral mRNAs. It is now known that an eIF4GI fragment (containing the binding sites for eIF4A and eIF3) and eIF4A are sufficient to recruit ternary 40S complexes to picornavirus IRES.

Substitution of PABP is the second method. It occurs in rotaviruses (genus *Rotavirus*) of *Reoviridae*. Rotaviruses contain eleven double-stranded RNA segments, each of which is transcribed into mRNA that possesses a 5`-terminal cap structure but lacks the 3`-terminal poly(A) tail. Instead, the 3`-end sequences contain a tetranucleotide motif that is conserved among different rotavirus groups. Rotaviruses disrupt host cell circularization, leading to severe inhibition of host mRNA translation. Viral NSP3 protein binds specifically to the conserved viral 3`-sequences and also interacts with a binding site in eIF4G that overlaps the PABP binding site. The eIF4G has higher affinity for NSP3 than for PABP so that interaction between PABP and eIF4G is disrupted in infected cells causing reduced efficiency of host mRNA translation. This example illustrates the way a virus, that encodes non-polyadenylated mRNAs, can usurp the host cell translation apparatus by encoding a protein that binds to 3`-ends of viral mRNAs, evicting PABP from eIF4G, which is the key player involved in recruitment of ribosomes to mRNAs. Thus, viruses use the closed-loop translation complex for redirecting the host translation machinery in favour of viral mRNA translation, by targeting PABP and disrupting interactions of ends of host mRNA resulting in attenuation of host mRNA translation.

Bypass of initiator tRNA is the third method. It occurs in insect paralysis-like viruses that are positive-strand RNA viruses with a bicistronic genome. The first large ORF encodes viral nonstructural proteins and is followed by a 200-nucleotide intergenic region (IGR) and an ORF that encodes viral structural proteins. These viruses contain IRES in its IGR sequence element, which can mediate translation initiation at a non-cognate CUU start codon without Met-tRNA. This is exceptional because only cognate (that is, AUG) or weak-cognate (that is, CUG or GUG) codons function as start codons. This IGR-IRES sequence themselves occupy the ribosomal P-site; a CUU triplet at start site base pairs directly with upstream IGR-IRES sequence. IRES can recruit both 40S and 60S subunits without any known canonical eIFs to form an 80S ribosome that can start protein synthesis from the next codon (a GCU), which is located in the ribosomal A-site. In this way, the first amino acid is alanine encoded by A-site located at GCU codon. This suggests that IGR-IRES elements can propel ribosome into elongation mode without prior formation of a peptide bond and that cricket paralysis-like viruses have evolved an RNA element that can initiate protein synthesis when intracellular amounts of ternary eIF2-GT-Met-tRNA complexes are low.

E. Translation of Viral RNAs

It is necessary for viral RNAs to mimic the properties of cellular mRNAs for being competent for translation. However, only a limited number of plant virus RNAs possess both the features [cap at 5`-end and poly(A) tail at 3`-end]. Some examples are: ASGV, BNYVV, PVX, GVB, and possibly still some more of the beny-, carla-, potex-, and trichoviruses (Héricourt *et al.*, 1999; Agranovsky and Morozov, 1999) and these plant viruses must be having canonical translation. However, viral templates of most of the plant viruses lack the cap, or poly(A), or both. Despite this, the 5`- and 3`-terminal structures in such viral RNAs function synergistically in translation because of the presence of alternative or additional elements. For example, TCV lacks both the 5`-cap structure as well as 3`-poly(A) tail but still it shows optimal and synergistic translational efficiency because of the interaction between 5`- and 3`-UTRs; this synergistic translation effect was fourfold greater than the sum of contributions of individual UTRs (Qu and Morris, 2000). Then, certain 3`-UTRs of nonpolyadenylated viral RNA enhance translation in a manner similar to that provided by a poly(A) tail, including a synergistic interaction with 5` cap as in plant viral RNAs terminating with a tRNA-like structure as in TMV, BMV (Gallie and Kobayashi, 1994) and TYMV (Matsuda and Dreher, 2004). Cap-independent translation also falls in this situation and is discussed later in detail.

The 5`-UTRs of some capped viral RNAs enhance translation expression as in *Potato virus S*, PVX. TMV (Gallie *et al.*, 1987), and TYMV (Matsuda *et al.*, 2004). This condition has been best studied in 5`-UTR enhancer of nonpolyadenylated TMV genomic RNA, the so-called Ω sequence, which enhances gene expression by elevated recruitment of eIF4F complex through the mediation of heat shock protein HSP101 (Gallie, 2002). The Ω sequence is a 68-nucleotide leader sequence that contains a 25-nucleotide poly (CAA) region responsible for enhancement of translational initiation (Gallie, 1996, 2002). The behaviour of 5`-TYMV sequences shares similarities with that of TMV Ω enhancer (Matsuda *et al.*, 2004). This suggests that, as hypothesized for TMV Ω element (Gallie, 2002), the TYMV 5`-sequences facilitate the earliest rounds of translation in a way that the cap is not capable of. The TYMV RNAs that contained all the three translation enhancing elements (5`-cap, 5`-TYMV sequences, and TYMV 3`-UTR) exhibited the highest gene expression (Matsuda *et al.*, 2004).

The RNA genomes of plant viruses undergo circularization/cyclization, which plays an essential role in replication and translation of viral RNAs. Formation of pan-handle structures by base-pairing between complementary sequences at 5`- and 3`-ends in several genera of negative-strand RNA viruses is known. Circularization of RNA genome may also occur essentially in positive-strand RNA viruses and appears to be mediated by protein-protein bridges between 5`- and 3`-ends of viral RNA. Many viral RNAs, like cellular mRNAs, have a 5`-cap-like structure and 3`-poly(A) tail. These structures in cellular mRNAs permit the formation of a protein bridge by interactions between a poly(A) stretch that communicates through PABP and eIFG4G with eIF4E bound to capped 5`-end (Gallie, 1991, 1996) leading to circularization of mRNAs. This is true of plant viruses also. However, in plant viruses, in the absence of 5`-cap structure, interaction between 5`-end VPg-Pro (a precursor of VPg) and eIF(iso)4E or eIF4F as well as with 3`-PABP could promote RNA circularization since formation of VPg-Pro-PABP complex could bring both ends of viral RNA in close proximity. Léonard *et al.* (2004) found that 5`-end VPg-Pro did actually interact with eIF(iso)4F and eIF4F and also with 3`-end PABP *in planta*. The VPg of TuMV interacts *in vitro* with eIF4E. The 3`-UTR (that contains tRNA-like structure) of TYMV enhances translation synergistically with 5`-cap; translational enhancement was maximum when aminoacylation and eIF1A binding properties stayed intact (Matsuda and Dreher, 2004). The circular form is generated in TYMV RNA through communication of 3`-tRNA-like structure with the start of ORF2 (Barends *et al.*, 2003). Translation of BYDV RNA requires base pairing between a stem-loop in 5`-UTR and a stem-loop in a 100-nucleotide translation enhancer element in 3`-UTR, presumably to deliver translation factors and/or ribosomes to 5`-end as well as to facilitate intramolecular RNA-RNA interactions leading to formation of a circular type of genomic RNA complex (Guo *et al.*, 2001). Some such thing happens in BMV as well (Barends *et al.*,

2004). Thus, the presence of 5`-cap structure or VPg and their interaction with translation initiation factors and with 3`-bound PABP could lead to formation of a circular genome complex in RNA plant viruses also. The CP of AMV and ilarviruses possibly also helps in circularization of RNA genomes. Either the CP or the exposed loop sequences of hairpin structure bind translation factors and establish 5`- to 3`-end communication of AMV RNA (Bol, 2003). In fact, efficient translation of AMV RNAs requires the binding of CP dimers to the nonadenylated 3`-termini of viral RNAs (Neeleman *et al.*, 2001, 2004). The circular form of genome is thought to assess integrity of mRNA as also to recycle ribosomes for multiple rounds of translation. In line with this conclusion is the observation that 5`-cap binding factors and PABP do synergistically stimulate translation (Wilkie *et al.*, 2003).

III. TRANSLATION STRATEGIES OF VIRAL RNAs

The translation strategies help the viral RNA to express their genes, to regulate gene expression, and enable the synthesis of more than one polypeptide per cistron. All this occurs through controlled and regulated synthesis of proteins. Gene expression of positive-sense RNA viruses can be regulated at four levels, with the strategies at each level that the viruses adopt being mentioned within parenthesis: first level is at the level of genome segments (unipartite or multipartite genome); second level is the transcription of RNA (the four such strategies are splicing of viral mRNAs to generate new ORFs, production of subgenomic RNAs from viral genomic RNA template, ambisense strategy, and cap-snatching); third level is that of the translation process [these strategies operate at translation initiation, translation elongation, and translation termination levels - the strategies at initiation step being scanning model (*cis*-acting and *trans*-acting factors), leaky scanning, shunting, and internal ribosome entry – that is, internal initiation on polycistronic mRNA; the only strategy at elongation step is frameshift or multiple reading frames; and the only strategy at termination step is the readthrough process], and the fourth level is at post-translational level [the only process at this level being proteolytic processing which happens because there are multiple initiation sites on monocistronic mRNA (independent gene translation) leading to synthesis of a polyprotein] (Drugeon *et al.*, 1999). Thus, some of the canonical mechanisms employed by viruses during translation of viral mRNAs are: internal translation initiation, ribosome shunt, leaky scanning, frameshifting, and control of termination and re-initiation.

Viruses have finite RNA genomes so that it is an ideal position if they could use the same nucleotide sequences for synthesis of different proteins. And that is precisely what the viruses do, as will be apparent in this chapter. Out of all the plant viruses, luteoviruses have achieved the highest level of sophistication for the expression of their genome. They employ multiple strategies: frameshift, readthrough, overlapping ORFs, subgenomic RNAs, and proteolytic processing. Then, in at least one luteovirus (BYDV-PAV which lacks a VPg) a motif within the 3`-region of mRNA facilitates initiation of translation by mimicking a 5`-cap structure. The bipartite PCV employs two translation techniques to produce its proteins. Protein p191 is produced by C-terminal extension of the replication protein p131 by readthrough of p131 stop codon on RNA1. The protein p39 is translated from full-length RNA2 by a leaky scanning mechanism starting from the second AUG. Many such examples commonly occur in plant viruses.

A. Regulation of Gene Expression at the Level of Genome Segments

The genomes of some plant virus taxons (bromo-, bymo-, como-, crini-, cucumo-, diantho-, furo-, hordei-, ilar-, idaeo-, nepo-, and tobraviruses) are divided among two (bipartite) or three (tripartite) RNA segments. The closely related viruses of a group can possess both monopartite and divided genomes as plant viruses in *Potyvirus* and *Bymovirus* genera of *Potyviridae*, and *Closterovirus* and *Crinivirus* genera in *Closteroviridae*. Each genome segment of a multipartite genome may contain one or more ORFs.

Genome splitting may occur even in the most conserved module – the replicative gene module. The conserved domains of methyltransferase (MT) and helicase (HEL), on the one hand, and polymerase (POL), on the other, are translated as distinct products of genomic RNAs α and γ, respectively, of the tripartite genomes of hordeiviruses (BSMV, *Lychnis ringspot*, and *Poa semilatent hordeiviruses*) (Gustafson *et al.*, 1989; Savenkov *et al.*, 1998). These domains in the closely related furo-, tobamo-, and tobraviruses are found in a single gene product. An identical division of the replicative domains amongst the RNA1 and RNA2 translation products also occurs in tripartite bromo-, cucumo-, and ilarviruses (Ahlquist, 1992).

The plant viruses having the TGB of movement protein genes also contain cases of complicated genome division patterns. The monopartite genomes of carla-, and potexviruses bear genes for replicase (TGB proteins) and CP while this gene array in BNYVV and PCV is divided between genomic RNA1 (encoding replicase) and RNA2 (encoding CP and TGB proteins) (Herzog *et al.*, 1994). Moreover, the genetic information on RNA2 of tripartite genomes of pomoviruses [*Beet soil-borne virus* (BSBV) and *Potato mop top virus* (PMTV)] is also split in such a way that the TGB and CP genes are located on two separate genomic RNAs (Kashiwazaki *et al.*, 1995; Koenig *et al.*, 1997).

Genome partitioning makes more genes accessible to ribosomes for translation. This also permits separate control of gene expression by *cis*-elements that may influence replication and translation of individual RNA components.

B. Regulation of Gene Expression at Transcription Level

Transcriptional control of translation can occur by following means: by splicing of viral mRNAs to generate new ORF; by production of subgenomic mRNA from genomic RNA template; by ambisense strategy in which both viral and viral complementary genomic RNAs code for viral proteins; by cap-snatching mechanism in which virus 'snatches' 5`-end of host mRNAs to prime synthesis of viral mRNAs; and by transcriptional editing which occurs in the *Paramyxoviridae* but is not reported among plant viruses (Maia *et al.*, 1996). Both editing and splicing as regulation mechanisms result from transcriptional and post-transcriptional modifications, lead to the production of two mRNAs and of two proteins that are identical in N-terminal regions but differ in C-terminal regions.

Splicing of mRNAs occurs only among plant DNA viruses. It occurs in some single-stranded DNA *Geminiviridae* [*Wheat dwarf geminivirus* (WDV) – Schalk *et al.*, 1989], and in double-stranded DNA-containing caulimovirus (CaMV - Kiss-László *et al.*, 1995), and badnavirus (*Rice tungro bacilliform badnavirus* - Fütterer *et al.*, 1994). Splicing is needed for any one of the several functions: for replication of WDV, for infectivity of CaMV, and for expression of the ordinarily silent ORF IV of *Rice tungro bacilliform badnavirus*. Both the spliced and unspliced forms of the corresponding mRNAs are present in the WDV-infected plants.

Formation of subgenomic RNAs ensures regulation of gene expression by preferential initiation of certain proteins. Genomic RNA of many, if not all, plant viruses contains at least two different initiation sites and preferential initiation from one or the other site can play an important role in regulation of gene expression. Preferential formation of CP subgenomic RNA followed by its preferential translation due to preferential initiation is one of the most common examples. Subgenomic RNAs of other cistrons are also known. Subgenomic RNAs and their translation products, at least the CP, have been detected both *in vivo* and *in vitro* systems. Internal genes in many plant viruses are expressed through sgRNAs that are formed by transcription from full-sized minus-sense RNA strand and represent the 5`-terminally truncated 3`-coterminal copies of genomic RNA. Each sgRNA, as a general rule, translates only one viral gene exposed at its 5`-end but some sgRNAs of hordeiviruses, luteoviruses, and tombusviruses express two genes that are nested in different reading frames. The number of sgRNAs varies in different virus groups - from one in bromo- and dianthoviruses to six to nine in closteroviruses. The subgenomic RNAs are treated in detail in a separate Chapter.

Cap snatching occurs in some negative-strand RNA viruses and entails cleaving of a host RNA generally at 10-20 nucleotides from its 5`-capped leader for priming viral transcription. Tospoviruses (genus *Tospovirus* of family *Bunyaviridae*; type tospovirus is *Tomato spotted wilt virus* - TSWV) (Schmaljohn *et al.*, 1996) are the multipartite minus-stranded RNA plant viruses consisting of three RNA segments. The L RNA has only one ORF that encodes a polymerase; this genomic RNA undergoes transcription for producing a specific positive-sense mRNA, which is then translated to produce the viral protein (polymerase). Both the M RNA and S RNA have ambisense coding strategy and possess an intergenic region having AU-rich inverted repeat that participates in termination of mRNA transcription. These three mRNAs of bunya-viruses are not polyadenylated while their capping occurs by the cap-snatching mechanism. Thus, transcription of segmented negative-strand RNA viruses is initiated by cap-snatching.

Tenuiviruses (genus *Tenuivirus*, type tenuivirus is *Rice stripe virus* – RSV) (Ramirez and Haenni, 1994) are flexuous thread-like negative-stranded multipartite RNA viruses, are closely related to genus *Tospovirus*, contain 4 to 6 RNA segments with conserved 5`- and 3`-terminal sequences (of about twenty bases) that are complementary to each other so that viral RNA forms an intramolecular secondary structure of panhandle shape. Genomic RNA1 (of RSV) encodes polymerase while RNAs 2, 3, and 4 have ambisense coding strategy. The genome of *Rice grassy stunt tenuivirus* contains six RNA segments, all having ambisense coding strategy.

In ambisense strategy, viral RNA contains an ORF in its 5`-region while another ORF is present in 5`-region of viral complementary RNA. The two ORFs do not overlap but are separated by an intergenic region (as in RNA4 of tenuiviruses) that generally has the capacity to adopt a hairpin structure. The ambisense coding strategy could provide independent regulation for synthesis of the proteins whose ORFs are located on opposite strands of a given RNA segment. Such a regulation could be important for virus development, if supposed synthesis of one mRNA in certain cells is specifically reduced or even inhibited as opposed to synthesis of complementary mRNA. In contrast to initiation by RdRps, initiation of synthesis of viral mRNAs is primer-dependent and the primers are derived from the 5`-end of host mRNAs by cap-snatching mechanism. As a consequence, the viral RNAs become capped, unlike the original uncapped viral RNA and viral complementary RNA. Viruses employing ambisense mode have developed a unique technique for producing sgRNAs.

The occurrence *in vitro* of "re-snatching" of viral mRNAs (i.e., the use of viral mRNAs as cap donors) has been demonstrated for TSWV, family *Bunyaviridae* (van Knippenberg *et al.*, 2005). Mechanisms of ambisense and cap-snatching do not operate in positive-sense RNA plant viruses (Drugeon *et al.*, 1999).

C. Regulation of Gene Expression at Translational Level

Gene expression can be regulated at all the stages of translation – at translation initiation step, at translation elongation step, and at translation termination step (Maia *et al.*, 1996; Drugeon *et al.*, 1999).

1. Regulation of Gene Expression at Translation Initiation Level

BMV polymerase expression is regulated at the level of translation initiation; both eIF4A and the second DEAD-box protein (Ded1p), and perhaps still other helicases, may be required for regulating translation initiation of particular mRNAs (Noueiry *et al.*, 2000).

a. Factors Affecting Translation Initiation

Translation initiation (that is, initiation of protein synthesis) in virus-infected eukaryotic cells is influenced by both the *cis*- and *trans*-acting viral factors.

i. *cis*-Acting Factors

Nature of 5`- and 3`-ends of RNA influence translation initiation. The structural elements [5`-terminal cap and 3`-terminal poly(A) tail] of cell mRNAs, that together influence their translational efficiency, are also present in many viral RNAs. But in viral RNAs, that lack either the cap or the poly(A) tail or both, these structures are substituted by other elements that are capable of recruiting cell translational machinery. The VPg is present at 5`-end of plant viruses of several groups like *Comoviridae, Potyviridae,* some luteoviruses and sobemoviruses – all of which belong to the picornavirus-like supergroup. In some of these plant viruses, the translation initiation appears to occur on IRES located in 5`-leader sequence. The 5`-terminus of genomic RNA of necroviruses and viruses of subgroup I luteoviruses bear neither a cap structure nor a VPg (Maia *et al.*, 1996).

Nature of initiator (initiation codon) is an important *cis* factor. The normal translation initiation triplet in all plant viruses is the conventional AUG codon at 5`-end. However, translation in a few plant viruses can also start from other distinct translation initiation codons that are non-conventional (or non-AUG). These non-AUG codon(s) could compete with the downstream in-frame AUG codon leading to the production of proteins having different N-terminal sequences. Use of non-AUG translation initiation codons by plant viruses is the non-canonical translation initiation. Such exceptions are reported in three plant viruses.

One exception is the N-terminal extension (25-kDa) protein of the 19-kDa CP cistron of RNA2 of bipartite SBWMV. An in-frame CUG codon precedes/is located upstream of the AUG codon of the 19-kDa capsid protein ORF so that the 25-kDa CP-related protein is suggested to be synthesized at this CUG codon (Shirako, 1998). The 25-kDa CP protein has N-terminal extension of 40 amino acids of the 19-kDa protein. Both these proteins are formed during *in vitro* translation studies. The CUG codon is conserved in several SBWMV isolates suggesting that the 25-kDa protein does play some role in virus life cycle. The second exception is the non-canonical translation initiation codon of AUU that exists in 5`-UTR of TMV in certain chimeric systems (Schmitz *et al.*, 1996). In fact, translation initiation in tobamovirus 5`-UTR may take place at several AUU codons that are upstream of and in-frame with AUG of replicase gene. This non-canonical initiation was highly efficient, at least for the AUU codon at positions 63-65, both *in vitro* and *in vivo* (Schmitz *et al.*, 1996). The third exception is the ORF2 in genome of SMYEAV; this ORF is deficient in AUG (Jelkmann *et al.*, 1992) and is the 5`-proximal gene of the conserved TGB encoding MPs in potex-viruses. Translation of this gene of *Strawberry mild yellow edge-associated potexvirus* (SMYEAV) is suggested to be initiated at a CUG or an AUU codon to produce a protein having the size and sequence comparable to those of the other MPs (25-26 kDa) of potexviruses. The AUG codons are also absent in the 3`-proximal TGB ORFs in genomes of the related *Lily virus X* and *Shallot virus X*.

The nucleotide context surrounding the AUG initiator codon matters a lot. Efficiency of translation is enhanced if initiation codon is surrounded by a favourable context. This condition is operative in all mRNAs irrespective of whether the initiation codon is AUG or non-AUG. When A of AUG is regarded as +1, then a favourable context is constituted by a purine at –3 and a guanine at +4 (Gallie, 1993). The sequence AACCAUGG is the most favourable for translation initiation (Drugeon *et al.*, 1999). The individual genes of a plant virus genome have variable contexts (Lehto and Dawson, 1990; Agranovsky *et al.*, 1994) so that plant viruses can choose the initiation codon(s) (with variable context) for quantitatively regulating the expression of their genes. Thus context of plant viral genes apparently has significant role in translation. Despite this, the expression level of a particular gene is achieved by a delicate interplay of different mechanisms so that a non-optimal context of an initiating codon may not always lead to poor expression of that gene.

Nature of the leader sequence is critical. Translation efficiency can also depend upon whether the 5`-UTR is highly or poorly structured and, if structured, then the location of this structure (Kozak, 1991, 1994). A poorly structured leader sequence is AU-rich and is dependent on cap to a lesser degree for translation initiation as in AMV and TMV. Moreover, position of the canonical initiation triplet AUG in relation to a hairpin structure can also influence translation initiation (Kozak, 1986b, 1990; Miller *et al.*, 1995). If the leader sequence is highly structured, then internal translation initiation can occur as in picornaviruses.

(ii) *trans*-Acting Factors

These factors are either cell factors or viral *trans*-activators like the gene VI protein product of CaMV. Virus-encoded proteins that specifically increase translation of a downstream ORF in a bicistronic or polycistronic mRNA are the virus-encoded transactivators. They possibly do so by stimulating reinitiation of translation. Several caulimoviruses exhibit transactivation activity. The transactivator of CaMV is most thoroughly worked out (Fütterer and Hohn, 1996). Only one example of *trans*-activation in RNA plant viruses has been published. While RNA2 of RCNMV is not required for replication of RNA1 in protoplasts, a 34-nucleotide sequence in RNA2 of this virus is required for transcription of subgenomic RNA from RNA1 (Sit *et al.*, 1998). It is the 20-nucleotide stem-loop structure SL2 (of RNA2) that functions as a *trans*-activator for subgenomic RNA synthesis (from RNA1) (Guenther *et al.*, 2004; Tatsuta *et al.*, 2005). Cellular initiation factors are very important for translation efficiency that can be increased in those viral mRNAs in which one of the *cis*-acting motifs in RNA is absent or mutated – as in STNV RNA mentioned above. Browning (1996) has reviewed this extensively.

b. Unconventional Translation Initiation

i. Leaky Scanning

During leaky scanning, some *cis*-acting elements may contribute to the strength of initiation at a particular start codon in some plant viruses. These *cis*-acting elements are - occurrence of secondary structures, length of leader sequence, and distance that separates the two AUGs (Kozak, 1991, 1992, 1994). Some specific examples are mentioned below. Translation of PPV genomic RNA starts at the second initiation codon at nucleotide 147 because the context of first AUG, at position 36, is poor in comparison to the second AUG. Upon translation of SBMV RNA, a suboptimal context of AUG in 5`-most ORF 1 permitted a part of the scanning ribosomes to reach the downstream ORF 2 encoding the replicase polyprotein (Sivakumaran and Hacker, 1998). The protein p39 of PCV is translated from full-length RNA2 by a leaky scanning mechanism starting from the second AUG. Two small ORFs in the middle of TCV RNA genome, that encode proteins p8 and p9 essential for cell-to-cell virus transport, are translated from a 1.7 kb subgenomic RNA by leaky scanning (Qu and Morris, 2000).

In luteovirus genomes, ORF encoding 17-kDa MP (p17) is nested in a different reading frame, within the CP ORF, and both the genes are expressed from one and the same sgRNA (Dinesh-Kumar and Miller, 1993). The 5`-proximal AUG of CP gene of BYDV is inefficient as it is in a suboptimal context and is masked by the RNA secondary structure elements. Consequently, most of the scanning 40S ribosomal subunits bypass this codon and initiate translation at the downstream AUG for p17 gene. Leaky scanning mechanism also has a feedback element: once a ribosome has initiated translation at the second AUG, it provokes initiation at the first AUG perhaps by melting RNA secondary structure and by stalling the following ribosomal subunits (Dinesh-Kumar and Miller, 1993).

The leaky or unconventional scanning mechanism of translation initiation shows several variations but there are only three main types in plant viruses (Maia *et al.*, 1996; Drugeon *et al.*, 1999). The three main types are: autonomous translation initiation at upstream ORF leading to termination-reinitiation i.e., translation initiation at two sites on one ORF; translation initiation on overlapping ORFs in which an upstream ORF overlaps the downstream ORF (that is, one gene is nested in another gene); and translation initiation on consecutive ORFs in which case the upstream ORF is in-frame with a downstream ORF (Fütterer and Hohn, 1996). The first type is common in prokaryotes; it has also been reported in CaMV but not in any plant RNA virus (Maia *et al.*, 1996). The remaining two types of leaky scanning (overlapping and in-frame ORFs) are extremely common in and a great asset to plant viruses since, by these means, viruses are able to maximize their translation capacity and also translationally control expression of their genes (Geballe and Morris, 1994; Rohde *et al.*, 1994; Kozak, 1995; Agranovsky and Morozov, 1999). Besides the above three modes of leaky scanning methods, some plant viruses possess internal translation initiation for initiating protein synthesis.

Autonomous ORFs occur and their existence leads to termination-reinitiation of translation. This envisages the presence of two consecutive ORFs on mRNA so that termination of translation of 5`-proximal ORF is possibly followed by reinitiation of the second ORF. This type of leaky scanning is found unambiguously only in CaMV but not in any of the RNA plant viruses (Fütterer and Hohn, 1996; Drugeon *et al.*, 1999). The viral transactivator, which is the product of CaMV ORF VI, facilitates translation of the downstream ORFs.

Overlapping of different reading frames occurs in the second type of leaky scanning. Translation initiation at overlapping reading frames/nested gene is common in plant viruses. It is found in translation of second ORF in RNA2 of PCV (Herzog *et al.*, 1995). This RNA bears two genes – the 5`-proximal CP gene and a 39-kDa-protein gene. The synthesis of 39-kDa protein occurs by efficient translation from another AUG codon which overlaps the terminator codon of the CP ORF. The translation initiation of the 39-kDa protein from the second AUG triplet *in vitro* is through a context-dependent leaky scanning mode even though an unusually very long distance of 620 nucleotides separates the two AUGs. The stop codon in the 5`-proximal CP gene and the initiating codon of the downstream 39-kDa gene overlap by two bases ....**A<u>UGA</u>**..., in which stop codon is underlined while start codon is shown in bold. About one third of translating ribosomes bypassed the AUG in the CP gene (that is in a poor context) to initiate translation at the AUG of 39-kDa gene. Thus, the second AUG codon acts as an efficient start codon for translation of the 39-kDa protein by a context-dependent leaky scanning mechanism.

Another example of overlap is translational control of the two ORFs of subgenomic RNA1 of BYDV. This subgenomic RNA in BYDV carries two ORFs – the shorter 17-kDa protein ORF and the longer about 23-kDa CP ORF. The ORF of 17-kDa protein is nested within the virus CP ORF but is located in a different frame. The 5`-end CP AUG is placed in an unfavourable sequence context (U at position –3 and A at position +4) so that it does not exert any inhibitory effect on translation initiation of 5`-distal 17-kDa protein AUG. The result is the preferential synthesis of the distal 17-kDa protein *in vivo* which is produced in two-to-seven-fold higher concentration than that of the CP (Dinesh-Kumar and Miller, 1993; Dinesh Kumar *et al.*, 1992). An identical situation prevails in PLRV: the shorter ORF is nested within the longer ORF but is placed in a different frame and is also preferentially synthesized through the overlap mechanism.

The single-stranded *Satellite Tobacco mosaic virus* has two overlapping ORFs (an upstream ORF and the downstream CP ORF) and both of these produce proteins of the expected size *in vitro*. The translation of the downstream CP ORF could be initiated by leaky scanning because the context surrounding its AUG is more favourable than that of the context of AUG of the upstream ORF. Leaky scanning also explains synthesis of overlapping 95- and 105-kDa proteins that are encoded in a single ORF of M RNA of CPMV. In CNV, there is coordinated translation of two nested ORFs encoding the proteins involved in the virus movement (Rochon and Johnston, 1991).

The three genes of TGB occur in carla-, furo-, hordei- and potexviruses, are concerned with virus transport within the plant, and generally overlap on the genome. Two 3`-coterminal subgenomic mRNAs encode the three proteins of the TGB in BSMV RNA ß (Zhou and Jackson, 1996) and PVX RNA (Verchot *et al.*, 1998). A functionally monocistronic subgenomic RNA encodes the firs ORF of TGB while a functionally bicistronic subgenomic RNA translates the second and third ORFs of TGB. Initiation codon of second ORF of TGB is located in a less favourable context for translation initiation than that of the third ORF in each bicistronic subgenomic RNA of BSMV and PVX. It results in leaky scanning for translation of the third ORF (Zhou and Jackson, 1996).

Translation of p22-nested p19 gene of TBSV is influenced by nucleotide sequence context of surrounding p22 start codon. This context-dependent leaky scanning leads to increased translation from the downstream p19 start codon. This co-translational regulation of gene expression results in different levels of p19 and p22 protein accumulation (Scholthof *et al.*, 1995, 1999). Genes p23 and p8 of *Hibiscus chlorotic ringspot virus* are nested, respectively, in genes p28 and p9 (Liang *et al.*, 2002). Two nested ORFs 4 and 5 of *Carnation Italian ringspot virus* encode 21-kDa MP and a 19-kDa protein involved in symptom expression in infected plants.

In-frame translation initiation is the third type of leaky scanning in which the upstream ORF is in-frame with a downstream ORF (i.e., the genome contains one ORF with two in-frame initiation codons). Two proteins are synthesized in this way and these proteins are identical over the total length of the shorter protein. This translation strategy is not common but is employed by the following plant viruses: CPMV for synthesis of MP and BSMV for *in vivo* translation of second ORF of RNA ß. The two in-frame codons encode two proteins.

The second ORF of RNA β of BSMV codes for two proteins, namely βb and βb`. These proteins are translated *in vivo* from two alternative in-frame initiator AUGs. The ribosomes initiate translation of the shorter protein βb` *in vivo* by bypassing start codon of the longer βb protein because of the context-dependent leaky scanning (Petty and Jackson, 1990). Thus, translation of 5`-proximal gene in TGB of BSMV could be initiated from two alternative in-frame AUG codons giving two forms of the same protein both of which are functionally active. Moreover, leaky scanning mechanism also results in expression of partially overlapping middle and 3`-proximal genes of the BSMV TGB.

Studies on ATC66 strain of BMV revealed a correlation between the length of 5`-UTR of viral RNA and occurrence of in-frame translation initiation. Formation of full-length CP and an N-terminally truncated CP was correlated with effect of codon context and probably also due to reduction in length of 5`-UTR because of elimination of two adjacent adenine residues immediately upstream of the first initiation codon.

Potyviruses in general employ internal initiation of translation. Translation of PPV genomic RNA is initiated at second in-frame AUG located at position 147 even though the first AUG codon is located at position 36 but is placed in a poor context in comparison to the second AUG and is hence ignored during translation (Riechmann *et al.*, 1991). The 5`-UTR of TEV and PSbMV (both potyviruses) RNAs contain an AU-rich zone which increases translation *in vivo* and *in vitro* (Carrington and Freed, 1990; Nicolaisen *et al.*, 1992). An IRES has been proposed for TEV RNA by Carrington and Freed (1990).

ii. Internal Translation Initiation OR Cap-independent Translation Initiation

This topic has been dealt with in detail later.

iii. Shunting

When scanning ribosomes 'ignore' certain regions within the leader sequence of an mRNA, that may contain short ORFs, and jumps to a downstream longer ORF, the mechanism is called shunting (Drugeon *et al.*, 1999). By this, ribosomes are transferred from a donor to an acceptor site on mRNA without involving mRNA scanning between these two sites. Presence of extensive secondary structure within the leader sequence could favour shunting. Synthesis of the protein encoded by ORF1 of transcripts of CaMV and of *Rice tungro bacilliform virus* is suggested to be by this mechanism.

2. Regulation of Gene Expression at Translation Elongation Level (Frameshifting)

The spontaneous switching of ribosomes to another reading frame on a single mRNA, during the elongation stage of translation, is called frameshifting or ribosomal frameshifting. It seems to occur through movement of ribosome by one nucleotide on the mRNA; is the only method of gene expression and regulation at elongation level of translation; is a very frequent strategy employed by RNA viruses to nearly always produce the polymerase (Table 1); and is detected in plant viruses either on the basis of *in vitro* (and sometimes *in vivo*) translation studies accompanied or not by mutations in the signals required for frameshifting, or is postulated on theoretical considerations. Two proteins, called frame and transframe proteins, are always produced during the frameshifting mechanism with frame protein always being in greater abundance than transframe protein. These two proteins are similar from the N-terminal to frameshift point but are different from that point onwards and hence have different C-terminal regions. The chances of frameshifting are very low; being only 10^{-4} (Lindsey and Gallant, 1993). But plant viruses overcome this limitation by incorporating certain

TABLE 1

Positive-sense RNA plant viruses showing frameshifting mechanism[1]
(Based on Table 2 of Maia *et al.*, 1996)

Virus	RNA genome segments	RNA[2]	Shifty sequence[3]	Proteins[4]	DS Sequence[5]
Carlavirus					
PVM	1	1	AAAAUGA	CP/12kDa	
Closterovirus					
BYV	1	1	GGGUUUA	295kDa/48kDa	PK
CCSV	1	1	GUUUGAC	ORF1a/b[6]	PK
CTV	1	1	GCGUUCG	349kDa/57kDa	
LIYV	2	1	AAAG	217kDa/55kDa	
Dianthovirus					
CRSV	2	1	GGAUUUU	27kDa/54kDa	HP
RCNMV	2	1	GGAUUUU	27kDa/57kDa	HP
*SCNMV	2	1	GGAUUUU	27kDa/57kDa	
Enamovirus					
PEMV	2	1	GGGAAAC	84kDa/67kDa	PK
		2	GAUUUUU	33kDa/65kDa	
Luteovirus					
BYDV-PAV	1	1	GGGUUUU	39kDa/60kDa	PK
BYDV-RPV	1	1	GGGAAAC	71kDa/72kDa	HP
Polerovirus					
PLRV	1	1	UUUAAAU	70kDa/67kDa	HP/PK
Sobemovirus					
BWYV	1	1	GGGAAAC	66kDa/67kDa	PK
CfMV	1	1	UUUAAAC	64kDa/56kDa	HP

[1]Frameshift is -1 except for the closteroviruses where +1 is specified. [2]RNA - The RNA whose proteins undergo frameshifting. [3]Shifty sequence - The hepata-nucleotide sequence in which frame- shift occurs. [4]Proteins - The CP or the size of the two ORFs. [5]DS sequence is the nature of the downsteam sequence involved in frameshifting either a hairpin (HP) or a pseudoknot (PK). [6] indicates that the size of the corresponding ORFs is unknown. *SCNMV - *Sweet clover necrotic mosaic virus.*

specific signals, including the so-called 'slippery sequences', in viral RNA templates. This strategy increases frameshifting frequency many times, up to as much as 2-3 x 10^{-1}. Table 1 also shows the nature of slippery sequence, whether a pseudoknot structure is/or may be potentially required for efficient frameshift, and the proteins that take part in the frameshift event.

a. Types of Frameshifting

The ribosome movement on an mRNA can be either in 5`- or in 3`-direction causing a -1 or a +1 frameshift, respectively; frameshift is -1 frameshift if the ribosomes move in 5`-direction and is +1 frameshift when the ribosomes move in 3`-direction. The structural motifs and the mechanisms resulting in the two types of frameshifts are distinct from each other.

i. -1 Frameshifting

The -1 frameshift requires the following three signals to be present in RNA. First signal is a heptanucleotide sequence, also known as the 'slippery' sequence, where the frameshift takes place. It mostly harbours XXXYYYZ-type two homopolymeric triplets in which X is A, G, or U; Y is A, or U; and Z is A, C, or U. The two ribosome-bound tRNAs occurring in one reading frame (X.XXX.YYZ), upon reaching the shifty heptanucleotide sequence, shift by one nucleotide towards the 5`-direction (XXX.YYY.Z) and retain two out of three base-paired nucleotides with the mRNA. Second signal comprises of the nucleotide sequence downstream of the point of frameshift. This signal nearly always consists of a hairpin structure that follows the heptanucleotide and in many cases may form a pseudoknot structure with downstream RNA sequences. The third signal consists of a spacer region between the slippery sequence (heptanucleotide region) and the hairpin structure. This spacer has a variable length between 4 to 9 nucleotides. All the three signals are essential since removal or mutations in any of them abolished frameshift. The above observations are also stated as: the -1 frameshifting process envisions a 'simultaneous slippage' of two tRNAs bound to the slippery sequence with the consensus XXX YYYZ, to decode it as XXX YYY. Nearby presence of a pseudoknot enhances the switching of reading frame at slippery site in many plant viruses since this possibly forces recognition of out-of-frame triplet by hindering progress of ribosomes along the template.

The occurrence of leftward (or –1) frameshifting (–1 frameshift) is far more common than the +1 frameshift; is frequently found in plant RNA viruses of positive-polarity as in carla-, diantho-, enamo-, luteo-, and sobemoviruses; also occurs in the small double-stranded RNA viruses and retroviruses; and is always required for synthesis of polymerase in the above-mentioned five positive-sense RNA plant virus groups, with the frame protein nearly always containing the RdRp. Viral RNAs of plus-polarity mostly exhibit -1 type of frameshift.

Luteoviruses are the best examined of all the plant viruses. The overlap zone found between the two BYDV ORFs is always several hundred nucleotides in length except in PAV isolate in which it is only 13 nucleotides in length. An UAG codon that terminates the 39-kDa ORF borders the slippery sequence in BYDV-PAV. A possible slippery sequence followed by a simple pseudoknot is proposed to be implicated in frameshift in BYDV-RPV. The slippery sequence in BWYV is bordered by a UAA that terminates the 66-kDa ORF. The slippery sequence in PLRV strain G (a German isolate) (PLRV-G) is followed by a stable hairpin but does not have a pseudoknot. However, in PLRV-P (Polish isolate) frameshift is pseudoknot-dependent. Expression of polymerase gene of BYDV in a protoplast system takes place through -1 frameshift (Brault and Miller, 1992). The minor 99-kDa fusion protein of ORFs 1 and 2 of BYDV is also produced due to frameshifting (Allen *et al.*, 1999).

RNA1 of CRSV (Ryabov *et al.*, 1994) and RCNMV (Xiong *et al.*, 1993; Kim and Lommel, 1994) (both dianthoviruses) contain replication elements including a potential hairpin structure downstream of slippery sequence but no potential pseudoknot. These two dianthoviruses, and also the SCNMV dianthovirus, possess identical slippery sequences followed immediately by the triplet, which terminates the frame protein. The protein p88 (that possesses the RdRp motif) of RCNMV is translated by extension of protein p27 by ribosome frameshifting. The efficiency of −1 frameshifting, leading to the synthesis of a fusion replicase, in dianthoviruses and luteoviruses is as low as ~1% (Garcia *et al.*, 1993; Kim and Lommel, 1994).

The two genomic RNAs of PEMV contain ORFs that encode proteins, which potentially contain RdRp-related sequences that are seemingly produced by −1 frameshift. The slippery motif in RNA1 may be followed by a pseudoknot structure. The RNA2 produces, during *in vitro* translation, a protein of the size of the frameshift product. This transframe protein harbours helicase- and RdRp-like elements. AMV RNA3 encodes a 35-kDa polypeptide and a high molecular weight polypeptide in wheat germ extract and in infected tobacco leaves. The high molecular weight protein could be produced either by two +1 frameshifts or by readthrough of the UGA termination codon present at the end of the 35-kDa polypeptide cistron followed by −1 frameshift. This conclusion is based on nucleotide sequence of AMV RNA3. A shifty heptanucleotide sequence in BYV (a closterovirus) contains the first two nucleotides of the codon that terminates the frame ORF. It is followed by a hairpin structure, which may potentially give rise to a pseudoknot (Agranovsky *et al.*, 1994). Expression of BYV ORF1b, that encodes polymerase, occurs by translational frameshift resulting in ORF1a/1b fusion protein.

The overlapping region of ORFs 2 and 3 (encoding proteinase and POL domains, respectively), in CfMV sobemovirus genome, bears a −1 frameshifting signal that directs synthesis of the replicase-containing polyprotein with high efficiency, up to 30% (Mäkinen *et al.*, 1995; Ryabov *et al.*, 1996; Lucchesi *et al.*, 2000). The related sobemoviruses also contain this conserved signal. However, the proteinase and POL motifs in other sobemoviruses, except CfMV, are encoded in one continuous ORF. Thus, only one member of sobemovirus group makes use of the frameshifting mechanism for expressing POL. Since ORF for protein p23 overlaps with ORF for protein p28 of *Hibiscus chlorotic ringspot carmovirus*, the expression of p23 may be due to a ribosomal frameshift (Liang *et al.*, 2002).

Conceivably genes, other than the replicase gene of plant viruses, may also be expressed by ribosomal frameshifting. Thus, the nucleic acid-binding zinc finger 12-kDa protein of PVM is translated by internal initiation as well as by ribosomal frameshifting involving a shifty stop codon (Gramstat *et al.*, 1994). In fact, the gene array encoding CP and 12-kDa nucleic acid-binding PVM protein may be translated by

-1 frameshifting yielding CP-12kDa fusion protein; moreover, the 12-kDa gene may also be expressed by internal initiation. An unusual -1 frameshifting signal, including UGA codon in CP gene and the upstream sequence of four A residues, is involved in this process while the secondary structure elements had no importance (Gramstat *et al.*, 1994).

ii. +1 Frameshifting

The rightward (or +1) frameshifting requires distinct signals and is reported only in prokaryotes and yeast *Ty* retrotransposons. Possibly the only plant viruses postulated (but not yet demonstrated) to make use of this strategy are the closteroviruses; no other plant virus does so (Maia *et al.*, 1996). The need for *cis* elements for +1 frameshift appears to be less rigorous than that of the −1 frameshift. Normally, it requires only a slippery sequence and a rare or 'hungry' termination codon on ribosomal A site. The genome of BYV contains a shifty sequence followed by a hairpin structure that could potentially form a pseudoknot. A +1 frameshift has also been suggested for CTV (Karasev *et al.*, 1995) and LIYV (Klaasen *et al.*, 1995) but no downstream structure, that might suggest frameshift, or even a shifty sequence could be detected. CTV produces two polyproteins: the shorter polyprotein (that contains two papain-like proteases and methyltransferase-like and helicase-like domains) is thought to produce the larger polyprotein (having an additional RdRp domain) by a +1 frameshift. The ORF 1a (encoding MT and HEL) and ORF 1b (encoding POL), in genomes of BYV and other closteroviruses, are placed in an unusual +1 configuration. The BYV POL is suggested to be expressed as an ORF 1a/1b fusion product as a result of +1 ribosomal frameshifting (Agranovsky, 1996; Agranovsky *et al.*, 1994).

3. Regulation Gene Expression at Translation Termination (Readthrough)

This topic has been reviewed (ten Dam *et al.*, 1990; Maia *et al.*, 1996; Drugeon *et al.*, 1999). Presence of an in-frame termination codon in an mRNA normally directs translation termination. But termination can be suppressed by suppression of a UAG or UGA termination codon and this leaky stop codon (termination site), with consequent readthrough of the message leading to the formation of readthrough protein, is employed as a translational strategy for controlling gene expression. Thus, termination/stop codons can act as modulators of gene expression resulting in regulation of synthesis/production of different proteins in a virus and may also regulate the relative amounts of various polypeptides synthesized as well as their sequence of formation. Thus, a leaky termination site immediately after CP cistron may regulate the relative synthesis of CP and some non-structural proteins. Readthrough results in the synthesis of two proteins; the stopped protein and the readthrough protein. The two proteins are identical over the entire length of the stopped protein, which is always produced far in excess of the readthrough protein. The readthrough is very common in plant viruses. The most common suppressible termination codons found in plant viruses are UAG and UGA while UAA has been proposed for BSBV (Koenig and Loss, 1997) and *Beet virus Q*, which is a furo-like virus (Koenig *et al.*, 1998).

Readthrough has been demonstrated or postulated to be present in carmoviruses, enamoviruses, furoviruses, luteoviruses, necroviruses, tobamoviruses, tobraviruses, tombusviruses, and *Oat chlorotic stunt virus* (Table 2). The protein p191 of PCV is a C-terminal extension of p131 produced by readthrough of p131 stop codon on RNA1.

TABLE 2

Positive-sense RNA plant viruses showing readthrough mechanism
(Based on Table 3 of Maia *et al.*, 1996)

Virus	RNA genome Segments	RNA[1]	Term codon[2]	Proteins[3]
Carmovirus				
CarMV	1	1	UAG, UAG	27kDa/86kDa/98kDa
TCV	1	1	UAG	28kDa/88kDa
CCFV	1	1	UAG	28kDa/87kDa
MCMV	1	1	UAG	50kDa/111kDa
			UGA	9kDa/33kDa
MNSV	1	1	UAG	29kDa/89kDa
			UAG	7kDa/14kDa
Enamovirus				
PEMV	2	1	UGA	CP/55kDa
Furovirus				
BNYVV	4	2	UAG	CP/75kDa
SBWMV	2	1	UGA	150kDa/209kDa
		2	UGA	
PCV	2	1	UGA	130kDa/191kDa
Luteovirus				
BYDV-PAV	1	1	UAG	CP/72kDa
Luteovirus				
BWYV	1	1	UAG	CP/74kDa
SDV	1	1	UAG	CP/80kDa
Necrovirus				
TNV	1	1	UAG	23kDa/82kDa
Polerovirus				
PLRV	1	1	UAG	CP/80kDa
Tobamovirus				
TMV	1	1	UAG	126kDa/183kDa
Tobravirus				
TRV	2	1	UGA	134kDa/194kDa
PEBV	2	1	UGA	141kDa/201kDa
Tombusvirus				
AMCV	1	1	UAG	33kDa/92kDa
CNV	1	1	UAG	33kDa/92kDa
CyRSV	1	1	UAG	33kDa/92kDa
TBSV	1	1	UAG	33kDa/92kDa
Miscellaneous				
OCSV	1	1	UAG	23kDa/84kDa

1RNA - The RNA whose proteins undergoes readthrough. [2]Term codon - The nature of thesuppressible codon. [3]Proteins - The stopped/readthrough protein; the stopped protein is iIndicated by CP or by the size when it is not CP and the readthrough protein is indicated by the total size. The readthrough protein always encompasses the polymerase when the stopped protein is not CP.

Protein p81 of HCRSV is a readthrough product of p28 gene (Liang *et al.*, 2002). The ORF2 of *Carnation Italian ringspot tombusvirus* encodes a 95-kDa polymerase by readthrough of ORF1 stop codon and this readthrough occurs at a frequency of about 10% in plant cells (Rubino *et al.*, 1995). CarMV is unique since its 98-kDa protein is produced through a double readthrough event, which also occurs in TCV (both are carmoviruses). SBWMV is possibly the only plant virus to have two readthrough signals, one signal being located on each of the genomic RNA segments (Agranovsky and Morozov, 1999). Thus two suppressible UGA codons are present in this furovirus; one each in RNA1 (that codes for polymerase) and RNA2 (that encodes capsid/fusion protein) (Shirako and Wilson, 1993). Readthrough product of TMV RNA increases at low, instead of high, viral RNA concentration. This shows that such a readthrough product is favoured during earlier stages of infection *in vivo*. The expression of βd protein, translated by the middle gene of BSMV TGB, occurs in two forms (as 14- and 23-kDa proteins). Of these, the larger form is a readthrough product of the amber stop codon (Zhou and Jackson, 1996). It is possible that other virus non-replicative and non-CP proteins could also be expressed in an identical way.

Maia *et al.* (1996) list the genera and plant viruses whose genome has been shown or postulated to resort to readthrough, the nature of the suppressible termination codon, the designation of the stopped and readthrough proteins, and other information including the mRNA motifs that favour readthrough.

a. Readthrough Mechanism

The readthrough mechanism has two aspects: motifs (present in mRNA) that favour readthrough in *cis* and the type of suppressor tRNAs involved.

i. mRNA Motifs Favouring Readthrough *in cis*

Besides the termination codon, some other *cis* features of mRNA are also required for efficient readthrough in a few plant viruses. For example, nature of two downstream TMV codons, following the UAG codon, is crucial for efficient readthrough *in vivo* and *in vitro*. On the other hand, situation in BYDV is completely different than that of TMV. For efficient suppression of leaky stop codon in CP gene, a composite signal consisting of two 3`-motifs situated downstream of suppressible UAG stop codon are essential for readthrough to occur *in vitro* and *in vivo*. The proximal element, out of the two, is placed 5 to 15 nucleotides downstream of the UAG, is made up of 16 repeats of CCN NNN (where N is any nucleotide), and readthrough is dramatically reduced upon deletion of 5`-proximal third of these 16 repeats. The distal motif/element is located 697 to 758 nucleotides downstream of the UAG stop codon in readthrough ORF and is thus nearly 60 nucleotides long. Readthrough is greatly reduced upon deletions within the distal element. It is not yet known if the proximal and the distal regions interact with each other through long-distance base-pairing and if the region downstream of suppressible termination codon is able to adopt a secondary structure like a hairpin or a pseudoknot for efficient readthrough. The distal element in luteoviruses is well conserved – suggesting that the distal element in these viruses may serve a role similar to that in BYDV.

ii. Presence of Suppressor tRNAs in Host Plants

The termination codon can be suppressed in the presence of an appropriate nonsense suppressor aminoacyl-tRNA that recognizes the nonsense (or termination) codon present at the end of a particular cistron. The tRNAs, that act as suppressor tRNAs and misread the termination codons in *trans*, have been isolated from various plant tissues, like the leaves of tobacco and wheat and from lupin (Maia *et al.*, 1996). Two Tyr-accepting tRNAs (tRNA^{Tyr}), isolated from tobacco leaves, suppressed the suppressible UAG codon present in TMV RNA. The suppression, accomplished during *in vitro* studies, was suggested to operate *in vivo* as well. Similarly, the host plant also contains a Trp- and a Cys-accepting tRNA for UGA codon present in TRV RNA. No suppressor tRNA has been isolated that specifically recognizes suppressible UAA codons although the tobacco host has been suggested to possibly contain a tRNA which can suppress UAA codon contained in TMV RNA. Two suppressor tRNA^{Tyr} species, present in the cytoplasm of tobacco cells, offset pairing of ribosomes with the UGA stop codon, thus facilitating readthrough (Zerfass and Beier, 1992a). The pseudouridine residue in the tRNA^{Tyr} GΨA anticodon is a host determinant of efficient UAG suppression.

Leaky stop codons of UAG, or of other types, terminate the 5`-proximal cistron in genomic RNA of AMV RNA1, potyvirus RNA, TMV RNA, TRV RNA1, and TYMV RNA. The readthrough of these codons is favoured in certain situations such as the addition of the corresponding (host-encoded) suppressor tRNAs or the presence of polyamines and their detection by a tRNA^{Tyr} isolated from tobacco leaves and from wheat germ or wheat leaves.

An identical but somewhat different readthrough mechanism synthesizes the closely related MT-HEL-POL-containing replicases encoded in RNA 1 of TRV and furoviruses (SBWMV and PCV) (Shirako and Wilson, 1993; Herzog *et al.*, 1994). In case of TRV, two tRNA^{Trp} species possessing methylated CmCA anticodons (that naturally occur in tobacco cytoplasm and chloroplasts) are necessary for suppression (Zerfass and Beier, 1992b). The POL expression in carmoviruses and tombusviruses also depends on readthrough of a leaky stop codon that is located far upstream of POL domain although replicases of these viruses lack MT and HEL while their POL domain is distantly related to that of the alpha-like virus superfamily (Koonin and Dolja, 1993).

b. Functions of Readthrough Mechanism

Readthrough performs several functions in plant viruses. Its most common function is the production of RdRp. An equally important function is that presence of such leaky stop codons regulates protein synthesis and makes it possible for synthesis of two or three overlapping polypeptide chains from a single cistron of an RNA. Thus, termination codons can act as modulators of gene expression. Still another function is that readthrough may induce the fusion of CP to the protein encoded by the readthrough domain to produce a fusion protein. In some plant viruses, the readthrough protein is required for encapsidation or for transmission of virus by its insect vector.

Synthesis of RdRp often occurs through readthrough mechanism. Several positive-sense RNA plant virus groups (carmo-, furo-, necro-, tobamo-, tobra-, and tombusviruses) employ readthrough of a leaky terminating codon for synthesis of RdRp. A long ORF of 87- to 111-kDa contains a UAG triplet in the monopartite RNA genome of carmo-, necro-, and tombusviruses and of MCMV and OCSV. The readthrough protein bears the conserved GDD motif characteristic of RdRps. A UAG codon present in PCV RNA1 produces the polymerase (Herzog *et al.*, 1994). The UAG stop codon separates the replicase gene in TMV genome into portions coding for MT plus HEL and POL, respectively. During translation of TMV genome, 95% of the translating ribosomes terminate at the stop codon while only 5% of them suppress it leading to the synthesis of the major 120-kDa and the minor 183-kDa products, respectively.

Fusion of CP with other proteins takes place through readthrough mechanism, In BNYVV, PEMV, PCV and luteoviruses, readthrough leads to fusion of CP to the protein produced by the readthrough domain. BNYVV has a UAG codon in its RNA2 that produces the fusion protein. In SBWMV (Shirako and Wilson, 1993) and *Beet virus Q* (BVQ) (Koenig *et al.*, 1998), both genomic RNAs possess a readthrough domain: RNA1 contains a readthrough domain encoding RdRp motifs and RNA2 carries the CP readthrough domain. The stop codons in CP gene in furoviruses, except of PCV, leads to formation of C-terminally extended forms of CP (Shirako and Wilson, 1993). Luteoviruses also employ an identical mechanism to express conventional and extended CP. A few copies of the aberrant CP are incorporated into the isometric virus particles of luteoviruses (Filichkin *et al.*, 1994) as well as into the rod-like particles of furoviruses (Haeberle *et al.*, 1994).

The protruding readthrough domains of proteins formed by fusion of CP with other proteins, in many of the above-mentioned plant viruses, ensure virus transmission by their respective vectors.

4. Regulation of Gene Expression at Post-Translational Level (Proteolytic Processing of Polyproteins)

The completion of gene translation does not stop the act of gene expression in many plant viruses. The newly synthesized protein chains may have to be processed by a post-translational mechanism by virus-encoded enzymes (proteinases) into mature structural and nonstructural proteins, called protein maturation. The polycistronic mRNA, in such cases, contains closed internal cistrons and leaky stop codons. The only translation initiation site available to ribosomes is the one adjacent to the 5`-terminus. The ribosomes then move along the chain from the 5`- to 3`-end, read and translate the various genes, and, in the absence of effective stop codons, synthesize a single large precursor polyprotein in which various polypeptides (gene products) encoded by various genes are strung along in a linear fashion. The readthrough polyprotein is later proteolytically processed by proteolytic enzymes (proteinases) to generate the various virus-specific functional proteins.

Proteolytic processing is a very important step of gene expression in many plant viruses, is regulated temporally through complex posttranslational cleavage, produces mature as well as intermediated-sized viral proteins, allows the virus to produce two or more proteins from one mRNA, and can occur in *cis* and/or in *trans* and cotranslationally or post-translationally. Only the positive-strand RNA viruses produce a virus-encoded proteolytic enzyme that cleaves a polyprotein precursor. Viruses of picorna-like superfamily (including plant como-, nepo-, and potyviruses) express their genome as a polyprotein precursor by contiguous translation of a single large ORF. There is no transcriptional control in such systems. The proteinases have been treated separately in the chapter 'PROTEINASES'.

IV. CAP-INDEPENDENT TRANSLATION

Many plant viruses lack one or both of the essential traditional regulatory elements [5` cap structure and/or poly(A) tail], yet they translate efficiently since they have evolved sequences that functionally replace 5`-cap and/or poly(A) tail. Thus these plant viruses have adopted alternative strategies of translation and for ensuring adequate levels of their encoded translation products (Gallie, 1996; Bailey-Serres, 1999; Kean, 2003). Cap-independent translation (CIT) is one of these strategies. Removal of cap structure from virion RNAs or from *in vitro* transcripts of genomic RNA reduces or abolishes infectivity in many viruses that normally have a cap structure [along with the presence of poly(A) tail] and the only known mRNAs naturally lacking a cap are found in viruses. Consequently, such viral mRNAs must undergo translation by a cap-independent mechanism for ensuring adequate levels of their encoded translation products (Gallie, 1996; Bailey-Serres, 1999). Two types of mechanisms ensure this.

One mechanism involves the presence of a special *cis*-acting nucleotide sequence motif called translational enhancer (3`-TE) or cap-independent translational enhancer (3`-CITE or CITE) domain located in the 3`-UTR of viruses. Absence of poly(A) tail in genomic RNAs of several viruses is compensated by the presence of CITE. Thus, plant virus RNAs in this case are naturally uncapped and nonpolyadenylated and carry out CIT through elements (CITEs) in their 3`-UTRs. In contrast to internal ribosome entry mechanism, the interaction between 3`-CITE and 5`-UTR appears to facilitate ribosome scanning from the 5`-end (Guo *et al.*, 2001), like normal capped mRNA (Kozak, 1991). For example, the BYDV CITE confers cap-independent translation by recruiting translation factors (E. Allen, personal communication, quoted in Shen and Miller, 2004) and interacting with the 5`-UTR via long-distance base pairing to ensure circularization of viral RNAs and transfer initiation factors to 5`-end (Guo *et al.*, 2001).

Second is the presence internal ribosome entry site(s) (IRES) in uncapped 5`-UTRs while 3`-poly(A) tail is present (potyviruses). Uncapped genomic RNAs are likely to possess a VPg at 5`-terminus and a long 5`-UTR that contains an IRES (Jackson and Kaminski, 1995). IRES are located upstream of the translated ORF, facilitate CIT, and recruit ribosomes to the mRNA through a variety of mechanisms (Pestova *et al.*, 1996; De Gregorio *et al.*, 1999). The mRNAs of *Potyviridae* family (including *Tobacco etch potyvirus*), like picornaviral RNAs, is polyadenylated and uncapped; its 5`-UTR is a functional alternative for a cap and shows moderate IRES activity (Carrington and Freed, 1990; Gallie *et al.*, 1995; Gallie, 2001). The translation initiation starts at IRES in 5`-UTR by promoting interaction between 5`-leader and 3`-poly(A) tail (Gallie, 1995). An interaction between eIF4G (bound to IRES) and PABP [bound to poly(A) tail] of potyviruses appears to be required for efficient translation of viral RNAs (Gallie, 2001). Like picornaviruses, the uncapped RNAs of *Hepatitis C virus*, and *Discistroviridae* family contain IRES (Hellen and Sarnow, 2001; Vagner *et al.*, 2001; Domier and McCoppin, 2003).

Shen and Miller (2004) have defined two classes of CITEs: the BTEs and non-BTEs. BTEs are BYDV-like CITE in its 3`-UTR, are present in at least three plant virus genera (*Dianthovirus*, *Luteovirus*, and *Necrovirus*) that can stimulate translation of uncapped mRNA; are located in the 3`-UTR; contain a highly conserved 18-nucleotide sequence (CGGAUCCUGGGAAACAGG); and possess similar secondary structure. However, notable differences exist between the CITEs of these three genera. Base pairing between 3`- and 5`-UTRs appears to necessary for luteovirus and necrovirus CITEs. The BTEs are present in all known or probable members of the *Luteovirus* genus but not in the two other genera of *Luteoviridae* family, and exist in only two of several genera of the *Tombusviridae*.

TNV-D RNA has a BYDV-like CITE in its 3`-UTR. The TNV-D CITE is at most 105-nucleotide-long, allows translation of uncapped viral mRNA efficiently, contains the 18-nucleotide sequence (CGGAUCCUGGGAAACAGG) that is well conserved in members of *Luteovirus* and *Necrovirus* genera, and depends on the viral 5`-UTR to function. Shen and Miller (2004) suggest, but do not prove, that long-distance base pairing between the TNV 3`-CITE and the 5`-UTR is required for cap-independent translation as in BYDV UTRs. Moreover, additional sequences of the 3`-UTR are required for cap-independent translation *in vivo*; these additional 3`-UTR portions may facilitate binding of translation initiation factor(s) and/or other *trans*-acting factor(s) to TNV-D CITE, enhance the interaction between UTRs, increase the stability of RNA, or all of the above. A double stem-loop structure at the extreme 3`-end of TNV-D functionally mimics a poly(A) tail (R. Shen, unpublished results). BYDV RNA also contains a "poly(A) mimic" function downstream of the 3`-CITE (Guo *et al.*, 2000). These elements are not needed *in vitro*.

Non-BTEs are non-BYDV-like 3`-CITEs, are present in other viruses of the *Tombusviridae* family and include TCV and *Hibiscus chlorotic ringspot virus* (HCRV) in the *Carmovirus* genus, TBSV in the *Tombusvirus* genus, and STNV. The RNAs of none of these genera harbour a 3`-UTR that bears sequence or structural similarity to a BTE. The 3`-element of STNV RNA stimulates cap-independent translation as efficiently as BTEs *in vitro* and *in vivo*, is about the same size, and is located at the 5`-end of a long 3`-UTR (Danthinne *et al.*, 1993) but its sequence and structure are entirely different from those of BTEs (Danthinne *et al.*, 1993; Meulewaeter *et al.*, 1998a, 1998b; Timmer *et al.*, 1993; van Lipzig *et al.*, 2002; Wang *et al.*, 1997). Cap-independent translation mediated by the TBSV CITE occurs only *in vivo* (Wu and White, 1999) and this sequence overlaps *cis*-acting replication elements. A 180-nucleotide sequence, including the essential hexanucleotide (GGGCAG) in 3`-UTR, of HCRV confers cap-independent translation (Koh *et al.*, 2002). The TCV CITE is located at the 5`-end of the 255-nucleotide 3`-UTR, is 150-nucleotide long and requires the 5`-UTR for optimal translation efficiency (Qu and Morris, 2000).

The 3`-UTR is thought to facilitate translation initiation at 5`-end by acting as a switch for preventing collisions between ribosomes and replicase on BYDV RNA (Barry and Miller, 2002). This speculation was based on the studies that showed that poliovirus negative-strand RNA synthesis was completely blocked by translating ribosomes (Barton *et al.*, 1998; Gamarnik and Andino, 1998), showing that RNA synthesis requires prior elimination of ribosomes from viral genome (Barton *et al.*, 1998).

A. Cap-Independent Translation Enhancer at 3` -End of Viral RNA

Cap-independent translational enhancers (3`-CITE) are located at 3`-UTRs of nonpolyadenylated positive-sense viral RNAs of several plant viruses, requirement of a 3`-CITE for efficient translation initiation of viral 5` ORF/at 5`-proximal AUG (of uncapped full-length RNA of BYDV-PAV and STNV) is related to the lack of either a cap or a VPg at 5`-end of genome, confer efficient translation initiation at 5`-proximal AUG on uncapped mRNA and so facilitate CIT, are the factors that enable viruses to conduct as well as enhance translation of viral RNAs, decrease concentration of eIF4F needed for maximum translation of capped RNA while destruction of translation capacity of STNV CITE increases amount of eIF4F required for translation, are necessary for viral replication since removal of CITE drastically decreases translation, and occur in viruses that lack a cap-like structure or a VPg at 5`-end but are also found in plant viruses possessing a cap-like structure, and appear to mimic the role of cap-like structure in translation initiation. Therefore, plant viruses having CITEs may show cap-independent translation as in luteo-, and potyviruses, or cap-dependent translation as in AMV, and potexviruses (Gallie, 1996).

Plant viruses showing CIT (because of the presence of 3`-CITE) belong to the carmo-like supergroup and include members (of genera *Carmovirus*, *Dianthovirus*, *Machlomovirus*, *Necrovirus*, and *Tombusvirus*) of *Tombusviridae* family, genus *Luteovirus* of *Luteoviridae* family, and STNV. The specific members of *Tombusviridae* showing CIT are: TCV (Qu and Morris, 2000), and HCRV (Koh *et al.*, 2002) of genus *Carmovirus*; TBSV (Wu and White, 1999) of genus *Tombusvirus*; RCNMV (Mizumoto *et al.*, 2003) of genus *Dianthovirus*; TNV (Shen and Miller, 2004) of genus *Necrovirus*; while specific member of genus *Luteovirus* is BYDV (Wang *et al.*, 1997; Guo *et al.*, 2000, 2001; van Lipzig *et al.*, 2002).

The 5`-ends of genomic RNAs of BYDV, STNV and TEV are uncapped and show CIT. The BYDV, STNV, and TBSV RNAs lack a poly(A) tail and possess the 3`-CITE in their respective 3`-UTRs. The 3`-CITEs of these three viruses do not have sequence homology (Meulewaeter *et al.*, 1998b; Allen *et al.*, 1999). The 3`-CITE of BYDV-PAV is located at 3`-UTR, is 5-kb downstream from 5`-end of genomic mRNA, is situated between nucleotides 4810 and 4920 (Allen *et al.*, 1999) and in intergenic region between ORFs 5 and 6 (the two distal ORFs on viral genome), is 109-nucleotide long, is required for replication of uncapped full-length BYDV genomic RNA and enables cap-independent translation of 5`-proximal ORF *in vivo* and *in vitro* (Wang *et al.*, 1997), and is required for translation of virion RNA from infected plants as well as uncapped genomic transcripts (Allen *et al.*, 1999). The 3`-CITE of **STNV** is located within 3`-UTR just downstream of the ORF, is 120-nucleotide long, forms an extended stem loop structure, *in vitro* translation of viral RNA requires both 5`-leader and a 3`-CITE, and possible base-pairing interactions between 5`-leader and 3`-enhancer have been proposed on the basis of complementarity between nucleotides within these regions. Similarly, a 167 nucleotide-long 3`-proximal sequence in **TBSV** RNA genome is an important part of a large 3`-terminal genome motif that acts as a 3`-CITE and facilitates CIT (Oster *et al.*, 1998; Wu and White, 1999). The TBSV 3`-CITE is distinct from 3`-CITEs of other positive-strand RNA plant viruses in several ways: significant increase of translation induced by 3`-CITEs (of BYDV and STNV) is not detectable *in vitro* in wheat germ extracts in case of TBSV 3`-CITE; 3`-CITE overlaps replication sequences of TBSV; position of 3`-CITE of TBSV is relatively more 3`-proximal than that of other plant viruses; and TBSV 3`-CITE does not show extensive sequence identity with any other known 3`-proximal enhancers of other plant viruses. Thus, a novel viral 3`-proximal CITE occurs in TBSV and, by implication, in the genus *Tombusvirus*.

TNV RNA has no cap and no 3`-poly(A) tail or tRNA-like structure (Meulewaeter, 1999) yet it translates efficiently. Shen and Miller (2004) report that a CITE exists in 3`-UTR of TNV strain D, is about 105-nucleotide-long, permits efficient translation of uncapped viral mRNA, and resembles the CITE of BYDV luteovirus. The CITEs of both these plant viruses (TNV and BYDV) resemble each other in following characters: function *in vitro* and *in vivo*; require additional sequence for cap-independent translation *in vivo*; have a similar secondary structure; share an 18-nucleotide sequence - the conserved sequence CGGAUCCUGGGAAACAGG that is well conserved in plant viruses of *Luteovirus* and *Necrovirus* genera; are inactivated by a four-base duplication in this conserved sequence; can function in the 5`-UTR; and when located in its natural 3`-location, the proper functioning of CITE depends upon the viral 5`-UTR since CITE can form long-distance base pairing with the conserved viral 5`-UTR that is required for CIT as in BYDV UTRs. However, TNV CITE differs from BYDV CITE by having only three helical domains instead of four. Similar structures exist in all members of the *Necrovirus* genus of *Tombusviridae* family, except *Satellite tobacco necrosis virus,* which harbours a different 3`-cap-independent translation domain. The presence of the BYDV-like CITE in select genera of different families indicates that phylogenetic distribution of CITEs does not follow standard viral taxonomic relationships. Shen and Miller (2004) proposed a new class of cap-independent CITE called BYDV-like CITE. (BTE). Shen and Miller (2004) suggest that RNAs of all necroviruses, but not STNV, initiate protein synthesis by highly similar CITE-mediated mechanisms, such as of BYDV RNA. Similar structures are present in the *Dianthovirus* genus (Mizumoto *et al.*, 2003) of *Tombusviridae* and *Luteovirus* genus of the *Luteoviridae* but are absent in other genera of these families. Because BYDV-like CITE is not limited to BYDV, Shen and Miller (2004) propose that it represents a new class of CITE called the BYDV-like CITE (BTE).

The 3` -UTR of TCV RNA plays a significant role in enhancement of translation (Qu and Morris, 2000) in line with the presence of CITE elements in 3` -UTRs of BYDV and STNV (Danthinne *et al.*, 1993; Wang *et al.*, 1997) although no similar sequence could be found in the 155-nncleotide CITE region of TCV.

1. Functions of 3` -CITE Domain

Different functions have been attributed to 3`-CITE. It is essential for translation of those genomic RNAs that naturally lack a cap structure but not for their capped genomic transcripts (Danthinne *et al.*, 1993; Timmer *et al.*, 1993; Wang *et al.*, 1997; Allen *et al.*, 1999) and efficiency of their translation is as good as that of the capped mRNAs (Wang *et al.*, 1997; Meulewaeter *et al.*, 1998a). The 3`-CITE motifs have been suggested to do so by playing a role in efficient translation initiation at 5`- proximal AUG on uncapped mRNAs (Gallie, 1991; Wang and Miller, 1995; Sachs *et al.*, 1997; Allen *et al.*, 1999).

It is possible that 3`-CITE increases the rate of replication initiation (Gallie, 1996). If so, this will require some type of communication between 3`-CITE and 5`-terminal of the RNA. Such an interaction between 5`- and 3`-ends does occur in eukaryotic cellular mRNAs and these interactions are mediated by a protein that binds 5`-cap and 3` poly(A) tail (Gamarink and Andino, 1998). Similarly, 3`-CITE of both BYDV and STNV requires the respective homologous viral 5`-terminus  for optimal replication activity indicating that a 5`-3` interaction does exist in these two plant viruses (Wang *et al.*, 1997; Meulewaeter *et al.*, 1998b). Similarly, the 3`-CITE may potentially regulate initiation of TBSV genome replication (Wu and White, 1999). Apart from the 3`-167 nucleotide long sequence, other parts of viral TBSV RNA could contain additional but less efficient CITE elements. Thus, the TBSV 3`-CITE extends into adjacent areas of viral genome that contain *cis*-acting replication motifs (Ray and White, 1999). This creates a localized arrangement of replication and translation sequences in 3`-region, which, in turn, could provide a possible mechanism that controls the use of viral genome as a template for replication or as an mRNA for translation (Wu and White, 1999).  Conceivably, 3`-CITE possesses properties like those of the poly(A) tail (Gallie, 1996; Wu and White, 1999). However, the precise mechanism(s) by which 3`-CITE motifs act in the above-mentioned functions is not known. Why these translation-stimulating elements are located at 3`-ends of RNAs of BYDV, STNV and other plant viruses and how they enhance recognition of 5`-proximal AUG in absence of cap? These questions also still need to be solved.

B. Cap-independent Translation Enhancer at 5` -End of Viral RNA

The 5` -leader of AMV RNA4 causes translation enhancement by a mechanism that leads to a reduced requirement for the translation initiation factors. The 5` -UTR of TMV RNA contains a CAA-rich CITE element (Ω leader), which dramatically enhances translation of downstream genes in both animal and plant cells (Gallie, 1996) and is made up of 68 nucleotides of which the key part is eight direct CAA repeats (Gallie *et al.*, 1987a, 1987b). Apart from TMV, presence of CA-rich motifs within 5` -UTR is a common feature in many plant viral RNAs like TEV, PVX, TCV, and TNV or CA-rich motifs are found in proximity of 5` -UTRs of both genomic and sgRNAs in TBSV, and in 5` -UTRs of CP sgRNAs of *Carnation mottle virus* and *Cardamine chlorotic fleck carmovirus* and may be performing identical function of enhancement of translation (Qu and Morris, 2000). The CA-rich motifs in 5` -UTR of TCV could possibly contribute to translational enhancement in both genomic and sgRNAs (like the CITE elements in 3` -UTRs of BYDV and STNV) but these motifs were essential for efficient translation of either message (Qu and Morris, 2000). Thus, different mechanisms may be operational for enhancement of translation in different plant viruses.

Both plant and viral mRNAs contain another type of regulatory elements in 5`-UTR – the small upstream ORFs (uORFs) which repress translation initiation at 'proper' AUG in the downstream gene (Geballe and Morris, 1994; Lovett and Rogers, 1996). An artificially created single uORF in 5`-UTR of bromovirus RNA3 reduced translation of MP gene *in vivo* up to 18 times (de Jong *et al.*, 1997). A natural uORF in RNAγ of BSMV strain CV17 induced great reduction in synthesis of γa protein (POL) and also caused host-specific inhibition of long-distance virus movement. The uORFs also exist in BSMV RNAα, and in genomic RNAs of some members of family *Bromoviridae*. LIYV (family *Closteroviridae*) is a bipartite virus; RNA 1 (its 5`-gene encodes replicase) contains a 97-nuleotide long 5`-UTR, which does not contain any uORFs while RNA2 (which possibly acts as mRNA for 5-kDa and 62-kDa non-structural proteins) contains a 326-nucleotide long 5`-UTR harbouring five uORFs (Klaassen *et al.*, 1995). However, the regulatory role of these uORFs in translation of these plant virus RNAs is still to be worked out.

C. Internal Ribosome Entry Sites (IRES)

The 5`-UTRs of viruses showing CIT (as picornaviruses) are well conserved, highly structured, long and varying in length from 610 to 1500 nucleotides, and contain multiple nonconserved AUG triplets located upstream of the initiation codon that should act as a strong barrier to scanning ribosomes. Jang *et al.* (1988) and Pelletier and Sonenberg (1988) discovered that animal picornaviral (EMCV and *Poliovirus,* respectively) mRNAs undergo translation by a mechanism that enables ribosomes to initiate translation effectively on highly structured areas located within the 5`-UTR. This internal translation initiation occurs because ribosomes and *trans*-acting factors bind to a complex secondary/tertiary structure, called the internal ribosome entry site

(IRES), within the 5`-UTR and is hence CIT initiation and was first reported in animal picornaviruses. Viral IRES elements have different structures in various viruses. Predictions of the folding patterns of these RNA motifs have been based primarily on sequence and biochemical analyses. Biochemical confirmation of the models has been achieved only for IRES of *Hepatitis C virus*, which adopts an open structure consisting of two major stems.

Such IRES elements also exist in genomes of plant picorna-like viruses (potyviruses and comoviruses) even though their 5`-leader sequences are not as long and as complex as those of animal picornaviruses (Fütterer and Hohn, 1996). Moreover, unlike animal picornaviruses, application of IRES by plant viruses seems to be optional. Thus, translation of comovirus M-RNA could be initiated either by internal initiation or by leaky scanning. The leaky scanning also operates in potyvirus TEV and could be due to the existence of a possible IRES within the last 63 nucleotides of its UTR. By now, an increasing number of plant viruses have been shown to initiate protein synthesis by a CIT mechanism involving IRES.

1. Translation Mechanism

The picornavirus IRES recruits 40S ribosomal subunits and various canonical and non-canonical translation initiation factors to promote translation in absence of 5`-cap structure. Because of the very different structures of IRES, 40S ribosomes can be recruited by at least three different mechanisms in animal viruses. First, interaction of eIF4G/eIF4A with IRES in EMCV is aided by canonical eIFs 2, 3, and 4B and is necessary to recruit 40S subunits to IRES. Second, the hepatitis C virus IRES can bind eIF-free 40S subunits without helicase eIF4A or other eIFs, yet recruit the ternary complex to position the initiator tRNA into ribosomal P site. Third, the cricket paralysis virus-like IRES elements can assemble 80S ribosomes without any eIF, including the tRNA. The above division indicates that structural features in IRES elements dictate the requirement for certain eIFs, such as eIF4A, that help in the recruitment of 40S ribosomes. Moreover, in all the three mechanisms, ribosomal recruitment involves non-canonical interactions with canonical components of translation apparatus. This shows that most IRES are likely to be refractory to mechanisms controlling 5`-end-dependent translation via dephosphorylation of eIR4E or sequestration of eIF4E by eIF4E-binding proteins. Svitkin *et al.* (2001) found that the initiation factors bind to an internal ribosome entry site in the 5`-UTR of picornaviral RNA and mediate interaction with PABP. No equivalent information is available for plant viruses.

All picornavirus IRESs contain an AUG triplet at their 3`-border, 25 nucleotides downstream from the beginning of pyrimidine-rich tract. In EMCV virus and *Foot-and-mouth disease virus*, this AUG codon is the initiation codon and ribosomes bind directly to it without scanning. The *cis*-signals that mediate direct access of 40S ribosomal subunits to initiation codon is an elaborate system of primary and secondary RNA structure elements in picornavirus genomic leaders. Translation by an internal ribosome entry mechanism does not require an intact eIF4F complex. The IRES-mediated translation requires *cis*-acting sequences within viral RNA that mediated interaction with *trans*-acting host factors.

2. Advantages of IRES

A process different from the usual canonical eukaryotic cap-dependent translation initiates translation in many RNA viruses having IRES. Many of the regulatory mechanisms, that control recruitment of most mRNAs to translation apparatus, are not applicable to IRES-containing mRNAs. One very important attribute of the IRES-mediated translation is that the IRES-mediated translation essentially eliminates competition for host factors from cap-dependent mRNA translation, favouring translation of viral mRNA. Few, if any definite, similarities in sequence, size, or structure, exist between different IRES elements except those between families of related viruses – implying that no universal mechanism occurs for the internal ribosome entry. Much is known about the sequence and factors responsible for recruiting ribosomal subunits to IRES of picornavirus, *Hepatitis C virus* and *Cricket paralysis virus* (Hellen and Sarnow, 2001).

D. Plant Viruses Showing Cap-Independent Translation

The mRNAs of many plant viruses lack a cap-like structure or poly(A) tail or both and yet they efficiently compete with host mRNAs for translational machinery. This is so because they have developed alternative strategies for translation regulation, the CIT mechanism based on internal ribosome entry of ribosomes, which is mediated by internal sequences in 5`-UTR, called IRES. Plant viruses showing CIT belong to several viral families like carmo-, como-, diantho-, luteo-, poty-, tobamo- and tombusviruses (Gallie *et al.*, 1995; Ivanov *et al.*, 1997). The internal translation initiation is reported to definitely occur in PVX (Hefferon *et al.*, 1997), even though it bears both the 5`-cap structure and 3`-poly(A). A bicistronic PVX subgenomic mRNA contains two non-overlapping ORFs; the 5`-proximal ORF codes for an 8-kDa protein and the distal ORF encodes CP *in vitro* as well as *in vivo* studies. The region upstream of the CP ORF possibly controls initiation of CP synthesis through an internal ribosome entry method (Hefferon *et al.*, 1997). The TMV CP was not expressed from the I_2 sgRNA in certain TMV strains but was expressed from this sgRNA of a crucifer-infecting TMV (cr-TMV) via an IRES upstream of the CP gene (Ivanov *et al.*, 1997). Canto *et al.* (2004) reported that TMV contains a sixth ORF (ORF6) that is a determinant of viral pathogenicity in *Nicotiana benthamiana* and potentially encodes a 4.8-kDa protein; this protein is perhaps expressed from an sgRNA associated with expression of 30-kDa MP and this possibly happens through IRES.

1. Plant Viruses Lacking Both 5`-Cap Structure and 3`-Poly(A) Tail

Plant viruses of genera *Carmovirus*, *Dianthovirus* (RCNMV – Mizumoto *et al.*, 2002, 2003), *Luteovirus*, *Necrovirus* and *Tombusvirus* contain neither cap structure nor poly(A) tail. Genomic RNAs of BYDV, RCNMV, STNV, TCV, TBSV (Wu and White, 1999; Qu and Morris, 2000), and *Carnation Italian ringspot tombusvirus* harbour sequences in 3`-UTR which confer cap-independent translation (Wang *et al.*, 1997; Wu and White, 1997; Meulewaeter *et al.*, 1998b; Mizumoto *et al.*, 2003; White and Nagy, 2004). The translation of TCV carmovirus RNA is co-ordinately enhanced by both 3`- and 5`-UTRs in a cap-independent manner (Qu and Morris, 2000). The 5`-end of RNA genome of STNV has neither a cap nor a VPg while the 3`-end does not have a poly(A) or a tRNA-like structure; instead, the 3`-UTR contains more than 600 nucleotides. Efficient translation of this virus needs 5`-leader sequence as well as a translational enhancer situated within 3`-UTR, immediately downstream of the ORF. The translational enhancer forms an extended stem-loop structure. Nucleotide complementarity exists between the 5`-leader and 3`-enhancer so that possibly base-pairing occurs between these regions. Elimination of 3`-enhancer reduces translation efficiency, which is restored by capping the RNA. The amount of eIF4F required for translation increases if the translation capacity of the enhancer is destroyed. Interaction between 5`-leader and a 3`-region of viral RNA genome stimulates translation of STNV RNA (Danthinne *et al.*, 1993; Timmer *et al.*, 1993). So, CIT has been proposed for STNV (Meulewaeter *et al.*, 1998a, 1998b).

The 5`-end of RNA genome of BYDV-PAV, like STNV RNA, has also neither a cap nor a VPg (Shams-baksh and Symons, 1997; Allen *et al.*, 1999) while the 3`-end does not have a poly(A) or a tRNA-like structure. Translation of BYDV RNA requires base pairing between a stem-loop in 5`-UTR and a stem-loop in a 100-nucleotide translation element in 3`-UTR, presumably to deliver translation factors and/or ribo somes to 5`-end (Guo *et al.*, 2001). The 3`-UTR of genomic BYDV RNA possesses 105-nucleotide CIT elements (CITE105), which facilitates CIT *in vitro* (Wang *et al.*, 1997; Guo *et al.*, 2000). CITE105 folds into a cruciform structure with three stem loops (SLI, SLII, SLIII) (Guo *et al.*, 2000). SLI stem-loop is conserved in 3`-UTRs of genomic RNAs of luteoviruses, STNV, and in RNA1 of dianthoviruses (Wang *et al.*, 1997). The 3`-CITE sequence also functions *in vivo* to significantly enhance translation initiation in BYDV-PAV in a cap-independent manner (Wang *et al.*, 1997; Allen *et al.*, 1999). The 3`-translational enhancer, located in intergenic region between ORFs 5 and 6 of BYDV-PAV RNA, is essential for translation of 5`-proximal ORF both *in vivo* and *in vitro* (Wang *et al.*, 1997). The enhancer decreases the concentration of eIF4F required for optimum translation of uncapped RNA. It thus seems to mimic the role of cap structure during translation initiation. It is possible, as in STNV, that the need for a 3`-translational enhancer for efficiently translating the 5` ORF of BYDV-PAV has some relationship with the absence of a cap structure or of a VPg at 5`-end of the genome (Allen *et al.*, 1999). Interaction between 5` leader and a 3` region of viral RNA genome stimulates translation of BYDV-PAV luteovirus (Shams-baksh and Symons, 1997; Allen *et al.*, 1999).

2. Plant Viruses Lacking 5`-Cap Structure but 3`-Poly(A) Tail Present

RNAs of potyviruses, including TEV, lack a 5`-cap structure but have a covalently-linked VPg at this end. In TEV RNA, the CIT initiation and efficient translation starts on IRES in 5`-UTR due to synergistic interaction between 143-nucleotide 5`-leader

(VPg) and poly(A) tail at 3`-end (Gallie *et al.*, 1995), similar to the interaction occurring between a cap and a poly(A) tail. Interaction between eIF4G (bound to IRES) and PABP [bound to poly(A) tail] of potyviruses (TEV) appears to be required for efficient translation of viral RNAs (Gallie, 2001).

Niepel and Gallie (1999) identified two specific elements within 5`-TEV leader that direct CIT. These two functionally distinct cap-independent regulatory elements (CIREs) (CIRE 1 and CIRE 2) are present within the central region of 143-base 5`-leader. The CIRE 1 is 5`-proximal, is present within a 39-nucleotide region but is more 5`-end independent than is the 5`-distal CIRE 2 element which is present within a 53-nucleotide region. Both elements are essential for CIT and so can be considered as components of a single regulatory motif. The combined effect of CIRE 1 and CIRE 2 is approximately multiplicative and is controlled by poly(A) tail. The two CIREs function optimally when placed in a 5`-proximal location but can also promote translation when they are located internally within a dicistronic mRNA. The last observation suggests that each CIRE can promote the translation of an internally located second cistron provided the viral leader sequence is present in intercistronic area of a dicistronic mRNA. In short, TEV 5`-leader contains two elements (CIRE 1 and CIRE 2) that have regulatory function and promote CIT but can also promote translation when located internally within a dicistronic mRNA.

3. Plant Viruses Lacking 3`-Poly(A) Tail but 5`-Cap Structure Present

Genomic RNAs of many plant viruses lack a poly(A) tail but have 5`-cap structure and elements in 5`- and/or 3`-UTR are believed to compensate for absence of a poly(A) tail (Gallie and Walbot, 1990; Gallie and Kobayashi, 1994; Gallie, 1996; Wells *et al.*, 1998; Qu and Morris, 2000; Lipzig *et al.*, 2001; Matsuda and Dreher, 2004). For example, functioning of STNV translational enhancer domain correlates with affinity with two wheat germ factors (Lipzig *et al.*, 2001). The 5`-end of RNA genomes of BMV, TMV, and TYMV is capped while the 3`-UTR is non-polyadenylated and has a complex 3`-UTR consisting of a pseudoknot domain and tRNA-like structure; the pseudoknot domain appears to substitute functionally for poly(A) tail to promote 5`-3` interaction and enhance CIT (Leathers *et al.*, 1993; Matsuda and Dreher, 2004). Thus, these terminal sequences stimulate viral RNA translation co-operatively similar to the cap and poly(A) in a cellular mRNA (Gallie, 1996).

The specific elements in TMV RNA were the CAA repeats in 5`-leader (Gallie and Walbot, 1992) and 3`-distal upstream pseudoknot motif that precedes the tRNA-like structure while no such activity was detected for tRNA-like structure alone; these elements (3`-pseudoknot and 5`-leader) bring the 5`- and 3`- ends into proximity (Gallie and Walbot, 1990; Gallie, 1996). A region within the TMV pseudoknot domain appears to functionally replace poly(A); this region in co-operation with cap structure at 5`-end lead to increased translation efficiency. Such a situation also obtains in BMV (Gallie and Kobayshi, 1994).

On the other hand, the pseudoknot upstream of tRNA-like structure in TYMV plays only a minor role in translation enhancement (Gallie and Kobayshi, 1994). However, Matsuda and Dreher (2004) identified a translational enhance in 3` -UTR of TYMV RNA and uncovered a novel type of translational enhancer and a new role for a plant viral tRNA-like structure. They found that the 3` -terminal 109 nucleotides comprising the tRNA-like structure and an upstream pseudoknot act as a translational enhancer that act in synergy with 5` -cap to enhance translation; that maximum translation enhancement requires that RNA be capable of aminoacylation but either the native valine or engineered methionine is acceptable; that mutations decreasing the affinity for translation elongation factor eEF1A (but also diminishing aminoacylation efficiency) strongly decrease translational enhancement, suggesting that eEF1A is mechanistically involved; and that the pseudoknot seems to act as an important, though nonspecific, spacer element ensuring proper presentation of a functional tRNA-like structure. Matsuda *et al.* (2004) concluded that translation from both AUG^{69} and AUG^{206} is strongly cap dependent; and that TYMV 3` -UTR is a general enhancer, with a principle mode of action being synergy with the cap (Matsuda and Dreher, 2004). Barends *et al.* (2003) have expressed a contrary view on the basis of their *in vitro* studies. They proposed that the 3` -UTR of TYMV communicates specifically with one or two AUG initiation codons to promote translation by a novel cap-independent mechanism that avoids the involvement of initiator $tRNA^{Met}$; and that the 3` -UTR is proposed to direct cap-independent translation initiation specifically at AUG^{206} and not at AUG^{69}.

The internal translation initiation on a bicistronic mRNA of a crucifer-infecting tobamovirus (cr-TMV) is reported to definitely occur (Ivanov *et al.*, 1997). The I_2 subgenomic RNA of crucifer-infecting TMV contains two genes, one gene is for virus MP and the second is for CP synthesis. The ORF for CP overlaps the 5`-proximal ORF for MP so that the CP is produced through internal translation initiation (Ivan *et al.*, 1997).

The RNAs 1 and 2 of RCNMV dianthovirus are capped at 5`-end but lack a 3`-poly(A) tail. However, uncapped *in vitro* transcripts of RCNMV RNA1 and RNA2 show high infectivity, implying that this virus possesses CIT mechanism of protein synthesis. The middle region (between nucleotides 3596 to 3732) in 3`-UTR, designated 3`-translation element of *Dianthovirus* RNA1 (3`TE-DR1), is the cap-independent element in RCNMV RNA1 and it plays an important role in CIT (Mizumoto *et al.*, 2003). A stem-loop is present within and is one of the functional structures in 3`TE-DR1. This stem-loop structure is conserved among members of genera *Dianthovirus* and *Luteovirus*.

E. Conclusions

Cap-independent translation occurs in plant viruses of various genera belonging to families *Luteoviridae, Potexviridae, Potyviridae, Tombusviridae,* and *Tubiviridae* of the three superfamilies (picorna-like viruses, carmo-like viruses, and alpha-like viruses). Plant viruses showing CIT belong to several viral families like luteo-, potex-, poty-, and tombusviruses. Out of these, BYDV, STNV, and TBSV have been studied extensively (Danthinne *et al.*, 1993; Timmer *et al.*, 1993; Rubino *et al.*, 1995; Wang and Miller, 1995; Wang, *et al.*, 1997; Allen *et al.*, 1999; Wu and White, 1999). All these viruses lack a 5`-cap structure but still are able to translate the encoded messages in their RNAs because of cap-independent translation.

The results of studies on RCNMV and BYDV suggest that conserved stem-loop structures can function in CIT initiation of distinct viruses belonging to different genera. This conserved loop is made up of GGAAA and might be involved in recruiting one or several proteins involved in CIT (Guo *et al.*, 2000). Despite similarities in CIT elements in 3` UTRs of RCNMV RNA1 and BYDV genomic RNA, there are differences in translation mechanism between the two viruses. For CIT of reporter mRNA, both 5`-UTR and 3`-UTR of genomic RNA of BYDV are required and direct base pairing between 5`-UTR and 3`-UTR is essential (Guo *et al.*, 2001) and possibly results in circularization of viral RNA. In contrast, the 5`-UTR of RCNMV RNA1 was not required for CIT of reporter mRNA conferred by viral 3`-UTR. Therefore, it is possible that RCNMV RNA1 forms circular structures by mechanisms that are different from that of BYDV. Possibly the sequence elements and/or secondary structures in RCNMV 3`-TE-DR1 could be involved in formation of a structure that includes a closed loop to facilitate CIT. This stem-loop structure is likely to be a *cis*-acting element required for viral RNA replication. Mizumoto *et al.*, (2003) suggest that the 3` stem-loop structure of RCNMV RNA1 is not important for translation of RdRp but could be important for replication of RNA1 by functioning as binding site for RdRp. Translation initiation of a downstream ORF, which in most of the cases is CP ORF, in a bicistronic subgenomic mRNA (as in PVX and cr-TMV) is by IRES elements resulting from internal ribosome entry. However, IRES is not conserved in the related potexviruses and tobamoviruses.

CIT is an effective viral strategy to compete aggressively with host mRNAs for translational machinery. By this means, virus is not subjected to cellular cap-mediated translational control mechanisms and cap-mediated translation is shut down by the virus in infected cells. It also eliminates the need to encode a capping enzyme or to acquire a cap through cellular enzymes. Thus, CIT continues.

Down-regulation of polymerase, with respect to synthesis of other replication proteins in many monocomponent RNA plant viruses, occurs by translational readthrough and frameshift mechanisms (Noueiry *et al.*, 2000). Regulation of ratio of 2a protein to 1a-like protein in TMV occurs by readthrough of leaky termination codons. Translation of BMV polymerase may also be down regulated with respect to other proteins. The 1a protein of BMV could play a similar role in regulating use of viral genome for its use as a template for replication or as an mRNA for translation since this protein stabilizes genomic RNA3 as well as down regulates its translation (Janda and Ahlquist, 1998; Sullivan and Ahlquist, 1999). Noueiry *et al.* (2000) suggested that this selective activation of mRNA by DEAD-box helicases may also be true of other mRNAs and may be a translational regulation mechanism by other members of DEAD-box RNA helicase family. Viral protein of RCNMV and *Cocksfoot mottle sobemovirus* are involved in viral replication and in down-regulation of other essential genes (Kim and Lommel, 1994; Lucchesi *et al.*, 2000).

The 5`-UTRs in plant mRNAs can affect translation both directly and indirectly. The direct influence depends on the length and degree of secondary structure of 5`-UTR while indirect affect is due to the presence of specific elements constituting the binding sites for protein factors (Gallie, 1996).

Shen and Miller (2004) suggest that the following proposed mechanism for BYDV RNA replication also applies to all viruses in *Tombusviridae* family. After translation of viral replicase (p82 in TNV) by the 3`-CITE, the replicase begins copying the viral RNA from the 3`-end and disrupts base pairing (or other form of interactions) of the 3`-CITE with the 5`-UTR as it proceeds in the 5`-direction on the viral template RNA. This shuts-off translation initiation at the 5`-end while the replicase is still in the 3`-UTR and clears the upstream ORFs of ribosomes by the time the replicase reaches them. This permits efficient replication of viral RNA, unimpeded by ribosomes. Later, after enough RNA accumulation has taken place, some RNA molecules become free of replicase and form the long-distance interactions that facilitate translation, and the cycle would begin again. This model shows the way a positive-strand virus RNA may achieve the potentially conflicting roles of acting as genome and mRNA.

V. REFERENCES

Agranovsky, A. A. 1996. Principles of molecular organization, expression and evolution of closteroviruses: Over the barriers. *Adv. Virus Res.* 47: 119-158.

Agranovsky, A. A., and Morozov, S. 1999. Gene expression in positive-strand RNA viruses. In: Mandahar, C. L. (Ed.). Molecular Biology of Plant Viruses. Kluwer Academic Publ. Boston/Dordrecht / London. p. 99-119.

Agranovsky, A. A., Koonin, E. V., Boyko, V. P., Maiss, E., Frötschl, R., Lunina, N. A., and Atabekov, J. G. 1994. *Beet yellows closterovirus*: Complete genome structure and identification of a leader papain-like thiol protease. *Virology* 198: 311-324.

Ahlquist, P. 1992. Bromovirus RNA replication and transcription. *Cur. Opin. Ge. Dev.* 2: 71-76.

Ahlquist, P., Noueiry, A. O., Lee, W.-M., Kushner, D. B., and Dye, B. T. 2003. Host factors in positive-strand RNA virus genome replication. *J. Virol.* 77: 8181-8186.

Allen, E., Wang, S., and Miller, W. A. 1999. Barley yellow dwarf virus RNA requires a cap-independent translation sequence because it lacks a 5`-cap. *Virology* 253: 139-144.

Atabekov, J. G., and Morozov, S. Yu. 1979. Translation of plant virus messenger RNAs. *Adv. Virus Res.* 25: 1-91.

Bailey-Serres, J. 1999. Selective translation of cytoplasmic mRNAs in plants. *Trends Plant Sci.* 4: 142-148.

Barends, S., Bink, H. H., van den Worm, S. H., Pleij, C. W. A., and Kraal, B. 2003. Entrapping ribosomes for viral translation: tRNA mimicry as a molecular Trojan horse. *Cell* 112: 123-129.

Barends, S., Rudinger-Thirion, J., Florentz, C., Giegé, R., Pleij, C. W. A., and Kraal, B. 2004. tRNA-like structure regulates translation of *Brome mosaic virus* RNA. *J. Virol.* 78: 4003-4010.

Barry, J. K., and Miller, W. A. 2002. A programmed −1 ribosomal frameshift that requires base-pairing across four kilobases suggests a novel mechanism for controlling ribosome and replicase traffic on a viral RNA. *Proc. Natl. Acad. Sci. USA* 99: 11133-11138.

Barton, D. J., Morasco, B. J., and Flanegan, J. B. 1999. Translating ribosomes inhibit poliovirus negative-strand RNA synthesis. *J. Virol.* 73: 10104-10112.

Bol, J. F. 2003. *Alfalfa mosaic virus*: Coat protein-dependent initiation of infection. *Mol. Plant Pathol.* 4: 1-8.

Brault, V., and Miller, W. A. 1992. Translational frameshifting mediated by a viral sequence in plant cells. *Proc. Natl. Acad. Sci. USA* 89: 2262-2266.

Browning, K. S. 1996. The plant translational apparatus. *Plant Mol. Biol.* 32: 107-144.

Bushell, M., and Sarnow, P. 2002. Hijacking the translation apparatus by RNA viruses. *J. Cell Biol.* 158: 395-399.

Canto, T., MacFarlane, S. A., and Palukaitis, P. 2004. ORF6 of *Tobacco mosaic virus* is a determinant of viral pathogenicity in *Nicotiana benthamiana*. *J. Gen. Virol.* 85: 3123-3133.

Carr, J. P. 2004. *Tobacco mosaic virus. Annu. Plant Rev.* 11: 27-67.

Carrington, J. C., and Freed, D. D. 1990. Cap-independent enhancement of translation by a plant potyvirus 5`-nontranslated region. *J. Virol.* 64: 1590-1597.

Danthinne, X., Seurinck, J., Meulewaeter, F., van Montagu, M., and Cornelissen, M. 1993. The 3`-untranslational region of satellite tobacco necrosis virus RNA stimulates translation *in vitro*. *Mol. Cell. Biol.* 13: 3340-3349.

De Gregorio, E., Preiss, T., and Hentze, M. W. 1999. Translation driven by an eIF4G core domain *in vivo*. *EMBO J.* 18: 4865-4874.

de Jong, W., Mise, K., Chu, A., and Ahlquist, P. 1997. Effects of coat protein mutations and reduced movement protein expression on infection spread by *Cowpea chlorotic mottle virus* and its hybrid derivatives. *Virology* 232: 167-173.

Dinesh-Kumar, S. P., and Miller, W. A. 1993. Control of start codon choice on a plant viral RNA encoding overlapping genes. *Plant Cell* 5: 679-692.

Dinesh-Kumar, S. P., Brault, V., and Miller, W. A. 1992. Precise mapping and *in vitro* translation of a trifunctional subgenomic RNA of *Barley yellow dwarf virus*. *Virology* 187: 711-722.

Domier, L. L., and McCoppin, N. K. 2003. *In vivo* activity of *Rhopalosiphum padi* virus internal ribosome entry sites. *J. Gen. Virol.* 84: 415-419.

Donahue, T. F. 2000. Genetic approaches to translation initiation in *Saccharomyces cerevisiae*. In: Sonenberg, N., Hershey, J. B. W., and Mathews, M. B. (Eds.). Translational Control of Gene Expression. Cold Spring Harbor Lab. Press, Cold Spring Harbor, USA. p. 487-502.

Drugeon, G., Urcuqui-Inchima, S., Milner, M., Kadaré, G., Valle, R. P. C., Voyatzakis, A., Haenni, A.-L., and Schirawski, J. 1999. The strategies of plant virus gene expression: Models of economy. *Plant Sci.* 148: 77-88.

Filichkin, S. A., Lister, R. M., McGarth, P. F., and Young, M. J. 1994. *In vivo* expression and mutational analysis of the barley yellow dwarf virus readthrough gene. *Virology* 205: 290-299.

Fraenkel-Conrat, H., Salvato, M., and Hirth, L. 1977. The translation of large viral RNAs. *Virology* 11: 201-235.

French, R., and Ahlquist, P. 1987. Intercistronic as well as terminal sequences are required for efficient amplification of brome mosaic virus RNA3. *J. Virol.* 6: 1457-1465.

Furuchi, Y., and Shatkin, A. J. 2000. Viral and cellular mRNA capping: Past and prospects. *Adv. Virus Res.* 55: 135-184.

Fütterer, J., and Hohn, T. 1996. Translation in plants – rules and exceptions. *Plant Mol. Biol.* 32: 159-189.

Fütterer, J., Potrykus, I., Valles-Brau, M. P., Dasgupta, I., Hull, R., and Hohn, T. 1994. Splicing in a plant pararetrovirus. *Virology* 198: 663-670.

Gale, M., Jr., Tan, S.-L., and Katze, M. G. 2000. Translational control of viral gene expression in eukaryotes. *Microbiol. Mol. Biol. Rev.* 64: 239-280.

Gallie, D. R. 1991. The cap and poly(A) tail function synergistically to regulate mRNA translational efficiency. *Genes Dev.* 5: 2108-2116.

Gallie, D. R. 1993. Post-translational regulation of gene expression in plants. *Annu. Rev. Plant Physiol. Plant Mol. Biol.* 44: 77-105.

Gallie, D. R. 1996. Translational control of cellular and viral mRNAs. *Plant Mol. Biol.* 32: 145-158.

Gallie, D. R. 1998. A tail of two termini: A functional interaction between the termini of an mRNA is a prerequisite for efficient translation initiation. *Gene* 216: 1-11.

Gallie, D. R. 2001. Cap-independent translation conferred by the 5`-leader of *Tobacco etch virus* is eukaryotic initiation factor 4G dependent. *J. Virol.* 75: 12141-12152.

Gallie, D. R. 2002. The 5`-leader of *Tobacco mosaic virus* promotes translation though enhanced recruitment of eIF4F. *Nucl. Acids Res.* 30: 3401-3411.

Gallie, D. R., and Browning, K. S. 2001. eIF4G functionally differs from eIFiso4G in promoting internal initiation, cap-independent translation and translation of structured mRNAs. *J. Biol. Chem.* 276: 36951-36960.

Gallie, D. R., and Kobayashi, M. 1994. The role of the 3`-untranslated region in non-polyadenylated plant viral mRNAs in regulating translational efficiency. *Gene* 142: 159-165.

Gallie, D. R., and Walbot, V. 1990. RNA pseudoknot domain of *Tobacco mosaic virus* can functionally substitute for a poly(A) tail in plant and animal cells. *Genes Dev.* 4: 1149-1157.

Gallie, D. R., and Walbot, V. 1992. Identification of the motifs within the *Tobacco mosaic virus* 5`-leader responsible for enhancing translation. *Nucl. Acids Res.* 17: 4631-4638.

Gallie, D. R., Sleat, D. E., Watts, J. W., Turner, P. C., and Wilson, T. M. A. 1987a. The 5`-leader sequence of tobacco mosaic virus RNA enhances the expression of foreign gene transcripts *in vitro* and *in vivo*. *Nucl. Acids Res.* 15: 3257-3273.

Gallie, D. R., Sleat, D. E., Watts, J. W., Turner, P. C., and Wilson, T. M. A. 1987b. A comparison of eukaryotic viral 5`-leader sequences as enhancers of mRNA expression *in vivo*. *Nucl. Acids Res.* 15: 8693-8711.

Gallie, D. R., Tanguay, R. L., and Leathers, V. 1995. The tobacco etch viral 5`-leader and poly(A) tail are functionally synergistic regulators of translation. *Gene* 165: 233-238.

Gamarink, A. V., and Andino, R. 1998. Switch from translation to RNA replication in a positive-stranded RNA virus. *Genes Dev.* 12: 2293-2304.

Garcia, A., van Duin, J., and Pleij, C. W. A. 1993. Differential response to frameshift signals in eukaryotic and prokaryotic translational systems. *Nucl. Acids Res.* 21: 401-406.

Geballe, A. P., and Morris, D. R. 1994. Initiation codons within 5`-leader of mRNAs as regulators of translation. *Trends Biochem. Sci.* 19: 159-164.

Gorbalenya, A. E., and Koonin, E. V. 1993. Helicases: Amino acid sequence comparisons and structure-function relationships. *Curr. Opin. Cell Biol.* 3: 419-429.

Gramstat, A., Prüfer, D., and Rohde, W. 1994. The nucleic acid-binding zinc finger protein of *Potato virus M* is translated by internal initiation as well as by ribosomal frameshifting involving a shifty stop codon and a novel mechanism of P-site slippage. *Nucl. Acids Res.* 22: 3911-3917.

Guenther, R. H., Sit, T. L., Gracz, H. S., Dolan, M. A., Townsend, H.-L., Liu, G., Newman, W. H., Agris, P. F., and Lommel, S. A. 2004. Structural characterization of an intermolecular RNA-RNA interaction involved in the transcription regulation element of a bipartite plant virus. *Nucl. Acids Res.* 32: 2819-2828.

Guo, L., Allen, E. M., and Miller, W. A. 2000. Structure and function of a cap-independent translation element that functions in either the 3`- or the 5`-untranslated region. *RNA* 6: 1808-1820.

Guo, L., Allen, E. M., and Miller, W. A. 2001. Base-pairing between untranslated regions facilitates translation of uncapped, nonpolyadenylated virus RNA. *Mol. Cell* 7: 1103-1109.

Gustafson, G. D., Armour, S. L., Gamboa, G. C., Burgett, S. G., and Shepherd, J. W. 1989. Nucleotide sequence of barley stripe mosaic virus RNA α: RNA α encodes a single polypeptide with homology to corresponding proteins from other viruses. *Virology* 170: 370-377.

Haeberle, A., Stussi-Garaud, C., Schmitt, C., Garaud, J.-C., Richards, K., Guilley, E., and Jonard, G. 1994. Detection by immunogold labeling of p75 readthrough protein near an extremity of *Beet necrotic yellow vein virus* particles. *Arch. Virol.* 134: 195-203.

Hefferon, K. L., Khalilian, H., Xu, H., and AbouHaidar, M. G. 1997. Expression of the coat protein of *Potato virus X* from a dicistronic mRNA in transgenic potato plants. *J. Gen. Virol.* 78: 3051-3059.

Hellen, C. U., and Sarnow, P. 2001. Internal ribosome entry sites in eukaryotic mRNA molecules. *Genes Dev.* 15: 1593-1612.

Héricourt, F., Jupin, I., and Haenni, A.-L. 1999. Genome of RNA viruses. In: Mandahar, C. L. (Ed.). Molecular Biology of Plant Viruses. Kluwer Academic Publishers, Boston / Dordrecht / London. p. 1-28.

Hershey, J. W. B., and Merrick, W. C. 2000. The pathway and mechanism of initiation of protein synthesis. In: Sonenberg, N., Hershey, J. W. B., and Mathews, M. B. (Eds.). Translational Control of Gene Expression. Cold Spring Harbor Laboratory Press, Cold Spring Harbor, New York, USA. p. 33-88.

Herzog, E., Guilley, H., Manohar, S. K., Dollet, M., Richards K., Fritsch, C., and Jonard, G. 1994. Complete nucleotide sequence of peanut clump virus RNA1 and relationship with other fungus-transmitted rod-shaped viruses. *J. Gen. Virol.* 75: 3147-3155.

Herzog, E., Guilley, H., and Fritsch, C. 1995. Translation of the second gene of peanut clump virus RNA2 occurs by leaky scanning *in vitro*. *Virology* 208: 215-225.

Ivanov, P. A., Karpova, O. V., Skulachev, M. V., Tomashevskaya, O. L., Rodionova, N. P., Dorokhov, Yu. L., and Atabekov, J. G. 1997. A tobamovirus genome that contains an internal ribosome entry site functional *in vitro*. *Virology* 232: 32-43.

Jackson, R. J. 2000. A comparative view of initiation site selection mechanisms. In: Sonenberg, N., Hershey, J. W. B., and Mathews, M. B. (Eds.). Translational Control of Gene Expression. Cold Spring Harbor Lab. Press, Cold Spring Harbor, USA. p. 127-183.

Jackson, R. J., and Kaminski, A. 1995. Internal initiation of translation in eukaryotes. The picornavirus paradigm and beyond. *RNA* 1: 985-1000.

Jackson, R. J., Hunt, S. L., Reynolds, J. E., and Kaminski, A. 1995. Cap-dependent and cap-independent translation: Operational distinctions and mechanistic interpretations. *Curr. Topics Microbiol. Immunol.* 203: 1-29.

Jang, S. K., Krausslich, H. G., Nicklin, M. J. H., Duke, A. C., Palmenberg, A. C., and Wimmer, E. 1988. A segment of the 5` nontranslated region of the *Encephalomyocarditis virus* RNA directs internal entry of ribosomes during *in vitro* translation. *J. Virol.* 62: 2636-2643.

Jelkmann, W., Maiss, E., and Martin, R. R. 1992. The nucleotide sequence and genome organisation of *Strawberry mild yellow edge-associated potexvirus*. *J. Gen. Virol.* 73: 475-479.

Karasev, A. V., Boykov, V. P., Gowda, S., Nikolaeva, O. V., Hilf, M. E., Koonin, E. V., Niblett, C. L., Cline, K., *et al.*, 1995. Complete sequence of the citrus tristeza virus RNA genome. *Virology* 208: 511-520.

Kashiwazaki, S., Scott, K. P., Reavy, B., and Harrison, B. D. 1995. Sequence analysis and gene content of potato mop top virus RNA3: Further evidence of heterogeneity in the genome organisation of furoviruses. *Virology* 206: 701-706.

Kean, K. M. 2003. The role of mRNA 5`-noncoding and 3`-end sequences on 40S ribosomal subunit recruitment and how RNA viruses successfully compete with cellular mRNAs to ensure their own protein synthesis. *Biol. Cell* 95: 129-139.

Kim, K.-H., and Lommel, S. A. 1994. Identification and analysis of the site of -1 ribosomal frameshifting in *Red clover necrotic mosaic virus*. *Virology* 200: 574-582.

Kiss-László, Z., Blanc, S., and Hohn, T. 1995. Splicing of cauliflower mosaic virus 35S RNA is essential for virus infectivity. *EMBO J.* 14: 3552-3562.

Klaassen, V. A., Boeshore, M. L., Koonin, E. V., Tian, T., and Falk, B. W. 1995. Genome structure and phylogenetic analysis of *Lettuce infectious yellows virus*, a whitefly-transmitted bipartite closterovirus. *Virology* 208: 99-110.

Koenig, R., and Loss, S. 1997. Beet soil-borne virus RNA1: Genetic analysis enabled by a starting sequence generated with primers to highly conserved helicase-encoding domains. *J. Gen. Virol.* 78: 3161-3165.

Koenig, R., Commandeur, U., Loss, S., Beier, C., Kaufmann, A., and Leseman, D. E. 1997. Beet soil-borne virus RNA2: Similarities and dissimilarities to the coat protein gene-carrying RNAs of other furoviruses. *J. Gen. Virol.* 78: 469-477.

Koenig, R., Pleij, C. W. A., Beier, C., and Commandeur, U. 1998. Genome properties of *Beet virus Q*, a new furo-like virus from sugar beet, determined from unpurified virus. *J. Gen Virol.* 79: 2027-2036.

Koh, D. C.-Y., Liu, D. X., and Wong, S.-M. 2002. A six-nucleotide segment within the 3` untranslated region of *Hibiscus chlorotic ringspot virus* plays an essential role in translational enhancement. *J. Virol.* 76: 1144-1153.

Koonin, E. V., and Dolja, V. V. 1993. Evolution and taxonomy of positive-strand RNA viruses: Implications of comparative analysis of amino acid sequences. *Crit. Rev. Biochem. Mol. Biol.* 28: 375-430.

Kozak, M. 1986a. Regulation of protein synthesis in virus-infected animal cells. *Adv. Virus Res.* 31: 229-292.

Kozak, M. 1986b. Point mutations define a sequence flanking the AUG initiator codon that modulates translation by eukaryotic ribosomes. *Cell* 44: 283-292.

Kozak, M. 1989. The scanning model for translation: An update. *J. Cell Biol.* 108: 229-241.

Kozak, M. 1990. Downstream secondary structure facilitates recognition of initiator codons by eukaryotic ribosomes. *Proc. Natl. Acad. Sci. USA* 87: 8301-8305.

Kozak, M. 1991. Structural features in eukaryotic mRNAs that modulate the initiation of translation. *J. Biol. Chem.* 266: 19867-19870.

Kozak, M. 1992. Regulation of translation in eukaryotic systems. *Annu. Rev. Cell Biol.* 8: 197-225.

Kozak, M. 1994. Determinants of translational fidelity and efficiency in vertebrate mRNAs. *Biochimie* 76: 815-821.

Kozak, M. 1995. Adherence to the first AUG rule when a second AUG codon follows closely upon the first. *Proc. Natl. Acad. Sci. USA* 92: 2662-2666.

Kozak, M. 1999. Initiation of translation in prokaryotes and eukaryotes. *Gene* 234: 187-208.

Le, H., Tanguay, R. L., Balasta, M. L., Wei, C. C., Browning, K. S., Metz, A. M., Goss, D. J., and Gallie, D. R. 1997. The translation initiation factors eIFiso4G and eIF-4B interact with the poly(A)-binding protein to increase its RNA binding affinity. *J. Biol. Chem.* 272: 16247-16255.

Le, H., Browning, K. S., and Gallie, D. R. 2000. The phosphorylation state of poly(A)-binding protein specifies its binding to poly(A) RNA and its interaction with eukaryotic initiation factor (eIF) 4F, eIFiso4F, and eIF4B. *J. Biol. Chem.* 275: 17452-17462.

Leathers, V., Tanguay, R., Kobayashi, M., and Gallie, D. R. 1993. A phylogenetically conserved sequence within viral 3` untranslated RNA pseudoknots regulates translation. *Mol. Cell. Biol.* 13: 5331-5337.

Lehto, K., and Dawson, W. O. 1990. Changing the start codon context of the 30a gene of *Tobacco mosaic virus* from "weak" to "strong" does not increase expression. *Virology* 174: 169-176.

Léonard, S., Viel, C., Beauchemin, C., Daigneault, N., Fortin, M. G., and Laliberté, J.-F. 2004. Interaction of VPg-Pro of *Turnip mosaic virus* with the translation initiation factor 4E and the poly(A)-binding protein *in planta. J. Gen. Virol.* 85: 1055-1063.

Liang, X.-Z., Lucy, A. P., Ding, S.-W., and Wong, S. M. 2002. The p23 protein of *Hibiscus chlorotic ringspot virus* is indispensable for host-specific replication . *J. Virol.* 76: 12312-12319.

Lindsey, D., and Gallant, J. 1993. On the directional specificity of ribosome frameshifting at a "hungry" codon. *Proc. Natl. Acad. Sci. USA* 90: 5469-5473.

Lipzig, R., van Montagu, M., Cornelissen, M., and Meulewaeter, F. 2001. Functionality of the STNV translational enhancer domain correlates with affinity for two wheat germ factors. *Nucl. Acids Res.* 29: 1080-1086.

Lovett, P. S., and Rogers, E. J. 1996. Ribosome regulation by the nascent peptide. *Microbiol. Rev.* 60: 366-385.

Lucchesi, J., Makelainen, K., Merits, A., Tamm, T., and Makinen, K. 2000. Regulation of –1 ribosomal frameshifting directed by cocksfoot mottle sobemovirus genome. *Eur. J. Biochem.* 267: 3523-3529.

Lukaszewicz, M., Feuermann, M., Jerouville, B., Stas, A., and Boutry, M. 2000. *In vivo* evaluation of the context sequence of the translation initiation codon in plants. *Plant Sci.* 154: 89-98.

Lütcke, H. A., Chow, K. C., Mickel, F. S., Moss, K. A., Kern, H. F., and Scheele, G. A. 1987. Selection of AUG codons differs in plants and animals. *EMBO J.* 6: 43-48.

Magnus, D. A., Evans, M. C., and Jacobson, A. 2003. Poly(A)-binding proteins: Multiplication scaffolds for the posttranscriptional control of gene expression. *Genome Biol.* 4: 223.

Maia, I. G., Séron, K., Haenni, A.-L., and Bernardi, F. 1996. Gene expression from viral RNA genomes. *Plant Mol. Biol.* 32: 367-391.

Mäkinen, K., Naess, V., Tamm, T., Truve, E., Aaspollu, A., and Saarma, M. 1995. The putative replicase of *Cocksfoot mottle sobemovirus is* translated as a part of the polyprotein by –1 ribosomal frameshift. *Virology* 207: 566-571.

Matsuda, D., and Dreher, T. W. 2004. The tRNA-like structure of turnip yellow mosaic virus RNA is a 3`-translatioinal enhancer. *Virology* 321: 36-46.

Matsuda, D., Bauer, L., Tinnesand, K., and Dreher, T. W. 2004. Expression of the two nested overlapping reading frames of turnip yellow mosaic virus RNA is enhanced by a 5`-cap and by 5`- and 3`-viral sequences. *J. Virol.* 78: 9325-9335.

McCarthy, J. 1998. Posttranscriptional control of gene expression in yeast. *Microbiol. Mol. Biol. Rev.* 62: 1492-1553.

Meulewaeter, F. 1999. Necroviruses (*Tombusviridae*). In: Granoff, A., and Webster, R. G. (Eds.). Encyclopedia of Virology. Second Ed., Academic Press, London, UK. p. 901-908.

Meulewaeter, F., van Montagu, M., and Cornelissen, M. 1998a. Features of autonomous function of the translational enhancer domain of Satellite tobacco necrosis virus. RNA 4: 1347-1356.

Meulewaeter, F., Danthinne, X., van Montagu, M., and Cornelissen, M. 1998b. 5`- and 3`-sequences of satellite tobacco necrosis virus RNA promoting translation in tobacco. Plant J. 14: 169-176.

Miller, W. A., Dinesh-Kumar, S. P., and Paul, C. P. 1995. Luteovirus gene expression. *Crit. Rev. Plant Sci.* 14: 179-211.

Mizumoto, H., Hikichi, Y., and Okuno, T. 2002. The 3`-untranslated region of RNA1 as a primary determinant of temperature sensitivity of *Red clover necrotic mosaic virus* Canadian strain. *Virology* 293: 320-327.

Mizumoto, H., Tatsuta, M., Kaido, M., Mise, K., and Okuno, T. 2003. Cap-independent translational enhancement by 3` -untraslated region of red clover necrotic mosaic virus RNA1. *J. Virol.* 77: 12113-12121.

Neeleman, L., Olsthoorn, R. C., Linthorst, H. J., and Bol, J. F. 2001. Translation of a nonpolyadenylated viral RNA is enhanced by binding of viral coat protein or polyadenylation of the viral RNA. *Proc. Natl. Acad. Sci. USA* 98: 14286-14291.

Neeleman, L., Linthorst, H. J. M., and Bol, J. F. 2004. Efficient translation of alfamovirus RNAs requires the binding of coat protein dimers to the 3`-termini of the viral RNAs. *J. Gen. Virol.* 85: 231-240.

Nicolaisen, M., Johansen, E., Poulsen, G. B., and Borkhardt, B. 1992. The 5`-untranslated region from pea seed borne mosaic potyvirus RNA as a translational enhancer in pea and tobacco protoplasts. *FEBS Lett.* 303: 169-172.

Niepel, M., and Gallie, D. R. 1999. Identification and characterization of the functional elements within the tobacco etch virus 5`-leader required for cap-independent translation. *J. Virol.* 73: 9080-9088.

Noueiry, A. O., Chen, J., and Ahlquist, P. 2000. A mutant allele of essential general translation initiation factor *DED1* selectively inhibits translation of a viral mRNA. *Proc. Nat. Acad. Sci. USA* 97: 12985-12990.

Okada, Y. 1999. Historical overview of research on the tobacco mosaic virus genome: Genome organization, infectivity and gene manipulation. *Phil. Trans. Roy. Soc. London* 354B: 569-582.

Osman, T. A. M., and Buck, K. W. 1997. The tobacco mosaic virus RNA polymerase complex contains a plant protein related to RNA-binding subunit of yeast eIF-3. *J. Virol.* 71: 6075-6082.

Oster, S. K., Wu, B., and White, K. A. 1998. Uncoupled expression of p33 and p92 permits amplification of tomato bushy stunt virus RNAs. *J. Virol.* 72: 5845-5851.

Pelletier, J., and Sonenberg, N. 1988. Internal initiation of translation of eukaryotic mRNA directed by a sequence derived from poliovirus RNA. *Nature (London)* 334: 320-325.

Pestova, T. V., Hellen, C. U. T., and Shatsky, I. N. 1996. Canonical eukaryotic initiation factors determine initiation of translation by internal ribosome entry *Mol. Cell. Biol.* 16: 6859-6869.

Pestova, T. V., Lomakin, I. B., Lee, J. H., Choi, S. K., Dever, T. E., and Hellen, C. U. 2000. The joining of ribosomal subunits in eukaryotes requires eIF5B. *Nature* 403: 332-335.

Pestova, T. V., Kolupaeva, V. G., Lomakin, I. B., Pilipenko, E. V., Shatsky, I. N., Agol, V. I., and Hellen, C. U. 2001. Molecular mechanisms of translation initiation in eukaryotes. *Proc. Natl. Acad. Sci. USA* 98: 7029-7036.

Petty, I. T. D., and Jackson, A. O. 1990. Two forms of the major barley stripe mosaic virus nonstructural proteins are synthesized *in vivo* from alternative initiation codons. *Virology* 177: 829-832.

Preiss, T., and Hentze, M. W. 1998. Dual function of the messenger RNA cap structure in poly(A)-tail-promoted translation in yeast. *Nature* (London) 392: 516-520.

Preiss, T., and Hentze, M. W. 1999. From factors to mechanisms: Translation and translational control in eukaryotes. *Curr. Opin. Genet. Dev.* 9: 515-521.

Prevot, D., Darlix, J. L., and Ohlmann, T. 2003. Conducting the initiation of protein synthesis: The role of eIF4G. *Biol. Cell* 95: 141-156.

Qu, F., and Morris, T. J. 2000. Cap-independent translational enhancement of turnip crinkle virus genomic and subgenomic RNAs. *J. Virol.* 74: 1085-1093.

Quadt, R., Kao, C. C., Browning, K. S., Hersberger, R. P., and Ahlquist, P. A. 1993. Characterization of a host protein associated with brome mosaic virus RNA-dependent RNA polymerase. *Proc. Natl. Acad. Sci. USA* 90: 1498-1502.

Ramirez, B.-C., and Haenni, A.-L. 1994. Molecular biology of tenuiviruses, a remarkable group of plant viruses. *J. Gen. Virol.* 75: 467-475.

Ray, D., and White, K. A. 1999. Enhancer-like properties of an RNA element that modulates tombusvirus RNA accumulation. *Virology* 256: 161-171.

Riechmann, J. L., Lain, S., and Garcia, J. A. 1991. Identification of the initiation codon of plum pox potyvirus genomic RNA. *Virology* 185: 544-552.

Rochon, D. M., and Johnston, J. C. 1991. Infectious transcripts from cloned cucumber necrosis virus cDNA: Evidence for a bifunctional subgenomic mRNA. *Virology* 181: 656-665.

Rohde, W., Gramstat, A., Schmitz, J., Tacke, E., and Prüfer, D. 1994. Plant viruses as model systems for the study of non-canonical translation mechanisms in higher plants. *J. Gen. Virol.* 75: 2141-2149.

Rubino, L., Burgyan, J., and Russo, M. 1995. Molecular cloning and complete nucleotide sequence of carnation Italian ringspot tombusvirus genomic and defective interfering RNAs. *Arch. Virol.* 140: 2027-2039.

Ryabov, E. V., Generozov, E. V., Kendall, T. L., Lommel, S. A., and Zavriev, S. K. 1994. Nucleotide sequence of carnation ringspot dianthovirus RNA-1. *J. Gen. Virol.* 75: 243-247.

Ryabov, E. V., Krutov, A. A., Novikov, V. K., Zheleznikova, O. V., Morozov, S. Yu., and Zavriev, S. K. 1996. Nucleotide sequence of RNA from the sobemovirus found in infected cocksfoot shows a luteovirus-like arrangement of the putative replicase and protease genes. *Phytopathology* 86: 391-397.

Sachs, A. B. 2000. Physical and functional interactions between the mRNA cap structure and poly(A) tail. In: Sonenberg, N., Hershey, J. W. B., and Mathews, M. B. (Eds.). Translational Control of Gene Expression. Cold Spring Harbor Laboratory Press, Cold Spring Harbor, N. Y., USA. p. 447-465.

Sachs, A. B., and Varani, G. 2000. Eukaryotic translation initiation: There are (at least) two sides of every story. *Nat. Struct. Biol.* 7: 356-361.

Sachs, A. B., Sanow, P., and Hentze, W. M. 1997. Starting at the beginning, middle, and end: Translation initiation in eukaryotes. *Cell* 89: 831-838.

Savenkov, B. I., Solovyev, A. G., and Morozov, S. Yu. 1998. Genome sequences of *Poa semilatent* and *Lychnis ringspot hordeiviruses*. *Arch. Virol.* 143: 1379-1393.

Schalk, H. J., Matzeit, V., Schiller, B., Schell, J., and Gronenborn, B. 1989. *Wheat dwarf virus*, a geminivirus of graminaceous plants, needs splicing for replication. *EMBO J.* 8: 359-364.

Schmaljohn, C. S. 1996. *Bunyaviridae*: The viruses and their replication. In: Fields, B. N., Knipe, D. M., Howley, P. M., *et al.* (Eds.). Fundamental Virology. Lippincott-Raven, Philadelphia, PA. p. 649-673.

Schmitz, J., Prüfer, D., Rohde, W., and Tacke, E. 1996. Non-canonical translation mechanisms in plants: Efficient *in vitro* and *in planta* initiation at AUU codons of the tobacco mosaic virus enhancer sequence. *Nucl. Acids Res.* 24: 257-263.

Scholthof, H. B., Scholthof, K.-B. G., Kikkert, M., and Jackson, A. O. 1995. *Tomato bushy stunt virus* spread is regulated by two nested genes that function in cell-to-cell movement and host-dependent systemic invasion. *Virology* 213: 425-438.

Scholthof, H. B., Desvoyes, B., Kuecker, J., and Whitehead, E. 1999. Biological activity of two tombusvirus proteins translated from nested genes is influenced by dosage control via context-dependent leaky scanning. *Mol. Plant-Microbe Interact.* 12: 670-679.

Shams-baksh, M., and Symons, R. H. 1997. *Barley yellow dwarf virus* – PAV RNA does not have a VPg. *Arch. Virol.* 142: 2529-2535.

Shen, R., and Miller, W. A. 2004. The 3`-untranslated region of tobacco necrosis virus RNA contains a barley yellow dwarf virus-like cap-independent translation element. *J. Virol.* 78: 4655-4664.

Shirako, Y. 1998. Non-AUG translation initiation in a plant RNA virus: A forty amino acid extension is added to the N terminus of the soil-borne wheat mosaic virus capsid protein. *J. Virol.* 72: 1677-1682.

Shirako, Y., and Wilson, 1993. Complete nucleotide sequence and organization of the bipartite RNA genome of *Soil-borne wheat mosaic virus*. *Virology* 195: 16-32.

Sit, T. L., Vaewhongs, A. A., and Lommel, S. A. 1998. RNA-mediated transactivation of transcription from a virus RNA. *Science* 281: 829-832.

Sivakumaran, K., and Hacker, D. L. 1998. The 105-kDa polyprotein of *Southern bean mosaic virus* is translated by scanning ribosomes. *Virology* 246: 34-44.

Svitkin, Y. V., Imataka, H., Khaleghpour, K., Kahvejian, A., Liebig, H.-D., and Sonenberg, N. 2001. Poly(A)-binding protein interaction with eIF4G stimulates picornavirus IRES-dependent translation. *RNA* 7: 1743-1752.

Tarun, S. J., and Sachs, A. B. 1995. A common function for mRNA 5`- and 3`-ends in translation initiation in yeast. *Genes Dev.* 9: 2997-3007.

Tatsuta, M., Mizumoto, H., Kaido, M., Mise, K., and Okuno, T. 2005. The *Red clover necrotic mosaic virus* RNA2 *trans*-activator is also a *cis*-acting RNA2 replication element. *J. Virol.* 79: 978-986.

Taylor, D. N., and Carr, J. P. 2000. The GCD10 subunit of yeast eIF-3 binds the methyltransferase-like domain of the 126- and 183-kDa replicase proteins of *Tobacco mosaic virus* in the yeast two-hybrid system. *J. Gen. Virol.* 81: 1587-1591.

ten Dam, E. B., Pleij, C. W. A., and Bosch, L. 1990. RNA pseudoknot: Translational frameshifting and readthrough on viral RNAs. *Virus Genes* 4: 121-136.

Thomas, A. A. M., ter Harr, E., Wellink, J., and Voorma, H. O. 1991. Cowpea mosaic virus middle component RNA contains a sequence that allows internal binding of ribosomes and that requires eukaryotic initiation factor 4F for optimal translation. *J. Virol.* 65: 2953-2959.

Timmer, R. T., Benkowski, L. A., Schodin, D., Lax, S. R., Metz, A. M., Ravel, J. M., and Browning, K. S. 1993. The 5`- and 3`-untranslated regions of satellite tobacco necrosis virus RNA affect translational efficiency and dependence on a 5`-cap structure. *J. Biol. Chem.* 268: 9504-9510.

Vagner, S., Galy, B., and Pyronnet, S. 2001. Irresistible IRES. Attracting the translation machinery to internal ribosome entry sites. *EMBO Rep.* 2: 893-898.

van Knippenberg, I., Lamine, M., Goldbach, R., and Kormelink, R. 2005. Tomato spotted wilt virus transcriptase *in vitro* displays a preference for cap donors with multiple base complementarity to the viral template. *Virology* 335: 122-130.

van Lipzig, R., Gultyaev, A. P., Pleij, C. W., van Montagu, M., Cornelissen, M., and Meulewaeter, F. 2002. The 5`-and 3`- extremities of the satellite tobacco necrosis virus translational enhancer domain contribute differentially to stimulation of translation. *RNA* 8: 229-236.

Verchot, J., Angell, S. M., and Baulcombe, D. C. 1998. *In vivo* translation of the triple gene block of *Potato virus X* requires two subgenomic mRNAs. *J. Virol.* 72: 8316-8320.

Wang, S., and Miller, K. S. 1995. A sequence located 4.5 to 5.0 kilobase from the 5`-end of the barley yellow dwarf virus (PAV) genome strongly stimulates translation of uncapped mRNA. *J. Biol. Chem.* 270: 13446-13452.

Wang, S., Browning, K. S., and Miller, W. A. 1997. A viral sequence in the 3`-untranslated region mimics a 5`-cap in facilitating translation of uncapped mRNA. *EMBO J.* 16: 4107-4116.

Wells, D. R., Tanguay, R. L., Le, H., and Gallie, D. R. 1998. HSP101 functions as a specific translational regulatory protein whose activity is regulated by nutrient status. *Gen. Develop.* 12: 3236-3251.

White, K. A., and Nagy, P. D. 2004. Advances in the molecular biology of tombusviruses: Gene expression, genome replication, and recombination. *Prog. Nucl. Acid Res. Mol. Biol.* 78: 187-226.

Wilkie, G. S., Dickson, K. S., and Gray, N. K. 2003. Regulation of mRNA translation by 5`- and 3`-UTR-binding factors. *Trends Biochem. Sci.* 28: 182-188.

Wilson, J. E., Powell, M. J., Hoover, S. E., and Sarnow, P. 2000. Naturally occurring dicistronic cricket paralysis virus RNA is regulated by two internal ribosome entry sites. *Mol. Cell. Biol.* 20: 4990-4999.

Wu, B., and White, K. A. 1999. A primary determinant of cap-independent translation is located in the 3`-proximal region of the tomato bushy stunt virus genome. *J. Virol.* 73: 8982-8988.

Xiong, Z., Kim, K. H., Kendall, T. L., and Lommel, S. A. 1993. Synthesis of the putative red clover necrotic mosaic virus RNA polymerase by ribosomal frameshifting *in vitro*. *Virology* 193: 213-221.

Zeenko, V. V., Ryabova, L. A., Spirin, A. S., Rothnie, H. M., Hess, D., Browning, K. S., and Hohn, T. 2002. Eukaryotic elongation factor 1A interacts with the upstream pseudoknot domain in the 3`-untranslated region of tobacco mosaic virus RNA. *J. Virol.* 76: 5678-5691.

Zerfass, K., and Beier, H. 1992a. Pseudouridine in the anticodon GΨA of plant cytoplasmic tRNA^{Tyr} is required for UAG and UAA suppression in the TMV-specific context. *Nucl. Acids Res.* 20: 5911-5918.

Zerfass K., and Beier, H. 1992b. The leaky UGA termination codon of tobacco rattle virus RNA is suppressed by tobacco chloroplast and cytoplasmic tRNA^{Trp} with CmCA anticodon. *EMBO J.* 11: 4167- 173.

Zhou, H., and Jackson, A. O. 1996. Expression of the barley stripe mosaic virus RNA β 'triple gene block'. *Virology* 216: 367-379.

10 ASSEMBLY OF VIRUS PARTICLES

I. INTRODUCTION

The first virus to be assembled from its constituents was TMV. It was performed way back in 1955 (Fraenkel-Conrat and Williams, 1955). Soon after the floodgates opened and the information on this topic has been reviewed many times starting from 1960-1963 (Klug and Caspar, 1960; Caspar, 1963) to 1999 (Butler, 1999; Klug, 1999) and 2002 (Culver, 2002).

The terms reconstitution, self-assembly, reassembly, and assembly have often been used interchangeably meaning a process in which biologically active structures are (re)constituted/(re)assembled *in vitro* and/or *in vivo* from their individual components (or sub-assemblies). Such studies have so far been conducted only on simple viruses (like TMV, BMV, PVX and others) that are constituted only by two macromolecular components – the RNA genome and CP and are the only two components involved in assembly. Still assembly of most plant viruses is not well characterized, although assembly of RNA viruses with only a single type of CP is commonly regarded to be a process that does not require scaffolding proteins. Interactions between viral RNAs and CPs are seemingly enough for viral assembly. No assembly studies have been reported on more complex plant viruses.

No cellular components or any cellular macromolecular catalytic factor(s) are involved or participate in viral assembly process although host plant does seem to exert some as yet unexplained influence. Thus, differential accumulation of defective (D) RNA3 of Fny strain of CMV (designated D RNA 3-1) in tobacco versus squash was due to its inability to become encapsidated in squash cotyledons – suggesting host-specific encapsidation of this defective RNA (Kaplan *et al.*, 2004) as also occurs in defective interfering (DI) RNA2 of BBMV (Romero *et al.*, 1993). Possibly, differences in interactions of genomic RNA, subgenomic RNA, and D/DI RNAs with host components in various plants could determine encapsidation or not of particular viral RNAs (Kaplan *et al.*, 2004).

Once the viral RNA and protein subunits have been synthesized, the two must aggregate to constitute the virus particles. Protein subunits fold up in a specific manner because of their intrinsic properties and aggregate by themselves to form the capsid/capsid sub-assemblies in which genome molecule(s) must be incorporated to give the complete virus particle. Assembly is a crystallization-like process in which, under suitable conditions, protein subunit, together with the nucleic acid, form complete virus particles. A nucleation step is thought to occur first. This entails the binding of a critical number of protein subunits to some specific point on the nucleic acid. Once this takes place or, in other words, once the assembly of a virus is initiated, the remaining protein subunits/sub-assemblies click into positions on nucleic acid where they forge maximum number of bonds with adjacent protein subunits resulting in the formation of a complete and fully infectious virus particle self-assembled from its two components. The stability of a protein aggregate is highest when maximum number of bonds is formed between the subunits. The initial fragment of the virus protein helix, the so-called protohelix, will have greater stability when the constituent protein subunits are able to form axial contacts besides the full complement of lateral contacts.

Viral RNAs are specifically selected for packaging during infection so that cellular RNAs are not packaged in progeny virus particles. Specific packaging occurs through interaction between viral RNAs possessing some sort of packaging signal and the structural units (CP molecules) of capsid. Viral elements involved in specific RNA packaging have been well characterised for several plant viruses but the best characterised RNA element acting as a packaging signal is the 'origin of assembly' of TMV (Turner *et al.*, 1988).

Genomes of majority of plant viruses, on which assembly studies are available, are monopartite but such studies are also available on multipartite plant viruses like BMV. Encapsidation of the single RNA genome molecule of a non-segmented plant virus is a simple affair compared to encapsidation of several RNA genome segments of a multipartite plant virus. There must be a means to ensure that all genome segments of such viruses are correctly packaged. The answer has been provided by work on BMV (Choi and Rao, 2000, 2003; Choi *et al.*, 2002).

The virus particles have to be disassembled first (during infection process of a host cell) before the self-assembly process (of progeny virus particles) could occur. The disassembly/uncoating of virus particles, which infect a cell, has already been treated in Chapter 3 (Infection by and Uncoating of Virus Particles).

II. ASSEMBLY OF ROD-LIKE VIRUS PARTICLES

A. *Tobacco mosaic tobamovirus*

TMV is the best worked out plant virus and some reviews on TMV self-assembly are Butler (1984, 1999), Okada (1986), Shaw (1999), Stubbs (1999), and Culver (2002) while little is known about the assembly of other rod-shaped plant viruses.

A brief recapitulation of information about TMV structure is in order here. The TMV CP structure has been studied at 2.8Å and 2.4Å for stacked 20S disks containing 34 subunits and at 2.9Å for the virion (Namba *et al.*, 1989; Wang *et al.*, 1997; Bhyravbhatla *et al.*, 1998). The CP has a central core that consists of a right-handed antiparallel helical bundle composed of four alpha-helices, the left and right slewed (LS and RS) and left and right radial (LR and RR). The LS and RS helices are connected through a short inner loop while the LR and RR helices are connected through a longer loop. The residues from both the loops and the LR helix constitute the

RNA binding site. Both the N and C termini are located along the outer surface of the virion. Four other tobamoviruses (*Cucumber green mottle mosaic virus, Ondontoglossum ringspot virus, Ribgrass mosaic virus*, and TMV strain U2) possess varying degrees of sequence homology with TMV but nevertheless all possess similar three-dimensional folds (Pattanayek and Stubbs, 1992; Wang and Stubbs, 1994; Wang *et al.*, 1997).

1. Assembly of Capsid Protein *in vitro*

Under *in vitro* conditions and depending upon solution conditions, TMV CP on degradation exists in a state of limited aggregation and forms a series of aggregates of increasing sedimentation coefficients. At least three/four such well-defined aggregates of TMV CP exist (4S, 20S disk or helix constituted by about 38 protein subunits, and 28-33S extended virion-like rod-like structures). All these aggregates are highly ordered structures; have extensive intermolecular interactions occurring both laterally (side to side) and axially (up and down); and involve intersubunit contacts between patches of hydrophobic or polar residues, including a salt-bridge network (Namba *et al.*, 1989). These intersubunit interactions are of much importance during virion assembly and virus-host interactions (Culver *et al.*, 1994; Lu *et al.*, 1998).

The lowest stable soluble aggregate is a 4S aggregate, known as the A-protein, and is a mixture of low-order aggregates (monomers, dimers, and trimers) of the chemical subunit. At weakly alkaline pH (~8.0) and moderate ionic strength (~0.05), the TMV CP exists in solution as small 4S aggregates. Some consider the A-protein to be a mixture of small aggregates consisting of one to five monomeric subunits. A double-layered disk (20S aggregate or disk) is the second stable aggregate; its diameter is the same as that of the virus and consists of two rings, each of 17 protein monomers; optimum conditions for its formation are identical to those of TMV assembly *in vitro* (pH 7.0, ionic strength 0.1, and temperature 20°C); has cylindrical symmetry; its formation is a slow process and takes several hours even under optimum conditions; is the form that is able to interact with viral RNA and participate in assembly process and thus plays important and essential role in initiation of TMV assembly, which cannot start in its absence primarily because of its recognition potential and stability; and differs significantly from intact TMV particles in antigenic specificity. The two layers of a disk are opposed in a head-to-tail fashion and show a pronounced 'pairing distortion' with respect to each other. This results in tilting of each subunit towards its neighbor of the other ring at the surface (high radius) so that the two subunits at low radius (at the inner surface towards the central channel) move away from each other. The disk thus has a loose structure, resulting in the formation of two open 'jaws' at the low radius. The third stable aggregate is the rod-like stacked-disk structure (stacked-disk rod) formed by linear stacking of the two-turn disks; it is the 28-33S aggregate; varies greatly in length from short rods of a few disks to rods many times longer than the virus particle; and has different antigenic properties to those of the intact TMV particles.

The fourth stable aggregate is a rod-like structure of helically polymerized protein subunits and so is a helical rod-shaped structure; its protein units are packed in the same way as in the virus particle; it exactly resembles a TMV particle in structure and helical arrangement of protein subunits but lacks RNA and so is non-infectious; possesses no distinct length; is presumed to be the usual end product of TMV protein polymerization; and is serologically identical to the intact TMV particles. Helical rods are formed when pH of protein solution containing disks is rapidly dropped to 5; and contain numerous nicks or gaps soon after the pH drop. However, with time, these gaps disappear due to slow annealing of the adjacent helical parts to give rise to a typical helix lacking interruptions. Helical rod is produced since each component disk undergoes a distortion (dislocation) to give rise to a transient two-turn helix or lock-washer configuration. Thus, this disk to helix transition is the most important step. Carboxyl-carboxylate groups regulate this transition. At acidic pH (≤ 6.0) and high ionic strength (I) (≥ 0.1), the TMV protein assembles (in the absence of RNA) into virus-like helical aggregates. This means that simple lowering of pH to ~6.0 can induce the same rearrangements in TMV protein subunits as interaction with homo-logous RNA at pH 7.0.

Trimer of A-protein is the basic unit, which may give rise either to stacked-disc rods or helical rods depending upon the conditions. The disk and the helical rods are more important for polymerization of TMV protein as well as TMV assembly.

Two sets of carboxyl-carboxylate pairs were initially suggested to be involved in the assembly process. However, later a certain number of interacting carboxyl groups were suggested to regulate the transition of stacked-disk rods to helical rods. Asp 88, Glu 95, Glu 97, Glu 106, Asp 109, Asp 115 and Asp 116 form a 'carboxyl cage', which is involved in this transition. Out of these, Asp 88, Asp 115 and Asp 116 are conserved in all known TMV strains. The carboxyls repel each other resulting in the formation of two-turn discs with loose structure in the interior near the central hole. These negative charges can be neutralised by multivalent cations (Ca^{2+}) binding or by the phosphates of the RNA backbone and then 'jaws' (open interior ends) close down dislocating the disks to transform them into helical structures. Thus, the abnormal cation binding sites act as a negative switch controlling and regulating the reversible disc – helix trans-formation. The role of carboxylate groups in the disassembly of TMV has been confirmed (Culver *et al.*, 1995). Capsid protein assembly is an entropy-driven reaction that is influenced by pH, divalent cations (Ca^{2+}), ionic strength, and other factors.

2. Assembly of Virus Particles *in vitro*

Infective TMV particles were reconstituted by reacting 1% TMV protein solution with $1/10^{th}$ its weight of TMV RNA at pH 6 in 0.03 *M* acetate at 3° for at least 24 hours (Fraenkel-Conrat and Williams, 1955). Later, it was demonstrated that mixing of TMV RNA and coat protein in the same ratio as in virus particle (1:20 by weight) and incubating in a 0.1 *M* pyrophosphate (ionic strength about 0.5 *M*) pH 7.2 solution approached completion in about 5 hours at 23°–25°C and in about 3 hours at 30°C leading to the formation of complete nucleoprotein particles. These particles sediment at the same rate as the intact virus, are identical to intact TMV in morphology and absorption spectra, are highly infectious, and their infectivity is resistant to RNase (as of native TMV). Thus, electron microscopy, X-ray diffraction, infectivity tests, nuclease resistance, etc. studies showed that reconstituted TMV particles are indistinguishable from the virus occurring in nature. Only correct pH, ionic strength, and temperature are required for reconstitution to occur; no additional factors were involved in this process. Assembly at pH 7.2 is comparatively specific so that homologous RNA is preferentially coated but, at lower pH of 6.6, foreign RNAs and synthetic polyribonucleotides could also be encapsidated within TMV protein.

Assembly of TMV particles occurs by the 'assembly from inside' mechanism because RNA enters the protein disk(s) from inside the central channel of the growing particle. This mechanism consists of two discrete steps: nucleation or assembly initiation step during which assembly is initiated and elongation step during which growth of the initiation complex occurs to yield a complete virus particle. The 'assembly from inside' mechanism is supported by all the available experimental results although it gives no insight into the fine structural mechanisms of RNA-protein and protein-protein interactions during TMV assembly.

a. Nucleation (Assembly Initiation)

Packaging of viral RNA within the protein disk starts from a specific site on viral RNA where the first protein disk interacts. This site has been variously called as reconstitution initiation site, assembly initiation site, nucleation sequence, nucleation core, nucleation site, or origin of assembly or origin of assembly sequence/site. Thus, TMV particle assembly is initiated by interaction between an origin of assembly sequence (OAS) in viral RNA and a stable two-layered 20S CP disk. Later, as subsequent disks are recruited, the nascent virus particle grows bi-directionally with both the free ends of TMV RNA being drawn up through the central hole as RNA complexes with CP disks.

The OAS is a stem-loop hairpin structure located in internal region of TMV RNA at a distance of 900 to 950 nucleotides from the 3`-end so that it exists internally on RNA. In TMV *vulgare* strain  (also known as common or U1 strain), the OAS is located about 900 nucleotides from the 3`-end, is formed by the sequence between nucleotides 5420 and 5546, is located (in this strain and in most tobamoviruses) within the ORF encoding movement protein, and its incorporation into nearly any recombinant RNA sequence permits that particular RNA to be encapsidated into virus-like particles both *in vitro* and *in vivo*.

Complete nucleotide sequence and the consequent possible secondary structure of the nucleation site containing OAS is known. The OAS sequence possesses certain peculiarities: it has low cytosine content, a regular pattern in which purines recur at every third position with high probability/ has a strong order 3 repeat of G residues, a 51-nucleotides long RNA stretch with –AAG-AAG-UUG- in its centre is assumed to constitute the OAS, and selective binding of multiple subunits of the disk to OAS during assembly initiation is suggested to indicate a little preference of one of the three RNA-binding sites of TMV CP for G residues (Turner *et al.*, 1988). The nucleotide sequence of the supposedly first turn of helix forms a double-stranded hairpin with a weakly base-paired stem and single-stranded loop at the top that contains the sequence – AAG-AAG-UUG- and possibly makes the first contact with the disk. This single-stranded loop is suggested to be the first to bind to the RNA-binding sites in the hole of the initiating disk; in other words, the single-stranded top of the hairpin loop inserts itself through the central hole of nucleating disk, and between the 'jaws' (loose ends) at low radius of subunits of the two layers and binds to the RNA binding site of the protein subunits of disk. The weakly base-paired part of the loop (that is, the stem) is then unwound (melted) to create single-stranded areas, which bind with other subunits of the disk. The process is continued till all subunits of the double disk have acquired RNA.

The CP 20S disk specifically recognises and interacts with the OAS in viral RNA to form a stable initiation complex called partially reconstituted RNA or intermediate nucleoprotein complex. This interaction between OAS in RNA and two-layered CP 20S disk is the nucleation step, which initiates the assembly process by conceivably inducing two far reaching effects: a rearrangement of the disk (having a cylindrical symmetry) into a so-called protohelix and packaging of the first turn(s) of RNA within the (now helical) protein disk.

The reorganization of the cylindrical double-disk into a two-turn helix fragment is called dislocation, which in turn generates several consequences. One, there is a displacement of the two layers of the disk by about half a subunit with respect to each other. This corresponds to a shift by about 1.6 nm at the high radius of the particle. Two, it brings some changes in the CP subunit structure like formation of the V-column, increase in the length of RR and LR α-helices, etc. Three, it also changes the position of protein subunits in relation to each other so that contacts between subunits also change. Axial contacts alter drastically while the lateral contacts change very little during change from the disk to helix conformation. Thus, structure of TMV protein subunits found in the disk and those in the helix is different, particularly at low radii. Four, neutralization of negative charges at low radius (inner surface) of the disk due to RNA incorporation draws the two layers together axially resulting in tight packing (that is, closing of the 'jaws') of subunits in the helix. Five, the inner loop connecting the LR and RR helices at low radius is flexible in disks or stacked-disk rods. This loop possibly bends aside as the RNA gets incorporated into the RNA-binding site located at low radius. But once the RNA is in position, the inner flexible loop is transformed into the rigid helical conformation of the virus particle. This effectively locks up the RNA in position. Thus, two factors (closing of the 'jaws' and transformation of inner flexible loop into a rigid helical structure) seal the RNA at its place of incorporation.

The manner of nucleation and internal location of nucleation region makes the 3`- and 5`-terminals of RNA to protrude from the same end of protohelix. The nucleation region, as already mentioned, is about 900 nucleotides from the 3`-end and thus about 5500 nucleotides from the 5`-end. This implies that the exposed 5`-end is more than five times longer than the exposed 3`-end and is looped back down the central channel of the initiation complex.

Another view about assembly initiation anticipates that, at pH 7.0 and low ionic strength of 0.2, two- or three-turn helices, called 'lock-washers' are also present in addition to the 20S disks (Schuster *et al.*, 1980). The number of protein subunits in each of these two-turn helices is 39 ± 2 (Correa *et al.*, 1985). Namba and Stubbs (1986) also suggested that, at the nucleating conditions of neutral pH and low ionic strength, the 20S-nucleating disk is a two-turn helical aggregate instead of the two-turn disk. The existence of the postulated two-turn helix as the nucleating aggregate suggests that initial RNA is not likely to be between the two turns. Instead the arginine groups on top surface of the aggregate may non-specifically bind to phosphates of RNA. However, on encountering the specific AAG-rich binding sequence at assembly initiation site, further binding of RNA could occur by interaction or by binding of bases to the bottom surface of the aggregate (Namba and Stubbs, 1986). This could happen because of the high affinity of CP for the AAG-rich binding sequence. This initiation of viral assembly is somewhat different from the one already given above but agrees with rest of the 'assembly from inside' mechanism.

b. Elongation Mechanism

As already mentioned, the two layers of the disk are in contact with each other at high radius (that is, toward outer periphery) but are loose at low radius (that is, towards the central hole) so that penetration of RNA between two open layers ('jaws') of the disk can occur only from inside via the central hole. Incorporation of RNA neutralises the negative charges of the carboxyl groups and immediately converts the two-layered disk into a two-layered nucleoprotein proto-helix in a manner similar to the proton (Ca^{2+})-mediated transformation of disks into two-turn protein helices upon rapid pH drop.

The assembly from inside mechanism is summarized below after Butler (1984) and Dobrov and Atabekov (1989). The internal region of TMV RNA molecule, containing OAS is a double-stranded hairpin with a single-stranded loop at its tip. This TMV RNA site, on approaching the initiating 20S double disk, penetrates its central channel. Then, the central part of the hairpin (containing OAS) is circumferentially packaged into the hole between the disk layers and, in one way or another, induces the dislocation of the initiating disk into a double-turn protohelix. Upon dislocation, one of the RNA tails (the shorter 3`-tail) comes out of the 3`-side of the protohelix at a radius of 40 Å, while the longer tail comes out of the 5`-side of the protohelix (also at 40 Å); it does not remain there, but folds back into the protohelix central channel and also appears on the 3`-side of the protohelix. So, both RNA tails of the growing TMV ribonucleoprotein are located on the same (3`) side of the particle. Further encapsidation (by disks or small aggregates) of long (about 5000 nucleotides) 5`-tail proceeds on the 5`-side of the growing particle, while the RNA sequences to be coated are delivered to the assembly site through the particle central channel. The 3`-side or 5`-side of a subunit or a disk is that side, which when within the virion, looks in the direction of the virion end that contains the 3`-end or 5`-end of RNA.

Mechanism of elongation towards the 5`-end is identical to the nucleation step (Butler, 1984). The 5`-end is folded back down the central hole of the growing rod (initiation complex) and is also bound to the upper surface of the complex to form a 'travelling loop'; that is, the 'travelling loop' is formed by a segment of RNA lying on the 5`-surface of the growing particle. The travelling loop interacts with the next incoming disk in exactly the same manner as the nucleation loop interacts with the first nucleating disk. More RNA then moves up through the channel of the growing rod, binds to upper surface, regenerates the traveling loop which then again gets encapsidated within a new incoming disk, and the process is repeated till whole of the 5'-end gets encapsidated (Butler, 1984).

The shorter (about 900 nucleotides) 3'-tail is encapsidated by 'normal' sequential coating on the 3'-side of the growing virus helix. Encapsidation of the 3'-end of RNA takes place only by A-protein and is hence slower (Butler, 1984); that is, the 4S-aggregates are used for elongation in the minor 3'-direction (in the 'assembly from inside' mechanism). However, in watermelon strain of *Cucumber green mottle mosaic virus* of TMV group, elongation in 3'-direction proceeds with the involvement of disks. Elongation in 3'-direction starts only after pulling in (of all the long 5'-tail) into the central channel.

c. Conclusions

Culver (2002) has excellently summed up the above discussion: For efficient nucleation with the viral RNA under cellular conditions, the 20S CP aggregate is required. However, it is still not finally settled whether it is the helix or the disk aggregate that initiates assembly (Correia *et al.*, 1985; Butler, 1999). Assembly of virion initiates at RNA OAS, which is located 876 nucleotides from the 3'-end of the RNA and is a stem-loop structure. RNA association with CP is predicted to take place by insertion of the OAS stem-loop into the central hole of the 20S aggregate (Butler, 1984). In the absence of RNA, the inner loop is disordered so that it allows access of the RNA to its binding site within the 20S aggregate. However, RNA binding causes structural ordering of residues within the inner loop that transforms the nucleoprotein complex into a virion-like helix and permits the assembly to proceed further (Bhyravbhatla *et al.*, 1998). Virion elongation in the 5' direction occurs through an inside-out mechanism in which the 5'-end of viral RNA is pulled up through the central hole of the helix. Assembly of the virion in 5' direction occurs rapidly with the addition of 20S aggregates to the growing nucleoprotein rod (Schuster *et al.*, 1980; Turpen *et al.*, 1995). Assembly of the virion in 3' direction is considerably slower and is suggested to involve the addition of single subunits or A protein aggregates (Lomonossoff and Butler, 1980). High CP concentrations *in vivo* most likely perform a critical role and control virion assembly by shifting the assembly-disassembly equilibrium towards assembly (Schuster *et al.*, 1980).

According to Namba and Stubbs (1986) the growing viral rod could elongate by incorporation of short helices at the bottom surface of the rod and top surface of the short helix. Disordered inner loop of the two-turn helix may be preventing its growth by itself but the specific binding of RNA induces this loop to fold, which in turn permits rod elongation and completion of assembly.

3. Assembly of Virus Particles *in vivo*

Carr (2004) has briefly dealt with this topic as below. It is not yet known for certain if *in vivo* TMV assembly occurs spontaneously without the involvement of any host factors, much in the same way as it occurs *in vitro*. However, some evidence shows that assembly process *in vivo* may not be identical to that occurring *in vitro*. This evidence comes from a comparison of protein composition of reconstituted virions with those purified from infected plants. TMV particles obtained from infected plants, unlike the reconstituted virus, contains an additional protein component called H-protein that is a monoubiquitinylated CP subunit. One copy of H-protein occurs in each virus particle at the end corresponding to the 3`-terminus of viral RNA. Ubiquitin is a highly conserved 76-amino acid eukaryotic protein that is covalently attached to proteins as monomers or lysine-linked chains of polyubiquitin. Carr (2004) speculates that possibly ubiquitination plays some role in either the *in vivo* assembly of virions or in their localization in the infected cell. However, the importance, if any, of H-protein is as yet unknown (Zaitlin, 1999).

B. Assembly of Other Rod-like Viruses

TRV CP forms small aggregates (like 4S TMV protein aggregates) in solution at pH < 5 but mainly forms 36 to 40S disks, along with some mixture of smaller aggregates, between pH 5 and 10. The 36 to 40S aggregates most probably correspond to three- or four-layer cylindrical structures and, at high temperature or protein concentration, these 36 to 40S disks tend to aggregate into long stacks of disks. No helical TRV aggregates have been observed. Thus, aggregation properties of TRV and TMV CPs differ strongly from each other. Assembly of particles of TRV (strains C and CAM) has been studied *in vitro*. The ribonucleoprotein reconstitution occurred at low ionic strength and 9°C in a broad pH range of 5 to 9. The short TRV RNA, which is encapsidated at a much greater efficiency than the long RNA, partially coated with the CP has been produced. These particles had the same morphology as native TRV, but with a tail of free RNA on one end of these particles, indicating that TRV reassembly *in vitro* starts near one of the RNA ends and proceeds by a mechanism that is different from the 'assembly from inside' employed by TMV. Moreover, short TRV RNA has a cap structure at the 5`-terminus so that assembly of these particles starts at 5`-end of RNA or very close to it. It is also suggested that assembly strategy is determined by the protein aggregate employed rather than by RNA.

The BSMV protein subunits assemble *in vitro* to form structures resembling the intact virus. The reassembled rod-like particles have helical organisation, indistinguishable from that of the intact BSMV, but are noninfective since they contain noninfective RNA. The BSMV polymerization proceeds step-wise through a number of stable intermediates of increasing size: 9.0 to 12.0S (called as 10S aggregates); 20 to 23S (called as 20S aggregates); 27 to 31S (called as 30S aggregates); 34 to 36S aggregate; and 40 to 42S (designated as 40S aggregates) that are serologically identical to each other and to BSMV, unlike TMV protein polymerization. The 10S aggregate, comparable to the TMV A-protein, is formed at ionic strength of about 0.01 and pH 7.5 to 10.0 while the 30S aggregate is possibly the double-disk structure. Divalent cations are extremely effective in inducing BSMV protein polymerization at low ionic strength (of about 0.01 or less) so that calcium ion induces formation of rod-like particles of helically polymerized protein (which are hollow tubes of helically packaged subunits). Possibly, calcium acts as a ligand that binds to viral protein and changes its conformation into a type favourable for protein polymerization.

The PCV *rigid rods superficially resemble those of the* TMV except *that* PCV *virions* possess bimodal length distribution (190 and 245 *n*m) and *its* genome is composed of two plus-sense RNAs. Hemme*r et* **al**. (2003) mapped RNA sequences required for assembly of PCV RNAs 1 and 2 into rod-shaped *virions*. Packaging of PCV RNA1 required a sequence domain in 5`-proximal part of protein p15 gene which is the 3`-proximal gene of RNA1. However, sgRNA that encodes p15 is not encapsidated. PCV RNA2 has two OAS sequences: a 5`-proximal OAS in CP gene and a 3`-proximal OAS in central part (p14 gene) of triple gene block. It is extremely unlikely that assembly of viral particles can take place at two different loci on the same viral RNA as this will have stearic problems during proper packaging of RNA at the point where the two growing helices meet. Instead, Hemmer *et al.* (2003) suggest on the basis of kinetic considerations that only one OAS is active in RNA2 while the second is a pseudo-OAS, which is not functional. The common strain of TMV also has pseudo-OAS that is located in CP cistron and is suggested not to be functional although it has structural and sequence homology with the authentic OAS. The three OASs - one in RNA1 and two in RNA2 – of PCV show no obvious sequence homologies. Hemmer *et al.* (2003) concluded that the first OAS on RNA2 resides somewhere within its 3`-terminal half and is situated between nucleotides 3370 and 4142 and that the second OAS on RNA2 is located within the CP gene. Like TMV, initiation of PCV assembly possibly also involves interaction of CP with a relatively short sequence that is presented by RNA secondary structure in a special configuration.

Encapsidation of RNA of *Indian Peanut clump virus* (IPCV) (regarded a strain of PCV) strain H does not require the presence of terminal 5`-UTR. Bragard *et al.* (2000) suggested, on this and on the basis of other experimental evidence, that OAS motif, if present, is not located in the 5`-UTR but is present within the CP gene. The OAS in multipartite plant viruses does not appear to be a particular nucleotide sequence (Bragard *et al.*, 2000) as no common nucleotide sequence was found in RNA1 (Miller *et al.*, 1996) and the CP gene (Wesley *et al.*, 1994) of IPCV strain H. Similarly, a nucleotide sequence at non-coding 5`-end of BNYVV RNA3 is essential for encapsidation of all viral RNAs *in vivo* even though no similarity occurs between this particular sequence and sequences of other BNYVV RNAs (Gilmer *et al.*, 1992). In a *Benyvirus* (BNYVV) (Schmitt *et al.*, 1992) and a *Pomovirus* (PMTV) (Cowan *et al.*, 1997), a minor CP, that is generated by readthrough of CP gene, is incorporated at one end of virion and is required for virion assembly and transmission.

III. ASSEMBLY OF FLEXUOUS VIRUS PARTICLES

Flexuous viruses possess slightly curved particles, due to a weakening of axial protein-protein interactions, in comparison to the rigid rod-like particles, particularly at high radii - suggesting that the relative contribution of RNA-protein interactions to virion stability is likely to be greater in filamentous viruses than that in the rigid viruses. Due to weak protein-protein interactions, CP of flexuous viruses has a low capacity to form different kinds of specific aggregates *in vitro*.

Optimum conditions for PVX assembly are pH 6.5 to 7.0, temperature of 20 to 25°C, and low ionic strength (0.01 *M*) at which it gives the highest yield of reconstituted virus. The reassembled PVX particles have low infectivity. The specificity of assembly is also low since RNAs from several plant viruses including TMV, BSMV and others were encapsidated with the same efficiency as homologous RNA. PVX reassembles very rapidly - more than 50% of the total increase in turbidity (OD_{310}) occurred during the first minute of the reconstitution. Non-specific ionic interactions between RNA and protein are mainly responsible for PVX assembly, although hydrophobic interactions may also make some contribution. Absence of helical PVX protein aggregates could be due to the probable little contribution of protein-protein interactions to formation of virions.

Several papaya mosaic virus (PapMV) protein aggregates, called 2-3S, 14S, and 27-30S aggregates, are found at pH 8.0, low ionic strength of 0.01, and 25°C. Their formation greatly depends upon protein concentration and their interconversion is rapid. Reassembly occurs with equal efficiency at all protein concentrations. The 14S aggregate corresponds either to the double-layer disks (18 subunits) or to the two-turn helices. PapMV reconstitution is best at pH 8.0, 25°C, and low ionic strength (0.01) leading to 90 per cent yield of reconstituted ribonucleoprotein, which, however, has low infectivity as compared to the infectivity of reconstituted PVX. The reconstitution initiation site is located on the 5`-terminal side (Sit *et al.*, 1994), possibly at a distance of about 200 nucleotides from the 5`-terminus of PapMV RNA. The first 138 nucleotides at the 5`-end including the assembly initiation site have been determined (Lok and Abou-Haidar, 1986). The first 40 nucleotides contain many adenosines, few uridines, and 8 consecutive repeated pentamers, which essentially are GCAAA. Possibly, these eight pentamers may interact with identical and repeated CP subunits based on one surface of the 14S double-layer protein disk. This could possibly induce conformational changes leading to formation of a protohelix and to subsequent nucleation. The first nucleotides of each pentamer, which have guanine in six nucleotides and U or A in the remaining two nucleotides, conceivably position the pentamers in relation to features and topology of 14S disk. Protein, and not RNA, determines the location of assembly initiation site.

Normally, PapMV protein-RNA interaction is specific so that virus-like 'tubular' particles are formed only with homologous RNA and with RNA of the closely related *Clover yellow mosaic virus* (CYMV) but not with RNAs of TMV and BMV. Thus, unlike PVX assembly, PapMV assembly *in vitro* exhibits some specificity. The elongation rate of homologously reconstituted PapMV particles under the above-mentioned optimal conditions is greatly dependent upon protein concentration. Virus reassembly was completed in 20 minutes at an RNA:protein weight ratio of 1:20 but was completed in 5 minutes at the RNA:protein ratio of 1:80. In all probability, monomers and/or small aggregates of protein participate during elongation of the growing virus particle, while the form of protein that participates in assembly initiation is not yet known. Thus, the nature of protein aggregates taking part in PapMV reassembly is not yet fully settled. However, under certain known conditions, the normal tubular PapMV particles can be constituted with homologous as well as heterologous RNAs. Encapsidation of both homologous and TMV RNA is very slow and encapsidation of both RNAs starts from the 5`-end. The short encapsidated RNA fragments contained 5`-cap structure since PapMV RNA has a cap structure at 5`-terminus. The reassembly of *Clover yellow mosaic potexvirus* (CYMV) is largely identical to that of PapMV, but has a somewhat lower optimum pH (7.0 to 7.5) and lacks specificity. CYMV protein formed normal tubular particles with both homologous RNA and with RNA of the completely unrelated spherical BMV at ionic strength 0.01 and pH 7.5. Causes for specificity of PapMV assembly and its absence in CYMV have not yet been worked out.

The OAS on RNA of both TVMV (Wu and Shaw, 1998) and PapMV (Sit et al., 1994) is located near 5`-extremity of viral RNA. However, this is not a general feature of *Potexviridae* since both genomic and 3`-coterminal sgRNAs of BaMV (V isolate) are packaged – suggesting that OAS is located somewhere in the terminal 1000 nucleotides of these viral RNAs (Lee et al., 1998).

IV. ASSEMBLY OF ICOSAHEDRAL VIRUS PARTICLES

Capsid of most of the spherical (icosahedral) viruses contains CP molecules arranged in quasi-icosahedral symmetry. Assembly of icosahedral viruses has been studied mostly in AMV, bromoviruses, cucumoviruses, sobemoviruses, tombusviruses, and tymoviruses. Of all the spherical viruses, protein-protein interactions make the greatest contribution to virion stability in tymoviruses while virions of bromoviruses, cucumoviruses, and AMV are stabilized chiefly by RNA-protein interactions.

Spherical viruses possess certain characteristics with relevance to their assembly. Firstly, many spherical plant viruses have low specific infectivity and low RNase sensitivity so that infectivity test is not very helpful for detection of a reconstituted virus in majority of these viruses. This made it necessary for employing analytical centrifugation (or sucrose gradient centrifugation) or CsCl or Cs_2SO_4 density gradient centrifugation and electron microscopy during studies of spherical virus assembly. In fact, they proved to be the most convenient methods. Secondly, despite the fact that the yield of reconstituted spherical virus ribonucleoprotein is rather high (up to 50 to 70%) in many cases, the reconstituted virus particles are generally somewhat different from the native virus in properties like sedimentation coefficient, density, or morphology. Consequently, the physiological character of the reconstitution is not a good and helpful feature. Thirdly, effective assembly of most spherical viruses is possible only under dialysis – either (most often) from high (1.5 to 1.0 *M*) to low ionic strength (about 0.01 *M*) solutions or from solutions containing a denaturing agent to solutions which do not contain this agent. Employing dialysis in these studies causes considerable problems for identifying the protein aggregates that participate in the assembly process. It also excludes the use of kinetic methods in such studies. Fourthly, in bromoviruses and cucumoviruses, virions contain either a single RNA molecule of molecular weight of 1.10 or 1.0 x 10^6 daltons or two molecules with molecular weights of 0.85 and 0.35 x 10^6 daltons. The encapsidation of the two (or more) RNA molecules into a single particle is a very serious problem. How the spherical viruses overcome it and what is the mechanism of this packaging? There is no answer yet to this pregnant question. Fifthly, single-stranded RNAs of both the helical and spherical viruses possess nearly the same total secondary structure in a solution. On encapsidation, the secondary structure of RNA genome of helical viruses is disrupted completely while no significant change in secondary structure of RNA of spherical viruses occurs. This ensures that no significant energy is spent on melting the RNA secondary structure during spherical virus assembly.

Because of the profound structural differences between elongated and spherical plant viruses, significant differences must also exist in the assembly strategies of these two types of viruses. Some of the relevant differences are as follows. Firstly, RNA interacts with the CP throughout the entire length of the virus particle. In contrast, the RNA molecule in spherical viruses is in direct contact with CP only at some place(s) with the result that RNA sequences that interact with the protein capsid could be located in different and distantly placed regions of RNA. Secondly, the particle length in helical viruses is dependent on RNA length, but dependence of virion size on RNA length in spherical viruses is much more complex. The number of protein subunits in viral capsid of spherical viruses and, consequently, the size (diameter) of the particle is determined by value of T (triangulation number). On geometrical considerations, only some definite values of T (1, 3, 4, 7) are permitted. In some spherical viruses, short RNAs (up to several hundred nucleotides) are packed into small $T = 1$ virions while long RNAs are packed into large $T = 3$ virions (Savithri and Erickson, 1983).

A. *Southern bean mosaic sobemovirus*, *Tomato bushy stunt tombusvirus*, and *Satellite tobacco necrosis virus*

Insights into assembly mechanisms of spherical plant viruses were generated by high-resolution X-ray diffraction studies of these three plant viruses. Detailed CP structure models based on X-ray diffraction data with 2.5 to 3.0 Å resolution were built for $T = 3$ SBMV, and TBSV and for small $T = 1$ STNV. The CP subunits of these three viruses possess identical secondary and tertiary structure and identical RNA packing modes. In capsids of $T = 3$ viruses (SBMV and TBSV) protein subunits existed in three different conformations designated A, B, and C. Small differences occur between A and B conformations but C-type subunits differ strongly from A and B subunits (Harrison, 1983; Rossmann *et al.*, 1983a, 1983b). The polypeptide chain is ordered along most of its length in all the three types of subunits while the N-terminal part of the subunit is disordered and is characterised by high content of basic amino acids. The length of disordered N-terminal 'arm' in SBMV is 62 residues in A-conformation, 64 residues in B-conformation, and only 38 residues in C-conformation. This implies that the C-subunits differ from A and B subunits by the ordering of some part of the N-terminal arm. The disordered N-terminal regions of all the three subunit types are located on the internal surface of the capsid and could possibly interact with intraviral RNA.

In vitro assembly of SBMV has been studied in detail. At pH 7.0 and 9.0 in the presence of Ca^{2+} or at pH 7.0 in absence of Ca^{2+}, intact SBMV RNA (of 1.4×10^6 daltons molecular weight) assembles into normal $T = 3$ virions (Savithri and Erickson,

1983) but fragmented SBMV RNA (about 0.5 x 10^6 daltons) produced small $T = 1$ particles under these conditions. The maximal yield was obtained at pH 9.0 with Ca^{2+} while no assembly occurred at pH 5.0. Rapid proteolysis of SBMV protein takes place during incubation in the absence of RNA and the unordered N-terminal region of the protein subunits is cleaved by proteolysis. However, no such degradation takes place in presence of RNA suggesting that SBMV RNA interacts in some way with the N-terminal arm of coat protein and protects it from proteolytic degradation.

For controlling assembly of several but not all spherical plant viruses, amino terminal arm (R domain) is an important factor. Proteolytic digestion of this arm from protein subunits invariably forms empty $T = 1$ capsids in several $T = 3$ icosahedral plant viruses like AMV, BMV, SBMV, and TCV. The R domain immobilizes the RNA, and is required for the formation of quasi-sixfold axes but the converse is not always true. Thus, formation of $T = 1$ nucleocapsids, as in STNV, is not compulsorily inhibited by the presence of amino terminal arm. It appears, therefore, that basic amino terminal arm-RNA interaction is the early step for forming an assembly initiation complex of both $T = 1$ and $T = 3$ virus particles. This possibly happens because phosphates of RNA backbone form non-specific electrostatic interactions with lysine and arginine side chains of the amino terminal arm of coat protein. This may be the nucleation step in which the initial RNA-protein binding generates an initiation complex that can produce, depending upon the conditions, both the $T = 1$ and $T = 3$ particles. Dimers are formed *in vitro*, at suitable coat protein concentration and ionic strength, by capsid protein of many $T = 3$ plant viruses – leading to the postulation that dimers are used for capsid assembly as in BMV, CCMV, SBMV, and TBSV. This happens since only the dimers allow the possible induction of a quaternary structural change during assembly of spherical viruses into $T = 1$ particles (Rossmann, 1984). Rossmann *et al.* (1983a, 1983b) and Rossmann (1984) proposed a detailed model of assembly of spherical viruses *in vitro*.

According to this model, SBMV coat protein dimeric molecules have this 'relaxed' AB-conformation with disordered N-terminal arms of both subunits. First stage of capsid assembly is the formation of a decamer assembly initiation complex from 5 AB dimers but subsequent addition of more AB dimers may only form small $T = 1$ capsids of 60 subunits. For formation of normal $T = 3$ capsids (180 subunits), 'tensed' CC dimers have to be added (the 'tensed' conformation of CC dimers comes from a partial ordering of their N-terminal arms). The conformational change of 'relaxed' AB dimers into the 'tensed' CC state is induced by interactions of AB dimers with the nucleating decamer. This transition is induced when some specific contact sites (three carboxylate clusters) on decamer subunits are in the high-charged state. But, if these sites are in a low-charged state, conformational transition of incoming dimers does not occur and they are added in the relaxed AB-state resulting in the formation of small $T = 1$ capsids. At the high-charged state of these sites (the three carboxylate clusters), the incoming CC dimers bind to periphery of the decamer initiation complex. This brings in a series of changes: creation of a cleft between B and C subunits, induction of ordering of βA arm and formation of β-annulus of C-subunits, and leading to the formation of a $T = 3$ capsid (Harrison, 1983). Thus, it is the switching of the dimers into a tensed state that determines the correct assembly.

The high- and low-charged condition of transition-inducing sites depends on protonation or de-protonation of some specific carboxyl groups in the subunit contact regions. The degree of protonation of these groups, in turn, depends on pH of a solution, Ca^{2+} concentration, and the presence of RNA. It is also suggested that residue His 132 may also be involved in 'switch-over' of the binding sites.

The above model, propounded by Rossmann *et al.* (1983a, 1983b) and Rossmann (1984) on the basis of high-resolution X-ray diffraction data on the three plant viruses, formulates the possible molecular mechanisms involved in assembly of spherical viruses *in vitro*. Such data are so far available on only the above-mentioned three spherical viruses so that it may be too optimistic to think that this model could be applicable to all spherical viruses. Nevertheless, some evidence does indicate that at least some concepts of this model are likely to be applicable to many other spherical viruses apart from SBMV. For example, nuclear magnetic resonance (NMR) studies showed that coat protein subunits of several other plant viruses (BMV, CCMV, and others) also possess an unordered N-terminal region facing into the internal cavity of the capsid. Moreover, CCMV coat protein also has a dimeric structure in solution so that in this virus certain essential characteristics of virion structure, RNA-protein interactions, and assembly are likely to be similar to those of the SBMV and TBSV.

B. *Alfalfa mosaic alfamovirus*

Efforts have been made to reconstitute AMV from a mixture prepared by the addition of previously separated AMV protein and nucleic acid and also from a heterogeneous mixture of AMV protein and RNA of TMV, BMV, BBMV, or TYMV or polyadenylic acid or sodium dextran sulphate. More or less spherical particles were formed in all cases except in the presence of sodium dextran sulphate in which case a significant proportion of elongated bacilliform particles, possessing lattice spacing identical to that of the native AMV, were constituted. Elongated particles were also obtained by using long-chain phospholipids or double-stranded DNA with AMV protein. This indicated that sodium dextran sulphate, long-chain phospholipids and double-stranded DNA have the desired configuration and can successfully act as nucleating agents to give rise to elongated particles. Thus, reconstitution of bacilliform AMV particles depends upon the conditions under which RNA possesses a suitable configuration. This crucial role of RNA is probably due to formation of salt links between the basic amino acids of the protein subunits and the ionized phosphate groups of RNA.

AMV coat protein dissociates into dimers *in vitro* in solution; such dimers, as in SBMV have two states: a relaxed and a tensed state. This two-state association of dimers is an important switch controlling AMV assembly. Only the relaxed dimers, with their tapering ends towards the interior of the particle, are located in the hemispherical ends (which have a $T = 1$ icosahedral symmetry) of bacilliform AMV particles. However, two different types of dimers occur in the cylindrical part (which has hexagonal surface lattice) of the virus particles: dimers (A type) on the hexagonal edges (along the lengthwise ridges) provide curvature, need less space in interior of virus and so are tapered towards the virus interior while dimers (B type) situated on the surface (faces) of the hexagonal lattice do not need to decrease in size and so need more space in the virus interior. The edge dimers are thought to be relaxed because of the absence of an ordered basic arm while the surface dimers might be tensed because of the presence of the ordered arm (Rossmann *et al.*, 1983a, 1983b). Tensing of some dimers is likely to control AMV assembly.

The N-terminal part of virus coat protein molecule possibly interacts with certain specific sequences, located near 3`-termini, of all the four viral RNAs to form nucleoprotein complexes. Formation of these complexes could have an important role in initiation of virus assembly as also in regulation of AMV replication. Stable complexes of AMV RNA4 with one or three coat protein dimers could be found in solution; in excess of protein in solution, these complexes may be changed into the ones that contain five or six protein dimers per RNA molecule. These bigger complexes supposedly have a pentagonal or hexagonal packing symmetry and act as a nucleus for further virion assembly.

In vitro, AMV CP binds to internal coding sequences so that internal binding sites may play a role in encapsidation of viral RNAs (Vlot *et al.*, 2001). In AMV- infected cells, replication of RNA3, synthesis of RNA4 and CP, and encapsidation of RNA3 are tightly coupled (Bol, 1999).

C. Bromoviruses

Fox *et al.* (1994) reviewed bromovirus virion structure, assembly and disassembly. The CCMV, BBMV, and BMV are small icosahedral bromoviruses that are constituted by 180 structural units clustered in 32 capsomeres. CCMV and BMV are serologically related and their capsids are 70 per cent identical in sequence, are functionally interchangeable *in vivo* (Osman *et al.*, 1997) and possess $T = 3$ quasi-icosahedral capsids. All the bromovirus capsid assemblies, isolated from infected plants, have 180-subunit, $T = 3$ virus particles containing viral RNA. All these viruses have been reconstituted from their respective proteins and RNAs. The reconstituted virus particles sediment at the same rate as intact virus, are about equally infectious, are about equally resistant to snake venom phosphodiesterase, are similar to the natural virus in serological properties, density, appearance under an electron microscope, and in ultraviolet spectrum.

RNAs of these viruses resemble each other in having the same molecular weight and in possessing almost identical base ratio. Early studies on reassembly of RNA of any of these viruses with protein of the other two viruses lead to the following conclusions. No specificity exists between particular RNAs and proteins for reconstitution, all the nine protein-RNA polymerization combinations of these three viruses are accomplished, sedimentation and electrophoretic properties of a reconstituted virus are the same as that of the virus supplying protein, the reassembled virus is infectious and resistant to diesterase, and infectivity of the reconstituted virus corresponds to the infectivity and host range of the native virus providing RNA. A series of homo- and hetero-polyribonucleotides also assemble with CCMV, BMV, and BBMV proteins to always form spherical particles. Particles reconstituted from protein mixtures of CCMV and BMV or CCMV and BBMV possess mixed coat, i.e., capsid is formed of proteins subunits obtained from two different viruses. These particles had normal spherical appearance, were infectious, and resistant to diesterase treatment.

The first 25 N-terminal amino acids of CPs of bromoviruses (BMV and CCMV) have a highly conserved arginine-rich motif (ARM) (Tan and Frankel, 1995; Rao and Grantham, 1996). Large number of basic amino acid residues is located in this N-proximal region: seven arginines and one lysine in BMV CP, and six arginines and three lysines in CCMV CP. These N-proximal basic amino acid residues in bromovirus capsid protein (BMV CP and CCMV CP) are thought to interact with negative phosphate groups in RNA during the encapsidation process (Speir *et al.*, 1995) so that deletion of N-terminal 25 amino acid residues eliminated the assembly of RNA containing virions but not of empty capsids (Zhao *et al.*, 1995). Zlotnick *et al.* (2000) speculated that an intact N terminus is essential to maintain the optimal structural conformation of the CP dimers that subsequently yield pentamers of dimers, which are essential for icosahedral virus assembly. Annamalai *et al.* (2005) support this conjecture.

Despite the CPs of BMV and CCMV being closely related and having apparently similar and highly conserved N-terminal ARMs, yet these viruses are not functionally analogous in terms of RNA packaging and interaction of ARM with respective genomic RNAs during packaging, which is distinct for each of the these two closely related plant viruses (Choi and Rao, 2000; Choi *et al.*, 2000; Annamalai *et al.*, 2005). Likewise, the interaction between RNA and CP leading to RNA encapsidation and role of the conserved N-terminal ARM in recognizing and packaging specific RNA appears to be distinct in each virus.

Hydrophobic interactions appear to be dominant in the assembly of TMV. In contrast, ionic interactions possibly play some critical part in assembly of these three viruses since they require low temperature, low ionic strengths and divalent cations for reconstitution. The proposed assembly model of bromoviruses is – association of protein dimer with RNA directs the protein conformational change from loose to compact form, which is then followed by further CP association into an icosahedral capsid.

1. *Brome mosaic bromovirus*

The three genomic and a single sgRNA of BMV are encapsidated by a single type of CP molecules to form three morphologically indistinguishable icosahedral virus particles possessing $T = 3$ quasi-symmetry. These virions are assembled from 180 identical subunits of a single CP type. Genomic RNAs 1 and 2 are packaged individually into separate particles while genomic RNA3 and sgRNA4 are copackaged together into a single virion. It is yet not known how does a single CP discriminate among four RNAs and packages them into three individual particles. BMV preparations obtained from infected leaves contain equal amounts of RNA1, RNA2, and RNA3 plus RNA4 (33% each) while during *in vitro* studies BMV capsid protein encapsidates lower molecular weight RNA3 and RNA4 more efficiently than higher molecular weight RNA1 and RNA2. However, no such preference for short RNAs is observed *in vivo*.

In vitro studies on competitive reconstitution of BMV protein with mixtures of homologous and heterologous RNAs showed that BMV protein has clear preference for its own homologous RNAs than the yeast tRNAs or the 12S AMV RNA showing that BMV CP possesses a high degree of specificity *in vitro* and also *in vivo* (Choi and Rao, 2000) to selectively package viral RNAs. Viral RNA may nucleate formation of pentamer-of-dimers which otherwise have weak hydrophobic interactions. Assembly of 180-subunit capsid is regarded to be initiated by a hexamer-of-dimers (Speir *et al.*, 1995), which is absent from the 120-subunit capsid. It has therefore been suggested that the two capsids may initiate assembly with the shared pentamer-of-dimers.

Amino acids located between residues 9 and 19 of CP N-terminal ARM region play crucial roles in BMV RNA packaging through RNA-CP interactions and also contain determinants specific for directing copackaging of BMV subgenomic RNA4 with genomic RNA3 into a single virion (Choi and Rao, 2000; Choi *et al.*, 2000). Moreover, the N-terminal ARM of BMV CP harbours determinants specific for selected viral RNAs so that RNA packaging in BMV is a highly specific process (Osman *et al.*, 1998; Choi and Rao, 2000; Choi *et al.*, 2002) resulting from the interactions between these determinants and a specific packaging signal(s) encoded within a given genomic RNA (Choi and Rao, 2003). RNA regions involved in packaging of BMV RNAs are assigned to coding region of BMV RNA1 (Duggal and Hall, 1993) as well as to 3`-proximal region of 3a ORF in RNA3 (Damayanti *et al.*, 2002).

a. Polymorphic Capsids

BMV can assemble *in vivo* as two distinct capsid types that have strong structural similarities (Krol *et al.*, 1999). The two capsid types are the 180-subunit capsid and the earlier unreported capsid formed by 120 subunits organised as 60 capsid protein dimers. The 180-subunit capsid has quasi-equivalent environment throughout and is similar in quasi-equivalence to the virus particles produced in natural infections while each of the two protein molecules of the BMV dimer (of the 120-subunit capsid), in contrast, is placed in distinct and non-equivalent environment while most of the other capsid protein interactions of the two BMV capsids are nearly the same. Krol *et al.*, (1999) concluded that capsid of BMV employs protein-protein interactions which may participate in assembly or disassembly of virus particles, that the two types of BMV capsids are likely to have common steps during their assembly, and that a common pentamer of capsid protein dimers could be an important intermediate. The presence of polymorphic capsids (the 180-subunit capsid and the 120-subunit capsid) during assembly of BMV capsid *in vivo* means that a single capsid protein molecule can form two distinct capsids by utilizing alternate interactions. Thus, single capsid protein molecules of BMV can switch between distinct capsids by using alternate interactions.

Both the capsid types are capable of selectively and specifically packaging viral RNA even if no RNA replication occurs (Krol *et al.*, 1999). A normal BMV genomic RNA is enclosed in the 180-subunit capsids while an engineered mRNA, that contains only the capsid protein gene, is enclosed in the 120-subunit capsid. RNA replication is not mandatory for this specific encapsidation of BMV RNAs *in vivo* – establishing that replication-independent methods have to operate *in vivo* for selecting BMV RNAs for coating. The encapsidated BMV RNA controls the *in vivo* switch between the two alternate viral capsids (Krol *et al.*, 1999). Thus, some features of RNA direct the assembly of a particular type of ribonucleoprotein complex *in vivo* – showing that the viral RNAs control interaction with capsid protein molecules.

The assembly of 180-subunit capsid is regarded to be initiated by a hexamer-of-dimers (Speir *et al.*, 1995), which is absent from the 120-subunit capsid. It has therefore been suggested that the two capsids may initiate assembly with the shared pentamer-of-dimers. Binding of structural proteins (that is, CP) to specific RNA structures in icosahedral viruses is proposed to produce an assembly *initiation* complex, which nucleates the subsequent addition of CP subunits to form a complete mature virus particle (Zlotnick *et al.*, 2000).

b. Capsid Protein-RNA Interactions

CP-RNA interactions are both nonspecific and specific interactions. Interaction between CP and 3`-tRNA-like structure (TLS) is non-specific and leads to virion assembly by compaction of RNA through neutralization of negative charge on phosphate backbone (Duggal and Hall, 1993). Choi *et al.* (2002) hypothesized that the highly conserved 3`-TLS of BMV RNAs serves as a chaperone, in a transient association with CP, and functions as a nucleating element to initiate assembly of viral RNA into BMV virions. The nucleating element promotes production of pentamers of dimers. Viral CP acts as a nucleation site for CP assembly and RNA encapsidation (Choi *et al.*, 2002).

The ARM located in N-terminal region of BMV CP is implicated in RNA binding during packaging. By binding to basic residues of N terminus, RNA (but specifically the 3`-TLS) could neutralize the charge repulsion between noncovalent CP dimers to produce pentamers of dimers for which viral TLS is hypothesized to act as a scaffold. Moreover, the non-specific interaction between CP and viral TLS possibly stabilizes these pentamers or productive complexes of pentamers. The non-specific interaction could stimulate encapsidation *in vivo* at early times when concentrations of viral components are comparatively low, or serve to chaperone productive capsid formation late in infection when high concentrations of CP could lead to non-productive aggregates.

BMV also has specific interactions since CP is highly specific for viral RNA packaging in vivo (Osman *et al.*, 1998) and *in vitro* (Choi and Rao, 2000). BMV RNA1 contains at least two regions within protein 1a-coding sequence that are capable of high-affinity sequence-specific interaction with CP. This suggests that these binding sites probably initiate encapsidation process (Duggal and Hall, 1993). In genomic RNA, the minimum sequence required for in vitro initiation of assembly is located in the first 38 to 47 nucleotides at 5`-terminus of RNA. This zone is rich in AC and lacks any discernible secondary structure (Sit *et al.*, 1994).

In contrast to the nonspecific interactions, RNA may have features that specifically interact with CP and are involved in assembly of icosahedral virions. Such RNA features have been reported in several animal viruses and in TCV plant virus (Qu and Morris, 1997). These specific interactions are thought to perform some significant role in selectively removing viral RNAs from the pool of heterologous host RNAs occurring in cytoplasm (Fox *et al.*, 1994). Choi and Rao (2003) thus proposed that a selective recognition of a domain specific for each of the four BMV RNAs occurs after or in concert with TLS/CP interaction. This specificity, along with size constraints enforced by capsid size, may help in isolating the four genomic RNAs into the three identical capsids. Thus, a particular region in BMV RNA3 may be regarded as a selective domain for CP interaction that facilitates specific packaging of RNA3. This indicates that, in addition to viral TLS, CP interacts with selective packaging signals located elsewhere on each packaged RNA.

Damayanti *et al.* (2003) delimited 866 *to* 934 *n*ucleo*ti*de sequence *that is* requi*red* f*or* eff*i*c*ient* packagi*ng of* BMV RNA3 a*n*d sho*w*ed *that* 69 *n*ucleo*ti*des *in* 3`-pro*x*i*m*al *region of* 3a ORF, espec*i*ally a p*re*dic*t*ed s*t*em-l*o*op s*t*ru*c*tu*re* (SL-II) (30 *n*ucleo*ti*des p*re*dic*t*ed f*ro*m 904 *to* 933), *i*s esse*nti*al f*or* eff*i*c*ient in vivo* packaging *of* BMV RNA3. They fu*rther* sugges*t*ed *that* TLS a*n*d SL-II may c*oo*pe*rate or* ac*t* *in*depe*nden*tly f*or* e*n*caps*i*da*ti*ng RNA3 *in vivo.*

Packagi*ng of* d*i*c*i*s*t*ro*n*ic RNA3 *re*qui*res* a b*i*par*ti*te s*i*g*n*al c*on*ta*i*ned *in* *t*he RNA - *t*he TLS a*n*d a *cis*-ac*t*i*n*g eleme*n*t (Ch*oi* a*n*d Ra*o*, 2003). The h*i*ghly c*on*se*rv*ed TLS *is* p*o*s*t*ula*t*ed *to* ac*t* as a *n*uclea*t*i*n*g eleme*n*t (NE) f*or* CP subu*n*i*t*s f*or* CP assembly a*n*d RNA e*n*caps*i*da*ti*on (Ch*oi et al.*, 2002) wh*i*le *t*he *cis*-ac*t*i*n*g eleme*n*t *is* *t*he p*o*s*i*t*i*on-depe*n*de*n*t packagi*ng* eleme*n*t (PE) *of* 187 *n*ucleo*ti*des *o*ccurr*i*ng *in* *t*he *n*o*n*s*t*ru*c*tu*r*al MP ge*n*e. These *t*wo packagi*ng* s*i*g*n*als (NE a*n*d PE) a*re* e*n*ough f*or* packagi*ng of* RNA3 a*n*d a*re* *in*tegral comp*on*e*n*ts *of* packagi*ng* co*re* (Ch*oi* a*n*d Ra*o*, 2003). Th*is* *is* *t*he f*i*rs*t* ev*i*de*n*ce *of* ex*i*s*t*e*n*ce *of* a b*i*par*ti*te packagi*ng* s*i*g*n*al *in* a*n* *i*cosahed*r*al v*i*ru*s*

Some mechanism must be regulating copackaging of RNAs 3 and 4 together. A regulatory element occurs in 3a ORF of BMV RNA3 from nucleotide 95 to 302 that regulates copackaging of RNA4 and suppresses packaging of RNA4 alone. This element seems to exist in 5`-proximal region of 3a ORF. Choi and Rao (2003) also give the first genetic evidence that *de novo* synthesized BMV RNA4 is incompetent for autonomous assembly and prior encapsidation of RNA3 is essential for RNA4 to be able to copackage in BMV virions. They hypothesize that copackaging of RNA3 and RNA4 into a single virion occurs either sequentially or in a concerted manner involving RNA-RNA interactions. Sequential packaging anticipates that binding of wild-type CP subunits to NE and PE causes prior encapsidation of RNA3 into a single virion with the basic N-arm of CP being on the surface of virion. The arginine residues in the exposed basic arm undergo specific interaction with RNA4 leading to copackaging of RNA3 and RNA4 into a single virion. Concerted packaging anticipates that PE may be acting in *trans* to form a complex with RNA4 resulting in copackaging. The complex may be formed through either RNA-RNA interactions or CP subunits (up on binding to PE) function directly or indirectly as chaperones to promote RNA3 and RNA4 copackaging.

2. *Cowpea chlorotic mottle bromovirus*

Unlike BMV, the N-terminal ARM of CCMV CP contains no specific determinants for viral RNA (Annamalai *et al.*, 2005); thus interaction between CCMV CP and an RNA species is nonspecific (Annamalai and Rao, unpublished results). Despite this, no cellular RNAs occur in mature CCMV virions. This suggests that CCMV possesses a mechanism that is distinct from that of BMV and acts as a selective filter during CCMV RNA encapsidation so that mature CCMV virions only package viral RNAs. Johnson *et al.* (2003) proposed such a mechanism, which is independent of viral RNA. CP binding slowly folds RNA into a compact structure (called CI complex) and, when CP concentration peaks during infection, CP preferentially binds to CI complex than to viral RNA, leading to RNA encapsidation.   However, Annamalai *et al.* (2005) proposed that binding of CP subunits to CCMV RNA1 does not lead to the formation of CI complex to which subsequently CP dimers are added leading to virion formation. Since no such CI complexes appear to be formed with CCMV genomic RNA2 or RNA3, it is not yet certain whether these genomic RNAs are encapsidated (Annamalai *et al.*, 2005). Thus, the closely related bromoviruses, BMV and CCMV, with apparently similar N-terminal ARMs, are not functionally analogous in terms of RNA packaging.

D. Other Spherical Plant Viruses

Protein-protein interactions stabilize tymoviruses [(TYMV) and *Eggplant mosaic virus* (EMV)] to a greater extent than all other groups of icosahedral plant viruses. This results in two important facts: presence of a large amount of empty RNA-free protein capsids in TYMV-infected cells and low solubility of coat protein subunits because of their high hydrophobicity. Thus reconstitution studies in TYMV were carried out on assembly of more or less intact empty capsids. Presence of spermidine or $MgCl_2$ in the reaction mixture was essential for re-association of TYMV and EMV capsids.

High cytosine content (38%) is present in RNAs of TYMV and EMV. Increased TYMV virion stability at pH 4.5 is due to interactions between carboxyl groups found on internal surface of CP and cytosines of RNA, which are protonated at this pH. TYMV reconstitution does not occur at pH 7.0 – showing the important role of cytosine protonation in RNA-protein interactions in virus particles. TYMV capsids reassociate with homologous RNA, EMV RNA (also containing 38% of cytosine), and with poly(C). Natural RNAs containing less than 30% of cytosine as well as poly(A) and poly (U) were not encapsidated. If 8% of cytosine in TYMV RNA was converted to uracil by bisulphite modification, this RNA was not recognised by TYMV capsids. These results confirm the importance of cytosine in RNA-protein recognition in TYMV. The above-described reconstitution conditions are absolutely unphysiological and the results obtained probably tell very little about mechanism of TYMV assembly in infected cells.

The 5`-UTR of tymoviruses contains conserved hairpins with protonatable internal loops, consisting of C-C and C-A mismatches (Hellendoorn *et al.*, 1996). The 5`-UTR of TYMV contains two such hairpins. These C-C and C-A mismatches have a functional role in replication of genomic RNA while 75% of 5`-UTR, including the two hairpins, seem to have a role in viral RNA packaging. In fact, the initiation signal for TYMV RNA encapsidation is possibly localized on the hairpin structure with characteristic C-C and C-A mismatches located in 5`-UTR close to polyprotein start codon, which is the contact area for 3`-TLS (Bink *et al.,* 2002).

Qu and Morris (1997) studied the assembly of TCV. Two regions of TCV interact specifically with CP: one region is made of two RNA motifs located within the 3`-proximal CP gene while the other region consists of three RNA motifs covering around 300 nucleotides in the putative RdRp-encoding gene and spans a suppressible termination codon. Four out of these five RNA motifs can adopt hairpin structures. Both of these regions are suggested to be involved in virus assembly. Thus, *cis*-acting sequences acting as packaging signal, for RNA packaging, have been demonstrated in this virus. Moreover, CP binding could switch viral RNA from translation/replication mode to encapsidation mode. The specific RNA-packaging signal is the 186-nucleotide region at 3`-end of P gene.

Lee *et al.* (2005) found that a surface loop of CP of PLRV is involved in virion assembly, systemic movement, and aphid transmission; that the structural model of the PLRV capsid predicted four amino acids are critical for virion assembly; and that these amino acids are located within and around the perimeter of a depression at the center of a coat protein trimer. These four amino acid residues are: Trp 171 that is oriented horizontally and forms the walls of a depression at the threefold axis of symmetry; Glu 170 that is projected into the center of the threefold axis of symmetry and forms a negatively charged depression; His 172 that surrounds the edge of a depression and is oriented toward Asp 177 in a manner that may allow their opposite charges to interact and stabilize the structure. Brault *et al.* (2003) reported that tryptophane residue in BWYV, corresponding to PLRV Trp 171, was projecting into the center of the threefold axis of symmetry but their results suggest that the negative charge is not important in assembly but may contribute to aphid-virus recognition.

ARM is also found in CPs of still other members of *Bromoviridae* and also in other plant virus genera like *Cucumovirus*, *Sobemovirus*, and *Tombusvirus*. But, keeping in view the different assembly mechanisms of the two closely related bromoviruses (BMV and CCMV), it can be safely assumed that assembly process in these icosahedral viruses may be somewhat different in each case.

V. SPECIFICITY OF VIRUS ASSEMBLY

Most of the experimental evidence shows that specificity of CP-RNA interactions is not of a high order. The most thoroughly recognised packaging signal is that of TMV in which specific interaction between CP and a nucleotide region of RNA leads to specificity of virion assembly. Such specificity of virion assembly can be very crucial in situations where a cell is infected by two different closely related plant viruses or strains of a virus. Two such phenomena that could exist in doubly infected plants are genome (genomic) masking and formation of phenotypically mixed virions. In genomic masking, virions contain RNA of one virus and CP of the other; i.e, CP of a particular virus encloses the RNA (masks the genome) of another virus. Phenotypically mixed virions possess a mosaic capsid made up of protein subunits of both the parental CPs (i.e., capsids of such particles contain coat protein subunits of both strains). Both these phenomena have been reported *in vivo* in several cases in plants infected by two viruses.

In doubly infected cells, the TMV CP *ts*-mutants Ni 118 and *flavum* are complemented by temperature-resistant (*tr*) strains *vulgare*, A14, and *Dolichos enation mosaic virus* (a bean strain of TMV distantly related serologically to the *vulgare* strain). At non-permissive (restrictive) temperature, CP *ts*-mutants are complemented by *tr*-helper virus by masking of *ts*-genome with the CP of *tr*-strain. Similarly, some CP-defective TMV strains are also complemented in plants doubly infected with a bean strain of TMV (Nigerian strain) and produce mature progeny. In these cases, genomic masking occurred in conditions when the infected cells contained functional RNAs from two strains and a functional protein from only one of them so that the mutant virus RNA did not have any access to an active homologous protein during the course of virus particle formation.

Genomic masking and phenotypically mixed virion formation was also studied in situations when infected cells contained both types of functional CPs as well as both types of active RNA; i.e., in conditions when homologous as well as heterologous RNA-protein and protein-protein interactions are possible in principle. Such *in vivo* studies were conducted in tobacco plants doubly infected with *vulgare* and U2 TMV strains. Virions having genomes masked by heterologous coat protein and phenotypically mixed capsids could not be produced *in vivo* in this system. – showing that specificity of RNA-protein and protein-protein interactions is rather high during TMV maturation *in vivo*.

Thus, when viral RNAs and CPs of both strains are present in active form in doubly infected cells, the homologous RNA and protein specifically recognise each other. However, joint infection of plants with *aucuba* and T TMV strains did produce many virus particles possessing a mosaic viral capsid but no genomic masking was observed. This also shows high specificity of RNA-protein interactions during TMV assembly *in vivo*. The occurrence of genomic masking suggests that the regulatory mechanism that enables a CP to specifically recognise the proper nucleic acid may be overcome. Many such experiments have been conducted. Hybrid TMV strains, showing genomic masking, have been reconstituted from protein of one strain and nucleic acid of another. Various combinations of RNAs and protein from different TMV strains like Holmes ribgrass, masked, yellow aucuba, and wild were conducted. Rod-like TMV particles were reconstituted in all cases. Reconstituted particles were infectious and disease symptoms invoked by the reconstituted virions correspond to the strain, which supplied RNA. TMV protein also reconstituted with RNA of other viruses (like TYMV, PVX, phage MS2 and synthetic polynucleotides) to produce hybrid viruses whose infectivity was always that of the virus providing RNA. However, rates and yields of heterologous reconstitution are much lower than reconstitution employing homologous TMV CP and RNA although this specificity is not absolute. The six triplets at the summit of RNA control this specificity and the discrimination between homologous and heterologous CP subunits possibly occurs at nucleation phase since the disk has increased affinity for specific nucleation site.

In vivo formation of TMV particles, possessing mixed coats in doubly infected-plant cells, has been widely reported. In fact, a correlation exists between the capacity of any two TMV strains to produce phenotypically mixed particles *in vivo* and the ability of the CPs of these particular strains to form hybrid 20S double disks *in vitro*. Thus, proteins of *aucuba* and T strains (which give phenotypically mixed particles *in vivo*) form hybrid 20S disks *in vitro*. In contrast, proteins of U2 and *vulgare* strains produced no mixed 20S disks *in vitro* and also no phenotypically mixed particles upon joint infection by these strains *in vivo*. TMV assembly initiation is characterized by greater specificity of RNA-protein and protein-protein interactions than in the elongation process. These results indicate that the degree of specificity upon TMV assembly *in vivo* is higher than expected on the basis of the *in vitro* reconstitution with TMV CP and heterologous RNAs or synthetic polynucleotides. Thus, some specific mechanism(s), which account for assembly with homologous RNA, possibly operates *in vivo*.

Most of the spherical viruses possess low specificity of assembly. Thus, CCMV capsid encapsidates homologous RNA, RNAs of other bromoviruses, TMV RNA, f2 phage RNA, and even ribosomal RNA with more or less the same efficiency. Small $T = 1$ virions were produced when CCMV protein was reconstituted with short poly(U) molecules but large $T = 4$ virions were produced when TMV RNA (which is longer than CCMV RNAs) was used for the assembly; the normal CCMV virions are $T = 3$. It establishes two facts: the size of RNA determines the size of the particles formed *in vitro* in bromoviruses and non-specific electrostatic interaction plays a major role in RNA-protein interactions during assembly of bromoviruses. Such studies have also been conducted on BMV. BMV RNA, on encapsidation in CCMV protein, was more infectious than when coated *in vitro* with homologous (BMV) protein. BMV protein also encapsidated TMV RNA to produce spherical particles containing TMV RNA and BMV CP; these particles were more infectious than free TMV RNA. Bromovirus protein also assembles with synthetic polyribonucleotides and polyvinylsulphate.

VI. CONCLUSIONS

Conditions required for assembly of different viruses vary widely. TMV is stable at high ionic strength and is reconstituted when the salt concentration of protein and RNA mixture is 0.1 *M*. Spherical viruses in contrast are disaggregated at high salt concentration and re-aggregate on decreasing the ionic concentration by dialysis of protein-RNA mixture against 0.01*M* KCl, 0.01*M* Tris and 5 x 10^{-3} *M* MgCl$_2$.

Degradation and reconstitution studies of spherical viruses indicate that ionic interactions or formation of salt links plays a dominant role in their reconstitution and that hydrophobic bonds between protein subunits of capsid are not the dominant bonds. This is unlike TMV and thus can be taken to indicate that different types of bonds are dominant in stabilizing the virions of different viruses.

Nucleic acid does play some definite role in assembly of viruses. In TMV, it is responsible for formation of initiation complex and to give a definite length to TMV particles. In AMV, nucleic acid appears to give the definite bacilliform shape to virions and thus serves to bind the protein subunits in a special configuration. This underlines the importance of nucleic acid in assembly and possibly also for the *in vivo* production of AMV. Thus, assembly *in vitro* seems to depend on the integrity and the necessary configuration of RNA.

It is conceivable that *in vivo* assembly of a virus may resemble the process of assembly *in vitro* and it occurs without the need for any morphopoietic factors. Enzymes play no role in the final aggregation of nucleic acid and protein to form mature virus particles. This is so because formation of new covalent bonds is not needed to build a stable protein coat or to affix it firmly to the nucleic acid core. Only weak secondary forces like hydrogen bonds, salt linkages and Van der Waals forces are needed for this. That is, reassembly of virus particles or formation of empty capsid can take place spontaneously without any external aid.

The specific recognition of initiation site on TMV RNA by TMV protein ensures that free host RNA is not encapsidated within TMV protein.

Carboxyl groups act as a negative switch for controlling viral assembly.

A double-layered disc in elongated viruses and a dimer in spherical viruses nucleate the RNA.

Large differences exist in our knowledge and understanding of assembly mechanisms *in vitro* and *in vivo* in different plant virus groups. In TMV, the endeavour is to work out the mechanisms of RNA-protein recognition at the level of individual functional groups of nucleotides and amino acid residues participating in self-assembly. On the other hand, even the general models of possible assembly mechanisms of many plant viruses are not yet enunciated.

Assembly initiation and elongation of the protohelix *in vitro* has two modes. One is functional in TMV in which CP subunits initiate TMV assembly from an internal site near the 3`-terminal of its RNA and the encapsidation of RNA is bidirectional. The second is in TRV, *Clover yellow mosaic virus* and PapMV in which *in vitro* assembly is polar and is initiated at the 5`-terminus of the respective RNA and encapsidation is unidirectional towards the 3`-terminus. In spherical viruses, tensing of the dimers appears to act as a switch for their assembly.

VII. REFERENCES

Annamalai, P., Apte, S., Wilkens, S., and Rao, A. L. N. 2005. Deletion of highly conserved arginine-rich RNA binding motif in cowpea chlorotic mottle virus capsid protein results in virion structural alterations and RNA constraints. *J. Virol.* 79: 3277-3288.

Bhyravbhatla, B., Watowich, S. J., and Caspar, D. L. D. 1998. Refined atomic model of the four-layer aggregate of the tobacco mosaic virus coat protein at 2.4 Å resolution. *Biophys. J.* 74: 604-615.

Bink, H. H. J., Hellendoorn, K., van der Meulen, J., and Pleij, C. W. A. 2002. Protonation of non-Watson-Crick base pairs and encapsidation of turnip yellow mosaic virus RNA. *Proc. Natl. Acad. Sci. USA* 99: 13465-13470.

Bol, J. F. 1999. *Alfalfa mosaic virus* and ilarviruses: Involvement of coat protein in multiple steps of the replication cycle. *J. Gen. Virol.* 80: 1089-1102.

Bragard, C., Duncan, G. H., Wesley, S. V., Naidu, R. A., and Mayo, M. A. 2000. Virus-like particles assemble in plants and bacteria expressing the coat protein gene of *Indian peanut clump virus*. *J. Gen. Virol.* 81: 267-272.

Brault, V., Bergdoll, M., Mutterer, J., Prasad, V., Pfeffer, S., Erdinger, M., Richards, K. E., and Ziegler-Graff, V. 2003. Effects of point mutations in the major capsid protein of *Beet western yellows virus* on capsid formation, virus accumulation, and aphid transmission. *J. Virol.* 77: 3247-3256.

Butler, P. J. G. 1984. The current picture of the structure and assembly of *Tobacco mosaic virus*. *J. Gen. Virol.* 65: 253-279.

Butler, P. J. G. 1999. Self-assembly of *Tobacco mosaic virus*: The role of a intermediate aggregate in generating both specificity and speed.. *Phil. Trans. R. Soc. London, Ser. B.* 354: 537-550.

Carr, J. P. 2004. Tobacco mosaic virus. Annu. Plant Revs. 11: 27-67.

Caspar, D. L. D. 1963. Assembly and stability of the tobacco mosaic virus particle. Adv. Protein Chem. 18: 37-121.

Choi, Y. G., and Rao, A. L. N. 2000. Molecular studies on bromoviruses capsid protein. VII. Selective packaging of BMV RNA4 by specific N-terminal arginine residues. Virology 275: 207-217.

Choi, Y. G., and Rao, A. L. N. 2003. Packaging of brome mosaic virus RNA3 is mediated through a bipartite signal. J. Virol. 77: 9750-9757.

Choi, Y. G., Grantham, G. L., and Rao, A. L. N. 2000. Molecular studies on bromovirus capsid protein. VI. Contributions of the N-terminal arginine-rich motif of BMV capsid protein to virion stability and RNA-packaging. *Virology* 270: 377-385.

Choi, Y. G., Dreher, T. W., and Rao, A. L. N. 2002. tRNA elements mediate the assembly of an icosahedral virus. *Proc. Natl. Acad. Sci. USA* 99: 655-660.

Correia, J. J., Shire, S., Yphantis, D. A., and Schuster, T. M. 1985. Sedimentation measurements of the intermediate-size tobacco mosaic virus protein polymers. *Biochemistry* 24: 3292-3297.

Cowan, G. H., Torrance, L., and Reavy, B. 1997. Detection of potato mop top virus capsid readthrough protein in virus particles. *J. Gen. Virol.* 78: 1779-1783.

Culver, J. N. 2002. Tobacco mosaic virus assembly and disassembly determinants in pathogenicity and resistance. Annu. Rev. Phytopath. 40: 287-308.

Culver, J. N., Stubbs, G., and Dawson, W. O. 1994. Structure-function relationship between tobacco mosaic virus coat protein and hypersensitivity in *Nicotiana sylvestris*. *J. Mol. Biol.* 242: 130138.

Culver, J. N., Daswon, W. O., Plonk, K., and Stubbs, G. 1995. Site-directed mutagenesis confirms the involvement of carboxylate groups in the disassembly of *Tobacco mosaic virus*. *Virology* 206: 724-730.

Damayanti, T. A., Nagano, H., Mise, K., Furusawa, I., and Okuno, T. 2002. Positional effect of deletions on viability, especially on encapsidation of brome mosaic virus D-RNA in barely protoplasts. *Virology* 293: 314-319.

Damayanti, T. A., Tsukaguchi, S., Mise, K., and Okuno, T. 2003. cis-Acting elements required for efficient packaging of brome mosaic virus RNA3 in barley protoplasts. J. Virol. 77: 9979-9986.

Dobrov, E. N., and Atabekov, J. G. 1989. Reconstitution of plant viruses. In: Mandahar, C. L. (Ed.). Plant Viruses. Volume I. Structure and Replication. CRC Press, Boca Raton, Florida, USA. p. 173-205.

Duggal, R., and Hall, T. C. 1993. Identification of domains in brome mosaic virus RNA1 and coat protein necessary for specific interaction and encapsidation. *J. Virol.* 67: 6406-6412.

Fox, J. M., Johnson, J. E., and Young, M. J. 1994. RNA/protein interactions in icosahedral virus assembly. Semin. Virol. 5: 51-60.

Fraenkel-Conrat, H., and Williams, R. C. 1955. Reconstitution of active *Tobacco mosaic virus* from its inactive protein and nucleic acid components. *Proc. Natl. Acad. Sci. USA* 41: 690-698.

Fukuda, M., and Ohno, Y. 1982. Mechanism of tobacco mosaic virus assembly: Role of subunit and larger aggregate protein. *Proc. Natl. Acad. Sci. USA* 79: 5833-5836.

Gilmer, D., Richards, K., Jonard, G., and Guilley, H. 1992. *cis*-Active sequences near the 5`-termini of beet necrotic yellow vein virus RNA 3 and 4. *Virology* 190: 55-67.

Goulden, M. G., Davies, J. W., Wood, K. R., and Lomonossoff, G. P. 1992. Structure of tobraviral particles: A model suggested from sequence conservation in tobraviral and tobamoviral coat proteins. *J. Mol. Biol.* 227: 1-8.

Harrison, S. C. 1983. Virus structure: High-resolution perspectives. *Adv. Virus Res.* 28: 175-240.

Hellendoorn, K., Michiels, P. J., Buitenhuis, R., and Pleij, C. W. A. 1996. Protonatable hairpins are conserved in the 5`-untranslated region of tymovirus RNAs. *Nucl. Acids Res.* 24: 4910-4917.

Hemmer, O., Dunoyer, P., Richards, K., and Fritsch, C. 2003. Mapping of viral RNA sequences required for assembly of peanut clump virus particles. J. Gen. Virol. 84: 2585-2594.

Johnson, J. M., Willitis, D. A., Young, M. J., Zlotnick, A. 2003. Interaction with capsid protein alters RNA structure and the pathway for *in vitro* assembly of *Cowpea chlorotic mottle virus*. *J. Mol. Biol.* 335: 455-464.

Kaplan, I. B., Lee, K.-C., Canto, T., Wong, S.-M., and Palukaitis, P. 2004. Host-specific encapsidation of a defective RNA3 of *Cucumber mosaic virus*. *J. Gen. Virol.* 85: 3757-3763.

Krol, M. A., Olson, N. H., Tate, J., Johnson, J. E., Baker, T. S., and Ahlquist, P. 1999. RNA-controlled polymorphism in the *in vivo* assembly of 180-subunit and 120-subunit virions from a single capsid protein. *Proc. Natl. Acad. Sci. USA* 96: 13650-13655.

Klug, A. 1999. The tobacco mosaic virus particle: Structure and assembly. *Phil. Tans. Roy. Soc. London* 354B: 531-535.

Klug, A., and Caspar, D. L. D. 1960. The structure of small viruses. *Adv. Virus Res.* 7: 225-325.

Lee, L., Kaplan, I. B., Ripoll, D. R., Liang, D., Palukaitis, P., and Gray, S. M. 2005. A surface loop of the potato leafroll virus coat protein is involved in virion assembly, systemic movement, and aphid transmission. *J. Virol.* 79: 1207-1214.

Lee, Y.-S., Lin, B.-Y., Hsu, Y.-H., Chang, B.-Y., and Lin, N.-S. 1998. Subgenomic RNAs of *Bamboo mosaic potexvirus* -V isolate are packaged into virions. *J. Gen. Virol.* 79: 1825-1832.

Lok, S., and Abou-Haidar, M. 1986. The nucleotide sequence of the 5`-end of papaya mosaic virus RNA: Site of *in vitro* assembly initiation. *Virology* 153: 289-296.

Lomonossoff, G. P., and Butler, P. J. G. 1980. Assembly of *Tobacco mosaic virus*: Elongation towards the 3`-hydroxyl terminus of the RNA. *FEBS Lett.* 113: 271-277.

Lu, B., Taraporewala, Z. F., Stubbs, G., and Culver, J. N. 1998. Inter-subunit interactions allowing a carboxylate mutant coat protein to inhibit tobamovirus disassembly. *Virology* 244: 13-19.

Miller, J. S., Wesley, S. V., Naidu, R. A., Reddy, D. V. R., and Mayo, M. A. 1996. The nucleotide sequence of RNA-1 of *Indian peanut clump furovirus*. *Arch Virol.* 141: 2301-2312.

Namba, K., and Stubbs, G. 1986. Structure of *Tobacco mosaic virus* at 3.6 Å resolution: Implications for assembly. *Science* 231: 1401-1406.

Namba, K., Pattanayek, R., and Stubbs, G. 1989. Visualization of protein-nucleic acid interactions in virus. Refinement of intact tobacco mosaic virus structure at 2.9 Å resolution by fiber diffraction. *J. Mol. Biol.* 208: 307-325.

Okada, Y. 1986. Molecular assembly of *Tobacco mosaic virus in vitro*. *Adv. Biophys.* 22: 95-145.

Osman, F., Grantham, G. L., and Rao, A. L. 1997. Molecular studies on bromovirus capsid protein. IV. Coat protein exchanges between *Brome mosaic virus* and *Cowpea chlorotic mottle virus* exhibit neutral effects in heterologous hosts. *Virology* 238: 452-459.

Osman, F., Choi, Y. F., Grantham, G. L., and Rao, A. L. N. 1998. Molecular studies on bromovirus capsid protein. V. Evidence for the specificity of brome mosaic virus encapsidation using RNA3 chimera of *Brome mosaic* and *Cucumber mosaic viruses* expressing heterologous coat proteins. *Virology* 251: 438-448.

Pattanayek, R., and Stubbbs, G. 1992. Structure of the U2 strain of *Tobacco mosaic virus* refined at 3.5Å resolution using X-ray fiber diffraction. *J. Mol. Biol.* 228: 516-528.

Qu, F., and Morris, T. J. 1997. Encapsidation of *Turnip crinkle virus* is defined by a specific packaging signal and RNA size. *J. Virol.* 71: 1428-1435.

Rao, A. L. N., and Grantham, G. L. 1996. Molecular studies on bromovirus capsid protein. II. Functional analysis of the amino terminal arginine rich motif and its role in encapsidation, movement and pathology. *Virology* 226: 294-305.

Romero, J., Huang, Q., Pogany, J., and Bujarski, J. J. 1993. Characterization of defective interfering RNA components that increase symptom severity of broad bean mottle virus infections. *Virology* 194: 576-584.

Rossmann, M. G. 1984. Constraints on the assembly of spherical virus particles. *Virology* 134: 1-11.

Rossmann, M. G., Abad-Zapatero, C., Hermodson, M. A., and Erickson, J. W. 1983a. Subunit interactions in *Southern bean mosaic virus*. *J. Mol. Biol.* 166: 37-83.

Rossmann, M. G., Abad-Zapatero, C., Murthy M. R. N., Liljas, L., Jones, T. A., and Strandberg, B. 1983b. Structural comparisons of some small spherical plant viruses. *J. Mol. Biol.* 165: 711-736.

Savithri, H. S., and Erickson, J. E. 1983. The self-assembly of the cowpea strain of *Southern bean mosaic virus*: Formation of $T = 1$ and $T = 3$ nucleoprotein particles. *Virology* 126: 328-335.

Schmitt, C., Balmori, E., Jonard, G., Richards, K., and Guilley, H. 1992. *In vitro* mutagenesis of biologically active transcripts of beet necrotic yellow vein virus RNA2: Evidence that a domain of the 75-kDa readthrough protein is important for efficient virus assembly. *Proc. Natl. Acad. Sci. USA* 89: 5715-5719.

Schuster, T. M., Scheele, R. B., Adams, M. L., Shire, S. J., Steckert, J. J., and Potschka, M. 1980. Studies on the mechanisms of assembly of *Tobacco mosaic virus. Biophys. J.* 10: 313-317.

Shaw, J. G. 1999. *Tobacco mosaic virus* and the study of early events in virus infections. *Phil. Trans. Roy. Soc. London* 354B: 603-611.

Sit, T. L., Leclerc, D., and AbouHaidar, M. G. 1994. The minimal 5`-sequence for *in vitro* initiation of papaya mosaic potexvirus assembly. *Virology* 199: 238-242.

Speir, J. A., Munshi, S., Wang, G., Baker, T. S., and Johnson, J. E. 1995. Structure of the native and swollen forms of *Cowpea chlorotic mottle virus* determined by X-ray crystallography and cryo-electron microscopy. *Structure* 3: 63-78.

Stubbs, G. 1999. Tobacco mosaic virus particle structure and the initiation of disassembly. *Phil. Trans. Roy. Soc. London* 354B: 551-557.

Tan, R., and Frankel, A. D. 1995. Structural variety of arginine-rich RNA-binding peptides. *Proc. Natl. Acad. Sci. USA* 92: 5282-5286.

Turner, D. R., Joyce, L. E., and Butler, P. J. 1988. The tobacco mosaic virus assembly origin. Functional characteristics defined by directed mutagenesis. *J. Mol. Biol.* 203: 531-547.

Turpen, T. H., Reinl, S. J., Charoenvit, Y., Hoffman, S. L., Fallarme, V., and Grill, L. K. 1995. Malarial epitopes expressed on the surface of recombinant *Tobacco mosaic virus. Biotechnology* 13: 53-57.

Vlot, A. C., Neeleman, L., Linthorst, H. J. M., and Bol, J. F. 2001. Role of the 3`-untranslated regions of alfalfa mosaic virus RNAs in the formation of a transiently expressed replicase in plants and in the assembly of virions. *J. Virol.* 75: 6440-6449.

Wang, H., and Stubbs, G. 1994. Structure determinants of *Cucumber green mottle mosaic virus* by X-ray fiber diffraction. *J. Mol. Biol.* 239: 371-384.

Wang, H., Culver, J. N., and Stubbs, G. 1997. Structure of *Ribgrass mosaic virus* at 2.9Å resolution: evolution and taxonomy of tobamoviruses. *J. Mol. Biol.* 269: 769-779.

Wesley, S. V., Mayo, M. A., Jolly, C. A., Naidu, R. A., Reddy, D. V. R., Jana, M. K., and Parnaik, V. K. 1994. The coat protein of Indian *Peanut clump virus*: Relationship among furoviruses and *Barley stripe mosaic virus. Arch. Virol.* 134: 271-278.

Wu, X. J., and Shaw, J. G. 1998. Evidence that assembly of a potyvirus begins near the 5`-terminus of the viral RNA. *J. Gen. Virol.* 79: 1525-1529.

Zaitlin, M. 1999. Elucidation of the genome organisation of *Tobacco mosaic virus. Phil. Trans. Roy. Soc. London* 354B: 587-591.

Zhao, X., Fox, J., Olson, N., Baker, T. S., and Young, M. J. 1995. *In vitro* assembly of *Cowpea chlorotic mottle virus* from coat protein expressed in *Escherichia coli* and *in vitro* transcribed viral cDNA. *Virology* 207: 486-494.

Zlotnick, A., Aldrich, R., Johnson, J. M., Ceres, P., and Young, M. J. 2000. Mechanism of capsid assembly for an icosahedral plant virus. *Virology* 278: 450-456.

11 HOST FACTORS AND VIRUS MULTIPLICATION

I. INTRODUCTION

Genetic and biochemical studies along with mutational studies in model genetic systems have unequivocally demonstrated that numerous cellular proteins of the host (host factors) are involved in replication and transcription processes of positive-strand RNA plant viruses (Buck, 1996; Lai, 1998; Strauss and Strauss, 1999; van der Heijden and Bol, 2002; Noueiry and Ahlquist, 2003; and Ahlquist *et al.*, 2003). In fact, genomes of all these viruses replicate *in vivo* in close association with existing or induced intracellular host membranes or membrane-bound replication (or RdRp/replicase) complexes. Virus-induced accumulation and/or proliferation of membranes is commonly employed by RNA viruses to increase the available membrane surface area and to provide for the membranous structures to be employed as scaffold to assemble their replication machinery.

Two approaches have been mainly used for studying the involvement and role of host proteins in plant virus RNA replication. The first approach is to purify the solubilized polymerase and to search for the host protein(s) that copurify with polymerase. Copurification of the host proteins along with RdRp has been detected in several plant viruses such as BMV, CMV, CPMV, RCNMV, and TYMV. The second approach is to search for host proteins that are specifically bound to terminal motifs of viral RNAs. Such specific binding of host proteins occurs in a wide variety of plant viruses (Hayes *et al.*, 1994).

Dzianott and Bujarski (2004) found that all cellular factors essential for BMV RNA replication, transcription, and RNA recombination are present in *Arabidopsis thaliana* so that studies employing this model genetic system should facilitate the identification of host factors governing BMV life cycle.

II. HOST PROTEINS AND MEMBRANES

Lai (1998) lists the following types of cellular proteins being involved in replication of RNAs of animal and plant viruses: the Pol III transcript binding proteins; heterogeneous nuclear ribonucleoproteins; cytoskeletal or chaperone protein; translation factors; and miscellaneous proteins like glyceraldehyde-3-phosphate dehydrogenase, calreticulin, and Sam 68. The concerned cellular proteins have been detected in three host types – infected plants, yeast genetic system, and *A. thaliana* genetic system. This shows that host factors (of the relevant host) are involved in plant viral

RNA replication whatever be the genetic system in which such studies were conducted. The host proteins could be bound to viral RNAs or exist as subunits of viral replicase.

Several studies have shown the involvement of plant factors/genes in the viral RNA replication. An unidentified cellular protein is reported in replicase preparations of CMV (Hayes and Buck, 1990). Similarly, replicase preparations obtained from CPMV- and RCNMV-infected tissues contained various unidentified host proteins. Abbink *et al.* (2002) identified two tobacco host proteins with a putative role in TMV RNA replication. One of these two proteins is designated #3. It belongs to a protein family of ATPases associated with various host activities. The second host protein is designated #13 and is the 33-kDa subunit of oxygen-evolving complex of photosystem II. Both these tobacco proteins interact with helicase region of 126-kDa TMV U1. The eIF3 is found in RdRp preparations of BMV and TMV, and is essential for activity of the relevant RdRp (Quadt *et al.*, 1993; Osman and Buck, 1996). The p41 subunit of eIF3 copurifies with BMV RdRp complex, is bound to 2a polymerase protein with great affinity and specificity irrespective of the presence or absence of 1a protein, and causes 3-fold stimulation of negative-strand RNA synthesis *in vitro*. Phosphorylation of replication cofactor p33 of CNV *in vivo* and *in vitro* by a membrane-bound plant kinase is a reversible process (Stork *et al.*, 2005; Shapka *et al.*, 2005). Phosphorylation of CNV p33 inhibited the ability of p33 to bind to viral RNA. Role of eIFs has already been discussed in an earlier Chapter. RNA replication of all plant viruses occurs in association with some specific host membrane. Thus, host factors and membranes are intimately associated with virus RNA replication.

Model genetic systems of *A. thaliana* plants (Lellis *et al.*, 2002; Hagiwara *et al.*, 2003; Tsujimoto *et al.*, 2003) and *Saccharomyces cerevisiae* yeast (Noueiry *et al.*, 2000; Diez *et al.*, 2000; Lee *et al.*, 2001; Tomita *et al.*, 2003) have been extensively employed for studying the host protein requirements for viral RNA replication and formation of replication complexes. Yeast two-hybrid system (Fields and Song, 1989; Chien *et al.*, 1991; Fields and Sternglanz, 1994) has particularly proved very useful and is now extensively employed in the study of various aspects of plant viruses. Host mutants of model genetic systems have been isolated in which intracellular multiplication of a virus is suppressed and the wild-type genes corresponding to some of these mutants have been cloned (Ishikawa *et al.*, 1997a, 1997b; Ohshima *et al.*, 1998; Noueiry *et al.*, 2000; Lee *et al.*, 2001). Such genetic studies have led to identification of *S. cerevisiae* genes essential for RNA replication and transcription of BMV RNA and an *A. thalaina* gene *TOM1* which is essential for efficient multiplication of TMV in which this gene encodes a putative transmembrane protein (Yamanaka *et al.*, 2000) (Table 1).

BMV successfully directs RNA replication, transcription of subgenomic RNA, and assembly of virus particles in yeast (Janda and Ahlquist, 1993). Some host genes are necessary for virus RNA replication (Ishikawa *et al.*, 1997a, 1997b). One such gene is *DED1*, another is *YDJ1* (Tomita *et al.*, 2003), and still another is *OLE1*. Mutations in any of these genes blocked BMV RNA replication in yeast. The blocking due to mutation in *OLE1* gene occurs after 1a and 2a replication proteins and viral RNA have associated with membrane but before start of initiation of negative-strand RNA synthesis (that is, between template recognition and RNA synthesis) and also because this mutated host gene disables still another host gene that encodes Δ9 fatty acid desaturase which is the key enzyme for converting saturated to unsaturated fatty acids and so interferes with membrane lipid composition (Lee *et al.*, 2001). Three more yeast genes (*MAB-1*, *MAB-2*, and *MAB-3*) seem to have a role in BMV RNA replication (Ishikawa *et al.*, 1997a). Efficient recruitment of BMV genomic RNAs from translation to replication mode is facilitated by viral 1a protein but it requires *LSM1* host yeast gene (Díez *et al.*, 2000) that encodes a protein Lsm1. The MAB-2 and MAB-3 yeast host proteins are bound to BMV RNA but their function in viral replication is not known.

Pantaleo *et al.* (2004) expressed both the p36 and p95 replicase proteins of *Carnation Italian ringspot virus* in yeast. These two proteins cooperated in dramatically increasing recruitment and stabilizing defective interfering RNA transcripts. However, p36 exerts its activity only after reaching a threshold below which the protein is unable to stabilize template RNA like the 'all-or-none' behaviour of ORF1 protein (p33) of TBSV during the *in vitro* studies of Rajendran and Nagy (2003).

TABLE 1

Proteins of host plants of *Arabidopsis thaliana*, and of *Saccharomyces cerevisiae* that interact with viral RNAs or RdRp subunits of plant alpha-like viruses (After van der Heijden and Bol, 2002, with additions)

VIRUS	HOST PROTEIN	VIRAL COMPONENT
	Viral RNA Genome	
Several plant viruses with 3` tRNA-like structure	eEF1a	3` aminoacylated tRNA- like structure
Brome mosaic virus	Lsm1p	Unknown
Brome mosaic virus	MAB2, MAB3	Unknown
Tobacco mosaic virus	TOM2	Unknown
	RdRp Subunits	
Brome mosaic virus	P41 subunit of eIF3	2a polymerase
Tobacco mosaic virus	GCD 10-like subunit of eIF3	126/183-kDa protein (MT)
	TOM1 #3 and 13 tobacco proteins	126/183-kDa protein (HE) 126-kDa (HE)

MT: Methyltransferase-like domain

HE: Helicase-like domain

A. thaliana TOM1 and TOM2A host proteins are thought to be essential for intracellular multiplication of TMV since inactivation of either of the two relevant genes (*AtTOM1* or *AtTOM2A*) causes decreased multiplication of the virus (Ishikawa *et al.*, 1993; Ohshima *et al.*, 1998). TOM1 host protein most likely is an essential component of TMV replication complex while TOM2A facilitates formation of replication complex without being its component (Hagiwara *et al.*, 2003). The *A. thaliana* protein TOM2 is possibly bound to TMV RNA.

III. FUNCTIONS OF CELLULAR FACTORS

Functions of host proteins (borrowed by plant viruses) within the host cell are not always known although most of the host proteins, that are essential for viral RNA replication, also have some essential role in the host (Noueiry *et al.*, 2000; Lee *et al.*, 2001). Moreover, host proteins do not necessarily perform the same functions during virus replication as they do in the host cell. Some of these factors have been identified but many still remain to be identified. Their identification has been slow primarily because purification of active replicases is very difficult and studies involving purified RdRps are not many.

Plant host proteins could function as subunits of viral replicase in many plant viruses; however, it is not yet settled whether these proteins are parts of RdRps or are mere contaminants. Similarly, yeast host proteins also function as subunits of viral RdRps. Yeast RNA-binding protein (GCD-10 subunit of eIF3) and two other host proteins copurified with RdRp complex obtained from TMV-infected tomato plants (Osman and Buck, 1997). The GCD-10-like protein interacts with methyltransferase domain (Taylor and Carr, 2000) of 126/183-kDa TMV replication proteins so that eIF3 performs a role in TMV replication. Functions of eIF3 are stabilization of binding of Met-tRNA to 40S ribosomal subunit, binding of mRNA to ribosome, dissociation of 80S ribosomes into 60S and 40S subunits, and is suggested to play a central role in assembly of initiation complex. Similarly, *A. thaliana* protein TOM1 is possibly involved with helicase domain of 126/183-kDa TMV replication proteins (Yamnanaka *et al.*, 2000).

An RdRp preparation could replicate any viral RNA without exhibiting any viral RNA template specificity. Thus, a basic function of the host proteins is to confer virus-specificity to RdRps. The first true replicase to be extracted from a eukaryotic positive-strand RNA virus-infected tissue was from CMV-infected plants (Hayes and Buck, 1990). It completely and specifically replicated viral genome *in vitro* and the final replicase preparation contained 1a and 2a viral proteins and a 50-kDa host protein. Elimination of the host protein from crude replicase extract during purification caused it to lose its replicase activity as well as template-specificity. Since then, all the purified RdRp preparations obtained from positive-strand RNA virus-infected tissues have been found to contain some host protein. The viral replicases of BMV, CMV, and TCV successfully replicate and transcribe viral RNAs in a specific manner only in the presence of the cellular protein (Lai, 1998). Moreover, many RdRps do not bind to viral RNA specifically or at all and their binding to template RNA must be mediated by cellular proteins that dock specifically to viral RNAs. Thus, cellular factors are necessary for template-specific RdRp synthesis and are a necessary pre-requisite for successful replication of the specific viral RNA (Lai, 1998).

Lai (1998) proposed two models in this connection. First model regards cellular proteins as being components of the replicase. The polymerase of *Vesicular stomatitis virus*, an animal virus, binds EF-1α very tightly. The complex so obtained then binds EF-1β and EF-1γ. Thus, all the three elongation proteins are required by this virus for *in vitro* synthesis of its RNA (Strauss and Strauss, 1999). In plant viruses, the elF3 binds to RdRps of BMV and TMV but the two RdRps contain different elF3 subunits. The elF3 binds directly to 2a protein (viral polymerase protein) of BMV but its mode of action in TMV replicase is not yet worked out. It is possible, therefore, that mechanistic and functional roles of elF3 translation factor are different in these two plant viruses.

Second model suggests that cellular proteins bind to viral RNA template and interact with *cis*-acting regulatory viral RNA sequences. It is necessitated because many viral replicases either do not seem to bind directly to *cis*-acting regulatory or promoter motifs on viral RNA or fail to do so in a specific manner - implying that viral replicases by themselves fail to initiate RNA synthesis at specific regions. Cellular proteins eliminate this deficiency. The cellular proteins by binding to RNA template direct the replicase to template. The protein that binds to viral RNA occurs in regions that regulate viral RNA synthesis. These sequences are located at 5`-end, 3`-end, and internal sequences of different RNA genomes of the positive-strand RNA viruses. Translation factor EF-1a binds to 3`-end of TYMV genome. The viral RNA synthesis, but not the viral RNA translation, seems to be directly linked to the translation factor. In short, these proteins are involved in RNA synthesis.

Interactions may also take place between viral RdRp and cellular factors to form an RdRp complex without the RNA template (as in BMV and TMV) or occur after the cellular factors have already docked with the RNA template. As a result, viral RdRps and cellular factors cooperate to form transcription or replication complexes on viral RNA.

The model of BMV RNA replication generated by Pogue *et al.* (1994) suggests that the cellular factors bind to internal control regions in double-stranded RNA replication intermediate. This may release a single-stranded 3`-terminus on negative-strand that could be essential for initiation of the positive-strand RNA synthesis. They further suggested that asymmetric synthesis of excess plus-strand over negative-strand RNA synthesis, which is typical of all positive-strand RNA viruses, could be possible by transition of RdRp competence from negative-strand to positive-strand synthesis by recruiting still more cellular elements.

Thus, host factors are being increasingly shown to be involved and play important roles during replication of positive-sense RNA plant viruses so much so that they participate in most, if not all, stages of infection including virus entry, selection and recruitment of viral RNA replication templates, viral gene expression, virus genome replication including assembling of viral RNA replication complexes and activation of replication complexes for RNA synthesis, virus particle assembly, and virion release. Initiation of negative-strand synthesis can be influenced in several ways by host factors: inhibition of viral RNA translation, activation of specific functions of replication complex, and recognition and utilization of 3`-end of RNA as initiation site

(Buck, 1996; Lai, 1998). Host factors could also participate in initiation of positive-strand RNA synthesis from negative-strand RNA templates (Ahlquist *et al.*, 2003).

Replication of tobamoviruses is inhibited by mutations in specific host genes and this regulation is achieved through interactions between viral proteins and a certain host factor(s) (Yamanaka *et al.*, 2002) – indicating that the limited host range of many plant viruses must be due to involvement of host-specific proteins in viral replication. In TBSV, host-specific symptom determination is through 19-kDa virus protein that enables systemic spread of virus in certain host plants through a mechanism associated with p19 suppression of gene silencing (Qiu *et al.*, 2002; Qu and Morris, 2002). The above virus-host interactions can be responsible for host and tissue specificity leading to specific infections.

Shapka *et al.* (2005) and Stork *et al.* (2005) arrived at the following conclusions: that posttranslational modification of replication proteins is a general feature among members of *Tombusviridae* family; that *in vitro* phosphorylation of threonine/serine residues adjacent to the essential arginine-proline-rich RNA-binding site in the auxiliary p33 protein likely plays a role in viral RNA replication and subgenomic RNA synthesis. They concluded that the primary function of reversible phosphorylation of p33 is to regulate the RNA binding capacity of p33, which could affect the assembly of new viral replicase complexes, recruitment of viral RNA template into replication and/or release of viral RNA from other processes like viral RNA encapsidation and cell-to-cell movement in infected hosts.

Host factors may be involved in still other steps. Ahlquist *et al.* (2003) concluded that the available results possibly have exposed merely a small part of the virus-host interactions that control viral replication. The repeated appearance of close integration of viral and host functions during the various reported studies has produced, according to Ahlquist *et al.* (2003), a more holistic view of virus-infected cells as a unified entity that constitutes the functional unit of infection.

IV. SITES OF VIRAL RNA REPLICATION

Different sites (membranes of different organelles) associated with viral RNA replication are cytoplasm, chloroplasts, nuclei, mitochondria, and vacuoles (tonoplast). Replication of no plant virus (unlike animal viruses) has so far been found to be associated with Golgi apparatus, endosomes, and lysosomes. The specific targeting of plant viruses to specific membranes of different organelles indicates that various plant viruses may require unique host factors that are only supplied by specific intracellular membranes or it may be related to other steps in viral life cycle as during viral protein translation and/or encapsidation. The membranes in general compartmentlize virus RNA replication.

A. Cytoplasm

It is the site of virus multiplication and/or assembly of a large number of viruses and all plant viruses present in cytoplasm generally belong to this category.

1. Cytoplasmic Inclusions (Viroplasms)

Viroplasms are electron-dense, proteinaceous, homogenous, amorphous bodies present in cytoplasm of host cells. They are not associated with any cellular organelle; contain virus particles; are the sites of virus replication/synthesis and/or assembly; and are formed by a large number of plant viruses including caulimoviruses, tospoviruses, plant reoviruses, plant rhabdoviruses, and several positive-sense RNA viruses including *Radish yellow edge virus*, *Sowbane mosaic virus*, TMV, and TSWV. Viroplasm replication complexes of TMV are the special cytoplasmic inclusions. They later enlarge during the course of infection to form the "X bodies", which are composed of aggregates of tubules that could be twisted around each other to form ropes, and are embedded in a ribosome-rich matrix. The tubules seem to be derived from ER. The viroplasms contain 126-kDa and/or 183-kDa replication proteins, are associated with tubules, and are possible sites of RNA replication.

Ultrastructurally, the X-bodies of CaMV are round or elliptical, lack a surrounding membrane, consist of compactly aggregated, dense, amorphous materials, and mostly possess within them one or more vacuole-like spaces that lack tonoplast membrane. Spherical virus particles, which often occur in masses but never in crystalline arrays, are scattered freely within vacuole-like spaces and also associated with inner portion and surfaces of X-bodies. These amorphous inclusions are the major and the most important replication site of caulimoviruses.

Viroplasms of *Fiji disease plant reovirus* are spherical bodies, are constituted by protein and double-stranded RNA, have vacuolate as well as reticulate appearance with light and dense staining areas, contain incomplete and/or complete virus particles, and are often the sites of virus replication and of the formation of complete virus particles. The presumptive viroplasms of plant rhabdoviruses are small or large structures, composed of granular and fibrillar material and contain a large number of naked nucleocapsids or complete virus particles at their periphery.

2. Endoplasmic Reticulum

Endoplasmic reticulum is involved in replication of picorna-like (including como-, nepo-, and potyviruses) and alpha-like (including bromoviruses, pecluviruses, and tobamoviruses) plant viruses.

Replication complexes of BMV (Sullivan and Ahlquist, 1997; Lee *et al.*, 2001) are targeted to ER membrane. The BMV 1a and 2a proteins and newly synthesized RNA colocalize in an ER-associated replication complex that is the site of BMV-specific RNA synthesis (Restrepo-Hartwig and Ahlquist, 1996, 1999; Dohi *et al.*, 2001). Localization of 2a protein on ER depends on interaction between 1a protein and an N-proximal 120-amino acid segment of 2a (Chen and Ahlquist, 2000). This segment of 2a contains one of the regions that interacts with 1a and is necessary and sufficient for 1a-directed localization of 2a to ER. This high affinity interaction between 1a and 2a N-terminal may be crucial to promote rapid assembly of replication complex that dramatically increases *in vivo* stability of genomic RNA3 before inoculum RNAs are degraded which can occur within minutes (Janda and Ahlquist, 1998).

Replication complexes of TMV (Más and Beachy, 1999) are targeted to ER membrane. They are nearly always assembled/localized on, and ultimately surround, strands or bodies of ER membranes (Heinlein *et al.*, 1998; Más and Beachy, 1999) as detected by immunofluorescent microscopy of infected cells while immunoelectron microscopy suggested them to be present in virus-induced cytoplasmic inclusion bodies that contained tubular membrane structures possibly of ER origin (Hills *et al.*, 1987). For TMV, ER alterations varied with stage of infection. TMV MP is an integral membrane protein, is responsible for ER redistribution, and both TMV viral replicase and MP localized to cytoplasmic face of cortical ER at initial stages of infection. At mid stage of infection, large cortical aggregates formed leading to complete disruption of cortical ER (Reichel and Beachy, 1998), as has been found in GFLV by Ritzenthaler *et al.* (2002).

Similarly, TEV employs ER-derived membranes for their replication and also induces modifications in ER network. TEV-encoded 6-kDa protein acts as an integral membrane protein that associates specifically with membranes derived from ER with which TEV replication complexes are associated. On the basis of these two observations, Schaad *et al.* (1997) proposed that direct interaction of 6-kDa protein with ER is involved in targeting TEV replication complexes to membranous sites of RNA synthesis. The 6-kDa protein acts as an anchor and secures part or all of the TEV replication complex apparatus to membrane sites in infected cells. The signal within the 6-kDa protein for the membrane-targeting of TEV replication complexes is located within the central 23 amino acid residues (amino acid residues 22 to 44) that consist of a central 19 amino acid hydrophobic domain and the flanking charged residues (Schaad *et al.*, 1997).

Infection by GFLV causes extensive cytopathic modifications of ER of host membrane system that finally forms a perinuclear 'viral compartment' (Ritzenthaler *et al.*, 2002). GFLV replication induces nearly complete depletion of cortical ER along with extensive redistribution of ER-derived membranes leading to formation of 'viral compartment' due to progressive build-up of perinuclear virus-induced ER aggregates/complex that are sites of GFLV replication as they contain virus RNA-encoded replication proteins, polyprotein precursors, newly formed viral RNA, and double-stranded viral replicative forms. It is significant that GFLV replication depends upon both ER-derived membrane recruitment and on *de novo* lipid synthesis (Ritzenthaler *et al.*, 2002). Thus, GLFV RNA replication on ER-derived membranes needs both continuous lipid biosynthesis leading to *de novo* synthesis of ER-derived membranes and a functional intracellular vesicle-dependent secretary transport system responsible for membrane vesicular trafficking. GFLV RNA1 alone is sufficient to cause complete cytopathic effects, unlike TMV in which viral MP is also involved.

CPMV also multiplies on ER (Carette *et al.*, 2000, 2002a, 2002b). Cortical ER is unaffected in cells infected with CPMV (Carette *et al.*, 2000) although this virus induced massive condensation of perinuclear ER on which replication occurred. The RNA1-encoded VPg and 110-kDa proteins of CPMV occur on condensed ER in plant cells (Carette *et al.*, 2000) and 60-kDa VPg induced membrane vesiculation in insect cells. PCV employs ER-derived membranes for replication and also induces modifications in ER network. Replication complex of PCV is associated with and occurs in modified vesiculated ER. Its replication proteins p131 and p191 along with double-stranded RNA colocalize in perinuclear region of ER (Dunoyer *et al.*, 2002). BYV, BWYV, CPMV, and many other plant viruses multiply on ER.

B. Chloroplasts

RNA replication of several plant viruses occurs on chloroplast membranes – AMV, BSMV, *Physalis mottle tymovirus*, TYMV, wild CMV and several other plant viruses. TYMV causes cytological abnormalities that are restricted to chloroplasts. These abnormalities include swelling and clumping of chloroplasts along with formation of peripheral structures consisting of membrane vesicles formed by invagination of chloroplast envelope into the organelle. The vesicles are of 50- to 100-nm diameter and are associated with TYMV RNA replication. Prod'home *et al.* (2001, 2003) and Jakubiec *et al.* (2004) show that TYMV 66-kDa protein localizes to virus-induced membrane vesicles present at chloroplast envelope; that 66-kDa protein shows cytoplasmic distribution and is targeted to chloroplast envelope by 140-kDa protein; that 140-kDa protein induces clumping of chloroplasts, is a key organizer of assembly of replication complexes and is the major determinant for virus localization and retention by chloroplasts. Jakubiec *et al.* (2004) further found that helicase domain of 140-kDa protein is unnecessary for proper recruitment of 66-kDa protein to chloroplast envelope while its proteinase domain appears to be essential for this process. Cytoplasmic invaginations of outer membrane of chloroplasts are sites of TYMV RNA synthesis. In AMV, p1 and p2 proteins (which act *in cis* for replication of RNAs 1 and 2, respectively) are targeted to outer membrane of chloroplasts where replication complexes are formed (de Graaff *et al.*, 1993; Bol, 1999). Thus, chloroplast outer membrane seems to be the probable site of AMV RNA replication.

Chloroplasts in all these cases develop vesicles, which are the sites of viral RNA synthesis or virus may be released by way of these vesicles. Vesicles contain numerous particles of *Physalis mottle tymovirus*, presumably synthesized in them, and are later budded off into cytoplasm from degenerating chloroplasts. TMV particles, observed by many workers to be present in chloroplasts, are enclosed within a double membrane of the chloroplast. They are regarded to be strictly within the chloroplasts but presumably occur in pinocytic-type cytoplasmic invaginations into the chloroplast membrane.

C. Other Cellular Organelles

Nuclei are sites of multiplication and/or assembly of many plant viruses; all viruses present in nuclei obviously multiply and/or assemble in nuclei. Nuclei may also be the exclusive sites of virus replication as in *Beet curly top curtovirus* and many other geminiviruses. Nucleoli are the specific sites of viral RNA replication of viruses multiplying in nuclei and have been found to be involved in replication of *Beet western yellows luteovirus*, *Pelargonium leaf curl tombusvirus*, and other plant viruses. Potyviruses cause invaginations in nuclear membrane. The 6-kDa protein expressed by

TEV potyvirus in transgenic plants appears to induce membranous proliferations associated with the periphery of nucleus. This protein also gets localized to these proliferations, which may be sites of virus replication in infected cells. Nuclear envelope is associated with replication of PEMV.

Mitochondria seem to be involved in replication of *Carnation Italian ringspot tombusvirus* (CIRSV), *Cucumber green mottle mosaic tobamovirus, Galinsoga mosaic carmovirus*, TRV, genus *Tombusvirus* (Burgyan *et al.*, 1996), TCV, and other plant viruses. Peripheral vesiculation of mitochondria leads to formation of multivesicular bodies (MVBs) in CIRSV. Both p36 and p95 of CIRSV, when expressed in yeast, localize to mitochondria independently of each other (Pantaleo *et al.*, 2004). The CIRSV ORF1 encodes a 36-kDa protein that localizes to mitochondria; the N-terminal part of ORF1 controls vesiculation of mitochondria. Weber-Lotfi *et al.* (2002) show that both the 36-kDa protein and complete replicase are targeted to mitochondria and anchor to outer mitochondrial membranes with the N-terminus while C-terminus is placed on cytosolic side and that anchor sequence appears to correspond approximately to amino acids 84 to 196 which contain two transmembrane domains. They suggest that membrane insertion of viral proteins is mediated through import of receptor-independent signal-anchor mechanism that relies on the two transmembrane segments and multiple recognition signals present in the N-terminal part of ORF1.

MVBs derived from peroxisome membrane are formed in carmo-like virus supergroup. These are characteristic of several tombusviruses (Burgyan *et al.*, 1996) like *Artichoke mottled crinkle virus, Cymbidium ringspot virus* (Russo *et al.*, 1983; Bleve-Zacheo *et al.*, 1997), CIRSV, *Eggplant mottled crinkle virus*, and TBSV. Double-stranded RNA is located in MVBs in some cases. Thus, MVBs are perhaps the sites of RNA replication. PCV infection is suggested to lead to partial redistribution of Golgi apparatus in ER but nothing is mentioned about its involvement in viral RNA replication (Dunoyer *et al.*, 2002). The subcellular origin of PCV MVBs is determined by N-terminal region of pre-readthrough protein encoded by the 5`-terminal ORF.

Tonoplast-associated vesicles protrude into vacuoles and are sites of replication of alfamo-, cucumo-, and ilarviruses (van der Heijden *et al.*, 2001). Such vesicles have been located in plant tissues infected by AMV, CMV, *Tomato aspermy cucumovirus*, occasionally TSV, and others. Vesicles induced by AMV are formed prior to appearance of virus particles in cytoplasm of infected cells and AMV p1 and p2 replication proteins are located in tonoplast. p1 is exclusively localized on tonoplast while p2 is localized on tonoplast as well as at other locations in the cell (van der Heijden *et al.*, 2001). CMV replication could also be associated with invaginations in vacuolar membrane (tonoplast) and these virus-induced vesicles contained double-stranded viral RNA. CMV replication proteins 1a and 2a and negative-stand RNAs colocalized predominantly to tonoplast while its plus-strand viral RNAs are distributed throughout cytoplasm in infected tobacco and cucumber hosts (Cillo *et al.*, 2002). In BMV, presence of membranes and/or phospholipids is essential for at least some steps of RNA replication both *in vivo* and *in vitro* (Lee *et al.*, 2001) while both plus- and negative-strand RNA synthesis occurs *in vivo* in membrane-associated complexes (Restrepo-Hartwig and Ahlquist, 1996, 1999).

V. REPLICATION COMPLEXES

Plant viruses induce extensive modifications and reorganization of intracellular membranes of infected cells leading to the formation of replication complexes known/regarded to contain and composed of viral template (replicating RNA), viral RdRp, cellular factors and/or eIFs, virus-encoded accessory proteins in some cases, and altered cellular membranes. Genome replication of all characterized positive-strand RNA plant viruses occurs in these large complexes – which are well-organised structures, perform many essential functions associated with virus multiplication in a highly coordinated manner, and are capable of synthesizing positive-strand, negative-strand, as well as subgenomic RNAs and are also able to cap viral RNAs. Therefore, methyltransferase-like, NTPase/helicase-like and polymerase-like proteins are also constituents of viral replication complexes as in plant viruses of alpha-like superfamily. Accordingly, some of the suggested and/or demonstrated functions of replication complexes are RNA polymerase activity, NTPase/helicase activity, capping functions, RNA replication, and assembly of virions. Transcription factors interact with viral polymerase to conduct RNA transcription or replication.

Replication complexes are also called polymerase (RdRp) complexes. Viral RdRps mostly fail to bind directly to *cis*-acting regulatory or promoter sequences on viral RNA. Ability of RdRps for initiation of RNA synthesis at specific sites appears to depend upon their interaction with cellular proteins that bind directly to viral templates. These interactions (between viral RdRp and cellular factors) may form an RdRp complex even in the absence of RNA template in situations when cellular factor is already a part of RdRp as in BMV and TMV. In other cases, cellular factors first dock with viral RNA template, as in TYMV, followed by interaction of this RNA template with viral RdRp. Thus, viral RdRps and cellular factors together form transcription or replication complexes on viral RNA.

The conclusion about the formation of replication complexes on intracellular membranes, which is generally coupled with formation of membrane vesicles, is based on four types of observations. Firstly, localization of viral replicase proteins and nascent viral RNA synthesis to intracellular membranes as revealed by immuno-fluorescence and immunoelectron microscopy studies on BMV (Restrepo-Hartwig and Ahlquist, 1996; Schwartz *et al.*, 2002), and TMV (Más and Beachy, 1999). Secondly, cofractionation of *in vitro* viral RdRp activity with cellular membranes (Schwartz *et al.*, 2002). Similarly, most of the viral replication complexes copurify with membrane extracts from infected cells (de Graaff and Jaspars, 1994). Thirdly, *in vitro* activities of viral replicase proteins are suppressed by non-ionic detergents but are enhanced by phospholipids in some cases. Fourthly, inhibitors of lipid synthesis (like brefeldin A) and mutations in lipid synthesis genes in BMV (Lee *et al.*, 2001) inhibit viral RNA replication. Brefeldin A inhibits formation of secretary vesicles and blocks RNA replication of *Poliovirus* and *Rhinovirus*. Similarly, multiplication of CPMV is sensitive to cerulenin that is a lipid synthesis inhibitor – indicating that CPMV RNA replication requires lipid and/or membranes synthesis (Carette *et al.*, 2000). Some positive-strand RNA viruses use membrane rearrangements for creating virus-specific, membrane-bounded compartments in which RNA replication occurs (Schwartz *et al.*, 2002).

Most plant viruses assemble their replication complexes on some specific membrane(s) but different viruses employ different membranes for this purpose. These membranes may be derived from endoplasmic reticulum, chloroplasts, mitochondria, nuclei, peroxisomes and vacuoles in plant viruses (Schaad *et al.*, 1997). This must be due to the establishment of specific interactions between such host membranes and virus-encoded proteins. However, the mechanism by which such viral replication complexes are targeted to and assembled on specific membrane sites remains poorly understood (Chen and Ahlquist, 2000). Many transcription factors are likely to be components of cellular RNA-processing pathways or translation machinery. These transcription factors are known to interact with each other and also with viral polymerase to conduct RNA transcription or replication (Lai, 1998). One or more viral proteins direct the assembly of virus replication complexes (Osman and Buck, 1996; Schaad *et al.*, 1997; Noueiry and Ahlquist, 2003). Animal viruses also employ Golgi bodies, endosomes, and lysomes (besides endoplasmic reticulum, mitochondria, nuclei, peroxisomes and vacuoles) for formation and function of replication complexes. The RdRp enzyme complex is synthesized/produced by vastly different strategies in different virus groups of animals and plants.

Recruitment of viral RdRp to replication complexes occurs in several plant viruses through an interaction with a membrane-anchored viral protein. Interaction of polymerase domain of one protein with helicase or proteinase domain of the other protein is essential for ensuring this. Polymerase-helicase interactions occur in BMV (Chen and Ahlquist, 2000) and the related AMV (van der Heijden *et al.*, 2001) and CMV (Kim *et al.*, 2002). It appears, therefore, that interactions of helicase and polymerase domains are essential for assembly of replication complexes in plant viruses of family *Bromoviridae* (AMV, BMV, and CMV) of alphavirus-like supergroup. A direct interaction of the nonconserved N terminus of polymerase protein 2a and the helicase domain of membrane-associated protein 1a is well documented in BMV (Kao and Ahlquist, 1992; O'Reilly *et al.*, 1995, 1997). An interaction of viral polymerase and proteinase domains occurs in potyviruses (TEV of picornavirus-like supergroup) for formation of replication complexes. The recruitment of NIb polymerase to replication initiation complexes occurs through protein-protein interactions with proteinase domain of the membrane-bound 6kDa/NIa precursor (Li *et al.*, 1997; Schaad *et al.*, 1997; Daros *et al.*, 1999); such interactions may represent a highly conserved core feature of the RNA replication apparatus of viruses within the picornavirus supergroup (Daros *et al.*, 1999).

TYMV (of alphavirus-like supergroup) seems to be a special case. Physical interaction has actually been detected between TYMV 140-kDa (that contains methyltransferase, proteinase, and NTPase/helicase domains) and 66-kDa (that contains polymerase domain) replication proteins by Jakubiec *et al.* (2004). The interaction domains mapped to the proteinase domain of the 140-kDa protein and to a large region encompassing the core polymerase domain within the 66-kDa protein. These protein-protein interactions in TYMV replication proteins differ from protein-protein interactions of some other alphavirus-like viruses in two aspects (Jakubiec *et al.*, 2004). One, in a number of alphaviruses (and several tripartite RNA plant viruses) the viral protein encompassing the polymerase domain contains a less conserved N-terminal segment in front of the core RdRp domain. In BMV 2a protein, this discrete N-terminal segment mediates interaction with the cognate helicase-like 1a protein (Kim and Ahlquist, 1992; O'Reilly *et al.*, 1995, 1997). The corresponding segment in 66-kDa protein is much smaller (~80 amino acids versus 220 and 280 amino acids in BMV 2a and AMV p2 proteins, respectively) and was not sufficient to promote binding to the 140-kDa protein. Instead, a large region of 66-kDa protein was required. Two, the interaction domain mapped within the TYMV 140-kDa protein and did not correspond to the helicase domain in other members of alphavirus-like supergroup (Kao and Ahlquist, 1992; O'Reilly *et al.*, 1995, 1997; van der Heijden *et al.*, 2001); instead   the helicase domain was unnecessary for recruitment of 66-kDa protein to chloroplast envelope while proteinase domain appeared to be essential for that process (Jakubiec *et al.*, 2004) as in potyviruses. This shows that different members of the alphavirus-like supergroup may have selected different pathways to assemble their replication complexes. Moreover, existence of similar polymerase-proteinase interactions in phylogenetically distinct TYMV and TEV viruses raises some interesting questions about the evolution of the assembly of replication complexes within positive-strand RNA plant viruses (Jakubiec *et al.*, 2004). Possibly interaction between the TYMV 140-kDa and 66-kDa proteins is regulated by conformational changes within the 66-kDa protein; these conformational changes could be caused by phosphorylation since phosphorylation of CMV 2a protein RdRp by a host kinase inhibited its interaction with 1a protein *in vitro* (Kim *et al.*, 2002).

A. Replication Complex of *Brome mosaic bromovirus*

This replication complex has been investigated in detail and is reviewed by Ahlquist *et al.* (2003). All BMV replication proteins are localized within the replicase complex. This replicase complex is an enzymatic complex that contains BMV-encoded 1a and 2a proteins, viral RNA, numerous 1a-containing intra-lumenal ER invaginations that form spherules (vesicles) in which BMV RNA templates are sequestered (Chen *et al.*, 2001; Schwartz *et al.*, 2002) and replicate, movement protein, a host factor eIF3, and possibly still other unidentified factors (Kao *et al.*, 1992; Quadt *et al.*, 1993; Schwartz *et al.*, 2002). BMV 1a is present in 25-fold excess over BMV 2a in membrane structures showing replicase activity. The BMV replicase complexes are electron-dense cytoplasmic inclusions similar to TMV viroplasms (Dohi *et al.*, 2001). The 5`-regions of BMV RNAs 1and 2 contain stem-loop structures with loop sequences resembling box B elements that are homologous to TΨC stem-loop of RNAs. These elements mediate recruitment of viral RNAs to replication complexes and are thus required for negative-strand RNA synthesis (Chen *et al.*, 2001; Schwartz *et al.*, 2002).

BMV proteins 1a and 2a together contain three domains (C-terminal helicase domain, N-terminal methyltransferase domain, and RdRp). Functions of 1a and 2a proteins are not entirely independent of each other; an intimate interaction between these proteins is essential for RNA replication and these proteins do interact both *in vivo* and *in vitro*. Their interaction is essential for successful completion of different steps of RNA replication *in vivo* including synthesis of negative-strand RNA, positive-strand genomic RNA, and subgenomic mRNA transcripts (Janda and Ahlquist, 1998). It has in fact been suggested that helicase-like 1a and polymerase-like 2a proteins together are most likely to form a well-organised replication complex in which helicase, polymerase, and conceivably capping functions are performed in an extremely coordinated manner.

BMV 1a is a multifunctional protein, has 109-kDa molecular weight, is involved in RNA replication, is required for formation of all BMV RNAs, controls template-specificity during RNA replication, seems to be part of catalytic subunit of RdRp and thus provides essential enzymatic functions, induces dramatic and specific stabilization of RNA3 template, plays key roles in assembly and functioning of replication complex, recruits viral RNA templates into replication mode (Chen and Ahlquist, 2000), and forms homodimers [through 1a-1a dimerization and involving guanylyltransferase/ methyltransferase-like domains as well as methyltransferase-helicase interactions (O'Reilley *et al.*, 1997, 1998)] that are suggested to interact with a 2a protein in replication complexes. Thus, it is suggested that each BMV replication complex contains two 1a molecules – one for eliminating secondary structure in template RNA strand and the other for unwinding the base pairs formed at the time of RNA synthesis (Buck, 1996). Template selection by 1a requires a box B-containing sequence present in 5`-non-coding regions of RNA1 and RNA2 or in intercistronic region of RNA3 (Chen *et al.*, 2001) and host protein Lsm1 in *trans* (Díez *et al.*, 2000). The 1a protein controls at least some stages of template-specificity in RNA replication by involvement of helicase-like motif in initiation of RNA replication. In short, 1a localizes to ER membranes, recruits BMV 2a polymerase and viral RNA templates, induces formation of membrane-bound spherules in which RNA replication occurs, and thus creates replication complexes.

Both C-terminal and N-terminal domains of 1a protein are involved in RNA replication and seem to be dependent upon each other. The C-terminal region contains DEAD box helicase domain, which seems to be involved in synthesis of all species and strands of BMV RNAs and is thus required for ongoing negative-strand, positive-strand, and subgenomic mRNA synthesis, and has key roles in assembly and functioning of RNA replication complex (Chen and Ahlquist, 2000). The N-terminal domain has 7-methylguanosine (m^{7}G) methyltransferase, and putative guanylyl-ransferase activities (Ahola *et al.*, 2000) and covalent GTP (m^{7}GMP) binding activities required for capping viral RNA.

The BMV 1a targets itself to ER membrane by signals residing in N-proximal half of the protein (den Boon *et al.*, 2001) and also recruits polymerase-like 2a protein to the ER membrane sites (spherules) of RNA replication (Chen and Ahlquist, 2000) through direct interaction between N-terminus of 2a and C-terminus of helicase-like domain of 1a (Schwartz *et al.*, 2002; Chen and Ahlquist, 2000). Thus, 1a ensures membrane association of BMV RNA replication complex.

Genetic studies indicate that 1a and RNA3 intergenic replication enhancer (RE) together perform important functions in viral RNA replication. They regulate selection of RNA3 templates for replication along with simultaneous inhibition of RNA3 translation, and strikingly and specifically increase stability and accumulation of RNA3 transcripts by blocking their degradation. Loss of RE and 1a-mediated events inhibit RNA3 negative-strand synthesis and thus of replication by about 100-fold. Similar mechanisms linking selection of viral RNA for replication along with translation inhibition are found in *Poliovirus*; and may be common or universal among positive-strand RNA viruses to avoid ribosome-polymerase collisions.

BMV 2a protein has 94-kDa molecular weight, is always present in active replication complex, is essential for the negative-strand RNA synthesis (Janda and Ahlquist, 1998), contains conserved sequence comprising GDD motif related to RdRp domain so that 2a is the putative polymerase that interacts with 1a helicase domain through a non-covalent bond, and is required for BMV replication. For successful BMV RNA replication, protein 2a must interact with protein 1a and/or it contributes to recognition of RNA 1 as replication template. The primary determinants for compatible and successful interaction of 2a protein with 1a protein are located within amino-terminal arm of 2a protein, which precedes the polymerase-like 2a protein sequence. These determinants on protein 2a are located within amino acid residues 25 to 140, and interact directly with C-terminal helicase-like domain of 1a.

B. Replication Complex of *Cucumber mosaic virus*

The first true replicase to be extracted from a eukaryotic plus-strand RNA plant virus-infected tissue was obtained from CMV-infected tissue and it completely and specifically replicated CMV genome *in vitro* (Hayes and Buck, 1990). This replicase in reality is regarded as replicase/replication complex since it contains viral proteins 1a and 2a (that perform polymerase and helicase functions) and one 50-kDa host protein, and possibly conducts capping functions by methyltransferase. Elimination of the host protein from crude replicase extract during purification caused the replicase to lose its activity as well as template-specificity. Interaction of CMV 1a protein with CMV 2a protein is disrupted by phosphorylation of 2a protein by a host kinase (Kim *et al.*, 2002) and occurs *in vivo*. This may either inhibit formation of a new replication complex late in infection so that virus is able to control its replication at a desired time or interaction between the two viral replication proteins helps the virus to maintain about 25:1 ratio of 1a to 2a protein in CMV replication complex (Schwartz *et al.*, 2002).

C. Replication Complex of *Tobacco mosaic virus*

Early cytological and immuno-electron microscope studies showed that cytoplasmic inclusion bodies, called X-bodies but now known as viroplasms, enlarge throughout infection, are sites of TMV RNA replication and/or assembly so that they are the viral replication complexes, and contain a large amount of 126- and 183-kDa replication proteins. The TMV replicase complex is associated with the cytoplasmic face of one or more types of intracellular membranes. Osman and Buck (1996, 1997) purified membrane-bound TMV RNA polymerase from infected plants and solubilized it from the virus-infected membrane fractions; Watanabe *et al.* (1999) also isolated TMV polymerase complex. Cytological, biochemical and electron microscopic studies (Osman and Buck, 1996, 1997; Reichel and Beachy, 1998; Watanabe *et al.*, 1999; Más and Beachy, 1999; Osman and Buck, 2003; Asurmendi *et al.*, 2004; Carr, 2004) generated the following information about TMV replication complexes: these bodies are not enclosed by a membrane; contain ribosomes, tubules, a number host-derived proteins including a plant protein related to RNA-binding subunit of yeast eIF3, heterodimer of 126-kDa and its readthrough 183-kDa viral replication proteins, viral RNA, CP, MP; the isolated TMV polymerase complex showed template-dependent and template-specific RNA polymerase activity and was virtually free from cellular RNA polymerase activity, and the polymerase for negative-strand RNA synthesis is perhaps composed of one molecule each of 126-kDa and 183-kDa proteins, possibly together with two or more host proteins, the polymerase is located within the movement protein bodies, and formation of replication complexes is regulated by CP. A region of 126- and 183-kDa proteins downstream of the core methyltransferase domain binds RdRp to 3`-terminal region of RNA *in vitro*. Two aromatic amino acids at positions 409 and 416 in this region of 126- and 183-kDa proteins are essential for this cross-linking to 3`-terminal region of RNA and for TMV RNA replication in tomato protoplasts (Osman and Buck, 2003). The two replication proteins have a common N-terminal methyltransferase-like domain that has guanylyltransferase activity, and a second domain that has been confirmed to be an ATP-dependent helicase capable of unwinding double-stranded RNA molecules (Goregaoker and Culver, 2003). The readthrough region of 183-kDa protein contains the polymerase domain. The specific regions required for heterodimerization of 126- and 183-kDa proteins and for full replicase activity have been identified and are located within the helicase domain and in the intervening region between the methyltransferase and helicase domains (Goregaoker *et al.*, 2001; Goregaoker and Culver, 2003).

In early stages of infection, virus replication complexes contain large amount of MP but small amount of polymerase but, at late stages of infection, amount of MP decreases while CP accumulation increases possibly as virions are formed (Asurmendi *et al.*, 2004). The TMV particles accumulate to a high level at periphery of viroplasms suggesting that virion assembly occurs there so that possibly TMV replicative intermediates are associated with these structures. The TMV CP is an integral membrane protein, is associated with host membrane with its N- and C-termini exposed to cytoplasm, and promotes formation of endoplasmic reticulum aggregates, and probably facilitates establishment of TMV replication complexes (Asurmendi *et al.*, 2004).

Asurmendi *et al.* (2004) give detailed information on structure and localization of TMV replication bodies. They found that viral replication complexes grow with time, that polymerase co-localizes with MP and CP, that polymerase is located within MP bodies/polymerase-rich bodies appear within MP-rich bodies or on periphery of replication complexes, and that CP surrounds the replication bodies leading to increase in their size. They proposed that virus replication complexes are anchored to endoplasmic reticulum by MP and polymerase, that viral RNA replication occurs in them, that CP performs a regulatory role in establishing virus replication complex, and that lack of formation of virus replication complexes restricts the efficiency of virus replication and formation of virus movement complexes leading to restriction of cell-to-cell spread of infection. S. Kawakami and Beachy (quoted in Asumendi *et al.*, 2004) calls structures, which are similar or identical to replication complexes and are transferred to and through plasmodesmata to adjacent cells, as virus movement complexes so that differences, if any, between replication and virus movement complexes are not yet known. Membrane interaction of host-encoded factors, that are part of viral replication complex, has also been detected in TMV-infected cells (Yamanaka *et al.*, 2000; Hagiwara *et al.*, 2003). Thus, MP and polymerase are primarily found in large bodies (virus replication complexes) existing throughout the cell but, at later stages of infection, CP becomes associated with these structures (Asurmendi *et al.*, 2004).

TMV multiplication, including the role of host proteins in formation and functioning of viral replication complexes, has also been studied in *Arabidopsis thaliana*. The *AtTOM1* and *AtTOM2A* genes are required for efficient TMV multiplication (Yamanaka *et al.*, 2000; Tsujimato *et al.*, 2003), their translation products [TOM1 (AtTOM1) and TOM2A (AtTOM2A)] are the host factors involved in virus multiplication (Ishikawa *et al.*, 1993; Ohshima *et al.*, 1998), both TOM1 and TOM2A host proteins play parallel and essential roles in TMV replication (Yamanaka *et al.*, 2000, 2002), *AtTOM1* gene encodes a 291-amino acid TOM1 polypeptide which is the transmembrane protein that acts as the membrane anchor for TMV replication complex *in vivo* and fixes the replication complex on host membrane (Yamanaka *et al.*, 2000, 2002), TOM1 host protein probably interacts with TMV 126-183-kDa proteins through conserved helicase domain; and TOM1 and TOM2A are the transmembrane proteins (Hagiwara *et al.*, 2003). TMV replication complexes (in *A. thaliana*) are formed on TOM1 protein-containing membranes and their formation is facilitated by TOM2A protein. Thus, these two integral transmembrane proteins are essential for efficient TMV multiplication, are constituents of and play important roles in formation of replication complexes on membranes on which they colocalize (Hagiwara *et al.*, 2003).

Hagiwara *et al.* (2003) also identified homologues of these proteins in *Nicotiana tabacum* (NtTOM1 and NtTOM2A), found that NtTOM1 interacts with tobamovirus-encoded 126/183-kDa replication proteins, that NtTOM1 and NtTOM2A are mainly localized on vacuolar membrane (tonoplast) and suggested that TMV replication complex is localized at least in part on tonoplast, that replication complex is formed on NtTOM1-containing tonoplast membranes so that NtTOM1 plays a critical role in formation of replication complex, and that formation of replication complex is facilitated by, but does not absolutely require, NtTOM2A. NtTOM1 interacts with helicase domain of TMV-encoded 126/183-kDa replication proteins as also with NtTOM2A so that both these host proteins are seemingly integral components of replication complex.

D. Replication Complex of *Cucumber necrosis virus*

Panaviene *et al.* (2004) confirmed that the functional CNV replicase complex contains both p33 and p92 replicase proteins and possibly also the RNA template. The p92 RdRp is functional *in vitro* only in presence of the auxiliary p33 protein. This was predicted earlier both in TBSV (Scholthof *et al.*, 1995) and CNV (Panaviene *et al.*, 2003) on the basis of biochemical and genetic studies. In addition, the endogenous defective interfering (DI) RNA is also an important factor in the CNV replicase complex. Presence of plus-strand DI-72 RNA in the same yeast cells gave a viral replicase preparation that had about 40-fold enhanced activity. The minus-strand DI-72 RNA failed to achieve the above enhancement in replicase activity. Possible function of the plus-stranded DI RNA templates is to promote assembly of functional replicase in cells by perhaps providing an assembly platform; similarly, viral RNA also possibly performs some role in assembly of the CNV replicase (Panaviene *et al.*, 2004).

E. Replication Complexes of Other Plant Viruses

Presence of multiple elements in replicase complex could also exist in replicases of several other plant viruses. On the contrary, RdRps obtained from heterologous systems in several viruses show them to be active without any auxiliary proteins. Such plant viruses are TVMV (Hong and Hunt, 1996), TCV (Rajendran *et al.*, 2002), BaMV (Li *et al.*, 1998) and some other viruses. In fact, the replicase complex of TCV, as also of tombus- and luteoviruses, is unusual since the helicase and NTP-binding domains, found in the more complete BMV replicase complex, is absent from it. The replicase of these small RNA viruses may be less efficient for synthesis of complementary RNA strands but show a corresponding increase in strand switching leading to recombined genomes. The AMV replicase complex requires several factors (like RdRp, helicase-like motif, a viral auxiliary protein, and viral RNA) in order to be functional *in vitro* (Vlot *et al.*, 2001, 2003).

The BYV replication complex has a complex ultrastructure consisting of multivesicular aggregates; each aggregate is composed of several 100-nm vesicles surrounded by a common membrane with cytoplasmic strands between them (Zinovkin *et al.*, 2003). The *in vivo* occurring methyltransferase-like (the 63-kDa product) and helicase-like (the 100-kDa product) proteins are co-localized *in situ* at membranes of BYV-induced multivesicular clusters (Erokhina *et al.*, 2000, 2001). Moreover, the papain-like cysteine proteinase is also associated with the induced membranous multivesicular aggregates (Zinovkin *et al.*, 2003).

Nepovirus-infected cells contain diffuse inclusions that often occur near nuclei and consist of complex membranous structures some of which form vesicles. Indeed, perinuclear membranous structures formed by massive proliferation of ER are the sites of replication of GFLV (Ritzenthaler *et al.*, 2002) and of CPMV comovirus (Carette *et al.*, 2000, 2002a). The 5`-region of PVX RNA forms complexes with a 54-kDa cellular protein which is bound to nucleotides 1 to 46 at 5`-terminus; presence of an ACCA sequence located at nucleotides 10 to 13 is important for optimum binding. CPMV infection of cowpea or *Nicotiana benthamiana* induces cytopathological structures consisting of amorphous matrix of electron-dense material traversed by rays of small membranous vesicles (Carette *et al.*, 2000).

F. Structure of Replication Complexes

Dohi *et al.* (2002) investigated molecular structure of active BMV RdRp complex. The intermediate area lying between N-terminal methyltransferase-like domain and

C-terminal helicase-like domain of 1a protein as well as N-terminus region of 2a protein are exposed on surface of the solubilized RdRp complex. But the intermediate region (between methyltransferase-like and helicase-like domains of 1a protein) in membrane-bound RdRp complex is located at border of a region buried within a membrane structure or within some membrane-associated material. Thus, BMV RdRp complex is comprised of RdRp and, nearly always, of an NTP-binding helicase motif located on the same or another protein (Kadarè and Haenni, 1997). BMV 1a localizes to outer nuclear envelope or perinuclear ER membrane (i. e., cytoplasmic face of ER), induces this membrane to invaginate into ER lumen to form 50- to 70-nm diameter vesicles (spherules), directs 2a polymerase and viral RNA template to these spherules which become sites of viral RNA synthesis (Restrepo-Hartwig and Ahlquist, 1999; Sullivan and Ahlquist, 1999; den Boon *et al.*, 2001; Schwartz *et al.*, 2002). The interior of these spherules is connected to cytoplasm through a membranous neck that is contiguous with ER membrane.

A yeast mutant, possessing the mutated essential yeast gene *OLE1*, was obtained by Lee *et al.* (2001). This gene encodes a fatty acid (FA) desaturase (ole1p) enzyme which is an integral ER membrane protein, catalyzes the synthesis of unsaturated FA (UFA) in yeast, and converts saturated palmitic acid and stearic acid into unsaturated palmitoleic acid and oleic acid, respectively. BMV RNA replication was inhibited in this yeast mutant indicating that BMV RNA replication depends on the presence of UFA, and that BMV RNA replication is highly sensitive to membrane lipid composition (Lee *et al.*, 2001). Lee and Ahlquist (2003) defined the relationship between BMV 1a protein, membrane lipid composition, and RNA synthesis. They showed that the numerous 1a protein-induced spherules are surrounded by invaginations of outer ER membrane, that 1a preferentially interacts with one or more types of membrane lipids, that 1a expression markedly increased membrane lipid accumulation in each cell, and that viral RNA synthesis is highly sensitive to membrane lipid composition.

Expression of BMV RNA replication factor 1a in yeast enhanced accumulation of membrane lipid per cell by 25 to 30 per cent possibly due to increased formation of intracellular membranes (Lee and Ahlquist, 2003). This supports the earlier observations of Schwartz *et al.* (2002) that 1a induces formation of vesicle-like invaginations, which greatly increase area of affected ER membrane and form compartments (spherules) in which RNA replication occurs. The BMV 1a replication protein accumulates in localized patches on ER membranes in high concentrations and locally modulates membrane lipid composition, and its numerous molecules are present in each vesicle (Schwartz *et al.*, 2002). Thus, local variations in lipid composition seem to be extremely significant for formation and properties of vesicular, spherular membrane invaginations that envelope replication complexes of BMV, and many other positive-strand plant viruses.

Host membrane performs several functions (Ahlquist *et al.*, 2003): it provides the surface on which replication factors are localized and concentrated for assembly; it surrounds and protects virus-induced replication compartments (spherules) in which RNA replication factors and genomic RNAs are sequestered from the competing RNA templates and competing processes such as translation; and spherules protect viral double-stranded RNA replication intermediate from double-stranded RNA-induced host defence responses like RNA interference or interferon-induced reactions. The replication complex, therefore, can be regarded a virus-specific organelle for viral RNA replication since it is new membrane-bound compartment that contains specific components and performs specific functions (Ahlquist *et al.*, 2003).

VI. MEMBRANE-TARGETING AND ANCHORING OF REPLICATION COMPLEXES

Viruses possess two types of specialized sequences, each encoding a specialized viral protein. One type of specialized proteins is thought to act as membrane anchors by either directly associating with intracellular membranes as integral membrane proteins and so become integral part of host membranes or by interacting with membrane proteins from host. The viral protein, which acts as membrane anchor, could interact with other viral proteins (like polymerase) to redirect them towards the membrane leading to the formation of a replication complex. The viral membrane anchor may associate with membranes as mature proteins or as larger polyprotein precursors as in picorna-like viruses. The 6-kDa protein of TEV is an integral membrane protein, is suggested to play a role as membrane anchor for replication complex, and its association with membrane is mediated by a transmembrane domain that consists of stretches of hydrophobic residues (Schaad *et al.*, 1997). Han and Sanfaçon (2003) and Wang *et al.*, (2004) studied the localization of replication complexes of bipartite *Tomato ringspot virus* (TomRSV) on ER. The RNA1-encoded polyprotein p1 contains domains including the putative nucleoside triphosphate (NTP)-binding protein (NTB), VPg, a proteinase, and a polymerase. The NTB protein contains a hydrophobic region at its C-terminus consisting of two adjacent stretches of hydrophobic amino acids separated by a few amino acids. Thus NTB-VPg polyprotein has characteristics of a transmembrane protein and so acts as a transmembrane protein, is an integral membrane protein, is associated with rough ER, and serves as a membrane anchor for TomRSV replication complex. In infected plants, the NTB-VPg polyprotein exists in association with ER membrane that are active in virus replication. It was hypothesized that one or several of the proteins, containing NTB domain, anchor replication complex to ER. The central portion of NTB protein is exposed to cytoplasmic face of membranes and an 8-kDa fragment is protected by membrane (i.e., is embedded in membrane and so is the anchor). This fragment probably consists of 3-kDa VPg and the 5-kDa stretch of hydrophobic residues at C terminus of NTB protein. Han and Sanfaçon (2003) accordingly proposed a model in which NTB-VPg (and NTB) is attached to membranes by both termini, with its central portion exposed to cytoplasm and accessible for protein-protein interactions with other viral or plant proteins. In the related CPMV comovirus, both the 66-kDa (NTB-VPg) polyprotein and 32-kDa protein are associated with ER. The 32-kDa protein is an hydrophobic protein, contains membrane-localization signal, can interact with membranes, and could target the 66-kDa proposed viral replicase and the replication complexes to ER membrane (Carette *et al.*, 2002b).

The second type of specialized viral proteins target viral replication complexes/ viral RNA/viruses/viral replication proteins to the specific host membrane. Some of these viral-encoded proteins have been identified. Membrane-targeting sequences exist in bromoviruses, nepoviruses, tombusviruses (Rubino and Russo, 1998), and tymoviruses. BMV replication protein 1a targets itself (Restrepo-Hartwig and Ahlquist, 1999) and interacts with and directs polymerase-like 2a protein (Chen and Ahlquist, 2000) to ER membrane sites of BMV RNA replication (Chen *et al.*, 2001). Targeting of 66-kDa viral replication protein, that contains RdRp domain, of TYMV to chloroplast envelope is mediated by 140-kDa viral protein. This 140-kDa protein interacts with cellular membranes, relocates 66-kDa viral protein from cytoplasm to membrane, and is the major protein that determines assembly of replication complexes and localization and retention of these replication complexes at chloroplasts, and the end encompassing methyltransferase-like domain in 140-kDa protein is the main determinant of membrane association (Prod'home *et al.*, 2003). The p15 protein of PCV contains at its C-terminus the triplet SKL, which corresponds to the canonical type1 peroxisomal targeting signal (Dunoyer *et al.*, 2001, 2002). Mitochondrial membrane-targeting and membrane-anchoring of viral replicase of *Carnation Italian ringspot virus* (CIRSV) occurs in plant and yeast cells. The CIRSV p36 (36-kDa) protein contains a signal targeting to outer mitochondrial membrane while its readthrough p95 protein contains conserved motifs of RdRps (Weber-Lotfi *et al.*, 2002). A 14-amino acid sequence, that corresponds to amino acids at positions 32 to 45 at N terminus of 36-kDa product of CIRSV ORF1, is the predicted mitochondrial membrane-targeting sequence (Weber-Lotfi *et al.*, 2002). No such sequence is present in ORF1 of other tombusviruses, which target peroxisomes. The anchor sequence possibly corresponds to amino acids at positions 84 to 196 and two transmembrane domains (Weber-Lotfi *et al.*, 2002). However, Pantaleo *et al.* (2004) found that p36 and p95 localize to mitochondria independently of each other and that p36, when expressed together with p95 in yeast, dramatically increased recruitment and stabilization of DI RNA transcripts. A possible mitochondrial targeting sequence also occurs in carmovirus ORF1 (*Galinsoga mosaic carmovirus*), which develop MVBs from mitochondria (Ciuffreda *et al.*, 1998).

Tethering of replication complexes to particular membranes is suggested to be controlled by either/or both of the following factors - a direct interaction between a helical peptide part and phospholipids as in BMV (Lampio *et al.*, 2000; den Boon *et al.*, 2001) and/or by specific interaction with membrane-bound host proteins as in TMV (Yamanaka *et al.*, 2000; Hagiwara *et al.*, 2003).

VII. VESICULATION

Innumerable reports show that vesicles, induced in membranes of various plant organelles, are intimately connected with viral RNA translation, replication, and even assembly of virus particles. Infection by positive-strand RNA plant viruses induces several pathological changes in morphology of intracellular host membranes of infected cells. Extensive proliferation of membranes (leading to increased surface area) of different organelles along with formation of small vesicles and/or redistribution of specific intracellular membranes are widely reported and are related to viral RNA replication. Vesicles occur in membranes of various organelles like cytoplasm, nucleus, mitochondria, chloroplasts, and vacuoles of virus-infected cells (Lesemann, 1991). Plant viruses can form either free smooth vesicles that accumulate in cytoplasm or membrane-associated invaginations of different organelles.

Vesicles (spherular vesicles or spherules) are spherical invagination of membranes of different organelles, are surrounded by double-layered or single-layered membranes, are presumed to assemble at cytoplasmic face of membranes, are connected to cytoplasm with a neck, and are 50- to 100-nm in diameter. The MVBs associated with tombusvirus infections consist of a main body surrounded by ovoid vesicles measuring 80- to 150-nm diameter. In early stages of infection, vesicles contain fibrillar networks, which most likely are the viral nucleic acids, while virus particles later appear in these vesicles. Vesicles originating in cytoplasm and/or from nucleus may ultimately degenerate and release virus particles in cytoplasm and/or nucleus.

Different plant viruses are specifically targeted to and modify different but specific intracellular membranes. This suggests two assumptions, either of which could be operative. First, these viruses require unique host factor(s) that can be supplied only by the specific intracellular membranes and specific interactions are formed between such host membranes and virus-encoded proteins. Second, many or all host intracellular membranes are potentially capable of providing the necessary requirements for formation and functioning of viral RNA replication complexes while specific intracellular localization of a particular virus is possibly dictated by some step(s) in viral life cycle (like viral protein synthesis or viral encapsidation) operating at a particular membrane.

Many plant viruses induce vesiculation. The 5`-terminal region of a tombusvirus genome determines the origin of MVBs (Burgyan *et al.*, 1996). TEV 6-kDa protein localizes to ER membrane and is proposed to cause modifications in infected cells (Schaad *et al.*, 1997). Induction of vesicular structures by CPMV is due to viral-induced proliferation of ER membranes and requires *de novo* synthesis of lipids (Carette *et al.*, 2000, 2002a, 2002b) while no *de novo* synthesis of lipids or *de novo* membrane synthesis takes place in PCV-infected cells and protoplasts (Dunoyer *et al.*, 2002). TMV also induces vesiculation of ER but its replication is not dependent on *de novo* lipid synthesis and membrane aggregates appear to be derived from pre-existing ER membranes (Reichel and Beachy, 1998; Carette *et al.*, 2000). Thus generation of ER-derived vesicles in CPMV-, PCV- (Dunoyer *et al.*, 2002), BMV-(Schwartz *et al.*, 2002) and TMV-infected cells is accompanied by drastic alterations of endomembrane system with or without *de novo* synthesis of lipids/membranes. MVBs, in PCV-infected protoplasts, are surrounded by a single membrane and strongly resemble MVBs induced by TBSV and CIRSV infections that are modified peroxisomes (Russo *et al.*, 1983: Rubino and Russo, 1998). Large number of membranous vesicles, that contain viral RNA and nonstructural proteins, are found in plant picorna-like CPMV-infected cells. Many other plant viruses act identically. TYMV replication is associated with virus-induced small invaginations of chloroplast membrane. On the other hand, insertion of CIRSV 36-kDa protein does not by itself trigger mitochondrial membrane vesiculation (Rubino *et al.*, 2000, 2001) so that the mechanisms or the molecular signal/viral protein, that enables CIRSV component to cause mitochondrial outer membrane to proliferate and form MVBs, is not yet known (Weber-Lotfi *et al.*, 2002). Small, unit membrane-bound vesicles containing fibrillar material also occur in mitochondria of cells infected with *Cucumber green mottle mosaic tobamovirus*. Prior to negative-strand RNA synthesis, BMV 1a protein induces the formation of spherules that bud into ER lumen so that BMV RNA templates are sequestered in these spherules.

A. Functions of Vesicles

Induction and accumulation of vesicles confers several advantages to the virus: vesicles isolate template RNA from cellular milieu of the host cell, compartmentalize viral RNA synthesis, are sites of initial or persistent viral RNA replication, increases surface area of the membrane and the increased surface area is used as scaffold for assembling replication machinery, and extensive membrane proliferation increases key lipid contents (Schwartz *et al.*, 2002; Prod'home *et al.*, 2003; Zinovkin *et al.*, 2003). Compartmentalization can have any one or more of the following positive effects on viral replication: it diverts viral RNA from translation machinery, increases local concentration of replication complexes and of the associated components, and avoids host defence responses that are directed against double-stranded RNA replicative intermediates.

Each spherule in BMV-infected cells contains numerous 1a protein molecules. The *cis*-acting BMV recognition elements act as virus-specific RNA packaging signals so that vesicles selectively package viral genomic RNAs. A few molecules of 2a polymerase, if expressed by this time, are also enclosed in each spherule, which then become sites of viral positive- and negative-strand RNA synthesis. RNA templates and 2a protein are sequestered into these vesicles and interact with 1a protein directly or indirectly to synthesize negative-strand RNA which is retained within these spherules and is used as template for synthesize positive-strand RNA molecules that are exported to cytoplasm (Schwartz *et al.*, 2002). Even double-stranded RNA replication intermediates have been found associated with these vesicles in BMV (Schwartz *et al.*, 2002) and CPMV (Carette *et al.*, 2002a).

Replication-associated proteins are also located at these vesicular sites in GFLV-infected cells (Ritzenthaler *et al.*, 2002). The methyltransferase-like and helicase-like proteins of BYV are associated with membrane compartments in infected cells and are identified *in planta* as 63-kDa and 100-kDa viral translation products, respectively (Erokhina *et al.*, 2000). Closterovirus infection induces characteristic membranous ultrastructures, consisting of BYV-type vesicles, which are implicated in virus replication (Coffin and Coutts, 1993). Replication of tombusvirus RNA in plant cells occurs in MVBs. This is true of CIRSV (Weber-Lotfi *et al.*, 2002), TBSV, and CymRSV. The presence of double-stranded RNA within these induced vesicles and association of replication proteins with these vesicles support this.

During *in vitro* studies, the isolated membrane complexes are able to conduct various steps of viral RNA replication like synthesis of replicative intermediates, elongation step, and release of genomic-length RNA. In a few such *in vitro* studies, RNA synthesis was also initiated on endogenous templates, which remain attached to the replication complex. Some plant viruses on which *in vitro* studies have been conducted are: AMV, CMV, *Foxtail mosaic potexvirus*, and TMV from alphavirus-like supergroup; CPMV and PPV from picornavirus-like supergroup; RCNMV and TCV from carmovirus-like supergroup; and *Velvet tobacco mottle virus* from sobemovirus-like supergroup.

B. Viral Signals Inducing Formation of Vesicles

Plant viral genomes possess definite and specific signals, and proteins encoded by these sequences induce formation of vesicles in membrane(s) of the host plant. The N-terminal half of the 36-kDa protein encoded by the ORF1 of genomic RNA of CIRSV contains information leading to vesiculation of mitochondria (Weber-Lotfi *et al.*, 2002). Membrane vesiculation in BMV-infected cells is induced by expression of methyltransferase-helicase-containing virus protein 1a (Schwartz *et al.*, 2002). The morphology of ER membrane is altered by expression of 6-kDa protein in TEV (Restrepo-Hartwig and Carrington, 1994), 32-kDa protein and 60-kDa protein in CPMV (Carette *et al.*, 2002b) and possibly by NTB and/or proteins containing NTB domain in case of TRSV-infected cells (Han and Sanfaçon, 2003).   Membrane vesiculation in TYMV-infected cells is regarded to be induced by 140-kDa protein (Prod'home *et al.*, 2003). This protein is proposed to interact with chloroplast envelope membrane and the intermolecular self-interaction between multiple 140-kDa proteins induces the formation and stabilization of chloroplast aggregates. The virus-encoded protein that harbours the above signals localizes to the relevant host membrane. Thus, the CIRSV 36-kDa protein localizes to mitochondria while TEV 6-kDa protein localizes to ER membrane.

Weber-Lotfi *et al.* (2002) have proposed the following hypothesis. The 36-kDa CIRSV protein as also the complete 95-kDa replicase (encoded by ORF2) protein contain the same sequence and might be targeted to mitochondria by 14 amino acid motif and subsequently anchored to organelle outer membrane by two hydrophobic transmembrane domains. The hydrophilic loop (the hinge) would protrude inside the intermembrane space, whereas N- and C-terminal regions would be cytosolic (Rubino and Russo, 1998). The N-terminal part of both 36-kDa and 95-kDa proteins contain multiple recognition-insertion signals for mitochondrial outer membrane within the two hydrophobic domains and their flanking regions; these together might constitute an efficient signal anchor sequence. The above hypothesis suggests a signal anchor pathway and eliminates the notion that 36-kDa and 95-kDa proteins are inserted into mitochondrial outer membrane by a stop-transfer mechanism. The import-independent signal anchor mechanism relies on two transmembrane segments and multiple recognition signals present in N-terminal part of ORF1 (Weber-Lotfi *et al.*, 2002).

VIII. REFERENCES

Abbink, T. E. M., Peart, J. R., Mos, T. N. M., Baulcombe, D. C., Bol, J. F., and Linthorst, H. J. M. 2002. Silencing of a gene encoding a protein component of the oxygen-evolving complex of photosystem II enhances virus replication. *Virology* 295: 307-319.

Ahlquist, P., Noueiry, A. O., Lee, W.-M., Kushner, D. B., and Dye, B. T. 2003. Host factors in positive-strand RNA virus genome replication. *J. Virol.* 77: 8181-8186.

Ahola, T., den Boon, J. A., and Ahlquist, P. 2000. Helicase and capping enzyme active site mutations in brome mosaic virus protein 1a cause defects in template recruitment, negative-strand RNA synthesis and viral RNA capping. *J. Virol.* 74: 8803-8811.

Asurmendi, S., Berg, R. H., Koo, J. C., and Beachy, R. N. 2004. Coat protein regulates formation of replication complexes during TMV infection. *Proc. Natl. Acad. Sci. USA* 101: 1415-1420.

Bleve-Zacheo, T., Rubino, L., Melillo, M. T., and Russo, M. 1997. The 33K protein encoded by *Cymbidium ringspot tombusvirus* localizes to modified peroxisomes of infected cells and of uninfected transgenic plant. *J. Plan. Pathol.* 79: 179-202.

Bol, J. F. 1999. *Alfalfa mosaic virus* and ilarviruses: Involvement of coat protein in multiple steps of the replication cycle. *J. Gen. Virol.* 80: 1089-1102.

Buck, K. W. 1996. Comparison of the replication of positive-stranded RNA viruses of plants and animals. *Adv. Virus Res.* 47: 159-251.

Burgyan, J., Rubino, L., and Russo, M. 1996. The 5`-terminal region of a tombusvirus genome determines the origin of multivesicular bodies. *J. Gen Virol.* 77: 1967-1974.

Carette, J. E., Stuiver, M., van Lent, J. W. M., Wellink, J., and van Kammen, A. 2000. Cowpea mosaic virus infection induces a massive proliferation of endoplasmic reticulum but not Golgi membranes and is dependent on *de novo* membrane synthesis. *J. Virol.* 74: 6556-6563.

Carette, J. E., Guhl, K., Wellink, J., and van Kammen, A. 2002a. Coalescence of the sites of cowpea mosaic virus RNA replication into a cytopathic structure. *J. Virol.* 76: 6235-6243.

Carette, J. E., van Lent, J., MacFarlane, S. A., Wellink, J. and van Kammen, A. 2002b. Cowpea mosaic virus 32- and 60-kilodalton replication proteins target and change the morphology of endoplasmic reticulum membranes. *J. Virol.* 76: 6293-6301.

Carr, J. P. 2004. *Tobacco mosaic virus. Annu. Plant Rev.* 11: 27-67.

Chen, J., and Ahlquist, P. 2000. Brome mosaic virus polymerase-like 2a is directed to the endoplasmic reticulum by helicase-like viral protein 1a. *J. Virol.* 74: 4310-4318.

Chen, J., Noueiry, A., and Ahlquist, P. 2001. Brome mosaic virus protein 1a recruits viral RNA2 to RNA replication through a 5`-proximal RNA2 signal. *J. Virol.* 75: 3207-3219.

Chien, C., Bartel, P. L., Sternglanz, R., and Fields, S. 1991. The two-hybrid system: A method to identify and clone genes for proteins that interact with a protein of interest. *Proc. Natl. Acad. Sci. USA* 88: 9578-9582.

Cillo, F., Roberts, I. M., and Palukaitis, P. 2002. *In situ* localization and tissue distribution of the replication-associated proteins *Cucumber mosaic virus* in tobacco and cucumber. *J. Virol.* 76: 10654-10664.

Ciuffreda, P., Rubino, L., and Russo, M. 1998. Molecular cloning and complete nucleotide sequence of galinsoga mosaic virus genomic RNA. *Arch. Virol.* 143: 173-180.

Coffin, R. S., and Coutts, R. H. A. 1993. The closteroviruses, capilloviruses and other similar viruses: A short review. *J. Gen. Virol.* 74: 1475-1483.

Daros, J. A., Schaad, M. C., and Carrington, J. C. 1999. Functional analysis of the interaction between VPg-proteinase (NIa) and RNA polymerase (NIb) of *Tobacco etch potyvirus*, using conditional and suppressor mutants. *J. Virol.* 73: 8732-8740.

de Graaff, M., and Jaspars, E. M. J. 1994. Plant viral RNA synthesis in cell-free systems. *Annu. Rev. Phytopathol.* 32: 311-335.

de Graaff, M., Coscoy, L., and Jaspars, E. M. J. 1993. Localization and biochemical characterization of alfalfa mosaic virus replication complexes. *Virology* 194: 878-881.

den Boon. J. A., Chen, J., and Ahlquist, P. 2001. Identification of sequences in brome mosaic virus replicase protein 1a that mediate association with endoplasmic reticulum membranes. *J. Virol.* 75: 12370-12381.

Díez, J., Ishikawa, M., Kaido, M., and Ahlquist, P. 2000. Identification and characterization of a host protein required for efficient template selection in viral RNA replication. *Proc. Natl. Acad. Sci. USA* 97: 3913-3918.

Dohi, K., Mori, M., Furusawa, I., Mise, K., and Okuno, T. 2001. Brome mosaic virus replicase proteins localize with the movement protein at infection-specific cytoplasmic inclusions in infected barley leaf cells. *Arch. Virol.* 146: 1607-1615.

Dohi, K., Mise, K., Furusawa, I., and Okuno, T. 2002. RNA-dependent RNA polymerase complex of *Brome mosaic virus*: Analysis of the molecular structure with monoclonal antibodies. *J. Gen. Virol.* 83: 2879-2890.

Dunoyer, P., Herzog, E., Hemmer, O., Ritzenthaler, C., and Fritsch, C. 2001. Peanut clump virus RNA-1-encoded p15 regulates viral RNA accumulation but is not abundant at viral RNA replication sites. *J. Virol.* 75: 1941-1948.

Dunoyer, P., Ritzenthaler, C., Hemmer, O., Michler, P., and Fritsch, C. 2002. Intracellular localization of the peanut clump virus replication complex in tobacco BY-2 protoplasts containing green fluorescent protein-labeled endoplasmic reticulum or Golgi apparatus. *J. Virol.* 76: 865-874.

Dzianott, A., and Bujarski, J. J. 2004. Infection and RNA recombination of *Brome mosaic virus* in *Arabidopsis thaliana*. *Virology* 318: 482-492.

Erokhina, T. N., Zinovkin, R. A., Vitushkina, M. V., Jelkmann, W., and Agranovsky, A. A. 2000. Detection of beet yellows closterovirus methyltransferase-like and helicase-like proteins *in vivo* using monoclonal antibodies. *J. Gen. Virol.* 81: 597-603.

Erokhina, T. N., Vitushkina, M. V., Zinovkin, R. A., Lesemann, D. E., Jelkmann, W., Koonin, E. V., and Agranovsky, A. A. 2001. Ultra-localization and epitope mapping of beet yellows closterovirus methyltransferase-like and helicase-like proteins. *J. Gen. Virol.* 82: 1983-1994.

Fields, F., and Song, O. 1989. A novel genetic system to detect protein-protein interactions. *Nature* 340: 245-246.

Fields, F., and Sternglanz, R. 1994. The two-hybrid system: An assay for protein-protein interactions. *Trends Gen.* 10: 286-292.

Goegaoker, S. P., and Culver, J. N. 2003. Oligomerization and activity of the helicase domain of the tobacco mosaic virus 126- and 183-kilodalton replicase proteins. *J. Virol.* 77: 3549-3556.

Goregaoker, S. P., Lewandowski, D. J., and Culver, J. N. 2001. Identification and functional analysis of an interaction between domains of the 126/183-kDa replicase-associated proteins of *Tobacco mosaic virus*. *Virology* 282: 320-328.

Hagiwara, Y., Komoda, K., Yamanaka, T., Tamai, A., Meshi, T., Funada, R., Tsuchiya, T., Naito, S., and Ishikawa, M. 2003. Subcellular localization of host and viral proteins associated with tobamovirus replication. *EMBO J.* 22: 344-353.

Han, S., and Sanfaçon, H. 2003. Tomato ringspot virus proteins containing the nucleoside triphosphate binding domain are transmembrane proteins that associate with the endoplasmic reticulum and cofractionate with replication complexes. *J. Virol.* 77: 523-534.

Hayes, R. J., and Buck, K. W. 1990. Complete replication of a eukaryotic virus RNA *in vitro* by a purified RNA-dependent RNA polymerase. *Cell* 63: 363-368.

Hayes, R. J., Pereira, V. C. A., and Buck, K. W. *et al.*, 1994. Plant proteins that bind to the 3`-terminal sequences of the negative-strand RNA of three diverse positive-strand RNA plant viruses. *FEBS Lett.* 352: 331-334.

Heinlein, M., Padgett, H. S., Gens, J. S., Pickard, B. G., Casper, S. J., Epel, B. L., and Beachy, R. N. 1998. Changing patterns of localization of the tobacco mosaic virus movement protein and replicase to the endoplasmic reticulum and microtubules during infection. *Plant Cell* 10: 1107-1120.

Hills, G. J., Plaskitt, K. A., Young, N. D., Dunigan, D. D., Watts, J. W., Wilson, T. M., and Zaitlin, M. 1987. Immunogold localization of the intracellular sites of structural and nonstructural tobacco mosaic virus proteins. *Virology* 161: 488-496.

Hong, Y., and Hunt, A. G. 1996. RNA polymerase activity by a potyvirus-encoded RNA-dependent RNA polymerase. *Virology* 226: 146-151.

Ishikawa, M., Naito, S., and Ohno, T. 1993. Effects of the *tom1* mutation of *Arabidopsis thaliana* on the multiplication of tobacco mosaic virus RNA in protoplasts. *J. Virol.* 67: 5328-5338.

Ishikawa, M., Díes, J., Restrepo-Hartwig, M., and Ahlquist, A. 1997a. Yeast mutations in multiple complementation groups inhibit brome mosaic virus RNA replication and transcription and perturb regulated expression of the viral polymerase-like gene. *Proc. Natl. Acad. Sci. USA* 94: 13810-13815.

Ishikawa, M., Janda, M., Krol, M. A., and Ahlquist, P. 1997b. *In vivo* DNA expression of functional brome mosaic virus RNA replicons in *Saccharomyces cerevisiae*. *J. Virol.* 71: 7781-7790.

Jakubiec, A., Notaise, J., Tournier, V., Héricourt, F., Block, M. A., Drugeon, G., van Aelst, L., and Jupin, I. 2004. Assembly of turnip yellow mosaic virus replication complexes: Interaction between the proteinase and polymerase domains of the replication proteins. *J. Virol.* 78: 7945-7957.

Janda, M., and Ahlquist, P. 1993. RNA-dependent replication, transcription and persistence of brome mosaic virus RNA replicons in *S. cerevisiae. Cell* 72: 961-970.

Janda, M ., and Ahlquist, P. 1998. Brome mosaic virus RNA replication protein 1a dramatically increases *in vivo* stability but not translation of viral genomic RNA3. *Proc. Natl. Acad. Sci. USA* 95: 2227-2232.

Kadarè, G., and Haenni, A.-L. 1997. Virus-encoded helicases. *J. Virol.* 71: 2583-2590.

Kao, C. C., and Ahlquist, P. 1992. Identification of the domains required for direct interaction of the helicase-like and polymerase-like replication proteins of *Brome mosaic virus. J. Virol.* 66: 7293-7302.

Kao, C. C., Quadt, R., Hershberger, R. P., and Ahlquist, P. 1992. Brome mosaic virus RNA replication proteins 1a and 2a form a complex *in vitro. J. Virol.* 66: 6322-6329.

Kim, S. H., Palukaitis, P., and Park, Y. I. 2002. Phosphorylation of cucumber mosaic virus RNA polymerase 2a protein inhibits formation of replicase complex. *EMBO J.* 21: 2292-2300

Lai, M. M. C. 1998. Cellular factors in the transcription and replication of viral RNA genomes: A parallel to DNA-dependent RNA transcription. *Virology* 244: 1-12.

Lampio, A., Kilpelainen, I., Pesonen, S., Karhi, K., Auvinen, P., Somerharju, P., and Kaariainen, L. 2000. Membrane binding mechanism of an RNA virus-capping enzyme. *J. Biol. Chem.* 275: 37853-37859.

Lee, W. M., and Ahlquist, P. 2003. Membrane synthesis, specific lipid requirements, and localized lipid composition changes associated with a positive-strand RNA virus RNA replication protein. *J. Virol.* 77: 12819-12828.

Lee, W. M., Ishikawa, M., and Ahlquist, P. 2001. Mutation of host Δ9 fatty acid desaturase inhibits brome mosaic virus RNA synthesis. *J. Virol.* 75: 2097-2106.

Lellis, A. D., Kasschau, K. D., Witham, S. A., and Carrington, J. C. 2002. Loss-of-susceptibility mutants of *Arabidopsis thaliana* reveal an essential role for eIF(iso)4E during potyvirus infection. *Curr. Biol.* 12: 1046-1051.

Lesemann, D. E. 1991. Specific cytological alterations in virus-infected plant cells. *In*: Mendgen, K., and Lesemann, D. E. (Eds.). Electron Microscopy of Plant Pathogens. Springer Verlag, K. G., Berlin. p. 147-159.

Li, X., Valdez, P., Olvera, R. E., and Carrington, J. C. 1997. Functions of the tobacco etch virus polymerase (NIb): Subcellular transport and protein-protein interactions with VPg/proteinase (NIa). *J. Virol.* 71: 1598-1607.

Li, Y.-L., Cheng, Y.-M., Huang, Y.-L., Tsai, C.-H., Hsu, Y.-H., and Meng, M. 1998. Identification and characterization of the *Escherichia coli*-expressed RNA-dependent RNA polymerase of *Bamboo mosaic virus. J. Virol.* 72: 10093-10099.

Más, P., and Beachy, R. N. 1999. Replication of *Tobacco mosaic virus* on endoplasmic reticulum and role of the cytoskeleton and virus movement protein in intracellular distribution of viral RNA. *J. Cell. Biol.* 147: 945-958.

Noueiry, A. O., and Ahlquist, P. 2003. Brome mosaic virus RNA replication: Revealing the role of the host in RNA-virus replication. *Annu. Rev. Phytopath.* 41: 77-98.

Noueiry, A. O., Chen, J., and Ahlquist, P. 2000. A mutant allele of essential general translation initiation factor *DED1* selectively inhibits translation of a viral mRNA. *Proc. Natl. Acad. Sci. USA* 97: 12985-12990.

Noueiry, A. O., Diez, J., Falk, S., Chen, J., and Ahlquist, P. 2003. Yeast Lsm1p-Tp/Pat1p deadenylation-dependent mRNA decapping factors are required for brome mosaic virus genomic RNA translation. *Mol. Cell Biol.* 23: 4094-4106.

Ohshima, K., Taniyama, T., Yamanaka, T., Ishikawa, M., and Naito, S. 1998. Isolation of a mutant of *Arabidopsis thaliana* carrying two simultaneous mutations affecting tobacco mosaic virus multiplication within a single cell. *Virology* 243: 472-481.

O'Reilley, E. K., Tang, N., Ahlquist, P, and Kao, C. C. 1995. Biochemical and genetic analyses of the interaction between the helicase-like and polymerase-like proteins of the *Brome mosaic virus. Virology* 214: 59-71.

O'Reilley, E. K., Paul, J. D., and Kao, C. C. 1997. Analysis of the interaction of viral RNA replication protein by using the yeast two-hybrid assay. *J. Virol.* 71: 7526-7532.

O'Reilley, E. K., Wang, Z., French, R., and Kao. C. C. 1998. Interactions between the structural domains of the RNA replication proteins of plant-infecting RNA viruses. *J. Virol.* 72: 7160-7169.

Osman, T. A. M., and Buck, K. W. 1996. Complete replication *in vitro* of tobacco mosaic virus RNA by a template-dependent membrane-bound RNA polymerase. *J. Virol.* 70: 6227-6234.

Osman, T. A. M., and Buck, K. W. 1997. The tobacco mosaic virus RNA polymerase complex contains a plant protein related to RNA-binding subunit of yeast elF-3. *J. Virol.* 71: 6075-6082.

Osman, T. A. M., and Buck, K. W. 2003. Identification of a region of the tobacco mosaic virus 126- and 183-kilodalton replication proteins which binds specifically to the viral 3`-terminal tRNA-like structure. *J. Virol.* 77: 8669-8675.

Panaviene, Z., Baker, J. M., and Nagy, P. D. 2003. The overlapping RNA-binding domains of p33 and p92 replicase proteins are essential for tombusvirus replication. *Virology* 308: 191-205.

Panaviene, Z., Panavas, T., Serva, S., and Nagy, P. D. 2004. Purification of the cucumber necrosis virus replicase from yeast cells: Role of coexpressed viral RNA in stimulation of replicase activity. *J. Virol.* 78: 8254-8263.

Pantaleo, V., Rubino, L., and Russo, M. 2004. The p36 and p95 replicase proteins of *Carnation Italian ringspot virus* cooperate in stabilizing defective interfering RNA. *J. Gen. Viol.* 85: 2429-2433.

Pogue, G. P., Huntley, C. C., and Hall, T. C. 1994. Common replication strategies emerging from the study of diverse groups of positive-strand virus. *Arch. Virol.* 9: 181-194.

Prod'homme, D., Le Panse, S., Drugeon, G., and Jupin, I. 2001. Detection and subcellular localization of the turnip yellow mosaic virus 66K replication protein in infected cells. *Virology* 281: 88-101.

Prod'homme, D., Jakubiec, A., Tournier, V., Drugeon, G., and Jupin, I. 2003. Targeting of turnip yellow mosaic virus 66K replication protein to the chloroplast envelope is mediated by the 140K protein. *J. Virol.* 77: 9124-9135.

Qiu, W. P., Park, J. W., and Scholthof, H. B. 2002. Tombusvirus p19-mediated suppression of virus-induced gene silencing is controlled by genetic and dosage features that influence pathogenicity. *Mol. Plant-Microbe Interact.* 15: 269-280.

Qu, F., and Morris, T. J. 2002. Efficient infection of *Nicotiana benthamiana* by *Tomato bushy stunt virus* is facilitated by the coat protein and maintained by p19 through suppression of gene silencing. *Mol. Plant-Microbe Interact.* 15: 193-202.

Quadt, R., Kao, C. C., Browning, K. S., Hersberger, R. P., and Ahlquist, P. 1993. Characterization of a host protein associated with brome mosaic virus RNA-dependent RNA polymerase. *Proc. Natl. Acad. Sci. USA* 90: 1498-1502.

Quadt, R., Ishikawa, M., Janda, M., and Ahlquist, P. 1995. Formation of brome mosaic virus RNA-dependent RNA polymerase in yeast requires coexpression of viral proteins and viral RNA. *Proc. Nat. Acad. Sci. USA* 92: 4892-4896.

Rajendran, K. S., and Nagy, P. D. 2003. Characterization of the RNA-binding domains in the replicase proteins of *Tomato bushy stunt virus. J. Virol.* 77: 9244-9258.

Rajendran, K. S., Pogany, J., and. Nagy, P. D. 2002. Comparison of turnip crinkle virus RNA-dependent RNA polymerase preparations expressed in *Escherichia coli* or derived from infected plants. *J. Virol.* 76: 1707-1717.

Reichel, C., and Beachy, R. N. 1998. Tobacco mosaic virus infection induces severe morphological changes of the endoplasmic reticulum. *Proc. Natl. Acad. Sci. USA* 95: 11169-11174.

Restrepo-Hartwig, M. A., and Ahlquist, P. 1996. Brome mosaic virus helicase- and polymerase-like proteins colocalize on the endoplasmic reticulum at sites of viral RNA synthesis. *J. Virol.* 70: 8908-8916.

Restrepo-Hartwig, M. A., and Ahlquist, P. 1999. Brome mosaic virus RNA replication proteins 1a and 2a colocalize and 1a independently localizes on the yeast endoplasmic reticulum. *J. Virol.* 73: 10303-10309.

Restrepo-Hartwig, M. A., and Carrington, J. C. 1994. The tobacco etch potyvirus 6-kilodalton protein is membrane associated and involved in viral replication. *J. Virol.* 68: 2388-2397.

Ritzenthler, C., Laporte, C., Gaire, F., Dunoyer, P., Schmitt. C., Duval, S., *et al.* 2002. *Grapevine fanleaf virus* replication occurs on endoplasmic-derived membranes. *J. Virol.* 76: 8808-8819.

Rubino, L., and Russo, M. 1998. Membrane targeting sequences in tombusvirus infections. *Virology* 252: 431-437.

Rubino, L., di Franco, A., and Russo, M. 2000. Expression of a plant virus non-structural protein in *Saccharomyces cerevisiae* causes membrane proliferation and altered mitochondrial morphology. *J. Gen. Virol.* 81: 279-286.

Rubino, L., Weber-Lotfi, F., Dietrich, A., Stussi-Garaud, C., and Russo, M. 2001. The open reading frame 1-encoded ('36K') protein of *Carnation Italian ringspot virus* localizes to mitochondria. *J. Gen. Virol.* 82: 29-34.

Russo, M., di Franco, A., and Martelli, G. P. 1983. The fine structure of cymbidium ringspot virus infections in host tissues. III. Role of peroxisomes in the genesis of multivesicular bodies. *J. Ultrastruct. Res.* 82: 52-63.

Schaad, M. C., Jensen, P. E., and Carrington, J. C. 1997. Formation of plant RNA virus replication complexes on membranes: Role of an endoplasmic reticulum-targeted viral protein. *EMBO J.* 16: 4049-4059.

Scholthof, K.-B. G., Scholthof, H. B., and Jackson, A. O. 1995. The tomato bushy stunt virus replicase proteins are coordinately expressed and membrane-associated. *Virology* 208: 365-369.

Schwartz, M., Chen, J., Janda, M., Sullivan, M., de Boon, J., and Ahlquist, P. 2002. A positive-strand RNA virus replication complex parallels form and function of retrovirus capsids. *Mol. Cell* 9: 505-514.

Shapka, N., Stork, J., and Nagy, P. D. 2005. Phosphorylation of the p33 replication protein of *Cucumber necrosis tombusvirus* adjacent to the RNA binding site affects viral RNA replication. *Virology*, in press.

Stork, J., Panaviene, Z., and Nagy, P. D. 2005. Inhibition of *in vitro* RNA binding replicase activity by phosphorylation of the p33 replication protein of *Cucumber necrosis tombusvirus. Virology*, in press.

Strauss, J. H., and Strauss, E. G. 1999. With a little help from the host. *Science* 283: 802-804.

Sullivan, M. L., and Ahlquist, P. 1997. *cis*-Acting signals in bromovirus RNA replication and gene expression: Networking with viral proteins and host factors. *Semin. Virol.* 8: 221-230.

Sullivan, M. L., and Ahlquist, P. 1999. A brome mosaic virus intergenic RNA3 replication signal functions with viral replication protein 1a to dramatically stabilize RNA *in vivo. J. Virol.* 73: 2622-2632.

Taylor, D. N., and Carr, J. P. 2000. The GCD10 subunit of yeast eIF-3 binds the methyltransferase-like domain of the 126- and 183-kDa replicase proteins of *Tobacco mosaic virus* in the yeast two-hybrid system. *J. Gen. Virol.* 81: 1587-1591.

Tomita, Y., Mizuno, T., Diez, J., Naito, S., Ahlquist, P., and Ishikawa, M. 2003. Mutation of host DNA1 homolog inhibits brome mosaic virus negative-strand RNA synthesis. *J. Virol.* 77: 2990-2997.

Tsujimoto, Y., Numaga, T., Ohshima, K., Yano, M. A., Ohsawa, R., Goto, D., Naito, S., and Ishikawa, M. 2003. *Arabidopsis* tobamovirus multiplication (TOM) 2 locus encodes a transmembrane protein that interacts with TOM1. *EMBO J.* 22: 335-343.

van der Heijden, M. W., and Bol, J. F. 2002. Composition of alphavirus-like replication complexes: Involvement of virus- and host-encoded proteins. *Arch. Virol.* 147: 875-898.

van der Heijden, M. W., Carette, J. E., Reinhoud, P. J., Haegi, A., and Bol, J. F. 2001. Alfalfa mosaic virus replicase proteins P1 and P2 interact and colocalize at the vacuolar membrane. *J. Virol.* 75: 1879-1887.

Vlot, A. C., Neeleman, L., Linthorst, H. J. M., and Bol, J. F. 2001. Role of the 3`-untranslated regions of alfalfa mosaic virus RNAs in the formation of a transiently expressed replicase in plants and in the assembly of virions. *J. Virol.* 75: 6440-6449.

Vlot, A. C., Laros, S. M., and Bol, J. F. 2003. Coordinate replication of alfalfa mosaic virus RNAs 1 and 2 involves *cis*-acting and *trans*-acting functions of the encoded helicase-like and polymerase-like domains. *J. Virol.* 77: 10790-10798.

Wang, A., Han, S., and Sanfaçon, H. 2004. Topogenesis in membranes of the NTB-VPg protein of *Tomato ringspot nepovirus*: Definition of the C-terminal transmembrane domain. *J. Virol.* 85: 535-545.

Wang, X., Ullah, Z., and Grumet, R. 2000. Interaction between zucchini yellow mosaic potyvirus RNA-dependent RNA polymerase and host poly(A) binding protein. *Virology* 275: 433-443.

Watanabe, T., Honda, A., Iwata, A., Ueda, S., Hibi, T., and Ishihama, A. 1999. Isolation from tobacco mosaic virus-infected tobacco of a solubilized template-specific RNA-dependent RNA polymerase containing a 126/183K protein heterodimer. *J. Virol.* 73: 2633-2640.

Weber-Lotfi, F., Dietrich, A., Russo, M., and Rubino, L. 2002. Mitochondrial targeting and membrane anchoring of a viral replicase in plant and yeast cells. *J. Virol.* 76: 10485-10496.

Yamanaka, T., Ohta, T., Takahashi, M., Meshi, T., Schmidt, R., Dean, C., Naito, S., and Ishikawa, M. 2000. *TOM1*, an *Arabidopsis* gene required for efficient multiplication of a tobamovirus, encodes a putative transmembrane protein. *Proc. Natl. Acad. Sci. USA* 97: 10107-10112.

Yamanaka, T., Imai, T., Satoh, R., Kawashima, A., Takahashi, M., Tomita, K., Kubota, K., Meshi, T., Naito, S., and Ishikawa, M. 2002. Complete inhibition of tobamovirus multiplication by simultaneous mutations in two homologous host genes. *J. Virol.* 76: 2491-2497.

Zinovkin, R. A., Erokhina, T. N., Lesemann, D. E., Jelkmann, W., and Agranovsky, A. A. 2003. Processing and subcellular localization of the leader papain-like proteinase of *Beet yellows closterovirus. J. Gen. Virol.* 84: 2265-2270.

SUBJECT INDEX

ATPase(s) / ATPase activity 151, 153, 154, 160, 225, 300

AUG initiation codon 12, 30, 80, 170, 176, 178, 215-216, 227-230, 237-242, 253, 255-257, 261

Aberrant strategies of translation 223, 224, 250

Adenine 14, 93, 242

Alfalfa mosaic alfamovirus (AMV) 11, 15, 22, 30, 32, 33, 35, 37, 39-40, 45, 51-54, 58-59, 79-80, 82, 87-88, 91-96, 99-100, 102, 105-110, 120, 123-129, 153-154, 159, 162, 196, 201-207, 209-210, 212-214, 218, 234, 239, 245, 249, 253, 256, 282, 284-286, 295, 307-308, 310-311, 316, 332

Alfamovirus 20, 22, 32, 35, 59, 126, 157, 196, 214, 285

Allexvirus 20

Alpha-like/ viruses/supergroup 35, 37, 120, 126, 157, 159, 167, 170, 181, 182, 202-203, 249, 261, 301, 305, 309

Alphavirus-like supergroup 20, 21, 37-38, 128, 136, 141-142, 157, 200, 202-203, 310-311, 322

Ambisense RNA/viruses 6, 136, 195, 234-237

Aminoacylation 45-46, 48-51, 54, 64-65, 233, 261

Apple chlorotic leaf spot trichovirus (ACLSV) 22, 33, 60, 122, 158, 167, 172, 181

Apple stem grooving capillovirus (ASGV) 20, 22, 32, 37, 59, 122, 126, 158, 167, 171, 181, 232

Apple stem pitting foveavirus (ASPV) 21, 171

Arabidopsis thaliana 43, 74, 141, 299-301, 315

Artichoke mottled crinkle tombus-virus 34, 308

Aspartic proteinases 169-170, 186-187, 190, 291

Assembly of viruses 6, 13, 53, 55, 73, 75-76, 78, 108, 189, 215, 271- 298, 300, 304, 307, 309, 319

Assembly origin 82-83, 109, 272, 275, 278

Aureusvirus 20

Avenavirus 20, 34

Badnaviridae 169

Badnavirus 186-187, 236

Bamboo mosaic potexvirus (BaMV) 22, 34, 37-38, 41, 54-56, 94-95, 119-120, 124-127, 129, 157, 200, 282, 316

Bamboo mosaic virus satellite RNA 127

Barley stripe mosaic hordeivirus (BSMV) 12, 21-22, 33, 37, 49, 58, 60, 73, 102-103, 122, 126- 127, 158, 197, 200, 206, 217, 235, 241-242, 248, 256, 279-280, 307

Barley yellow dwarf luteovirus (BYDV) 9, 13, 20, 22, 31, 33-34, 41, 60, 88, 103, 121, 126, 141, 156, 167, 197, 200-201, 206-209, 211-213, 217-218, 233-234, 239- 240, 243-249, 251-256, 258-259, 261-262

Barley yellow mosaic bymovirus (BYMV) 22, 32, 41, 44, 59, 167, 175, 183-184

Bean common mosaic bymovirus (BCMV) 22

Bean pod mottle comovirus 141, 163

Beet cryptic alphacryptovirus 1 22

Beet cryptic alphacryptovirus 3 22

Beet curly top curtovirus 307

Beet furo-like virus Q (BVQ) 22, 247, 250

Beet necrotic yellow vein benyvirus (BNYVV) 20, 22, 31-32, 37, 55- 56, 60, 74, 103, 122, 126, 141, 153, 157-158, 167, 171, 181, 183, 197, 203, 206-207, 209, 212, 232, 235, 247, 250, 280

Beet soil-borne pomovirus 126, 235

Beet virus Q – see *Beet furo-like virus Q*

Beet western yellows luteovirus (BWYV) 22, 41, 121, 156, 180, 243, 244, 247, 291, 307

Beet yellows closterovirus (BYV) 20, 22, 31-32, 34, 37, 59, 120, 122, 126-128, 158, 167, 170, 181-185, 187-190, 199-202, 205, 217, 243, 245-246, 307, 316, 321

Benyvirus 20, 31, 34, 169, 280

Benyviruses 171, 181

Bipartite genome/viruses 15, 62, 101, 171, 173-174, 201, 204, 212, 234, 238, 256, 318

Birnaviridae 91, 156

Bluberry scorch carlavirus (BlSV) 22, 181

Bovine diarrhea virus 136

Broad bean mottle bromovirus (BBMV) 20, 22, 34, 49, 62, 271, 285-286

Broad bean wilt fabavirus 20

Brome mosaic bromovirus (BMV) 1, 4, 7-14, 20, 22, 30-32, 34-38, 40, 45-59, 62, 76, 82, 88-96, 98-99, 101, 104, 105, 107-109, 120, 122-131, 132-134, 136-138, 140, 153, 158, 161-163, 195-196, 201-214, 226, 233, 237, 242, 260, 262, 271-272, 282, 284-290, 292-294, 299, 300-305, 308-313, 316-317, 319-322

Brome streak mosaic rymovirus (BrSMV) 22, 33, 60, 167, 175, 184

Bromoviridae 12, 14, 19, 21, 34, 37, 92, 120, 122, 195, 209, 211, 214, 256, 292, 310

Bromovirus 20, 32, 35, 46, 59, 157, 196, 214, 287

Bromoviruses 35, 50, 54, 99, 106, 162, 200, 210, 218, 282, 286, 292-294

Bunyaviridae 39, 195, 236-237

Bymovirus 20, 32, 41, 44, 55, 59, 159, 167, 169, 173-174, 184, 235

Bymoviruses 44, 142, 182-183

Cap/cap-like structure 16, 30-41, 43, 56, 80, 88, 90, 107-109, 127, 135, 196-201, 215, 224-234, 237-238, 242, 251-262, 279, 281, 309, 312-313

Cap-dependent translation 18, 40, 225-229, 253, 258

Cap-independent translation 13, 16, 31, 43, 52, 227-228, 232, 242, 251-255, 258-261

Cap-independent translational enhancer 45, 52, 251, 253, 259-262

Cap snatching 36, 39, 234-237

Caliciviridae 134

Capillovirus 20, 32, 55, 59, 121, 126, 157, 167, 169

Capilloviruses 141, 171, 181

Capsid/coat 6, 53, 55, 59, 62, 79, 80, 82, 168, 271-272, 282-294

Capsid protein / coat protein (CP) 2, 3, 5, 6, 11, 12, 14-15, 33, 53, 58, 61-62, 75, 77, 79-82, 88, 100-111, 175-176, 178-179, 186-187, 190, 199, 200, 216-218, 223, 240-241, 248, 273-274, 283-288, 292-293

Capsid protein-binding sites 33, 106

Cardamine chlorotic fleck carmo-virus 256

Carlavirus 20, 31, 32, 37, 55, 59, 121, 157, 167, 169, 195-196, 243

Carlaviruses 142, 171, 181, 183

Carmo-like viruses/supergroup 20-21, 34-35, 126, 167, 253, 261, 308, 322

Carmovirus 20, 22, 32, 35, 59, 89, 103, 124, 127, 196, 212, 247, 252-253, 258, 319

Carmoviruses 8, 102-104, 137, 142, 247-249

Carnation Italian ringspot tombus-virus (CIRSV) 31, 75, 125-126, 128, 241, 248, 258, 301, 308, 319

Carnation latent carlavirus 20, 121

Carnation mottle carmovirus (CarMV) 20, 22, 32, 34, 45, 58-59, 121, 156, 196, 247-248, 256

Carnation ringspot dianthovirus (CRSV) 20, 22, 31

Carrot mottle umbravirus 21, 121, 126

Cauliflower mosaic caulimovirus (CaMV) 22, 74, 186, 190, 229, 235, 239, 240, 242, 305

Caulimoviridae 169, 186

Caulimovirus 156, 186, 235

Caulimoviruses 168, 239, 305

Cell-to-cell movement 58, 140, 157, 189, 201, 217, 304

Cherry leaf roll nepovirus (CLRV) 22, 198, 200, 218

Chenopodium amaranticolor 73

Chloroplasts 110, 249, 304, 307, 310-311, 319, 320, 322

Chymotrypsin-like proteinase 168, 170-177, 179-181

cis-Acting factors/sequences/elements/ activity/cleavage 4-5, 7-15, 29, 35, 55, 58, 62, 87-89, 92-93, 103, 110, 129, 136-137, 140, 162, 167, 168, 171, 173-179, 181-183, 187, 199, 203-208, 214, 226, 234-237, 239, 246, 248, 251, 253, 255, 257, 262, 290, 292, 303, 307, 309, 321

Citrus leaf blotch virus 34, 199-200, 202

Citrus tatter leaf capillovirus (CTLV) 22, 32

Citrus tristeza closterovirus (CTV) 22, 31, 34, 59, 94, 99, 100, 122, 126, 181-185, 196, 199, 200, 203-205, 211, 214, 217, 243, 246

Closteroviridae 20, 34, 126, 183, 185, 195, 199, 235, 256

Closterovirus 20, 32, 35, 59, 122, 126, 141, 157, 167, 169, 182, 234, 243, 321

Closteroviruses 31, 34, 181, 183, 185-186, 200, 205, 208, 236, 243, 246

Clover yellow mosaic potexvirus (CYMV) 22, 34, 200, 282

Clover yellow vein potyvirus 41-43, 187

Coat - see Capsid

Coat protein - see Capsid protein

Cocksfoot mottle sobemovirus (CfMV) 22, 41, 121, 126, 243, 245, 262

Comoviridae 20, 34, 121, 169, 175, 195, 238

Comovirus 20, 32, 41, 55, 59, 159, 167, 175-177, 180, 290

Comoviruses 141, 176, 178, 257

Core promoter(s) 4, 6-8, 12-14, 92- 93, 102-103, 135, 138, 208-214, 217

Cowpea chlorotic mottle bromovirus (CCMV) 22, 32, 49, 62, 73, 82, 89, 96, 123-124, 205, 211, 284-287, 290-292, 294

Cowpea mosaic comovirus (CPMV) 4, 20, 22, 30, 32, 35, 41-44, 55-56, 59, 71, 73, 85, 96-97, 102-103, 119, 121, 123-124, 129, 153, 158-159, 167, 170, 176-177, 241, 290, 299, 300, 306-307, 309, 316, 318, 320-322

Cricket paralysis virus 19, 230-232, 257-258

Crinivirus 20, 185, 205, 235

Cucumber chlorotic spot closterovirus (CCSV) 22, 243

Cucumber green mottle mosaic tobamovirus 215, 273, 277, 308, 321

Cucumber mosaic cucumovirus (CMV) 1, 4, 8, 11-12, 20, 22, 32, 37, 40, 49, 54, 57-59, 73-74, 90, 92-93, 97, 123, 129, 130, 132, 140, 153, 156, 158, 196, 203, 205-207, 212-214, 271, 291, 299, 300, 302, 307-308, 310-311, 313, 322

Cucumber necrosis tombusvirus (CNV) 8, 22, 101, 119-127, 129, 136-137, 140, 156, 195, 198, 200-201, 206-207, 212-213, 216, 218, 241, 247, 300, 315-316

Cucumovirus 20, 32, 35, 46, 59, 101, 126, 157, 196, 214, 292, 308

Cucumoviruses 46, 50, 106, 218, 282

Cymbidium ringspot tombusvirus (CymRSV) 4, 22, 56, 302, 321

Cysteine proteinases 127, 169, 171-174, 176, 180-185, 187, 316

Cysteine-rich protein(s) 103, 217

Cystoviridae 91, 133, 136

Cytoplasmic inclusion(s) 124, 185, 306, 313

Cytosine 275, 291

Decapsidation – see Uncoating

Defective interfering RNA 139, 301, 316

Dengu 2 virus 136

Dianthovirus 20, 31, 32, 34, 59, 121, 156, 243, 245, 252-253, 255, 258, 261

Dianthovirus(es) 141, 200, 236, 245, 259

Disassembly 43, 75-76, 78-83, 278, 286, 288, see also Uncoating

Discistroviridae 252

Double-stranded DNA genome/ viruses 155, 235

Double-stranded RNA genome/ viruses 4, 29, 36, 72, 90-92, 96, 98, 100, 131, 137, 154-155, 206, 231, 244, 303, 305, 307-308, 314, 321

Eggplant mosaic tymvovirus 49, 122, 291

Eggplant mottled crinkle tombusvirus 308

Enamovirus 20, 32, 41-42, 59, 121, 167, 243, 247, 248

Encephalomyocarditis virus (EMCV) 229, 231, 256-257

Endoplasmic reticulum (ER) 23, 44, 61, 95, 108, 124, 162-163, 305, 310, 314

Erysimum latent virus 47-49

Eukaryotic translation initiation factor(s) (eIFs) 18-19, 31, 40-44, 49, 53, 56, 66, 107, 152, 155-156, 160-162, 224-233, 237, 252-253, 257, 259, 263, 300, 302, 309, 311, 314

Fabavirus 20

Fiji disease reovirus 305

Flavivirdae 134

Foot-and-mouth disease virus 230, 257

Foveavirus 21, 31, 172

Foxtail mosaic potexvirus 122, 124, 154, 322

Frameshift(ing)/frameshift 13, 59, 100, 120, 122, 126-128, 167, 195, 223, 230, 234, 243-246, 262

Furovirus 21, 32, 34-35, 37, 46, 49, 59, 122, 126, 157, 167, 195-196, 247

Furoviruses 49, 247-250

GB virus-B 125

Galinsoga mosaic carmovirus 8, 308, 319

Geminivirus(es) 157, 236, 307

Gene silencing 89, 137, 190, 201, 304

Genome activation 107-108, 110

Genome-linked protein (VPg) 18, 30, 32-36, 41-44, 126, 134, 168, 170, 174, 176-179, 187, 197-200, 233-234, 237, 252-253, 259, 306, 318

Grapevine chrome mosaic virus 158, 178

Grapevine fanleaf nepovirus (GFLV) 22, 41, 60-61, 96, 170-171, 176- 179, 306, 316, 321

Grapevine fleck virus 30

Grapevine trichovirus B (GVB) 22

Groundnut rosette umbravirus (GRV) 22, 33, 60, 74, 120-121

Grapevine vitivirus A (GVA) 21, 22

Guanine 36-37, 224, 228, 238, 281

Guanylyltransferase activity/domain 36-39, 58, 61, 312, 314

Heat shock protein (HSP) 16, 23, 59, 201-202, 217, 233

Helicase (HEL) 1, 2, 4, 5, 8, 37-38, 40, 58-62, 87, 90-93, 96, 125, 127-128, 133, 140-142, 151-165, 181-185, 218, 224, 225, 230, 235, 237, 245-246, 257, 262, 300-302, 307, 309-317, 321-322

Helper component proteinase (HC-Pro) 23, 42, 170, 173-174, 181-190

Hepatitis C virus 1, 125, 132-134, 136, 139, 151, 161, 229-230, 252, 257-258

Hepatitis E virus 38

Herpes simplex virus-1

Hibiscus chlorotic ringspot carmovirus (HCRSV) 7, 89, 94, 101, 124, 200, 216, 241, 245, 248

Hordeivirus 21, 33-35, 37, 46, 49, 50-51, 60, 62, 92, 122, 126, 157, 195, 197

Hordeiviruses 49, 142, 157, 195, 218, 235-236

Host factors 2, 4, 7-8, 15, 17, 43, 55, 75, 82, 87, 89, 99, 102, 104, 129-130, 135, 226, 257-258, 278, 299-322

Host gene shut-off 15-17, 225

Human immunodeficiency virus 139

Hungarian grapevine chrome mosaic virus 158

Idaeovirus 21, 33, 60, 122, 126, 157, 197

Ilarvirus / ilarviruses 20, 22, 33, 51, 93, 105-108, 110, 197, 214, 218, 233, 235, 308

Indian peanut clump virus 49, 280

Influenza orthomyxovirus 39

Internal control region 11, 303

Internal initiation of gene expression/translation 12, 52, 136, 203, 228-229, 234, 242, 257

Internal ribosome binding/entry site (IRES) 18-19, 31, 215, 223, 229-232, 234, 237, 242, 251-252, 256-260, 262

Ipomovirus 20, 183-184

Kennedya yellow mosaic tymovirus 37, 49, 158

Kinase(s) 41-42, 300, 311, 313

Leader proteinases 181-183, 185-186, 190

Leader sequence 12, 76, 80-82, 195, 200, 232, 236-239, 242, 252, 254, 256-260

Leaky scanning/gene expression 195, 216-217, 227-228, 234, 239-242, 257

Leaky stop codon/termination 61, 126, 246, 248-250, 262

Lettuce infectious yellows closterovirus (LIYV) 15, 20, 22, 99, 101, 122, 183, 185, 199, 243, 246, 256

Lettuce mosaic virus (LMV) 43-44, 130, 183-184, 188

Lily virus X 238

Luteoviridae 19, 20, 42, 252-253, 261, 292

Luteovirus 20, 31, 33-34, 41, 60, 121, 126, 156, 167-169, 195, 197, 239, 243, 247, 252-255, 258, 261, 307

Luteoviruses 141, 180, 195, 207, 234, 236-237, 244-245, 248-250, 259, 316

Lychnis ringspot hordeivirus 235

Machlomovirus 20, 33-35, 60, 121, 197, 253

Maclura mosaic macluravirus 20

Macluravirus 20

Maize chlorotic dwarf waikavirus 179

Maize chlorotic mottle machlomovirus (MCMV) 20, 22, 33-34, 60, 121, 156, 197, 200, 247, 250

Maize rayado fino virus 21-22

Maize rough dwarf fijivirus 22

Maize stripe tenuivirus 22

Marafivirus 21-22, 31, 33-34, 60, 167, 169

Marafiviruses 172, 181

Methyltransferase 5, 8, 36-39, 58, 61, 90, 125, 127-128, 133, 181-182, 185, 201, 217, 235, 246, 301-302, 309-310, 312-317, 319, 321-322

Monocomponent/monopartite genome/viruses 29-30, 74, 179, 195, 200, 234-235

Movement complex 43, 314-315

Movement protein 2, 15, 23, 74-75, 80, 111, 124, 175-176, 178, 187, 199, 200-201, 212-213, 215-217, 235, 275, 311, 314

Multicomponent/multipartite genome/viruses 13, 29-30, 34, 41, 48, 58, 96-97, 102, 200, 218, 234, 236, 272, 280

Narcissus mosaic potexvirus 158, 214

Necrovirus 20, 31, 33, 60, 121, 156, 247, 252-254, 258

Necroviruses 141, 238, 247, 255

Negative-sense RNA viruses/plant viruses/template 3, 4, 7, 13-14, 17, 19, 29, 51, 72, 101, 119

Nepovirus 20, 21, 33, 41, 43-44, 55, 60, 127, 159, 167, 169, 172, 175, 177-178, 180, 198, 316, 318

Nepoviruses 141, 171, 176, 179, 319

Nicotiana benthamiana 61, 119, 122-213, 258, 316

Nicotiana tabacum var. Xanthi-nc 71, 315

Nuclear inclusion 16, 41, 58, 173

Nuclear localization 42, 74-75, 141

Nucleo-cytoplasmic shuttling 17, 40, 73-75, 88

Nucleoside 5`-triphosphate (NTP) and NTPase activity 8, 44, 132-133, 135-136, 151, 153-157, 159-163, 177, 182, 309-310, 316-318

Nucleus 17, 36, 39, 73-75, 174, 186, 308, 319, 320

Nucleotide binding activity/sites 161

Oat blue dwarf marafivirus (OBDV) 22, 31, 33, 60, 167, 171, 181

Oat chlorotic stunt avenavirus 20, 34, 200, 247

Odontoglossum ringspot tobamo-virus 136

Oleavirus 20, 214

Olive latent oleavirus 2 20

Origin of assembly sequence (OAS) 109, 272, 275, 277-280, 282

Ourmia melon ourmiavirus 21

Ourmiavirus 21

Panicovirus 20

Panicum mosaic panicovirus 20, 22, 34, 200

Papain-like proteinase(s) 59, 127, 168, 170-173, 180-185, 187, 190, 246

Papaya mosaic virus 82, 158, 281

Pararetroviruses 36, 167, 186

Parsnip yellow fleck sequivirus (PYFV) 20, 22, 33, 41, 60, 121, 141, 167, 171, 179

Pea early browning tobravirus (PEBV) 22, 122, 247

Pea enation mosaic enamovirus (PEMV) 20, 22, 32, 59, 67, 199, 243, 245, 247, 250

Pea enation mosaic enamovirus-1 42, 121, 126

Pea enation mosaic enamovirus-2 42

Pea seed-borne mosaic potyvirus (PSbMV) 22, 16-17, 41-43, 92, 101, 158, 184, 242

Peanut clump pecluvirus (PCV) 21, 22, 45, 48-49, 96, 103, 200201, 234, 235, 239-240, 347-250, 279-280, 306, 308, 319-320

Pelargonium leaf curl tombusvirus 307

Pecluvirus 21, 22

Pecluviruses 305

Pepper mild mottle virus 122

Pepper mottle virus 130

Pepper ringspot virus 122

Pepper vein banding virus 41, 126

Physalis mottle tymovirus 307

Picorna-like supergroup/picorna virus (es) 18, 20, 41, 141-142, 159, 167, 169, 170-172, 174, 180, 195, 229, 231, 237-238, 251-252, 257-258, 265, 305, 310, 318, 320, 322

Picornaviridae 35, 133

Plasmodesmata 43, 71, 190, 315

Plum pox potyvirus (PPV) 22, 41, 44, 56, 91, 121, 151, 153-156, 158-162, 175, 183, 239, 242, 322

Poa semilatent hordeivirus 235

Polerovirus 20, 22, 41, 126, 195, 198, 243, 247

Poliovirus 1, 17-18, 21, 97, 125, 127, 131-134, 156, 159, 253, 256, 309, 313

Poly(A) binding protein (PABP) 18-19, 23, 31, 44, 53, 56, 107-109, 224-228, 231, 233, 252, 257, 260

Poly(A) tail/end 4, 10, 16, 18, 23, 30-34, 40, 44-45, 53, 55-56, 88, 90, 94-95, 107, 109, 112, 153, 200-201, 210, 212, 214, 225-228, 231-233, 251-255, 258-261, 291

Polymerase(s) (POL) – see RNA dependent RNA polymerase

Pomovirus 21, 22, 126, 280

Pomoviruses 235

Potato leaf roll polerovirus (PLRV) 20, 22, 41, 60, 121, 126, 156, 180, 198, 241, 243, 245, 247, 292

Potato mop top pomovirus (PMTV) 22, 49, 235, 280

Potato carlavirus M (PVM) 22, 32, 37, 59, 158, 171, 196, 243, 245-246

Potato virus A (PVA) 41, 42-43, 184

Potato virus S 232

Potato potexvirus X (PVX) 11, 22, 33-37, 43, 45, 60, 71-72, 120, 124, 131, 157-158, 162, 189, 198, 201-203, 206-207, 212, 232, 241, 256, 258, 262, 271, 280-281, 293, 316

Potato potyvirus Y (PVY) 22, 33, 41-43, 60, 92, 96, 126, 183-184, 188

Potexviridae 20-21, 31, 34, 261, 282

Potexvirus 21-22, 31, 33, 35, 55, 60, 122, 126, 154, 195, 198, 214, 282, 322

Potexviruses 94, 120, 141, 153, 157, 200, 206, 232, 235, 238, 241, 253, 261-262

Pothos latent aureusvirus 20

Potyviridae 20, 22, 33-34, 44, 121, 172, 175, 183, 195, 235, 237, 261

Potyvirus 20, 33, 41, 55, 60, 157, 159, 167, 169, 190, 226, 234

Potyvirus(es) 16-17, 22, 31, 42-44, 74, 92, 101, 121, 126, 130, 136, 142, 151, 153, 159, 167, 171-175, 178-189, 221, 242, 249, 251-253, 257-261, 305, 307, 310-311

Proteases 17, 59, 75, 126, 140, 163, 167, 246

Proteinases 1-2, 23, 41, 57-58, 60, 62, 127, 167-194, 245, 250-251, 307, 310-311, 316, 318

Proteolytic activity/processing 43, 120, 126-128, 167-181, 184, 186-187, 190, 231, 234, 250-251, 283, 284

Pseudoknot 4, 10, 31, 47-48, 50-54, 62, 93-95, 106, 108-110, 244-246, 250, 260-261

Prune dwarf ilarvirus (PDV) 22, 60, 93

Qβ virus 90

RNA-binding activity/domain/site 18, 42, 59, 100, 105, 122, 134-137, 151, 154-155, 159, 188-189, 210, 227, 275-276, 302, 304, 314
RNA-dependent RNA polymerase (RdRp) 1-5, 7-10, 12, 15, 22-23, 36-40, 42-43, 51-53, 56-62, 81, 87, 89-106, 110, 119-142, 148, 153-159, 161-163, 170, 177-178, 181-183, 187, 193, 202-206, 210-214, 217, 222-223, 226, 236-237, 242, 244-250, 262, 291, 299-303, 309-319, 321
Rabbit haemorrhagic disease virus 132-134
Radish yellow edge virus 305
Raspberry bushy dwarf idaeovirus (RBDV) 21-22, 33, 37, 60, 122, 126, 141, 158, 197, 201
Readthrough domain/protein/process 6, 23, 61, 75, 120, 122, 126-128, 136, 167, 195, 199, 215, 217, 223, 234, 245-246, 251, 262, 308, 314, 319
Red clover necrotic mosaic diantho-virus (RCNMV) 15, 22, 31, 32, 34, 45, 97, 103, 121, 124, 127, 133, 200, 204, 206, 207, 209, 212, 239, 243, 253, 258, 261, 262, 299-300, 322
Reoviridae 18, 72, 91, 134, 231
Reovirus 134
Reovirus(es) 132, 152, 305
Replicase/replicase activity/replicase gene 2-4, 8, 11, 13, 14, 29, 38, 45, 48, 51, 55-57, 61, 75, 80-81, 83, 87, 90-91, 94, 99-100, 102, 104-105, 107-108, 120-142, 174, 176, 178, 185, 200, 202-203, 205-212, 215, 226, 235, 238-239, 245, 249-250, 256, 262-263, 299-303, 306, 308-309, 313-314, 316, 318, 319, 322

Replicase complex 3, 5, 12, 53, 82, 90, 104, 203-205, 210, 305, 311, 313, 315-316
Replication complex(es) 3, 8, 10, 13, 35, 42, 44, 52, 61, 90, 95-96, 105, 111, 124, 127-130, 133, 136, 162-163, 205, 227, 300, 302-307, 309-322
Replication enhancer(s) 3, 11-12, 14, 15, 102-103, 137, 140, 312
Replicative form (RF) RNA 4-6, 91-92, 97-98, 206
Replicative intermediate (RI) RNA 90, 97-98, 100
Retroviridae 133, 156
Retroviruses 36
Rhabdoviridae 72, 156
Rhabdoviruses/plant rhabdoviruses 36, 305
Rhinovirus 309
Ribgrass mosaic tobamovirus 273
Ribosomal frameshifting 26, 246
Rice dwarf phytoreovirus 22
Rice grassy stunt tenuivirus 22, 237
Rice stripe tenuivirus (RStV) 22, 237
Rice tungro bacilliform badnavirus 187, 236, 242
Rice tungro spherical waikavirus (RTSV) 20, 22, 33, 41, 60, 121, 158, 161, 167, 171, 179, 199
Rice yellow mottle sobemovirus (RYMV) 22, 60
Rotavirus 231
Rotaviruses 18-19, 136, 231
Ryegrass mosaic rymovirus 20, 41
Rymovirus 20, 22, 33, 41, 55, 60, 167, 173, 184

Saguaro cactus virus 34
Satellite maize white line mosaic virus 14
Satellite panicum mosaic virus (SPMV) 14-15, 88
Satellite tobacco mosaic virus (STMV) 14, 241

Satellite tobacco necrosis virus (STNV) 14, 22, 31, 34-35, 239, 252-256, 258-261, 283-284

Semliki Forest virus (SFV) 37-38, 63, 162

Sequiviridae 20, 121, 169, 179, 195

Sequivirus 20, 33, 41, 60, 159, 167, 169, 179, 195

Sequivirus(es) 22, 169

Serine/serine-like proteinases 17, 168-174, 176-177, 180, 187

Shallot allexvirus X 20, 37, 158, 238

Simian virus 40 (SV40) 156, 159

Sindbis virus 1, 37, 209

Sobemo-like supergroup 20-21, 34, 41, 121, 203, 245, 322

Sobemovirus 21, 33-34, 41, 60, 121, 156, 167, 169, 195, 198, 243, 245, 283, 292

Sobemovirus(es) 31, 126, 142, 167, 169, 172, 180, 195, 200, 238, 244-245, 282

Soil-borne wheat mosaic furovirus (SBWMV) 21-22, 32, 37, 45, 49, 51, 60, 122, 126, 158, 196, 238, 247-250

Solanum commersoni 43

Sonchus yellow net virus 74

Southern bean mosaic sobemovirus (SBMV) 21-22, 33, 41, 45, 60, 82, 121, 126, 156, 167, 180, 198, 215, 239, 283-285

Sowbane mosaic virus 305

Soybean dwarf luteovirus (SbDV) 22, 156

Stem-loop region/structure 7, 10-11, 13-15, 31, 34-35, 46, 53, 57, 86, 93-96, 101-102, 106, 110, 138, 204, 261-262, 275, 278, 290, 311

Strawberry mild yellow edge-associated virus 158

Subgenomic RNA promoter 5, 9, 12, 88, 93, 99, 203, 206-214, 217-218

Sunnhemp mosaic tobamovirus 49-50

Sweet potato chlorotic stunt crini-virus 34, 200, 205

Sweet potato mild mottle ipomovirus 20, 183

Switch 4-5, 8-10, 52, 54, 75-76, 78, 96, 99-100, 110, 138, 253, 274, 285-286, 292, 295

Tamarillo mosaic potyvirus 151

Template switching 139

Tenuivirus(es) 22, 195, 237

Tobacco etch potyvirus (TEV) 16, 22, 31, 41-44, 56, 73-74, 101, 120-121, 130, 170, 173-175, 181, 183-184, 187-188, 243, 252, 255, 257-258, 261, 308-309, 312-313, 320, 322, 326

Tobacco mosaic tobamovirus (TMV) 1, 4, 7-8, 11-15, 21-22, 30-31, 33-34, 37-38, 45-55, 58, 61-63, 71-83, 88-89, 92, 94-96, 99-101, 107, 111, 119, 120, 122-124, 126, 129-130, 136-137, 140-141, 157-158, 162, 189, 195, 198, 200-201, 208, 215-219, 225-227, 234, 239-240, 249-251, 257, 260, 262-264, 272-282, 288, 294-296, 302-309, 311, 313, 315-317, 321-322, 324

Tobacco necrosis necrovirus (TNV) 20, 22, 33-34, 60, 77, 197, 121, 206, 208, 249, 254-257, 264, 283

Tobacco rattle tobravirus (TRV) 21-22, 33, 37, 45, 48, 59, 61-62, 71, 73, 76-78, 122, 158, 198, 200- 201, 217, 249-51, 280, 296, 310

Tobacco ringspot virus 22, 121

Tobacco streak ilarvirus (TSV) 20, 22, 33, 45, 60, 122, 197, 308

Tobacco vein mottling potyvirus (TVMV) 22, 41-44, 119, 125, 130, 158, 173, 183-184, 282, 316

Tobamovirus 4, 34-35, 37, 46, 92, 122, 126, 158, 195, 198, 247, 272

Tobamovirus(es) 12, 21-22, 49-50, 60-61, 136, 142, 206, 238, 247, 261-262, 272, 275, 304-305, 308, 315, 321

Tobravirus 21, 33-35, 37, 46, 60, 122, 158, 195, 198, 247

Tobraviruses 103, 129, 142, 234-235, 247

Tomato aspermy cucumovirus 48, 50, 130, 308

Tomato black ring nepovirus (TBRV) 22, 33, 41, 60, 127, 158, 167, 178-179

Tomato bushy stunt tombusvirus (TBSV) 10-11, 20, 22, 31, 33-35, 60, 89, 101-104, 120-124, 127, 129, 134-135, 137, 140, 200, 202-204, 207, 212, 216, 241, 247, 252-258, 261, 283-285, 301, 304, 308, 315, 320-321

Tomato mosaic virus 92, 96, 104

Tomato ringspot nepovirus (TomRSV) 22, 41, 43-44, 96-97, 177-179, 318

Tomato spotted wilt tospovirus (TSWV) 22, 39, 189, 236-237, 305

Tomato yellow leaf curl virus 74

Tombusviridae 20-21, 31, 34, 103, 105, 121, 156, 195, 252-253, 255, 263, 304

Tombusvirus 20, 31, 60, 105, 198, 247, 252-254, 258, 292, 307

Tombusvirus(es) 4, 12, 31, 56-58, 75, 101, 105, 120, 122, 125, 128, 137, 139-141, 195, 200, 206, 236, 248-250, 258, 261, 282, 308, 319-321

Tospovirus 39, 236

Tospovirus(es) 236

Transcription factor 12, 17, 73, 309-310

Transcription initiation/start site 4, 9, 51, 209, 213

trans-Acting nucleotides/factors 9-10, 14-15, 87, 99, 237-239, 252, 256-257

Transcriptional enhancer(s) 5, 103

Translation enhancer 88, 207, 211, 233, 253, 256

Translation initiation factors – see eukaryotic translation initiation factors (eIFs)

Trichovirus 21-22, 31, 33-34, 37, 55, 60, 122, 167, 169

Trichovirus(es) 120, 142, 172, 181, 232

Triple gene block (TGB) 153-154, 196-199, 217, 235, 238, 241-242, 248

Tritimovirus 20, 173

Turnip crinkle carmovirus (TCV) 11-12, 22, 34, 56-57, 88, 92, 94, 102-103, 111, 120, 124-127, 129, 136-138, 200-203, 206-207, 209, 211-212, 216-217, 232, 239, 247-248, 252-253, 255-256, 258, 284, 289, 291, 302, 308, 316, 322

Turnip mosaic potyvirus (TuMV) 12, 18, 22, 41-44, 233

Turnip rosette sobemovirus 31, 41, 45

Turnip yellow mosaic tymovirus (TYMV) 11, 21-22, 33, 37, 46-55, 57, 60, 73, 76-80, 82, 94-95, 98, 112, 120, 122, 124-125, 127, 129-131, 136, 157-158, 162, 167, 170, 181-182, 198, 203, 209, 213, 216, 226, 232-233, 249, 260-261, 285, 291, 293, 299, 303, 307, 309-311, 319-320, 322

Tymovirales 20-21

Tymovirus 21, 33-35, 37, 46, 60, 92, 122, 167, 169, 195, 198

Tymoviruses 37, 47, 48-50, 54, 79, 120, 142, 157, 181, 183, 200, 206, 213, 282, 291, 301, 319

Umbravirus 21, 33, 42, 60-62, 121, 127

Umbraviruses 74, 126

Uncoating of virus particles/RNA 3, 75-82, 107, 123, 272, see also Disassembly

Untranslated 5` region (5`-UTR) of RNA 10, 15, 30, 31, 34-35, 45, 82, 88, 93, 101, 108, 207, 226-227,

232-233, 238, 242, 251-259, 262-263, 280, 291
Untranslated 3` region (3`-UTR) of RNA 10, 12, 14-15, 31, 45, 51, 53, 55, 88, 93-95, 106, 108-111, 127, 130, 226, 232-233, 251-263
Uracil 291

Velvet tobacco mottle virus 35, 322
Vesicular stomatitis virus 17, 136, 303
Vesicles/vesiculation of membranes 72, 76, 78, 108, 185, 307-311, 316-317, 319-322
Viroplasm 305, 311, 313-314
Virus assembly – see Assembly of viruses
Virus movement/transport in plants 43, 58, 88-89, 106, 153, 175, 186, 190, 241, 239, 241, 256, 314-315
Vitivirus 21

Waikavirus 20, 22, 33, 41, 55, 60, 159, 167, 169, 179
Waikaviruses 172
Wheat dwarf geminivirus 236
Wheat streak mosaic tritimovirus 20, 173-174
White clover mosaic potexvirus (WClMV) 22, 37, 55-56, 158
Wound tumor phytoreovirus 22, 72-73

X-bodies 305, 313
Xenopus 17, 40, 125